Heat and Mass Transfer for Chemical Engineers

About the Author

Giorgio Carta is Lawrence R. Quarles Professor of Chemical Engineering at the University of Virginia. He received his Laurea in Chemical Engineering from the University of Cagliari, Italy, in 1980 and his PhD in Chemical Engineering from the University of Delaware in 1984. He has authored/co-authored 170 refereed journal articles mainly in the area of adsorption and ion exchange. He has co-authored the first and second editions of the book *Protein Chromatography—Process Development and Scale-Up*, Wiley-VCH, 2010 and 2020, and has served as section co-editor of Section 16 "Adsorption and Ion Exchange" of *Perry's Chemical Engineers' Handbook*, McGraw-Hill, 7th, 8th, and 9th editions. He has served as a member of the Organizing Committee of the International Symposium on Preparative and Process Chromatography—PREP since 1997 and as the PREP Symposium Chair annually since 2009. In addition to his university teaching, he has also been involved extensively in professional education and has been a consultant for a number of chemical and pharmaceutical companies. He has served as chair of AIChE Area 2e Adsorption and Ion Exchange, as chair of AIChE Area 2g Bioseparations, as co-chair of the 2007 Fundamentals of Adsorption Conference—FOA9, and as a Member of the Board of the International Adsorption Society. He is a Fellow of AIChE and has been recognized with the 2020 AIChE Area 2e Honorary Session.

Heat and Mass Transfer for Chemical Engineers

Principles and Applications

Giorgio Carta

McGraw Hill

New York Chicago San Francisco
Athens London Madrid
Mexico City Milan New Delhi
Singapore Sydney Toronto

Library of Congress Control Number: 2021938714

McGraw Hill books are available at special quantity discounts to use as premiums and sales promotions or for use in corporate training programs. To contact a representative, please visit the Contact Us page at www.mhprofessional.com.

Heat and Mass Transfer for Chemical Engineers: Principles and Applications

1 2 3 4 5 6 7 8 9 CCD 26 25 24 23 22 21

ISBN 978-1-264-26667-8
MHID 1-264-26667-7

The pages within this book were printed on acid-free paper.

Sponsoring Editor
Robin Najar

Editorial Supervisor
Donna M. Martone

Production Supervisor
Lynn M. Messina

Acquisitions Coordinator
Elizabeth M. Houde

Project Manager
Nitesh Sharma,
KnowledgeWorks Global Ltd.

Copy Editor
Girish Sharma

Proofreader
Upendra Prasad

Indexer
Edwin Durbin

Art Director, Cover
Jeff Weeks

Composition
KnowledgeWorks Global Ltd.

Contents

Preface

The goal of this textbook is to provide a conceptual understanding of heat and mass transfer phenomena along with the tools needed to quantitatively apply these concepts to solve practical problems. The unified analysis of transport phenomena based on differential balances has been a centerpiece of chemical engineering for decades. This book follows this approach but seeks to provide a more physically based understanding of the underlying molecular phenomena by relating their mathematical description to practical situations in many different areas. Heat and mass transfer are covered assuming that the reader already has a knowledge of momentum transfer. Rather than attempting to provide a comprehensive coverage of all three transport phenomena, the goal here is to provide what can be covered in a typical one-semester, junior-level course. Heat and mass transfer are separated in two major sections, with considerably more space dedicated to the latter. Mass transfer is not only often more relevant for the chemical engineering professional, but it is also inherently more complex since it deals with mixtures. Mass transfer is also, in many ways, less intuitive than heat transfer and more challenging to relate to practical, everyday observations. After all, we can, for example, easily sense temperature gradients and the effects of insulation on heat losses, but we cannot as easily sense concentration gradients or appreciate mass transfer resistances in series. Thus, covering heat transfer first provides a primer to better understand mass transfer later. Each major section starts with simple problems that demonstrate the extent to which macroscopic energy and material balances can be used for their solution and why analysis of the underlying heat and mass transfer phenomena becomes necessary for their complete solution.

Chapter 1 introduces heat transfer mechanisms and their descriptions in terms of Fourier's law and convective heat transfer coefficients. Chapter 2 explores the magnitude of the thermal conductivity of different materials and its estimation for homogenous fluids and composites. Chapter 3 covers steady-state, one-dimensional heat conduction with a focus on understanding resistances in series and in parallel. Chapter 4 addresses systems where conductive heat transfer in one dimension is coupled with convective heat transfer in a perpendicular direction. The chief example here is heat transfer to/from finned surfaces. Chapter 5 introduces generalized multi-dimensional energy balances in different coordinate systems. This chapter emphasizes solutions to transient conductive and convective problems providing the mathematical tools needed to understand and predict the evolution of temperature within objects. Chapter 6 addresses the prediction of convective heat transfer coefficients. The goal here is for students to understand how a velocity gradient at an interface affects the temperature gradient and, thus, the heat transfer flux. A second goal is for the reader to become comfortable using engineering correlations to predict heat transfer coefficients by understanding which physical properties are relevant and their role in these predictions. Finally, Chap. 7 covers the application of heat transfer concepts to the analysis and design of heat exchangers. This chapter reconnects to Chap. 1 where the design of a heat exchanger was first proposed as a motivational example. Upon completing Chap. 7, students will realize that they now have the knowledge needed to solve this problem in full. I decided not to include a chapter covering radiant heat transfer. This mechanism is covered only with illustrative examples in Chap. 1. While very important at extreme temperatures, this topic is really distinct and does not fit well with the general treatment of transport analogies that is emphasized in this book. In any case, many current texts are available in the mechanical engineering literature covering this topic in depth.

The mass transfer section begins with Chap. 8 with an introduction to what we can and cannot determine just based on material balances and phase equilibrium relationships. An illustrative example shows that we need to understand and apply mass transfer concepts if we are to fully design a gas absorption column. Chapter 9 introduces Fick's law as a phenomenological description of diffusion, while Chap. 10 explores the diffusivities in different phases and how we can predict their values based on molecular properties. A view of diffusion based on Brownian motion of particles is also provided. Chapter 11 introduces one-dimensional diffusion problems with special emphasis on pseudo-steady state situations. Chapter 12 mirrors Chap. 5 by introducing general differential material balances in different coordinate systems and exploring the solution of various transient diffusion problems. This chapter draws on the analogous solutions presented in Chap. 5 for heat transfer but extends the analysis to other situations encountered when dealing with mass transfer between phases and with the interplay of transport and phase equilibrium at the interface. Chapter 13 also mirrors Chap. 6 dealing with convective mass transfer coefficients. The reader learns about the appropriate dimensionless transport numbers that allow us to use heat transfer correlations to predict mass transfer coefficients. Examples of convective mass transport in laminar flow are also covered in detail to emphasize the interplay of momentum and mass transport phenomena. Chapter 14 provides the reader with the opportunity to apply the fundamental concepts and tools learned to the analysis and design of gas-absorption columns. Not only is this subject of practical significance, but it also provides a readily accessible opportunity for students to test their ability to apply concepts to practical engineering calculations. Finally, Chaps. 15 and 16 provide coverage of additional, more advanced mass transfer topics. It may be difficult to cover these topics in their entirety within the scope of a one-semester course. Nevertheless, I think it is important that students are at least exposed to the concept of coupling of mass transfer fluxes, which is covered in Chap. 15, and to the coupling of diffusive and reactive processes, which is covered in Chap. 16. Neither of these chapters provides comprehensive coverage of these topics. The intent is to provide enough so that students can get an appreciation of what lies beyond the simpler cases of mass transfer in binary gas mixtures or in dilute solutions.

Two important features of this book are worth describing in detail. The first is the extensive reference to *Perry's Chemical Engineers' Handbook*. Rather than providing extensive data tables, only enough physical property data are included in this book to understand trends and work out most of the chapter problems. Exceptions are the properties of air and liquid water for which more extensive tables are provided in App. A.

Specific references to *Perry's Chemical Engineers' Handbook* are made throughout the book, particularly to Section 2 in *Perry's Chemical Engineers' Handbook*, which covers extensively the values and prediction of physical properties, and to Section 5, which covers many of the properties specifically relevant to heat and mass transfer. This section also provides extensive coverage of correlations to predict heat and mass transfer coefficients. As a result, only a limited number of these correlations are included here with the main goal of illustrating trends. The second feature is the inclusion of Wolfram Mathematica notebooks, available in the companion website, to facilitate calculations. While not all equations presented in the book are available as Mathematica scripts, the more complicated ones are, including many with interactive graphical output. I feel that relieving the students from the burden of coding complex equations or reading dense graphs will provide more time and pedagogically more effective and gratifying ways of exploring what these equations actually predict. Students can, of course, easily modify and extend these notebooks to handle different cases. Equations, figures, tables, and examples for which Mathematica notebooks are provided are marked in the book with the corresponding notation (M.E), (M.F), (M.T), and (M.EX).

My own engineering education was shaped to a large extent by three giants in this field—Prof. Arthur B. Metzner, Prof. Gianni Astarita, and my own PhD advisor Prof. Robert L. Pigford, with whom I was fortunate to interact with as a graduate student at the University of Delaware and beyond. Art had an uncanny ability to dissect and explain the most complex of transport problems. He also had a wonderful sense of humor. In fact, if any humor is to be found in transport phenomena, he is the one who could find it. Gianni had an amazing sense for the magnitude of things. His oft celebrated quote "chemical engineers really only need to know three numbers—0, 1, and ∞—if they just learn to properly scale the physical variables" is not to be forgotten. For Bob, mathematics was just like words, or, shall we say, like a musical score that can be played to beautifully reveal the symphony of transport phenomena at play in physical systems. Although a long time has passed and they have not been with us for many years, their teachings have remained and I have done my best to reflect in this book what I have learned from them.

Those of us who have chosen academic careers know well that we learn from our own students at least as much as we learn from our mentors. Over 37 years of teaching chemical engineering at the University of Virginia, much of what I have learned has come from my own students. For this I am very grateful to them. In particular, I want to thank the students in my Spring 2021 Heat and Mass Transfer class at UVa who have provided invaluable comments on a draft version of this book.

Please go to www.mhprofessional.com/CartaHeatandMassTransfer for online materials that include Wolfram Mathematica notebooks to facilitate calculations.

Giorgio Carta

General References

Bennett, C.O., Myers, J.E., Momentum, Heat, and Mass Transfer, 3rd Edition, McGraw-Hill, 1982.

Bergman, T.L., Lavine, A.S., Fundamentals of Heat and Mass Transfer, 8th Edition, John Wiley & Sons, 2017.

Bird, R.B., Stewart, W.E, Lighfoot, E.N., Transport Phenomena, 2nd Edition, John Wiley & Sons, 2007.

Geankoplis, C.J., Hersel, A.A., Lepek, D.H., Transport Processes and Separation Principles, 5th Edition, Pearson, 2018.

Green, D.W., Southard, M.Z., Perry's Chemical Engineers' Handbook, 9th edition, McGraw-Hill, 2019. Section 2: Physical and Chemical Data; Section 5: Heat and Mass Transfer; Section 6: Fluid and Particle Dynamics; Section 11: Heat-Transfer Equipment; and Section 14: Equipment for Distillation, Gas Absorption, Phase Dispersion, and Phase Separation.

McCabe, W.L., Smith, J.C., Harriott, P., Unit Operations of Chemical Engineering, 7th Edition, McGraw-Hill, 2005.

Middleman, S., An Introduction to Mass and Heat Transfer—Principles, Analysis, and Design, John Wiley & Sons, 1998.

Sherwood, T.K., Pigford, R.L., Wilke, C.R., Mass Transfer, McGraw-Hill, 1975.

Welty, J.R., Rorrer, G.L., Foster, D.G., Fundamentals of Momentum, Heat, and Mass Transfer, 7th Edition, John Wiley & Sons, 2019.

Nomenclature

a Gas–liquid interfacial area per unit column volume, m^2/m^3

a_p Specific surface area parameter for random packing, m^2/m^3

a_s Surface area of a porous material per unit volume of material, m^2/m^3

A Cross-sectional area perpendicular to the heat transfer flux, m^2

A_o Heat exchanger area based on outer radius of pipe, m^2

B Thickness or half-thickness of a slab, m

c Molar density of a liquid or solid phase, mol/m^3

c_i Molar concentration of component i in a liquid or solid mixture, mol i/m^3

C Molar density of a fluid phase, mol/m^3

C_i Molar concentration of component i in a fluid mixture, mol i/m^3

C_p Mass-based heat capacity, J/kg K

$\tilde{C}_p$ Molar-based heat capacity, J/mol K

d_p Packing size, m

d_{pore} Pore diameter, m

D Diameter, m

$\mathcal{D}$ Self-diffusivity, m^2/s

$\mathcal{D}_{A,B}$ Binary diffusion coefficient in a mixture of components A and B, m^2/s

$\mathcal{D}^0_{A,B}$ Diffusion coefficient of solute A in solvent B at infinite dilution, m^2/s

$\mathcal{D}^K_A$ Knudsen diffusivity of component A in a pore, m^2/s

$\mathcal{D}_{A,\text{pore}}$ Overall diffusivity of component A in a pore, m^2/s

$\mathcal{D}_{A,e}$ Effective pore diffusivity of component A in a porous material, m^2/s

$\bar{\mathcal{D}}_A$ Diffusivity of ion A in an ion exchange resin, m^2/s

$\bar{\mathcal{D}}_{AB}$ Effective diffusivity for counter-diffusion of ions A and B in an ion exchange resin, m^2/s

$\mathcal{D}'_A$ Limiting ionic diffusivity of ion A in solution, m^2/s

E_D Eddy mass diffusivity, m^2/s

E_H Eddy thermal diffusivity, m^2/s

E_M	Eddy kinematic viscosity, m^2/s
f	Friction factor
F	Force, N
F_{LG}	Mass velocity ratio for predicting flooding $[F_{LG} = \frac{L'}{G'}\left(\frac{\rho_G}{\rho_L}\right)^{1/2}]$
F_p	Packing factor for random packing, m^2/m^3
F_T	Correction factor to log-mean temperature difference for a shell and tube heat exchanger with multiple passes
$\mathcal{F}$	Faraday's constant [$\mathcal{F}$ = 96,500 Coulomb/mol]
g	Gravity constant, m/s^2
G	Gas molar velocity, mol/m^2 s, or power generated per unit volume, W/m^3
G'	Gas mass velocity, kg/m^2 s
G_M	Gas molar flow rate, mol/s
G_s	Molar velocity of inert gas, mol/m^2 s
G'_s	Mass velocity of inert gas, kg/m^2 s
h	Average heat transfer coefficient, W/m^2 K
h_x	Local heat transfer coefficient at a position x, W/m^2 K
$\hat{h}$	Specific enthalpy of humid air, J/kg dry air
$\hat{h}_s$	Specific enthalpy of humid saturated air, J/kg dry air
H	Height, m
H_G	Height of a gas-phase mass transfer unit [$H_G = G/ak_y$], m
H_L	Height of a liquid-phase mass transfer unit [$H_L = L/ak_x$], m
$H_{o,G}$	Height of an overall gas-phase mass transfer unit [$H_{o,G} = G/aK_y$], m
$H_{o,L}$	Height of an overall liquid-phase mass transfer unit [$H_{o,L} = L/aK_x$], m
$\hat{H}$	Specific enthalpy on a mass basis, J/kg
j_i	Mass flux of a component i relative to the mass-average velocity, kg i/m^2 s
J_i	Molar flux of a component i relative to the molar-average velocity, mol i/m^2 s
k	Thermal conductivity, W/m K, or second-order reaction rate constant, m^3/mol s
k_B	Boltzmann constant [$k_B = 1.30 \times 10^{-23}$ J/molecule K]
k_c	Average mass transfer coefficient based on a concentration driving force, m/s
$k_{c,x}$	Local mass transfer coefficient based on a concentration driving force at a position x, m/s
h_d	Fouling coefficient, W/m^2 K
k_p	Average mass transfer coefficient based on a partial pressure driving force, mol/m^2 s Pa
k_W	Average mass transfer coefficient based on a mass ratio driving force, mol/m^2 s

k_x	Average mass transfer coefficient based on a liquid-phase mole fraction driving force, mol/m^2 s
k_y	Average mass transfer coefficient based on a gas-phase mole fraction driving force, mol/m^2 s
k_I	First-order or pseudo-first-order reaction rate constant, 1/s
K	Equilibrium constant describing partitioning of a component i between two phases as a mole fraction ratio $[K = y_i/x_i]$
K_c	Equilibrium constant describing partitioning of a component i between two phases as a concentration ratio $[K_c = c_i/C_i]$
$K_{D,i}$	Partition coefficient of a component i in a pore defined as the ratio of accessible and total pore volumes
K_x	Overall liquid-side mass transfer coefficient based on a liquid-phase mole fraction driving force, mol/m^2 s
K_y	Overall gas-side mass transfer coefficient based on a gas-phase mole fraction driving force, mol/m^2 s
L	Length, m, or liquid molar velocity, mol/m^2 s
L'	Liquid mass velocity, kg/m^2 s
L_M	Liquid molar flow rate, mol/s
L_s	Molar velocity of solute-free absorbent liquid, mol/m^2 s
m	Mass, kg, or mass flow rate, kg/s
m_i	Mass of component i, kg i, or mass flow rate or mass transfer rate of component i, kg i/s
M	Total moles, mol, or total molar flow rate, mol/s
M_i	Moles of component i, mol i, or molar flow rate or molar mass transfer rate of component i, mol i/s
M_r	Molecular mass, kg/mol or kg/kmol
$M_{t,i}$	Total or cumulative mass transfer rate of component i, mol i/s
n_i	Mass flux of component i, kg i/m^2 s
N	Fin modulus or reaction-diffusion modulus for mass transfer with reaction
N_i	Molar flux of component i, mol of i/m^2 s
N_G	Number of gas-phase mass transfer units
N_L	Number of liquid-phase mass transfer units
$N_{G,\hat{h}}$	Number of gas-phase enthalpy transfer units
$N_{o,G}$	Number of overall gas-phase mass transfer units
$N_{o,L}$	Number of overall liquid-phase mass transfer units
N_t	Number of tubes per pass in a shell and tube heat exchanger
$\dot{N}$	Revolution rate, rotations per second
p	Perimeter of the cross-section of a fin, m

p_i	Partial pressure of component i [$p_i = y_i P$], Pa
p_i^*	Vapor pressure of component i, Pa
P	Absolute pressure, Pa, or power, W
P_c	Critical pressure, Pa
P_r	Reduced pressure [$P_r = P/P_c$]
q	Heat transfer flux, W/m^2
q_e	Total thermal energy flux, W/m^2
Q	Heat transfer rate, W
Q_{ideal}	Heat dissipation rate in an ideal fin at a uniform temperature equal to the base temperature, W
Q_t	Total or cumulative heat transfer rate, W
r	Radial coordinate, m
r_A	Molecule radius, m, or net rate of mass generation of component A, kg A/m^3 s
r_H	Hydrodynamic radius of a molecule defined as the radius of a sphere having the same diffusivity, m
r_i	Pipe inner radius, m
r_o	Pipe outer radius, m
r_{pore}	Pore radius, m
R	Heat transfer resistance, K/W, or radius of a cylinder or sphere, m
R_A	Net molar rate of generation of component A, mol A/m^3 s
R_{overall}	Overall heat transfer resistance, K/W
R_e	Electrical resistance, Ω
$\mathcal{R}$	Ideal gas law constant [$\mathcal{R}$ = 8.314 J/mol K or 82.06 cm^3 atm/mol K]
S	Area perpendicular to the mass transfer flux or cross-sectional area of an absorption column, m^2
t	Time, s
T	Temperature, K
T_b	Normal boiling point, K
T_c	Critical temperature, K
T_r	Reduced temperature [$T_r = T/T_c$]
T_{sat}	Saturation temperature, K
T_{WB}	Wet bulb temperature, K
u	Dimensionless velocity
U_o	Overall heat transfer coefficient based on the outer radius of pipe, W/m^2 K
$U_{o,d}$	Overall heat transfer coefficient based on the outer radius of pipe for a fouled heat exchanger, W/m^2 K
$\hat{U}$	Specific internal energy on a mass basis, J/kg

v	Fluid velocity, m/s
v_i	Velocity of a component i in a mixture, m/s
v_{avg}	Average velocity of a fluid in a pipe or in a falling film, m/s
v^*	Molar-average velocity of a mixture, m/s
v_i'	Velocity of component i in a mixture relative to the mass-average velocity, m/s
v_i''	Velocity of component i in a mixture relative to the molar-average velocity, m/s
V	Applied voltage, V, or volume, m^3
V_f	Volume of fluid, m^3
$\dot{V}$	Volumetric flow rate, m^3/s
$\hat{V}$	Specific volume on a mass basis, m^3/kg
$\hat{V}_A$	LaBas volume or molar volume of component A in the liquid phase at the normal boiling point, cm^3/mol
w_i	Mass fraction of component i in a mixture
W	Width of a flat plate, m
W_i	Mass ratio of component i in a mixture with an inert component B, kg i/kg B
x	Axial coordinate, m
x_i	Mole fraction of component i in a liquid mixture
X_i	Mole ratio of component i in a liquid mixture with an inert component B, mol i/mol B
y	Axial coordinate, m
y_i	Mole fraction of component i in a gas mixture
Y	Dimensionless temperature or dimensionless concentration
Y_{avg}	Dimensionless average temperature or dimensionless average concentration
Y_i	Mole ratio of component i in a gas mixture with an inert component B, mol i/mol B
z	Axial coordinate, m
Z	Height of packing, m

Greek Symbols

α	Thermal diffusivity $[\alpha = k/\rho C_p]$, m^2/s, or ratio of ion diffusivities $[\alpha = \bar{\mathcal{D}}_A/\bar{\mathcal{D}}_B]$
β	Concentration-diffusivity ratio for mass transfer with reaction $[\beta = \mathcal{D}_B c_{B,b}/v\mathcal{D}_A c_{A,s}]$
$\hat{\beta}$	Thermal expansion coefficient $[\hat{\beta} = -\frac{1}{\rho}\frac{d\rho}{dT}]$, K^{-1}
δ	Thickness of the momentum transfer boundary layer, m
δ_D	Thickness of the mass transfer boundary layer, m
δ_T	Thickness of the heat transfer boundary layer, m

Δ	Slab or film thickness, m
$\Delta\hat{H}_v$	Latent heat of vaporization, J/kg
ε	Porosity or void fraction of a packed bed
ε_e	Emissivity
ε_i	Lennard-Jones parameter, J/molecule
ε_p	Porosity or void fraction of a porous material
$\bar{\varepsilon}$	Agitator power input per unit mass, W/kg
γ	Stripping factor $[\gamma = KG_M/L_M = KG/L]$
λ	Mean free path, m
λ_i	Limiting equivalent conductance of ion i in solution, Coulomb cm^2/s equivalent V
$\lambda_{m,i}$	Limiting molar conductance of ion i in solution, Coulomb cm^2/s mol V
ϕ	Angular coordinate in the spherical coordinate system, rad, or enhancement factor for mass transfer with reaction, Eq. (16.9)
ϕ_B	Solvent association parameter in the Wilke–Chang equation, Eq. (10.24)
φ_A	Ratio of molecule radius and pore radius $[\varphi_A = r_A/r_{\text{pore}}]$
Λ	Capacity ratio for heat transfer $[\Lambda = \rho_f C_{p,f} V_f/\rho C_p V]$ or for mass transfer $[\Lambda = V_f/VK_c]$
η	Fin efficiency $[\eta = Q/Q_{\text{ideal}}]$ or Boltzmann transformation variable $[\eta = x/2\sqrt{\alpha t}$ for heat transfer or $\eta = x/2\sqrt{\mathcal{D}^0_{A,B}t}$ for mass transfer]
μ	Viscosity, Pa s
ν	Kinematic viscosity $[\nu = \mu/\rho]$, m^2/s
θ	Angular coordinate in cylindrical and spherical coordinate systems, rad, or dimensionless temperature
θ_i	Dimensionless concentration of component i
ρ	Density, kg/m^3, or dimensionless radial coordinate
ρ_e	Electrical resistivity, Ω m
ρ_i	Mass concentration of component i in a mixture, kg i/m^3
σ	Lennard-Jones parameter, Å, Stefan-Boltzmann constant $[\sigma = 5.68 \times 10^{-8}$ W/m^2 K^4], or surface tension, N/m
τ	Dimensionless time $[\tau = \alpha t/\ell^2$ for heat transfer or $\tau = \mathcal{D}^0_{A,B}t/\ell^2$ for mass transfer where $\ell = B$ for slabs and $\ell = R$ for cylinders and spheres]
τ_p	Tortuosity factor for diffusion in a porous material
τ_w	Shear stress, Pa
υ	Stoichiometric coefficient in the bimolecular reaction $A + \upsilon B \rightarrow P$
ξ	Dimensionless distance $[\xi = x/B$ for a slab or $\xi = r/R$ for a cylinder or for a sphere]
Ω_D	Collision integral for diffusivity based on the kinetic theory of gases, Eq. (10.16)

Ω_k	Collision integral for thermal conductivity based on the kinetic theory of gases, Eq. (2.6)
Ω_μ	Collision integral for viscosity based on the kinetic theory of gases, Eq. (2.6)
Ψ	Function defined by Eq. (15.26)
ψ_A	Hindrance factor for macropore diffusion, Eqs. (10.32)–(10.33)
ζ	Dimensionless distance from the entrance of a pipe with radius R [$\zeta = z/(R\,\mathrm{Re}\,\mathrm{Pr})$ for heat transfer or $\zeta = z/(R\,\mathrm{Re}\,\mathrm{Sc})$ for mass transfer]
Z	Dimensionless length of a pipe with radius R [$Z = L/(R\,\mathrm{Re}\,\mathrm{Pr})$ for heat transfer or $Z = L/(R\,\mathrm{Re}\,\mathrm{Sc})$ for mass transfer]

Dimensionless Transport Groups

Bi	Biot number [$\mathrm{Bi} = h\ell/k_{\mathrm{solid}}$ for heat transfer or $\mathrm{Bi} = k_c B/(\mathcal{D}_{A,B})_{\mathrm{solid}}$ for mass transfer where $\ell = B$ for heat or mass transfer in a slab with thickness B or $\ell = R$ for heat or mass transfer in a cylinder or in a sphere with radius R. k_{solid} and $(\mathcal{D}_{A,B})_{\mathrm{solid}}$ are the thermal conductivity of the solid and the diffusivity in the solid, respectively]
Gr	Grashof number [$\mathrm{Gr} = g\hat{\beta}L^3\Delta T/\nu^2$]
$\mathrm{J_D}$	Chilton-Colburn factor for mass transfer [$\mathrm{J_D} = \mathrm{Sh}/(\mathrm{Re}\,\mathrm{Sc}^{1/3})$]
$\mathrm{J_H}$	Chilton-Colburn factor for heat transfer [$\mathrm{J_H} = \mathrm{Nu}/(\mathrm{Re}\mathrm{Pr}^{1/3})$]
Le	Lewis number [$\mathrm{Le} = \mathrm{Pr}/\mathrm{Sc}$]
Nu	Average Nusselt number [$\mathrm{Nu} = hL/k$ for flow over a flat plate, $\mathrm{Nu} = hD/k$ for flow over a cylinder or over a sphere with diameter D, $\mathrm{Nu} = hD/k$ for flow in a pipe with inner diameter D]
Nu_x	Local Nusselt number [$\mathrm{Nu_x} = h_x x/k$ for flow over a flat plate, $\mathrm{Nu_x} = h_x D/k$ for flow in a pipe with inner diameter D]
Ra	Rayleigh number [$\mathrm{Ra} = \mathrm{Gr}\,\mathrm{Pr}$]
Re	Reynolds number [$\mathrm{Re} = v_\infty L/\nu$ for flow over a flat plate of length L, $\mathrm{Re} = \rho v_\infty D/\mu$ for flow over a cylinder or over a sphere with diameter D, $\mathrm{Re} = \rho v_{\mathrm{avg}} D/\mu$ for flow in a pipe of inner diameter D]
Re_x	Local Reynolds number [$\mathrm{Re_x} = \rho v_\infty x/\mu$ for flow over a flat plate]
Pe_L	Peclet number [$\mathrm{Pe}_L = Lv/\mathcal{D}_{\mathrm{A,B}}$]
Pr	Prandtl number [$\mathrm{Pr} = \nu/\alpha = \mu C_p/k$]
Sc	Schmidt number [$\mathrm{Sc} = \nu/\mathcal{D}_{A,B} = \mu/\rho\mathcal{D}_{A,B}$]
Sh	Sherwood number [$\mathrm{Sh} = k_c L/\mathcal{D}_{A,B}$ for flow over a flat plate, $\mathrm{Sh} = k_c D/\mathcal{D}_{A,B}$ for flow over a cylinder or over a sphere with diameter D, $\mathrm{Sh} = k_c D/\mathcal{D}_{A,B}$ for flow in a pipe with inner diameter D]
Sh_x	Local Sherwood number [$\mathrm{Sh}_x = k_{c,x} x/\mathcal{D}_{A,B}$ for flow over a flat plate, $\mathrm{Sh}_x = k_{c,x} D/\mathcal{D}_{A,B}$ for flow in a pipe with inner diameter D]
$\mathrm{St_D}$	Stanton number for mass transfer [$\mathrm{St_D} = k_c/v_\infty$]
$\mathrm{St_H}$	Stanton number of heat transfer [$\mathrm{St_H} = h/\rho C_p v_\infty$]

Subscripts

a	Ambient value
b	Value in bulk fluid
f	Fluid value
lm	Logarithmic mean value
max	Maximum value
min	Minimum value
s	Surface value
0	Initial or boundary value
∞	Value away from the surface

PART I

Heat Transfer

CHAPTER 1

Introduction to Heat Transfer

1.1 Introduction

Most readers of this textbook are likely to be junior-level chemical engineering students, who are fresh from their experience with material and energy balances and with the calculation of thermodynamic properties. Thus, we will start here with an example of application of these concepts and tools to highlight both their value and their limitations by contrasting energy balance problems with heat transfer problems.

Example 1.1. Heat exchanger design problem. Let's consider the task of designing a heat exchanger to heat 12.0 kg/s of liquid toluene at 1 atm from 30°C to 95°C using saturated steam at 75 psia as the energy source. It is specified that the steam will leave the heat exchanger as subcooled liquid at 70°C. The problem at hand encompasses two different dimensions. The first is about flow or quantities of energy and materials, while the second is about equipment size. We can solve for the first dimension using what we already know from mass and energy balances. However, we will need to learn heat transfer concepts before we can address the second dimension. We begin with a simple process flow diagram (PFD) showing all labeled input and output streams as shown in Fig. 1.1. The PFD also includes the quantity Q_t, which is the rate of energy transfer (J/s or BTU/h) between the condensing steam and the liquid toluene. We assume that the heat exchanger itself is adiabatic; i.e., there are no heat losses or gains to/from the surrounding. This is a reasonable assumption for a process unit that we expect will be quite large in size. Note that the particular arrangement of mass flows in and out of the unit is not important for the time being.

Next, we list the unknowns in this process that we can solve for based on material and energy balances and thermodynamics alone. These unknowns are:

1. The temperature of the inlet steam, T_3;
2. The flow rate of the inlet steam, m_3, which is equal to the condensate flow rate; and
3. The total rate of energy transfer between two streams, Q_t.

We can find the inlet steam temperature from a steam table. For example, from Table 2-135 in *Perry's Chemical Engineers' Handbook*, 9th edition, we find that the saturated steam temperature at 75 psia (0.52 MPa) is T_{sat} = 153°C. Fortunately, this is higher than the temperature of the toluene stream leaving the unit, which means that the process can actually work as specified. If the steam temperature were lower than 95°C, the process could not work since it would be impossible to heat the toluene to the specified temperature. Next, we can find the steam flow rate from an overall energy balance. To do this, we need to determine what the phase of the outlet toluene will be. From Table 2-1 in *Perry's Chemical Engineers' Handbook*, we find that the normal boiling

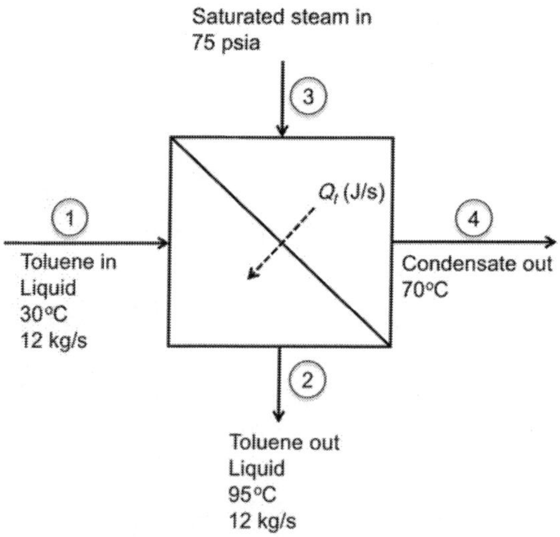

FIGURE 1.1 Process flow diagram for a heat exchanger unit.

point of toluene is 110.8°C. Thus, at 95°C, toluene will be liquid. The overall energy balance can be written as follows:

$$m_1\hat{H}_1 + m_3\hat{H}_3 = m_2\hat{H}_2 + m_4\hat{H}_4 \tag{1.1}$$

where m_i is the mass flow rate (kg/s) of stream i and $\hat{H}_i$ is the corresponding specific enthalpy on a mass basis (kJ/kg). Since $m_1 = m_2 = m_{\text{toluene}}$ and $m_3 = m_4 = m_{\text{steam}}$, from Eq. (1.1), we obtain:

$$m_{\text{steam}} = \frac{\hat{H}_2 - \hat{H}_1}{\hat{H}_3 - \hat{H}_4} m_{\text{toluene}} \tag{1.2}$$

We can find $\hat{H}_3$ and $\hat{H}_4$ again from the steam tables in *Perry's Chemical Engineers' Handbook*. Their values are 2,748 and 293 kJ/kg, respectively. Next, we find $\hat{H}_2 - \hat{H}_1$ from the following integral:

$$\hat{H}_2 - \hat{H}_1 = \int_{T_1}^{T_2} C_p(T)\,dT \tag{1.3}$$

where $C_p(T)$ is the heat capacity of liquid toluene. From Perry's Table 2-72, we find:

$$C_p = 1.52 - 1.65 \times 10^{-3}T + 7.55 \times 10^{-6}T^2$$

with C_p in kJ/kg K and T in K. Integrating Eq. (1.3) gives:

$$\hat{H}_2 - \hat{H}_1 = \int_{30+273}^{95+273} (1.52 - 1.65 \times 10^{-3}T + 7.55 \times 10^{-6})\,dT$$

$$= 1.52(368 - 303) - \frac{1.65 \times 10^{-3}}{2}(368^2 - 303^2) + \frac{7.55 \times 10^{-6}}{3}(368^3 - 303^3) = 118 \text{ kJ/kg}$$

Inserting these quantities into Eq. (1.2,) we obtain the final result:

$$m_{\text{steam}} = \frac{118}{2{,}748 - 293} 12.0 = 0.58 \text{ kg/s}$$

Note that this calculation indicates that a small mass flow rate of steam will be sufficient to heat a large flow of toluene. This is because of the large latent heat of vaporization which is transferred to the toluene as the steam condenses. We can now find the rate of energy transfer, Q_t, from an energy balance around the toluene stream, which gives:

$$m_1\hat{H}_1 + Q_t = m_2\hat{H}_2 \tag{1.4}$$

Solving this equation for Q_t we finally obtain:

$$Q_t = m_2\hat{H}_2 - m_1\hat{H}_1 = m_{\text{toluene}}(\hat{H}_2 - \hat{H}_1) = 12.0 \times 119 = 1{,}430 \text{ kJ/s}$$

At this point, we have exhausted what we can do with energy balances and the first law of thermodynamics. To go further, we need to turn our attention to heat transfer concepts in order to determine the following:

1. The flow configuration of the heat exchanger (i.e., the direction in which the streams will actually go in and out); and
2. The physical size of the heat exchanger.

Let's assume that the heat exchanger will consist of pipes for the toluene and steam to flow in and out. The pipes will keep the two streams separate while allowing thermal contact for the transfer of energy between the two streams. For example, the toluene could flow inside the pipes while steam will be condensing outside the pipes. Heat transfer will occur through the walls of the pipes. We can consider two possible configurations as shown in Fig. 1.2: co-current and counter-current. Both configurations are consistent with the PFD in Fig.1.1. However, in the co-current configuration, the

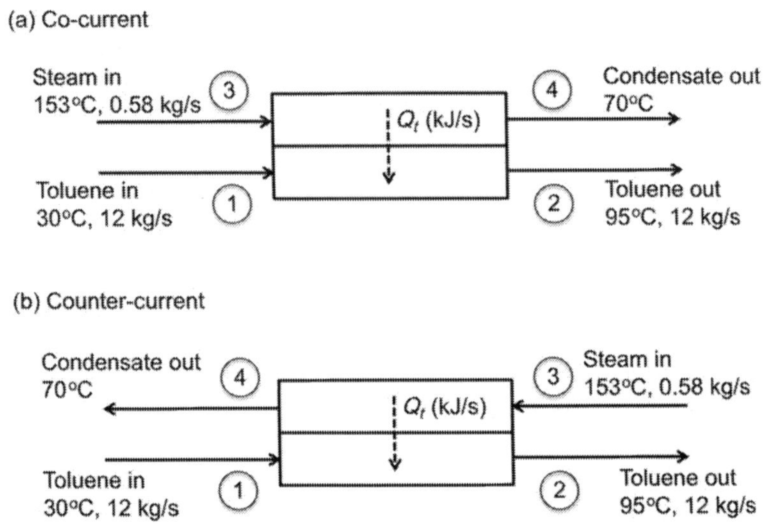

FIGURE 1.2 Co-current and counter-current configurations for Example 1.1.

two streams flow in the same direction, while in the counter-current configuration, the two streams flow in opposite directions. In this case, we can easily determine that one of these two configurations will not work—the co-current one.

To make this determination, we recall from the second law of thermodynamics that heat will flow from a body at a higher temperature to one at a lower temperature. Let's follow what happens to the stream temperatures as we move along the heat exchanger from left to right. For the co-current case in Fig. 1.2, initially, heat will transfer from the hotter steam, starting at 153°C, to the cooler toluene, starting at 30°C. However, as the steam condenses and then cools along the tube and the toluene heats up, we can see that we would reach a point where the steam and toluene temperatures will become equal to each other. At that point, heat transfer between the two streams will stop, making it impossible for the outflowing condensate to reach a temperature that is lower than that of the outflowing toluene. As seen in Fig. 1.2, the counter-current configuration does the trick. As the steam condenses and then cools in one direction, the toluene heats up in the opposite direction, so that the two stream temperatures will never become equal. While, in this particular case, only one configuration is possible, we will see that in other cases, both co-current and counter-current can work. Which configuration will be chosen in those cases will then depend on other considerations, including the efficiency of the heat transfer process and the local rate of heating or cooling.

Having selected the counter-current configuration for this process, the next step will be to determine the physical size of the unit. We are initially interested in determining the heat transfer area, A, that is required to achieve the necessary rate of energy transfer, Q_t. If the heat exchanger consists of pipes, A will be the surface areas of the pipes, perpendicular to the direction of heat transfer. Determining A will, in turn, require that we describe the local rate of heat transfer, which, in general, will not be uniform along the length of the pipes. To this end, we define the heat transfer flux q, typically in units of J/m^2 s or W/m^2, as the local rate of heat transfer per unit cross-sectional area. We obtain this quantity by dividing the rate of energy transfer by the area through which heat transfer occurs. On the average, $q_{avg} = Q_t/A$. In reality, however, q will vary along the length of heat exchanger. Intuitively, the larger the temperature difference between the two streams is, the larger the heat transfer flux will be. At the point where the steam goes in, the temperature difference between the steam and the toluene is $\Delta T = T_3 - T_2 = 153 - 95 = 58$°C. Conversely, at the point where the condensate comes out, the temperature difference between the streams is $\Delta T = T_4 - T_1 = 70 - 30 = 40$°C. Thus, we will expect a larger heat flux where the steam goes into the heat exchanger compared to the point where the condensate comes out. We will learn that, in practice, other factors, including the physical properties of the two fluids, will determine the actual flux. Since the heat flux varies along the heat exchanger, finding the area will require that we calculate the average flux accounting for all of these factors giving $A = Q_t/q_{avg}$. Once the area is known, we will then be able to calculate the length of the pipes, L, if their radius r_o is known since $A = 2\pi r_o L$. How exactly this calculation will be done will be shown later once we have mastered the mechanisms through which heat transfer occurs and how we can calculate the local value of q and its average value q_{avg}.

1.2 Heat Transfer Rate and Heat Transfer Flux

As discussed in Example 1.1, we define the rate of heat transfer across a certain area A as Q (or Q_t), typically measured in units of J/s or W or in units of BTU/h. Accordingly, the heat transfer flux $q = Q/A$ will typically be measured in units of J/s m^2 or W/m^2 or in units of BTU/h ft^2. Note that q is actually a vector. For example, in rectangular coordinates q can be written as:

$$q = q_x\vec{i} + q_y\vec{j} + q_z\vec{k} \tag{1.5}$$

where q_x, q_y, and q_z are the x, y, and z components and $\vec{i}$, $\vec{j}$, and $\vec{k}$ are the corresponding unit vectors. A is the area perpendicular to the flux vector. For example, in Example 1.1,

A is the pipe surface area through which heat transfer takes place. The sign of the q-vector components will be taken to be positive when heat transfer occurs in the positive direction of the corresponding axis.

1.3 Heat Transfer Mechanisms

There are three distinct mechanisms through which heat is transferred from one region of space to another across a surface area A:

1. Conductive heat transfer;
2. Convective heat transfer; and
3. Radiant heat transfer.

These mechanisms can occur individually or in combination. Their relative importance will vary widely depending on the physical system and on the range of temperatures that are involved in the process.

Conductive heat transfer can occur in the absence of fluid motion and is driven by a spatial temperature gradient according to Fourier's law that relates the heat transfer flux to the temperature gradient within a certain phase. The latter can be a fluid or a solid. Along the x-coordinate of a rectangular coordinate system, Fourier's law is given by the following equation:

$$q_x = -k\frac{dT}{dx} \tag{1.6}$$

where k is the thermal conductivity of the medium where conductive heat transfer occurs. The minus sign is included so that the flux q_x will be positive in the direction of decreasing temperature; i.e., when dT/dx is negative. k is a physical property that, in general, will depend on temperature and pressure. Since we can write

$$k = -\frac{q_x}{dT/dx}, \tag{1.7}$$

the dimensions of k are (power/area)/(temperature/distance). Thus, k will typically be expressed in units of J/s m K or W/m K or in units of BTU/h ft°F. In gases and liquids, conductive heat transfer depends on intermolecular collisions, which transfer kinetic energy from one atom or molecule to another. Thus, in general, heat conduction will be faster in higher density fluids. In solids, conductive heat transfer depends on atomic or molecular vibrations. As a rule, heat conduction will be faster in solids with higher electrical conductivity.

We can see that although the physical basis is completely different, Eq. (1.6) is formally analogous to Newton's law for momentum transport in a fluid. This law defines the momentum transport flux, as follows:

$$\tau_{y,x} = \mu\frac{dv_y}{dx} \tag{1.8}$$

where μ is the viscosity of the fluid and dv_y/dx is the velocity gradient. We will see that beside the mathematical equivalence, this analogy between momentum and heat transfer has important practical implications.

System	k (W/m K)*	Dependence on Temperature
Gases at 1 atm	0.005–0.2	~Linear
Liquids (except molten metals)	0.1–1	Weak
Solids	1–500	Weak

*divide by 1.731 to get BTU/h ft°F.

TABLE 1.1 Ranges of Thermal Conductivity Values Encountered in Practice

For a homogenous or "isotropic" material, the thermal conductivity is the same in all directions but may vary with position as a result of temperature and/or density changes. Conversely, for non-isotropic materials, k may depend on the direction of heat transfer. This can be the case for liquid crystals or for composite materials, which may be designed to conduct heat more readily in one direction than another. Experimental values of the thermal conductivity and estimation methods are discussed in Chap. 2. The values of k encountered in practice vary widely, but they typically fall in the ranges indicated in Table 1.1. The lowest values are found for gases at low pressure, higher values are usually found for liquids, and even higher values are found for many solids. Most metals, which have high electrical conductivity, will also have very high thermal conductivity. Note that, for gases, the thermal conductivity typically varies linearly with temperature, while for liquids and solids it is relatively constant.

Convective heat transfer is associated with fluid motion at the interface between two phases, such as the interface between a moving fluid and a stationary solid or the interface between two fluids in relative motion. The flow can be generated with a pump, a fan, a stirrer, or another mechanical device. In this case, we refer to the process as *forced convection*. Or it can be generated in a fluid because of the existence of a density gradient, which, in turn, results from a temperature gradient. In this case, we refer to the process as *natural* or *free convection*, since heat transfer, along with the ensuing temperature gradient, is itself responsible for fluid motion. Convective heat transfer can also occur when a vapor condenses on a cold surface or when a liquid is vaporized on a hot surface.

Figure 1.3 shows an example of forced convection, where a fluid at temperature T_∞ with free stream velocity v_∞ flows past the edge of a wide flat plate kept at a temperature T_s. At the plate surface, the fluid velocity will be zero while the fluid temperature will be

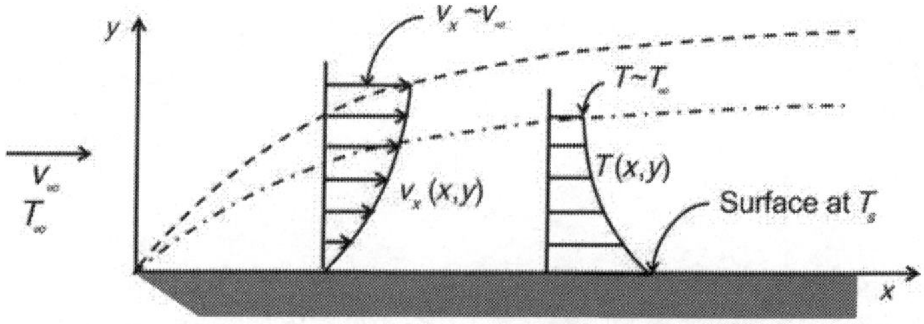

FIGURE 1.3 Velocity and temperature profiles near the edge of a flat plate kept at temperature T_s.

equal to T_s. Beyond the surface, velocity and temperature gradients will develop starting at the leading edge of the plate and evolving along the x-coordinate with velocity and temperature reaching the corresponding free-stream values v_∞ and T_∞, respectively, at increasing distances δ_M and δ_T from the plate surface. These distances define the momentum and thermal boundary layers, respectively. Both will depend on the free-stream velocity and on the fluid properties. Within the boundary layer, heat transfer will occur from the plate surface to the fluid (or vice-versa depending on whether T_s is higher or lower than T_∞) as a result of the combination of conduction, which arises because of the temperature gradient, and fluid motion, which arises from the existence of a velocity gradient at the surface. This combination is what we call convective transport. In this case, we can express the heat-transfer flux according to the following equation:

$$q_y = h_x(T_s - T_\infty) \tag{1.9}$$

where h_x is the local convective heat transfer coefficient at the fluid-solid interface. The subscript x indicates that this is the coefficient found at a position x along the plate surface. Since we can write

$$h_x = \frac{q_y}{T_s - T_\infty} \tag{1.10}$$

the dimensions of h_x are (power/area)/(temperature). Thus, h_x will typically be measured in units of J/s m^2 K or W/m^2 K or in units of BTU/h ft^2°F. h_x and, thus, q_y, will depend on v_∞ and on the physical properties of the fluid and will vary over the distance x from the edge of the plate. As a result, it is useful to define a heat transfer coefficient, h, that is averaged over a distance L as follows:

$$h = \frac{1}{L}\int_0^L h_x(x)dx \tag{1.11}$$

Accordingly, the rate of heat transfer over a plate of length L and width W will be given by:

$$Q = Ah(T_s - T_\infty) \tag{1.12}$$

where $A = L \times W$ is the heat-transfer area. Note that Eq. (1.12) is analogous to the following equation expressing the force exerted by the fluid on the plate:

$$F = A\tau_w = Af\frac{\rho v_\infty^2}{2} \tag{1.13}$$

where τ_w is the shear stress at the plate surface, f is the friction factor, ρ is the density of the fluid, and $\rho v_\infty^2/2$ is the specific kinetic energy of the fluid. Comparing Eqs. (1.12) and (1.13), we can see that, mathematically, f and $\rho v_\infty^2/2$ play roles in determining the force on the plate that are analogous to those played by h and $T_s - T_\infty$ in determining the heat transfer rate. We will refer to the difference $T_s - T_\infty$ between the temperature of the fluid at the surface and the bulk fluid temperature as the driving force for heat transfer.

In practical applications, convective heat transfer coefficients will vary widely depending on the fluid properties, whether forced or free convection occurs, whether condensation or boiling takes place, the geometry of the surface, and the flow conditions (e.g., whether the flow is laminar or turbulent). Methods for the prediction of these coefficients are, thus, critically important and are presented in Chap. 6.

System/Process	h (W/m² K)*
Air/free convection	5–50
Air/forced convection	25–250
Water/forced convection	250–15,000
Boiling	2,500–25,000
Condensation	5,000–100,000

*divide by 5.68 to convert to BTU/h ft² °F.

TABLE 1.2 Ranges of Convective Heat Transfer Coefficient Values Encountered in Practice

Typical values of the heat transfer coefficients encountered in practice usually fall in the ranges indicated in Table 1.2. As seen in Table 1.2, h is usually the lowest for gases when only natural convection occurs and the highest for systems where boiling or condensation occurs. Gases and liquids under forced convection will result in intermediate values of h.

Radiant heat transfer is associated with the propagation of electromagnetic waves and can occur through transparent fluids, through transparent solids, or in vacuum. It is, of course, the mechanism through which energy is transported from the sun to the earth. Radiant heat transfer depends on the temperatures of the emitting and receiving surfaces between which heat transfer occurs, the nature of the materials of which these surfaces are made, and the geometric exposure of the two surfaces to each other. In general, radiant heat transfer between a surface at temperature T_1 and another at temperature T_2 is described by the Stefan-Boltzmann equation:

$$Q = AF_v \sigma \varepsilon_e (T_1^4 - T_2^4) \tag{1.14}$$

where A is the surface area, F_v is a geometric viewing factor that takes into account the different exposure of one surface to the other, σ is a universal constant with a value of 5.68×10^{-8} W/m² K⁴, and ε_e is the emissivity, a material surface property that determines the fraction of the incident electromagnetic wave that is absorbed. ε_e has a value of 1 for a so-called "black body," which is highly absorbing, and values near zero for highly reflective surfaces. Consider, for example, the case in Fig. 1.4 depicting radiant heat transfer between the flat wall of a furnace and a metal pipe through which a fluid may flow. In this case, A will be the pipe surface, T_1 the furnace wall temperature, and T_2 the pipe surface temperature. F_v will account for the fact that different portions of the pipe surface will have a different exposure to the furnace wall.

Note that T_1 and T_2 in Eq. (1.14) are absolute temperatures. Since these temperatures are to the fourth power, radiant heat transfer is especially important at elevated temperatures or at very low temperatures, near absolute zero, when the fourth power makes the driving force $T_1^4 - T_2^4$ large. As shown in Example 1.2, radiant heat transfer can be combined with other mechanisms.

Example 1.2. Combined convective and radiant heat transfer—Freezing of a puddle. We sometimes observe that early in the morning, after a very clear night, when the air is very still, water held in shallow pools over insulated surfaces, such as a roadway, can freeze even though the air temperature is

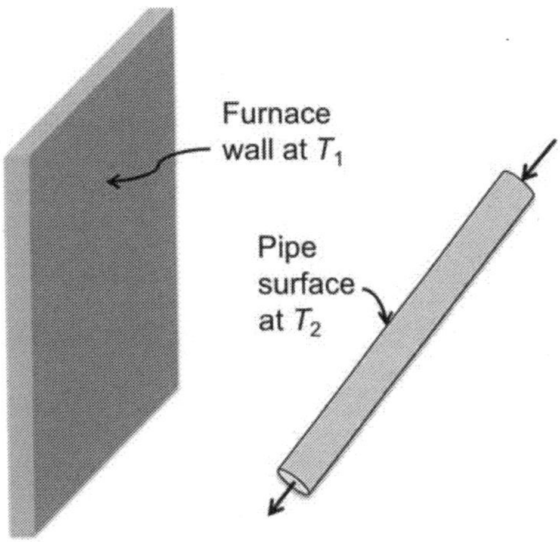

Figure 1.4 Radiant heat transfer between a furnace wall at T_1 and the surface of a pipe at T_2.

above 0°C. This phenomenon can be explained by the combination of convective and radiant heat transfer. Consider the situation in Fig. 1.5. If we assume that the shallow pool of water is insulated (i.e., there is no heat transfer to/from the roadway), heat will be absorbed only from the surrounding air at the rate:

$$Q_{\text{in}} = Ah(T_a - T_s)$$

where A is the surface area of the puddle, h is the convective heat transfer coefficient, and T_a and T_s are the air temperature and the water temperature, respectively. The value of the heat transfer coefficient, h, will depend on the air movement. If there is no wind, only natural convection will occur, which will lead to a relatively low value of h, say 40 W/m^2 K. While heat is gained from the air as a result of convective heat transfer, heat will also be lost to the surrounding by radiation

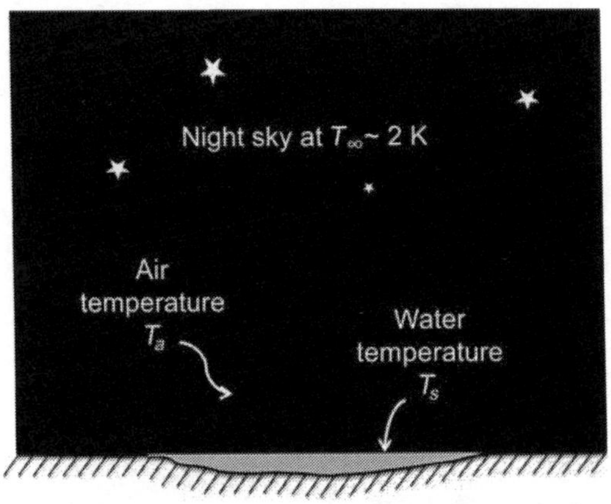

Figure 1.5 Freezing of an insulated shallow pool of water when the air temperature is above freezing according to Example 1.2.

according to Eq. (1.14). Assuming that $F_v = 1$ and taking $\varepsilon_e = 1$ for water, from Eq. (1.14) we will have:

$$Q_{\text{out}} = A\sigma(T_s^4 - T_\infty^4)$$

where T_∞ is the temperature of the night sky. The latter can be assumed to be 2 K. At steady state, we will have $Q_{\text{in}} = Q_{\text{out}}$ and, thus,

$$h(T_a - T_s) = \sigma(T_s^4 - T_\infty^4)$$

If the air temperature is known, we can solve this equation by trial and error to find T_s. For example, if $T_a = 5$°C (278 K), we obtain $T_s = 270$ K or −3°C. Of course, air movement will increase h, which, in turn, will increase T_s. Likewise, conductive heat transfer from the roadway to the puddle will also increase T_s. Thus, freezing with an air temperature of 5°C may not occur in certain practical situations, but it is obviously theoretically possible.

1.4 Problems

P.1.1. The thermal conductivity of a certain solid material is $k = 25$ W/m K. At a certain point and time in a block of this material sketched in Fig. P.1.1, the temperature is 65°C, the temperature gradient in the x direction is 0.20°C/m, and the cross-sectional area is 0.3 m^2. (a) Determine the magnitude and sign of the heat transfer flux, q_x, in W/m^2. (b) Determine the heat transfer rate, Q_x, in W. (c) Convert the values of the temperature gradient to °F/ft and the values of q_x and Q_x to BTU/h ft^2 and BTU/h, respectively.

P.1.2. A condenser at the top of a distillation column operated at 1 atm absolute pressure is to be designed to condense 0.50 kg/s of essentially pure isopropyl alcohol using cooling water entering the condenser at 25°C and leaving at 35°C. As a rough approximation, the heat transfer rate in the condenser is given by the equation:

$$Q \sim Ah(T_{iPA} - T_{w,\text{avg}})$$

where A is the heat transfer area, T_{iPA} is the temperature of the condensing isopropyl alcohol, and $T_{w,\text{avg}}$ is the average temperature of the cooling water. (a) Determine the heat transfer rate and the flow rate of the cooling water. (b) Determine the heat transfer area required for the heat exchanger if the heat transfer coefficient is $h = 1{,}000$ W/m^2 K. (c) Suppose that the condenser is

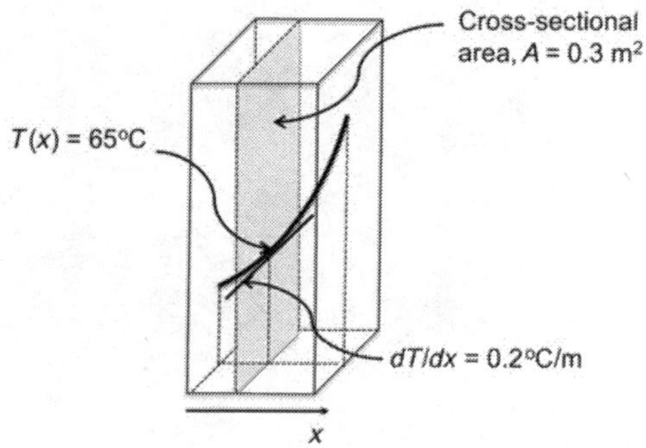

FIGURE P.1.1

made up of tubes that are 2.5 cm in diameter and 1.5 m long. How many tubes will be required for the heat exchanger?

P.1.3. Floor heating is used to keep a room at a toasty 25°C. On a cold day, the room loses heat through the walls, ceilings, door, and windows at the rate of 2,000 W. The heated floor area is 30 m^2. The heat transfer coefficient for convective transport from the heated floor area to the room air is estimated to be 5.0 W/m^2 K. (a) Determine the floor temperature. (b) If the maximum allowable floor temperature is to be 30°C, what should the value of the heat transfer coefficient be?

P.1.4. *Bona fide* Italian pizza is cooked in a domed brick oven. A wood fire is first lit inside the dome, which brings the brick temperature to something around 800°F. The fire is then pushed aside prior to putting the pizza in. Heat transfer to the pizza is primarily by conduction from the hot brick on the bottom and by a combination of convective and radiant heat transfer from the top. Assume that the pizza is 10 in. in diameter, that it starts at 60°F, and that the temperature of the air in the oven is 600°F (lower than the brick temperature because of the draft). Estimate the initial rate of heat transfer to the top of the pizza; i.e., the rate when the pizza is first put in the oven.

CHAPTER 2

Thermal Conductivity

2.1 Introduction

As discussed in Chap. 1, the thermal conductivity of the medium through which heat transfer occurs is a critical physical property that will determine the rate of heat conduction. Even in the case of convective heat transfer, the thermal conductivity will still play a key role. Thus, determination of the thermal conductivity will almost always be necessary to solve heat transfer problems. The thermal conductivity, k, has dimensions of power per unit length and temperature and is typically expressed in units of W/m K or BTU/h ft°F. According to Fourier's law of heat conduction, Eq. (1.6), k multiplied by the temperature gradient gives the heat transfer flux, q. As a result, k in W/m K can be regarded as the rate of energy transfer per unit area obtained with a 1 K/m temperature gradient. Obviously, for a given gradient, the higher is the thermal conductivity, the greater the heat transfer rate will be.

As shown in Table 1.1, the thermal conductivity varies widely from one material to another. For homogenous materials, k is smallest for gases at low pressure, intermediate for liquids, and greatest for metals. Composite materials, such as porous solids, foams, or laminated products, will have conductivities intermediate between those of the constituent materials. Some of these materials (e.g., laminates) can have different conductivity in different directions.

In many cases, actual heat transfer calculations will depend on experimentally determined values of thermal conductivity, including its temperature dependence. A convenient source of such values is Section 2 of *Perry's Chemical Engineers' Handbook*, 9th edition. Estimation methods are also available and can be used when no data can be found or to correlate experimental data as a function of temperature and pressure. However, their accuracy is often limited, and using experimentally determined values of the thermal conductivity is recommended to the extent possible for accurate calculations.

2.2 Experimental Thermal Conductivity Values

2.2.1 Gases

Table 2-145 in *Perry's Chemical Engineers' Handbook*, 9th edition, provides data for a substantial number of inorganic and organic vapors at 1 atm absolute pressure. In most cases, the data are correlated as a function of absolute temperature T in K by the following empirical equation:

$$k = \frac{c_1 T^{c_2}}{1 + c_3/T + c_4/T^2} \tag{2.1}$$

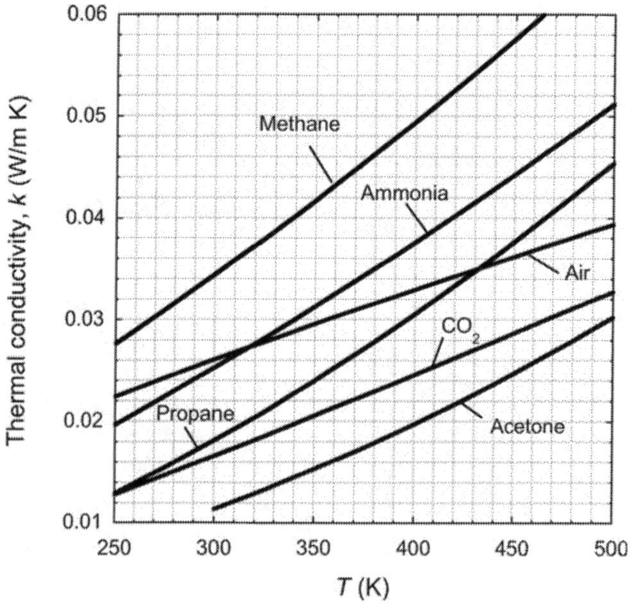

FIGURE 2.1 Thermal conductivity of representative gases and vapors at 1 atm absolute pressure. (Data from Section 2 of *Perry's Chemical Engineers' Handbook*, 9th edition.)

where c_1, c_2, c_3, and c_4 are constants specific for each component found in Table 2-145 in *Perry's Chemical Engineers' Handbook*, 9th edition. Figure 2.1 shows the values of k for some representative gases. As shown in Fig. 2.1, these k values are in the range of 0.01 to 0.1 W/m K. Hydrogen (not included in Fig. 2.1) is an exception having a thermal conductivity in the range of 0.2 to 0.6 W/m K, which is substantially higher than that of other gases. In most cases, over modest ranges of temperature, the thermal conductivity of gases at low pressures increases approximately linearly with temperature.

2.2.2 Liquids

Table 2-147 in *Perry's Chemical Engineers' Handbook*, 9th edition, provides data for a substantial number of inorganic and organic liquids. In most cases, the data are correlated empirically as a function of absolute temperature T in K by the following equation:

$$k = c_1 + c_2T + c_3T^2 + c_4T^3 + c_5T^4 \tag{2.2}$$

where c_1, c_2, c_3, c_4, and c_5 are constants specific for each component found in Table 2-147 in *Perry's Chemical Engineers' Handbook*, 9th edition. In a majority of cases, liquid thermal conductivities are only weakly dependent on temperature and generally decrease as the temperature is increased. Figure 2.2 shows the k values for some representative liquids. As shown in Fig. 2.2, typical values for organic liquids are in the range of 0.08 to 0.18 W/m K. Water has a very high thermal conductivity compared to organic liquids, averaging around 0.65 W/m K, with a slight maximum around 400 K. We know that water conducts heat very well when, for example, on a hot day, we take a dip in a pool or in the ocean. We get cooled quickly because water conducts heat rapidly. Liquid ammonia also has a high thermal conductivity with an unusually strong dependence on temperature.

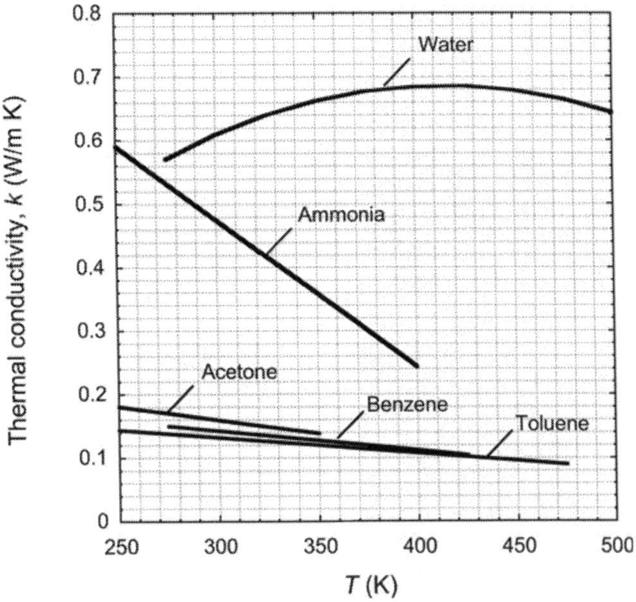

FIGURE 2.2 Thermal conductivity of representative liquids. (Data from Section 2 of *Perry's Chemical Engineers' Handbook*, 9th edition.)

2.2.3 Solids

Tables 2-149, 2-150, and 2-152 in *Perry's Chemical Engineers' Handbook*, 9th edition, provide data for a number of solids, metal alloys, and miscellaneous materials, respectively. In most cases, the thermal conductivities of solids are only weakly dependent on temperature and generally decrease as the temperature is increased. Figure 2.3 shows the k values for some representative solids. As shown in Fig. 2.3, k varies over very ample ranges from one material to the other, being highest for metals like aluminum and copper and lowest for non-metallic solids like silica. In general, there is a direct correlation between the thermal conductivity of a solid and its electrical conductivity. Typically, materials that have higher electrical conductivity will also have higher thermal conductivity. The thermal conductivity in most solids is an isotropic property; i.e., it is independent of the direction in which heat conduction takes place. There are exceptions, however, and the graphite data included in Fig. 2.3 provide an example. We can see that because of the larger k, heat conduction will be much easier when heat transfer is perpendicular to the basal plane of the graphite structure compared to when it is parallel to it.

Table 2.1 provides thermal conductivity data for some miscellaneous solid materials along with their density and heat capacity. Obviously, some of the materials included in Table 2.1 are essentially composites, such as paper, pine wood, or plywood, which may contain trapped air. Conductivity values in these cases will vary substantially depending on their exact structure and on whether the material is dry or wet. Table 2.1 is provided only for illustrative purposes, and original sources should be consulted when accurate values are needed.

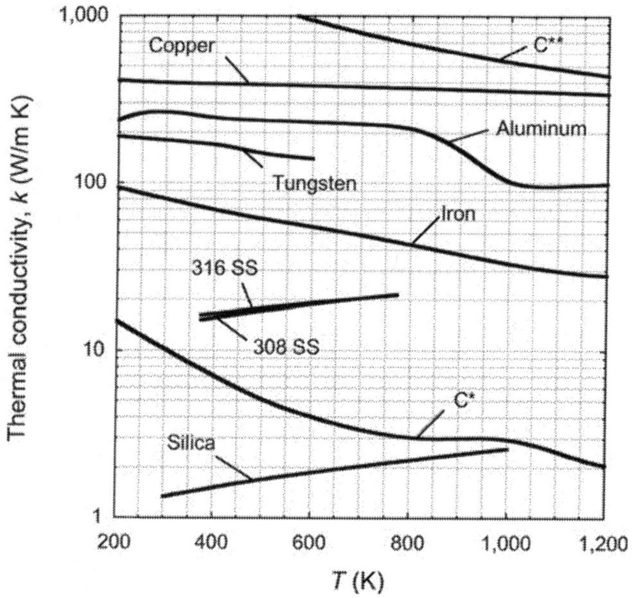

FIGURE 2.3 Thermal conductivity of representative solids. The curve labeled C* is the thermal conductivity of graphite for heat conduction parallel to the basal plane of the graphite structure, while the curve labeled C** is for heat conduction perpendicular to the basal plane. 308 SS and 316 SS are examples of nickel-chromium stainless steel alloys. (Data from Section 2 of *Perry's Chemical Engineers' Handbook*, 9th edition.)

Material	Density, ρ (kg/m^3)	Heat Capacity, C_p (J/kg K)	Thermal Conductivity, k (W/m K)
Brick	1,930	835	0.72
Carbon (amorphous)	1,950	724	1.6
Ice (0°C)	916	2,050	2.2
Paper	930	1,340	0.011
Pine wood	530	2,750	0.12
Plywood	540	1,220	0.12
Pyrex glass	2,250	840	1.4
Rubber	1,150	2,000	0.2
Teflon	2,200	350	0.26
Window glass	2,800	800	0.8

TABLE 2.1 Thermal Properties of Selected Solid Materials

2.3 Prediction and Correlation of Thermal Conductivities

2.3.1 Estimates for Gases

The Chapman–Enskog relationship, based on the kinetic theory of gases, provides the thermal conductivity of monoatomic gases at 1 atm absolute pressure according to the following dimensional equation:

(M.E.2.3) $$k(\mathrm{W/m\ K}) = 0.083 \times \frac{\sqrt{T/M_r}}{\sigma^2 \Omega_k} \tag{2.3}$$

where T is the absolute temperature in K, M_r is the atomic mass in g/mol, σ is in Å, and Ω_k is the collision integral. The latter is a function of the dimensionless temperature $T^* = k_B T/\varepsilon$ where $k_B = 1.30 \times 10^{-23}$ J/molecule and K is the Boltzmann constant. σ and ε are the Lennard–Jones parameters and are available for many species from many different sources. For the purpose of predicting transport properties, including the thermal conductivity, σ and ε can be estimated from the critical temperature, T_c, and the critical pressure, P_c, of the gas based on the following equations:

$$\sigma = 2.44\left(\frac{T_c}{P_c}\right)^{1/3} \tag{2.4}$$

$$\frac{\varepsilon}{k_B} = 0.75 T_c \tag{2.5}$$

with units of K, Å, and atm. Table 5-12 in *Perry's Chemical Engineers' Handbook*, 9th edition, provides alternative estimation methods. Over the range of T^* of 0.3–100, Ω_k is very well approximated by the following equation from Neufeld et al. [J. Chem. Phys. 57 (1972) 1100]:

$$\Omega_k = \Omega_\mu = 1.16145(T^*)^{-0.14874} + 0.52487\exp(-0.77320T^*) + 2.16178\exp(-2.43787T^*) \tag{2.6}$$

For the estimate of k for monoatomic and nonmonoatomic gases at 1 atm absolute pressure, *Perry's Chemical Engineers' Handbook*, 9th edition, recommends the following alternative dimensionless expressions:

(M.E.2.7) $$\frac{kM_r}{\mu(\tilde{C}_p - \mathcal{R})} = 2.5 \quad \text{for monoatomic gases} \tag{2.7}$$

(M.E.2.8) $$\frac{kM_r}{\mu(\tilde{C}_p - \mathcal{R})} = 1.30 + \frac{\mathcal{R}}{\tilde{C}_p - \mathcal{R}}\left(1.76 - \frac{0.352}{T_r}\right) \quad \text{for linear molecules} \tag{2.8}$$

(M.E.2.9) $$\frac{kM_r}{\mu(\tilde{C}_p - \mathcal{R})} = 1.15 + 2.03\frac{\mathcal{R}}{\tilde{C}_p - \mathcal{R}} \quad \text{for nonlinear molecules} \tag{2.9}$$

In these equations, μ is the viscosity of the gas in Pa s, $\tilde{C}_p$ is the gas molar heat capacity in J/mol K, $\mathcal{R} = 8.314$ J/mol K is the ideal gas law constant, M_r is the gas molecular mass in kg/mol, $T_r = T/T_c$ is the reduced temperature, and T_c is the gas critical temperature in K. These equations require knowledge of the gas viscosity at the same conditions at which k is being estimated. If not available experimentally, μ can also be estimated from the Chapman–Enskog theory, which gives

$$\mu(\text{Pa s}) = 2.67 \times 10^{-6} \frac{\sqrt{M_r T}}{\sigma^2 \Omega_\mu} \tag{2.10}$$

where, as for Eq. (2.3), M_r is in g/mol, T is in K, and σ is in Å. $\Omega_\mu = \Omega_k$ is approximated by Eq. (2.6).

Figure 2.4, based on data in Lenoir et al. [Chem. Eng. Progr. 49 (1953) 539], provides an approximate means of predicting the thermal conductivity of a gas at a pressure P, if the value of k is known at the temperature of interest and 1 atm. In this graph, the normalized thermal conductivity is defined as $k' = k/k_{P=1}$, where k is the conductivity at T and P and $k_{P=1}$ is the thermal conductivity at the same temperature T and $P = 1$ atm. Example 2.1 illustrates the use of these relationships.

Example 2.1. Estimation of the thermal conductivity of a gas. Let's use the equations presented so far in this section to estimate the thermal conductivity of CO_2 gas at 400 K and 150 atm absolute pressure. For CO_2, the critical temperature and critical pressure are $T_c = 304$ K and $P_c = 72.8$ atm, respectively. We don't expect the Chapman–Enskog theory to be accurate since CO_2 is not monoatomic, but let's use it anyway to see what we get. From Eqs. (2.4) and (2.5), we have

$$\sigma = 2.44\left(\frac{304}{72.8}\right)^{1/3} = 3.929 \text{ Å}$$

$$\frac{\varepsilon}{k_B} = 0.75 \times 304 = 228 \text{ K}$$

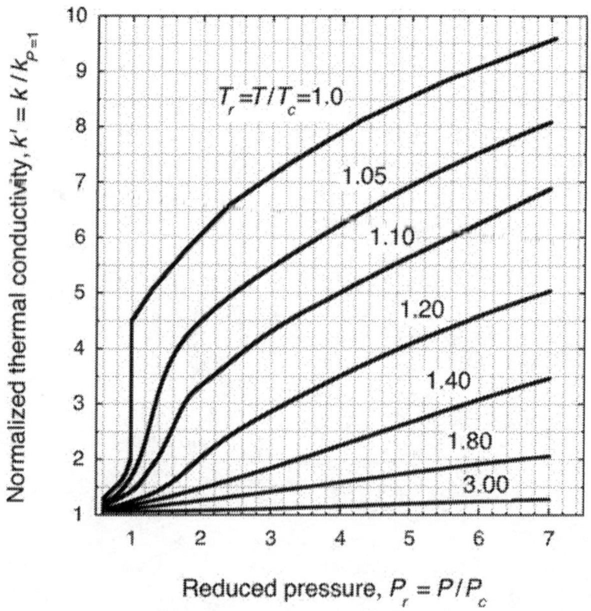

FIGURE 2.4 Normalized thermal conductivity as a function of reduced pressure and reduced temperature. [Data from Lenoir et al., Chem. Eng. Progr. 49 (1953) 539.]

Thus, $T^* = 400/228 = 1.75$. With this value of T^*, Eq. (2.6) gives $\Omega_k = 1.23$. Finally, from Eq. (2.3) we obtain

$$k = 0.083 \times \frac{\sqrt{400/44}}{3.929^2 \times 1.23} = 0.013 \text{ W/m K}$$

Although of the same order of magnitude, this value is quite different from the experimental value of $k = 0.024$ W/m K at 400 K and 1 atm. Let's now use Eq. (2.9) to estimate k at 1 atm pressure. The experimental value of the viscosity of CO_2 gas at 400 K and 1 atm is $\mu = 19.7 \times 10^{-6}$ Pa s [note that Eq. (2.10) gives $\mu = 18.7 \times 10^{-6}$ Pa s, which is pretty close to the experimental value], and its molar heat capacity is $\tilde{C}_p = 41.3$ J/mol K. With these values of μ and $\tilde{C}_p$, taking CO_2 as a linear molecule, Eq. (2.8) yields

$$k = \frac{19.7 \times 10^{-6}(41.3 - 8.314)}{44/1{,}000}\left[1.30 + \frac{8.314}{41.3 - 8.314}\left(1.76 - \frac{0.352}{400/304}\right)\right] = 0.0247 \text{ W/m K}$$

which is very nearly the same as the experimental value. In order to account for the effect of pressure, we find $T_r = 400/304 = 1.3$ and $P_r = 150/72.8 = 2.06$. From Fig. 2.4 we read $k' \sim 1.6$. Thus, $k = 1.6 \times 0.0246 = 0.039$ W/m K at 150 atm.

Gas Mixtures In the case of gas mixtures at low pressure, a simple weighted average of the thermal conductivities using the mole fractions of the mixture components as the weights is often sufficient for preliminary estimates. A more accurate estimation can be obtained using Wilke's mixing rule [Wilke, J., Chem. Phys. 18 (1950) 517], which is an empirical extension of the Chapman–Enskog theory. According to this rule, the thermal conductivity of a gas mixture is given by the following equation:

(M.E.2.11)

$$k = \sum_i^n \frac{y_i k_i}{\sum_{j=1}^n y_j \Phi_{i,j}} \tag{2.11}$$

where

$$\Phi_{i,j} = \frac{\left[1 + \left(\frac{\mu_i}{\mu_j}\right)^{0.5}\left(\frac{M_{r,j}}{M_{r,i}}\right)^{0.25}\right]^2}{\sqrt{8}\left(1 + \frac{M_{r,i}}{M_{r,j}}\right)^{0.5}} \tag{2.12}$$

In these equations, k_i and μ_i are, respectively, the thermal conductivity and viscosity of component i, and y_i is the corresponding mole fraction in the mixture. Example 2.2 illustrates the application of this method.

Example 2.2. Thermal conductivity of a gas mixture. Let's estimate the thermal conductivity of a gas containing 20 mol% CO_2 and 80 mol% N_2 at 400 K and 1 atm. From Example 2.1, we already have $k_{CO_2} = 0.025$ W/m K. For N_2, using the same method as in Example 2.1, we obtain $\sigma = 3.78$ Å, $\varepsilon/k_B = 94.7$ K, and $\mu = 2.06 \times 10^{-5}$ Pa s. Since for N_2 the molar heat capacity is $\tilde{C}_p = 29.1$ J/mol K, we have

$$k_{N_2} = \frac{20.6 \times 10^{-6}(29.1 - 8.314)}{28/1{,}000}\left(1.15 + 2.03\frac{8.314}{29.1 - 8.314}\right) = 0.030 \text{ W/m K}$$

The $\Phi_{i,j}$ terms are as follows (setting CO_2 =1 and N_2 = 2):

$$\Phi_{CO_2,CO_2} = 1$$

$$\Phi_{CO_2,N_2} = 0.734$$

$$\Phi_{N_2,CO_2} = 1.27$$

$$\Phi_{N_2,N_2} = 1$$

Finally, with these values, Eq. (2.11) yields

$$k = \frac{y_{CO_2} k_{CO_2}}{y_{CO_2}\Phi_{CO_2,CO_2} + y_{N_2}\Phi_{CO_2,N_2}} + \frac{y_{N_2} k_{N_2}}{y_{CO_2}\Phi_{N_2,CO_2} + y_{N_2}\Phi_{N_2,N_2}}$$

$$= \frac{0.2 \times 0.024}{0.2 \times 1 + 0.8 \times 0.734} + \frac{0.8 \times 0.030}{0.2 \times 1.27 + 0.8 \times 1} = 0.025 \text{ W/m K}$$

Note that a linear combination would give $k = 0.2 \times 0.024 + 0.8 \times 0.030 = 0.029$ W/m K, which is not too different from the value obtained from Wilke's method.

2.3.2 Estimates for Liquids

For organic liquids, *Perry's Chemical Engineers' Handbook*, 9th edition, recommends the following dimensional equation:

(M.E.2.13)
$$k(\text{W/m K}) = c\frac{(M_r)^m}{\hat{V}_{293}} \frac{3.0 + 20(1 - T_r)^{2/3}}{3.0 + 20(1 - T_{r,293})^{2/3}} \tag{2.13}$$

where M_r is the molecular mass of the liquid in units of g/mol, $\hat{V}_{293}$ is the molar volume of the liquid at 293 K in units of cm³/mol, $T_r = T/T_c$ is the reduced temperature, $T_{r,293} = 293/T_c$, and T_c is the critical temperature in K. The parameters c and m are empirical constants with $c = 0.181$ and $m = 1.0$ for linear hydrocarbons and $c = 0.441$ and $m = 0.772$ for cyclic and aromatic hydrocarbons. A slightly simpler relationship is the following Sato–Riedel equation [Poling et al., *The Properties of Gases and Liquids*, 5th edition, McGraw-Hill, 2001]:

(M.E.2.14)
$$k(\text{W/m K}) = \frac{1.10}{\sqrt{M_r}} \frac{3.0 + 20(1 - T_r)^{2/3}}{3.0 + 20(1 - T_{r,b})^{2/3}} \tag{2.14}$$

where M_r is again in g/mol and $T_{r,b} = T_b/T_c$ where T_b is the normal boiling point in K.

Example 2.3. Thermal conductivity of liquid benzene. Let's estimate the thermal conductivity of liquid benzene at 30°C. For benzene, we have the following data: $M_r = 78$ g/mol, $T_c = 562$ K, and $T_b = 353$ K. Based on the density of liquid benzene, $\rho = 0.88$ g/cm³, we can estimate $\hat{V}_{293}$ as $\hat{V}_{273} = 78/0.88 = 90$ cm³/mol. With $T_r = 303/562 = 0.54$ and $T_{r,293} = 293/562 = 0.52$, Eq. (2.13) gives:

$$k = 0.441\frac{(78)^{0.772}}{90} \frac{3.0 + 20(1 - 0.54)^{2/3}}{3.0 + 20(1 - 0.52)^{2/3}} = 0.138 \text{ W/m K}$$

The slightly simpler Sato–Riedel equation, Eq. (2.14), gives the following result:

$$k = \frac{1.10}{\sqrt{78}} \frac{3.0 + 20(1 - 0.54)^{2/3}}{3.0 + 20(1 - 0.63)^{2/3}} = 0.140 \text{ W/m K}$$

From Fig. 2.2, the experimental value of k is about 0.13 W/m K, so the agreement with both equations is pretty good.

Liquid Mixtures The Sato–Riedel equation, Eq. (2.14), can be extended to predict the thermal conductivity of a hydrocarbon mixture using the pseudo-component concept. In this case, mixture values of M, T_r, and $T_{r,b}$ are calculated according to the following mixing rules [Arikol and Gürbüz, Can. J. Chem. Eng. 70 (1992) 1157]:

$$T_r' = \frac{T}{\sum x_i T_{c,i}} \tag{2.15}$$

$$T_{r,b}' = \frac{\sum x_i T_{b,i}}{\sum x_i T_{c,i}} \tag{2.16}$$

$$M_r' = \sum x_i M_{r,i} \tag{2.17}$$

where x_i, $M_{r,i}$, $T_{c,i}$, and $T_{b,i}$ are, respectively, the mole fraction, the molecular mass, the critical temperature, and the normal boiling point of component i.

Example 2.4. Estimate the thermal conductivity of mixtures of benzene and 2,2,4 tri-methyl pentane (TMP) containing (a) 75 mol% benzene and (b) 25 mol% benzene, both at 313 K and 1 atm. The following data are readily available from the *NIST Chemistry WebBook* [https://webbook.nist.gov/chemistry/]:

Component	M_r (g/mol)	T_c (K)	T_b (K)
Benzene	78	562	353
TMP	114	544	372

If the mixture contains 75 mol% benzene, we will have:

$$T_r' = \frac{313}{0.75 \times 562 + 0.25 \times 544} = 0.56$$

$$T_{r,b}' = \frac{0.75 \times 353 + 0.25 \times 372}{0.75 \times 562 + 0.25 \times 544} = 0.64$$

$$M_r' = 0.75 \times 78 + 0.25 \times 114 = 87$$

Inserting these values in Eq. (2.14) in lieu of M_r, T_r, and $T_{r,b}$ yields $k = 0.13$ W/m K. The same procedure yields $k = 0.11$ W/m K if the mixture contains 25 mol% benzene. The experimental values for these mixtures from Wakeham et al. [Int. J. Thermophys. 11 (1990) 987], are 0.12 and 0.10 W/m K, respectively, which are not too different from the estimated values.

2.3.3 Estimation for Porous Materials

Porous materials are found in many applications ranging from catalysts and adsorbent pellets to thermal insulation. For catalysis and adsorption, porosity provides high surface area that gives high reaction rates and high adsorption capacity, respectively. For thermal insulation, the porosity is designed to trap air, which reduces the overall thermal conductivity without adding weight. The porous material will have conductivity intermediate between the conductivity of the solid matrix and the conductivity of the fluid that fill the pores. The following Russell equation [Russell, J. Am. Chem. Soc. 18 (1935) 1] is useful to estimate the conductivity, k, of such materials:

(M.E.2.18)
$$\frac{k}{k_{\text{solid}}} = \frac{\gamma\varepsilon_p^{2/3} + 1 - \varepsilon_p^{2/3}}{\gamma(\varepsilon_p^{2/3} - \varepsilon_p) + 1 + \varepsilon_p - \varepsilon_p^{2/3}} \tag{2.18}$$

where

$$\gamma = \frac{k_{\text{pore}}}{k_{\text{solid}}} \tag{2.19}$$

is the ratio of the thermal conductivity of the material that fills the pores, k_{pore}, and of the solid, k_{solid}. ε_p is the porosity or void fraction of the porous material. This quantity is defined as the pore volume divided by the volume of porous material. Typical units of ε_p will, thus, be m^3 of pore volume per m^3 of pore volume plus solid.

Example 2.5. Thermal conductivity of a porous solid. Synthetic cork board can be produced by gluing together pine wood particles to form a porous network. When dry, the pores are filled with air. We can estimate the thermal conductivity of dry cork at 300 K using the following data from Fig. 2.1 and Table 2.1, respectively:

$$k_{\text{pore}} = k_{\text{air}} = 0.026 \text{ W/m K}$$

$$k_{\text{solid}} = k_{\text{pine}} = 0.12 \text{ W/m K}$$

These values give $\gamma = 0.026/0.12 = 0.217$. The value of k will vary with porosity as shown in Fig. 2.5 for these conditions. For cork board with a density of 10 lb/ft^3, Bennett and Myers, *Momentum, Heat, and Mass Transfer*, 1982, Table A-14, give $k = 0.025$ BTU/h ft °F or 0.043 W/m K. Since the density of pine wood is 34.4 lb/ft^3, also according to Bennett and Myers, we can estimate the porosity of cork board from the following equation:

$$34.4 \times (1 - \varepsilon_p) = 10$$

which gives $\varepsilon_p = 0.71$. With this porosity, from Eq. (2.18) or Fig. 2.5, we estimate $k = 0.048$ W/m K, which, given the uncertainties about the actual type of wood and the cork board structure, is remarkably close to the experimental value of Bennett and Myers. Of course, if the cork board were wet and the pores were saturated with water instead of air, quite a different value of k would be obtained because of the much higher conductivity of liquid water compared to air. For instance, if we were to take $k_{\text{pore}} = k_{\text{water}} = 0.65$ W/m K for wet cork, with $\varepsilon_p = 0.71$, Eq. (2.18) would yield $k = 0.39$ W/m K. Thus according to this estimate, we would predict that the wet cork board would lose almost all of its thermal insulating ability. In general, avoiding condensation of water vapor in insulation is obviously critical and it is one of the reasons why, for example, exposed home insulation is often lined with a vapor barrier.

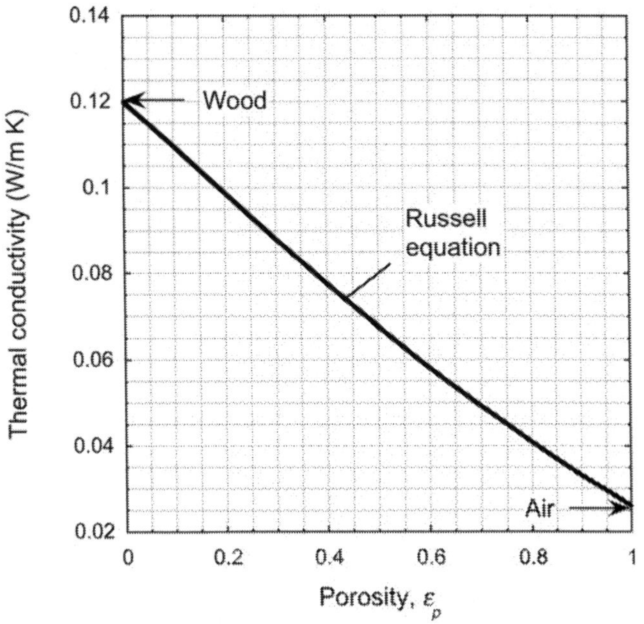

FIGURE 2.5 Thermal conductivity of cork board for Example 2.5.

2.4 Problems

P.2.1. Estimate the thermal conductivity of NH_3 gas at 450 K and 1 atm absolute pressure and compare your estimate with the experimental value.

P.2.2. Estimate the thermal conductivity of CO_2 at 365 K and 1 atm absolute pressure. Then use this value to find the thermal conductivity at 365 K and 200 atm absolute pressure.

P.2.3. Estimate the thermal conductivity of a gas mixture containing 40 mol% ammonia and 60 mol% N_2 at 400 K and 1 atm absolute pressure.

P.2.4. Estimate the thermal conductivity of liquid isopropyl alcohol at 300 K and compare your estimates with the experimental value available in Section 2 of *Perry's Chemical Engineers' Handbook*.

P.2.5. Estimate the thermal conductivity of liquid mixtures of *n*-hexane and *n*-decane at 295 K and 335 K and 1 atm absolute pressure as a function of the mole fraction of *n*-hexane and compare your estimate with the experimental results reported in Vasquez and Briano, Ind. Eng. Chem. Res. 32 (1993) 194 with 20% and 80% *n*-hexane.

P.2.6. A silica-based catalyst particle has a porosity $\varepsilon_p = 0.60$. Estimate its thermal conductivity at 25°C if the pores are (a) filled with air at 1 atm and (b) filled with liquid water. The thermal conductivity of solid silica can be found in Table 2-152 of *Perry's Handbook Chemical Engineers' Handbook*, 9th edition.

Tissue	Density, ρ (kg/m^3)	Heat Capacity, C_p (J/kg K)	Thermal Conductivity, k (W/m K)
Fat	911	2,350	0.21
Muscle	1,090	3,421	0.49
Skin	1,110	3,391	0.37

TABLE P.2.6 Density and Thermal Properties of Animal Tissue

P.2.7. The thermal properties of animal and human tissue have been extensively characterized, and an ample database is available at https://itis.swiss/virtual-population/tissue-properties/database/database-summary/. Some representative data are given in Table P.2.6. Based on these data, estimate the density, heat capacity, and thermal conductivity of ground beef containing 23.8% (w/w) fat. Compare your estimates with the experimental values given by Pan and Singh, LWT—Food Science Technol. 34 (2001) 437.

CHAPTER 3

Steady-State, One-Dimensional Heat Conduction

3.1 Introduction

Fourier's law was introduced in Sec. 1.3 to describe conductive heat transfer occurring within a region of space occupied by a stagnant fluid or by a solid. This law predicts that the heat transfer flux is proportional to the temperature gradient. As a result, the solution of steady-state, conductive heat transfer problems will typically involve the determination of the temperature profile and, thus, the temperature gradient in a certain domain in order to predict the rate of heat transfer given conditions that are defined at the boundaries of this domain. These boundary conditions can be specified in two different ways, depending on the particular problem. The first way is to specify the temperature. The second way is to specify the heat transfer flux and, thus, the temperature gradient. In general, at a boundary between two solids, between two immiscible fluids, or between a solid and a fluid, the temperature will be a continuous function; i.e., the temperature will be the same on the two sides of the boundary. Additionally, the rate of heat transfer on one side of such a boundary will also be the same as the rate of heat transfer on the other side provided that no thermal energy is generated or consumed right at the boundary. Thus, in such cases, both the temperature and the heat fluxes will be identical on the two sides of such a boundary.

The exact form of Fourier's law will depend on the coordinate system that is chosen to describe the spatial dependence of temperature in the domain of interest. Table 3.1 provides the expressions for the components of the heat flux q in rectangular, cylindrical, and spherical coordinate systems. The choice will depend on the particular geometry of the system where heat conduction occurs. In many cases, problems that are three-dimensional in rectangular coordinates become one-dimensional in cylindrical or spherical coordinates resulting in major simplification of the mathematical treatment.

In the following, we consider the application of Fourier's law to a number of one-dimensional, steady-state heat conduction problems, where the temperature is a function of a single geometric variable.

<table>
<tr><td colspan="3">Rectangular</td></tr>
<tr><td>$T = T(t, x, y, z)$</td><td>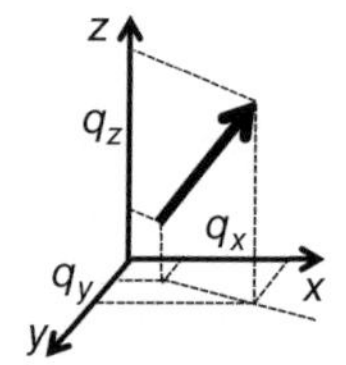
</td><td>$q_x = -k\frac{\partial T}{\partial x}$
$q_y = -k\frac{\partial T}{\partial y}$
$q_z = -k\frac{\partial T}{\partial z}$</td></tr>
<tr><td colspan="3">Cylindrical</td></tr>
<tr><td>$T = T(t, r, z, \theta)$</td><td>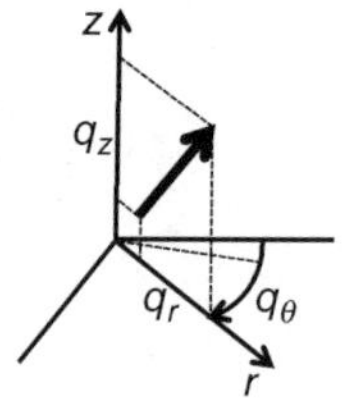
</td><td>$q_r = -k\frac{\partial T}{\partial r}$
$q_z = -k\frac{\partial T}{\partial z}$
$q_\theta = -k\frac{1}{r}\frac{\partial T}{\partial \theta}$</td></tr>
<tr><td colspan="3">Spherical</td></tr>
<tr><td>$T = T(t, r, \theta, \phi)$</td><td>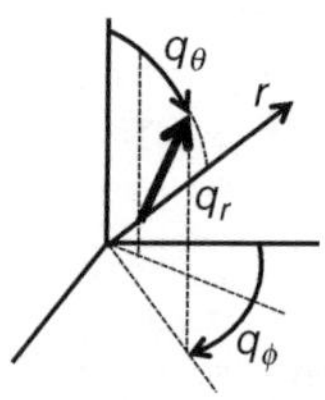
</td><td>$q_r = -k\frac{\partial T}{\partial r}$
$q_\theta = -\frac{k}{r}\frac{\partial T}{\partial \theta}$
$q_\phi = -k\frac{1}{r\sin\theta}\frac{\partial T}{\partial \phi}$</td></tr>
</table>

TABLE 3.1 Expressions for the Components of the Heat Transfer Flux in Rectangular, Cylindrical, and Spherical Coordinates

3.2 Flat Walls and Slabs

In general, the solution of conductive heat transfer problems is facilitated by going along the following five-step procedure:

1. Select a differential control volume;
2. Write a differential thermal energy balance;
3. Combine the differential energy balance with Fourier's law;
4. Identify the boundary conditions; and
5. Integrate the resulting differential equation.

The control volume is the volume over which we will write our energy balance. It will be chosen based on the coordinate system used to describe the temperature profile and on the geometric variable along which the temperature changes. Consider, for example, the problem of steady-state heat conduction across a large slab or a wall of thickness Δ and cross-sectional area A as shown in Fig. 3.1. Let's assume that the surface

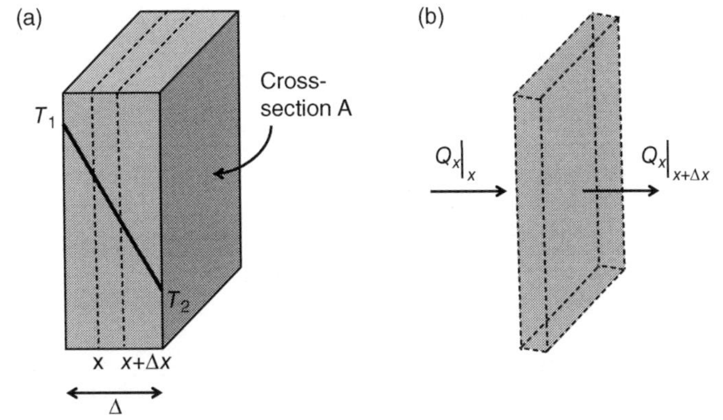

FIGURE 3.1 Heat conduction across a single-layer slab or flat wall (a) and control volume used to derive the differential energy balance (b).

temperatures T_1 and T_2 are specified on the sides of the slab and that these temperatures are uniformly distributed over each surface. If the wall length and width are large compared to its thickness, the situation can be treated as a one-dimensional heat conduction problem where the temperature within the slab will be just a function of x, or $T = T(x)$. Since our goal is to determine the function $T(x)$, we will choose a control volume that is differential along x as shown in Fig. 3.1b. For simplicity, we start with a volume of finite thickness Δx. After writing the energy balance for this volume, we will then take the limit $\Delta x \to 0$ to obtain a differential description of the process. In general, the thermal energy balance will be as follows:

$$\text{Energy accumulated} = \text{Energy in} - \text{Energy out} + \text{Energy generation}$$

At steady state, there is no accumulation of energy. Thus, the first term will be zero. Likewise, if there is no energy generation within the slab, the last term will be zero. In this case, for our control volume, the energy balance will simply be:

$$Q_x\Big|_x = Q_x\Big|_{x+\Delta x}$$

where $Q_x\big|_x$ and $Q_x\big|_{x+\Delta x}$ are the rates of energy transfer in and out of the control volume. Dividing both terms of this equation by Δx and taking the limit for $\Delta x \to 0$ yield:

$$\lim_{\Delta x \to 0}\left(\frac{Q_x\big|_{x+\Delta x} - Q_x\big|_x}{\Delta x}\right) = 0$$

or

$$\frac{dQ_x}{dx} = 0 \tag{3.1}$$

This result implies that Q_x will be a constant and, thus, the same at each x position from one side of the slab to the other. Further, since, in this case, the area A perpendicular to the heat flux vector does not vary with x, the heat transfer flux $q_x = Q_x/A$

will also be a constant across the thickness of the slab. Next, we write Fourier's law as follows:

$$q_x = -k\frac{dT}{dx} \tag{3.2}$$

Since, from the energy balance, we have determined that q_x is constant, we can separate the variables T and x in Eq. (3.2) and integrate as follows:

$$q_x \int_0^{\Delta} dx = -\int_{T_1}^{T_2} k\,dT \tag{3.3}$$

where the integration limits are based on the temperatures specified on the two sides of the slab. The integral on the right-hand side of this equation is easy if k is independent of temperature. In this case, we obtain:

$$q_x = \frac{k}{\Delta}(T_1 - T_2) \tag{3.4}$$

If k is temperature-dependent, we would need to first insert the function describing the temperature dependence, $k(T)$, into Eq. (3.3) and then integrate. We can also find the temperature profile in the slab by changing the upper integration limits of Eq. (3.3) to arbitrary values of x and T, which yields:

$$q_x \int_0^{x} dx = -\int_{T_1}^{T} k\,dT \tag{3.5}$$

If k is constant, from this equation we obtain:

$$T = T_1 - \frac{q_x}{k}x \tag{3.6}$$

This result shows that with k constant, the temperature profile will be a straight line with slope inversely proportional to k. This means that with a higher k, we will obtain the same heat transfer flux with a smaller temperature difference $T_1 - T_2$. In other words, with a higher thermal conductivity, a smaller driving force will be sufficient to get the same heat flux. Finally, we can find the heat transfer rate, which is given by the following equation for the constant k case:

$$Q_x = Aq_x = \frac{Ak}{\Delta}(T_1 - T_2) \tag{3.7}$$

or

$$Q_x = \frac{\Delta T}{R} \tag{3.8}$$

where $\Delta T = T_1 - T_2$ and $R = \Delta / Ak$. We can recognize that this expression for Q_x is analogous to the expression for the electrical current, I, passing through an electrical conductor with electrical resistance R_e as a result of an applied voltage difference ΔV. In this case, Ohm's law gives:

$$I = \frac{\Delta V}{R_e} \tag{3.9}$$

This equation is analogous to Eq. (3.8) with the applied voltage difference ΔV playing the role of ΔT and the electrical resistance R_e playing the role of R. We thus refer to the quantity $R = \Delta / Ak$ as the *heat transfer resistance* offered by the slab. Its dimensions will be temperature over power and typical units will be K/W or °F/(BTU/h). Obviously, the heat transfer resistance will be greater for a thicker slab or for a slab made of a material with a smaller thermal conductivity.

We can now extend this analysis to the problem of heat transfer across a multi-layer slab or wall that consists of multiple layers in series. Each layer can be made of a different material with a different thermal conductivity and can have a different thickness. With reference to Fig. 3.2, consider a two-layer slab. We can apply Eq. (3.4) to each layer with the corresponding thermal conductivity and thickness. At the interface between the two layers, we can assume that the temperature will be the same on both sides and equal to T'. In this case, working directly in terms of heat transfer fluxes, we obtain the following two equations:

$$q_{x,1} = \frac{k_1}{\Delta_1}(T_1 - T') \qquad q_{x,2} = \frac{k_2}{\Delta_2}(T' - T_2) \tag{3.10}$$

Since heat cannot accumulate at the interface between the two layers, the heat fluxes on the two sides of the interface must also be equal, or $q_{x,1} = q_{x,2} = q_x$. We can thus combine these two equations to find an expression for T'. The algebra is a bit lengthy, but yields the following result:

$$T' = \frac{\frac{k_1}{\Delta_1}T_1 + \frac{k_2}{\Delta_2}T_2}{\frac{k_1}{\Delta_1} + \frac{k_2}{\Delta_2}} \tag{3.11}$$

This result shows that the intermediate temperature at the boundary between the two layers, T', will be a weighted average of the two temperatures T_1 and T_2 that are

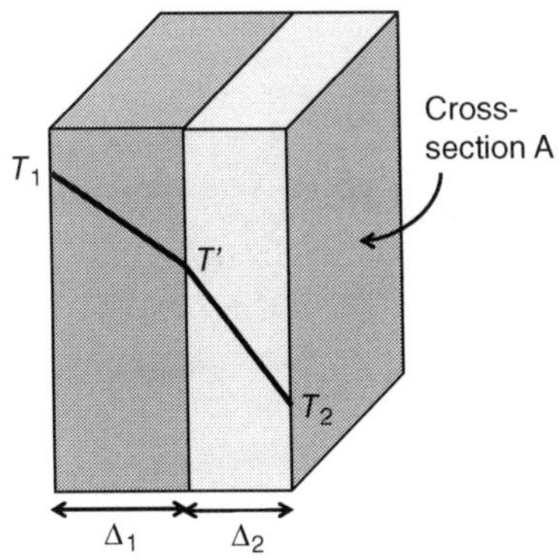

FIGURE 3.2 Heat conduction across a two-layer slab or flat wall. The temperature is continuous at the interface between the two layers but the slope of the temperature profile varies from one layer to the other. The slope is greater in the layer with smaller thermal conductivity.

specified at the sides or the two-layer slab. Alternatively, and with a little bit less effort, we can rewrite Eq. (3.10) as follows:

$$\begin{aligned}\frac{q_{x,1}}{k_1/\Delta_1} &= T_1 - T' \\ \frac{q_{x,2}}{k_2/\Delta_2} &= T' - T_2\end{aligned} \tag{3.12}$$

Adding these two expressions together and remembering that $q_{x,1} = q_{x,2} = q_x$ yields:

$$q_x\left(\frac{\Delta_1}{k_1} + \frac{\Delta_2}{k_2}\right) = T_1 - T_2 \tag{3.13}$$

which no longer contains the unknown intermediate temperature T'. Having eliminated this unknown temperature is advantageous as we can now calculate the heat transfer flux directly as a function of the specified temperatures T_1 and T_2 on the sides of the two-layer slab. The following result is obtained from Eq. (3.13):

$$q_x = \frac{T_1 - T_2}{\dfrac{\Delta_1}{k_1} + \dfrac{\Delta_2}{k_2}} \tag{3.14}$$

Once q_x is known, we can go back to Eq. (3.12) and find the intermediate temperature T'. Finally, we can write an expression for the heat transfer rate. Since the area A is the same for both layers, we will have:

$$Q_x = q_x A = \frac{T_1 - T_2}{\dfrac{\Delta_1}{Ak_1} + \dfrac{\Delta_2}{Ak_2}} \tag{3.15}$$

In terms of heat transfer resistances, Eq. (3.15) can also be written as follows:

$$Q_x = \frac{T_1 - T_2}{R_1 + R_2} = \frac{T_1 - T_2}{R_{\text{overall}}} \tag{3.16}$$

where $R_1 = \Delta_1/Ak_1$ and $R_2 = \Delta_2/Ak_2$ are the heat transfer resistances of the two layers and the quantity:

$$R_{\text{overall}} = R_1 + R_2 \tag{3.17}$$

is the overall heat transfer resistance. Note that, based on Ohm's law, we would obtain an equivalent expression for the case of two electrical resistors in series, with electrical resistances $R_{e,1}$ and $R_{e,2}$. In this case, the overall electrical resistance would be the sum of the two electrical resistances in series, just like the overall heat transfer resistance in our two-layer slab is the sum of the two heat transfer resistances in series.

We consider next the problem of heat transfer through multiple layers in either a series or a parallel arrangement. Consider the two scenarios in Fig. 3.3 where a laminated block is formed by pressing together multiple layers of two different materials one with thickness Δ_1 and the other with thickness Δ_2.

In both arrangements, there are n_1 layers of material 1 and n_2 layers of material 2. In one case (a), heat transfer is perpendicular to the layers. This case is an extension of the

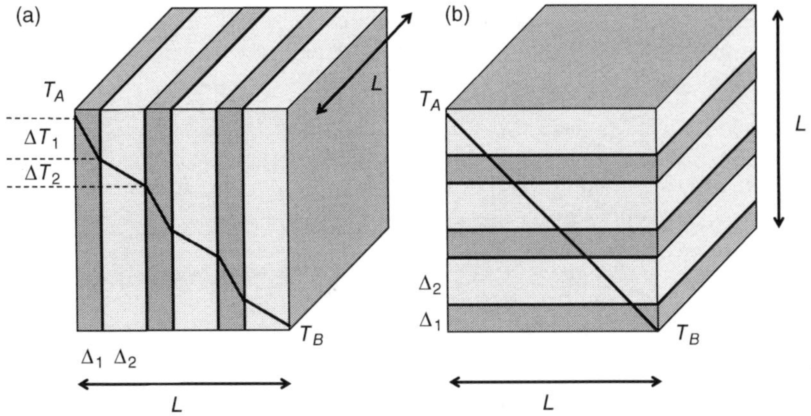

FIGURE 3.3 Series (a) and parallel (b) arrangements of two materials 1 and 2 in a composite block.

two-layer slab with resistances in series. The steady-state heat transfer rate will be the same for all the layers and equal to:

$$Q_x = \frac{Ak_1}{\Delta_1}\Delta T_1 = \frac{Ak_2}{\Delta_2}\Delta T_2 \tag{3.18}$$

where ΔT_1 is the temperature difference across each layer of material 1 and ΔT_2 is the temperature difference across each layer of material 2. Since the layers are in series, the overall resistance will be the sum of the resistances of each individual layer giving the following result:

$$Q_x = \frac{T_A - T_B}{\dfrac{n_1\Delta_1}{Ak_1} + \dfrac{n_2\Delta_2}{Ak_2}} \tag{3.19}$$

where T_A and T_B are the temperatures on the two sides of the multi-layer block. In terms of heat transfer resistances, this equation can be written as follows:

$$Q_x = \frac{T_A - T_B}{n_1R_1 + n_2R_2} = \frac{T_A - T_B}{R_{\text{overall}}} \tag{3.20}$$

where $R_1 = \Delta_1/Ak_1$ and $R_2 = \Delta_2/Ak_2$ are the resistances of the individual layers and $R_{\text{overall}} = n_1R_1 + n_2R_2$ is the overall resistance. In the second case (b), the layers are in parallel to the direction of heat transfer. If assume one-dimensional heat transfer, which may be justified if the layers are thin, each layer will be subject to the same temperature difference $T_A - T_B$. In this case, the rate at which heat is conducted through the layers of material 1 will be given by:

$$Q_{x,1} = \frac{n_1\Delta_1Wk_1}{L}(T_A - T_B) = \frac{A_1k_1}{L}(T_A - T_B) \tag{3.21}$$

while the heat conducted through the layers of material 2 will be:

$$Q_{x,2} = \frac{n_2\Delta_2Wk_2}{L}(T_A - T_B) = \frac{A_2k_2}{L}(T_A - T_B) \tag{3.22}$$

In both cases, W is the width of the block, A_1 is the area of material 1, and A_2 is the area of material 2. The total rate of heat transfer will be the sum of the two rates. Thus:

$$Q_x = Q_{x,1} + Q_{x,2} = \left(\frac{A_1 k_1}{L} + \frac{A_2 k_2}{L}\right)(T_A - T_B) \tag{3.23}$$

Since in this case, $R_1 = L/A_1 k_1$ and $R_2 = L/A_2 k_2$ are the resistances offered by the two materials, we can write:

$$Q_x = \frac{T_A - T_B}{R_{\text{overall}}} \tag{3.24}$$

The quantity:

$$R_{\text{overall}} = \left(\frac{1}{R_1} + \frac{1}{R_2}\right)^{-1} \tag{3.25}$$

is the overall heat transfer resistance for the layers in parallel. So, just like in an electrical circuit, the overall electrical resistance of multiple resistors in parallel is equal to the reciprocal of the sum of the reciprocals of the individual resistances.

A final consideration is the possibility that a convective heat transfer resistance exists at a boundary of our domain where heat conduction occurs. For example, let's return to the problem of heat conduction across a flat wall or slab as depicted in Fig. 3.1. We can envision a situation where rather than specifying the surface temperatures at $x = 0$ and at $x = \Delta$, we specify, instead, the temperatures of a fluid that flows past those surfaces as shown in Fig. 3.4. In this case, heat transfer from each slab surface to the moving fluid will be described by Eq. (1.9) based on a convective heat transfer coefficient, h. In general, the surface temperature will be different from the temperature of the fluid and will be unknown. The problem is now described by the following equations:

$$\begin{aligned} q_x &= h_1(T_{a,1} - T_1) \\ q_x &= \frac{k}{\Delta}(T_1 - T_2) \\ q_x &= h_2(T_2 - T_{a,2}) \end{aligned} \tag{3.26}$$

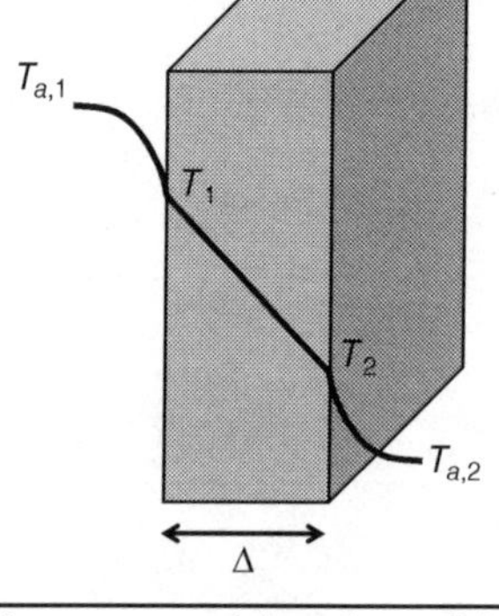

FIGURE 3.4 Conductive heat transfer across a slab or flat wall including convective heat transfer resistances at the boundaries.

where h_1 and h_2 are the convective heat transfer coefficients on the two sides of the slab. The first and last of these equations describe convective heat transfer on the two sides of the slab, while the middle equation describes conductive heat transfer within the slab itself. At steady state and with no heat generation within the slab, the three heat fluxes will be exactly the same. Rearranging these equations, we obtain:

$$\frac{q_x}{h_1} = T_{a,1} - T_1$$

$$\frac{q_x}{k/\Delta} = T_1 - T_2 \tag{3.27}$$

$$\frac{q_x}{h_2} = T_2 - T_{a,2}$$

Finally, since q_x is the same in all three equations, adding these expressions together gives the result:

$$q_x = \frac{T_{a,1} - T_{a,2}}{\dfrac{1}{h_1} + \dfrac{\Delta}{k} + \dfrac{1}{h_2}} \tag{3.28}$$

In terms of the heat transfer rate, this result can be written as follows:

$$Q_x = \frac{T_{a,1} - T_{a,2}}{\dfrac{1}{Ah_1} + \dfrac{\Delta}{Ak} + \dfrac{1}{Ah_2}} = \frac{T_{a,1} - T_{a,2}}{R_{\text{overall}}} \tag{3.29}$$

where

$$R_{\text{overall}} = \frac{1}{Ah_1} + \frac{\Delta}{Ak} + \frac{1}{Ah_2} \tag{3.30}$$

is the overall heat transfer resistance. The quantities $1/Ah_1$ and $1/Ah_2$ represent the convective heat transfer resistances at the slab surfaces. As seen from Eq. (3.30), these resistances are added to the conductive resistance of the slab, Δ/Ak, to give the overall heat transfer resistance. In terms of resistances, the overall result is the same that we obtained for multiple conductive resistances in series in a multilayer slab. However, an important difference between conductive and convective resistances is that the former depends on the thickness of the layer while the latter will depend on the flow conditions. Example 3.1 demonstrates the application of these relationships.

Example 3.1. Heat transfer across a double-pane glass window. A double-pane glass window consists of two glass panes that sandwich a thin, air-filled gap designed to increase the overall heat transfer resistance and, thus, reduce heat losses. Let's evaluate the heat losses from a window considering the following three cases:

(a) A single pane glass window 3/8 in. thick neglecting convective resistances;

(b) A single pane glass window 3/8 in. thick considering internal and external convective resistances; and

(c) A double pane glass window comprising two 1/8 in. glass panes sandwiching a 1/8 in. air gap considering internal and external convective resistances.

For this example problem, we will assume that the internal and external air temperatures are 70°F and 32°F, respectively. The thermal conductivity of window glass is $k_{glass} = 0.80$ W/m K (*cf.* Table 2.1) or 0.46 BTU/h ft °F. The thermal conductivity of the air in the gap between glass panes at the average temperature of 51°F (284 K) is $k_{air} = 0.025$ W/m K (*cf.* Fig. 2.1) or 0.014 BTU/h ft °F. For this problem, let's assume that the internal and external convective heat transfer coefficients are $h_i = 5$ and $h_e = 20$ BTU/h ft² °F, respectively. The actual values will, in fact, depend on air flow and natural convection effects. Let's base our calculations on 1 ft² of window area. For case (a), if convective resistances are neglected, the internal and external glass surface temperatures will be equal to, respectively, the internal and external air temperatures. Thus, the heat transfer rate will be:

$$Q = \frac{Ak_{glass}}{\Delta}(70 - 32) = \frac{1 \times 0.46}{0.375/12}(70 - 32) = 560 \text{ BTU/h}$$

For case (b), we will have three heat transfer resistances in series. Thus, from Eq. (3.30) we will obtain the following for the overall resistance:

$$R_{overall} = \frac{1}{Ah_i} + \frac{\Delta}{Ak_{glass}} + \frac{1}{Ah_e}$$

$$= \frac{1}{1 \times 5} + \frac{0.375/12}{1 \times 0.46} + \frac{1}{1 \times 20}$$

$$= 0.20 + 0.068 + 0.05 = 0.32 \text{ h °F/BTU}$$

In this case, the heat transfer rate will be:

$$Q = \frac{70 - 32}{0.32} = 120 \text{ BTU/h}$$

For case (c), we will have five resistances in series, including the two convective resistances, the conductive resistance of the two glass panes, and the conductive resistance of the air gap. In this case, we will obtain the following for the overall resistance:

$$R_{overall} = \frac{1}{Ah_i} + \frac{2\Delta_{glass}}{Ak_{glass}} + \frac{\Delta_{air}}{Ak_{air}} + \frac{1}{Ah_e}$$

$$= \frac{1}{1 \times 5} + 2 \times \frac{0.125/12}{1 \times 0.46} + \frac{0.125/12}{1 \times 0.014} + \frac{1}{1 \times 20}$$

$$= 0.20 + 0.045 + 0.74 + 0.05 = 1.04 \text{ h°F/BTU}$$

As a result, the heat transfer rate will be:

$$Q = \frac{70 - 32}{1.04} = 36.5 \text{ BTU/h}$$

Based on these calculations, we can see that the convective heat transfer resistances reduce heat losses by about 79% compared to the case where only the conductive resistance of a single glass pane is considered. Obviously, reducing air flow past the glass surface will help considerably. Adding a 1/8 in. air gap will reduce heat losses by (120–36.5)/120 × 100% = 70% compared to the single glass pane case, although the overall thickness of the double pane glass window is the same as that of the single pane window. This reduction is due to the heat transfer resistance added by the air gap. This gap provides the greatest contribution to the overall resistance (0.74 h °F/BTU compared to the total resistance of 1.04 h °F/BTU) and is, thus, the greatest contributor to the reduced heat transfer rate.

Example 3.1 introduced the very important concept of which resistance is dominant when resistances are in series. Since the individual resistances are additives, the dominant resistance is the largest one. For instance, for Example 3.1, in the case of the double pane window, the resistance of the air gap is 0.74/1.04 or 71% of the total. On the other hand, in a system where resistances are in parallel, since the overall resistance is the reciprocal of the sum of the reciprocals of the resistances, the smallest resistance will tend to dominate the overall heat transfer process. Heat will flow mainly through the path of least resistance.

3.3 Cylindrical Walls

Heat conduction across cylindrical walls is important in many applications, such as those involving pipes. Pipes are often insulated with layers of insulating materials and can be described as multilayer cylindrical walls. Let's consider first an un-insulated pipe as shown in Fig. 3.5a. The inner and outer surfaces of the pipe wall are at temperatures T_i and T_o, respectively, and conductive heat transfer occurs across the pipe wall. If the pipe is long compared to its radius, we can treat the problem as one-dimensional heat conduction along the radial coordinate r with T being a function of r only. In order to find the function $T(r)$, following the steps outlined in Sec. 3.1, we will base our energy balance on a control volume defined by a cylindrical shell with thickness Δr and length L with its faces located at r and at $r+\Delta r$ as sketched in Fig. 3.5b. Assuming steady state and no heat generation within the pipe wall, the energy balance for this control volume yields:

$$Q_r\big|_r = Q_r\big|_{r+\Delta r}$$

where $Q_r\big|_r$ and $Q_r\big|_{r+\Delta r}$ are the rates of energy transfer in and out of the control volume. Dividing this equation by Δr and taking the limit $\Delta r \to 0$, we obtain:

$$\lim_{\Delta r \to 0}\left(\frac{Q_r\big|_{r+\Delta r} - Q_r\big|_r}{\Delta r}\right) = 0$$

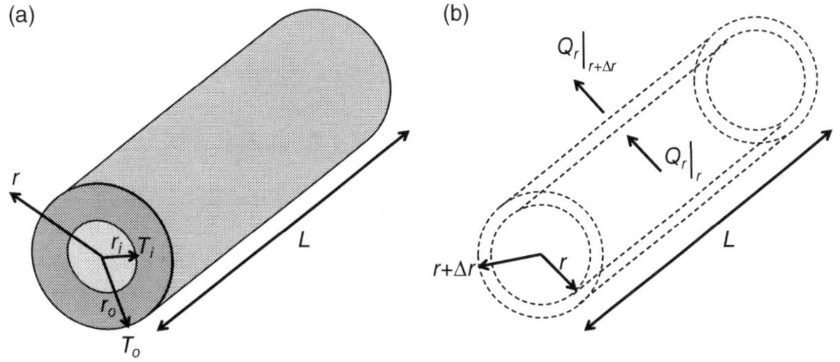

Figure 3.5 Heat conduction across a pipe or a cylindrical wall (a) and control volume used to derive the differential energy balance (b).

or

$$\frac{dQ_r}{dr} = 0 \tag{3.31}$$

This result shows that Q_r will be constant across the cylindrical wall, just like it was for a flat wall. However, in this case, unlike the flat wall situations, the heat transfer flux, $q_r = Q_r/A$ will vary along the radial coordinate since the surface area A varies with r according to the following equation:

$$A = 2\pi rL \tag{3.32}$$

Since A increases with r and Q_r is constant, it is evident that the heat flux q_r will decrease as we move from the inner surface of the pipe wall to its outer surface. In order to find the heat transfer rate, we need to combine the differential energy balance with Fourier's law. In cylindrical coordinates, Fourier's law is (see Table 3.1, item B):

$$q_r = -k\frac{dT}{dr} \tag{3.33}$$

Thus,

$$\frac{Q_r}{2\pi rL} = -k\frac{dT}{dr} \tag{3.34}$$

Since Q_r is constant, we can separate the variables r and T in this equation to obtain:

$$\frac{Q_r}{2\pi L}\frac{dr}{r} = -kdT \tag{3.35}$$

or, integrating across the wall,

$$\frac{Q_r}{2\pi L}\int_{r_i}^{r_o}\frac{dr}{r} = -\int_{T_i}^{T_o} k\,dT \tag{3.36}$$

Note that the integration limits are based on the temperatures T_i and T_o that are specified at the inner and outer surfaces of the pipe wall. If k is assumed to be a constant, integrating this equation yields:

$$\frac{Q_r}{2\pi L}\ln\frac{r_o}{r_i} = k(T_i - T_o) \tag{3.37}$$

or, after some rearrangement:

$$Q_r = \frac{2\pi Lk}{\ln\frac{r_o}{r_i}}(T_i - T_o) = \frac{T_i - T_o}{R} \tag{3.38}$$

The quantity:

$$R = \frac{\ln(r_o/r_i)}{2\pi Lk} \tag{3.39}$$

is the heat transfer resistance of the cylindrical wall.

The treatment above for a cylindrical wall can be extended to more complicated situations involving multiple cylindrical layers by adopting the same approach that we used for a multi-layer flat wall. The difference is that, in this case, the surface area perpendicular to the heat transfer flux will vary with radial position and, thus, from layer to layer. For instance, if fluids at bulk temperatures $T_{b,i}$ and $T_{b,o}$ flow, respectively, inside and outside the pipe shown in Fig. 3.5a, we can write the following equations:

$$Q_r = 2\pi r_i L h_i (T_{b,i} - T_i)$$
$$Q_r = \frac{2\pi Lk}{\ln\frac{r_o}{r_i}}(T_i - T_o) \tag{3.40}$$
$$Q_r = 2\pi r_o L h_o (T_o - T_{b,o})$$

where h_i and h_o are the convective heat transfer coefficients at the inner and outer pipe surfaces, respectively. As we saw for the flat wall, the first and last of these equations describe convective heat transfer at the pipe surfaces, while the middle equation describes conductive transport through the pipe wall itself. At steady state, Q_r will be the same in each of these equations. Rearranging these equations gives the following expressions:

$$\frac{Q_r}{2\pi r_i L h_i} = T_{b,i} - T_i$$
$$\frac{Q_r}{\frac{2\pi Lk}{\ln r_o/r_i}} = T_i - T_o \tag{3.41}$$
$$\frac{Q_r}{2\pi r_o L h_i} = T_o - T_{b,o}$$

Finally, adding these expressions together we obtain:

$$Q_r = \frac{T_{b,i} - T_{b,o}}{\frac{1}{2\pi r_i L h_i} + \frac{\ln r_o/r_i}{2\pi Lk} + \frac{1}{2\pi r_o L h_o}} = \frac{T_{b,i} - T_{b,o}}{R_{\text{overall}}} \tag{3.42}$$

The quantity:

$$R_{\text{overall}} = \frac{1}{2\pi r_i L h_i} + \frac{\ln r_o/r_i}{2\pi Lk} + \frac{1}{2\pi r_o L h_o} \tag{3.43}$$

is the overall heat transfer resistance. As expected, since the resistances are in series, R_{overall} is the sum of the two convective resistances at the pipe surfaces plus the conductive resistance of the pipe wall. Examples 3.2 and 3.3 illustrate some applications of these relationships.

Example 3.2. Heat transfer across a pipe wall. Let's calculate the heat transfer rate across the wall of a 1-m-long pipe with inner and outer radii $r_i = 0.03$ m and $r_o = 0.04$ m, respectively. The pipe wall is made of a steel alloy with $k = 19$ W/m K. Internal and external heat transfer coefficients are $h_i =$

500 W/m² K and h_o = 1,200 W/m² K, respectively, and the internal and external bulk fluid temperatures are $T_{b,i} = 120°C$ and $T_{b,o} = 30°C$, respectively. From Eq. (3.43), we have:

$$R_{\text{overall}} = \frac{1}{2\pi \times 0.03 \times 1 \times 500} + \frac{\ln(0.04/0.03)}{2\pi \times 1 \times 19} + \frac{1}{2\pi \times 0.04 \times 1 \times 1{,}200}$$

$$= 0.011 + 0.0024 + 0.0033 = 0.017 \text{ K/W}$$

Thus, the heat transfer rate will be

$$Q_r = \frac{120 - 30}{0.017} = 5{,}290 \text{ W}$$

The concept of a dominant resistance among resistances in series applies also in this case. Since the resistances are additive, the largest resistance will tend to be dominant. For this example, based on the numbers above, the internal convective resistance is dominant, accounting for 0.011/0.017 = 0.65 or 65% of the total resistance. The conductive resistance through the pipe wall is the least important, in this case, accounting for only 0.0024/0.017 = 0.14 or 14% of the total. The external convective resistance is intermediate, accounting for about 21% of the total. Note that, beyond this particular example, there are certainly cases where no one resistance is dominant with each resistance providing comparable contributions to the total.

Example 3.3. Heat losses from an insulated pipe. Let's consider the heat losses to the surrounding air for an insulated copper pipe whose cross-section is shown in Fig. 3.6. The pipe itself has inner and outer radii r_i = 0.015 m and r_o = 0.02 m, respectively, and a length L = 1 m. Internal and external bulk fluid temperatures are 100°C and 0°C, respectively, while the internal and external convective heat transfer coefficients are h_i = 5,000 W/m² K and h_o = 100 W/m² K, respectively. The larger value of h_i suggests that the internal fluid is likely a liquid, while the smaller value of h_o is consistent with forced convection to air (*cf.* Table 1.2). The thermal conductivity of copper is k = 390 W/m K (*cf.* Fig. 2.3). We want to insulate this pipe with an insulating material having thermal conductivity k_e = 1.2 W/m K and wish to determine the thickness of insulation needed to reduce heat losses to the surrounding by 50%

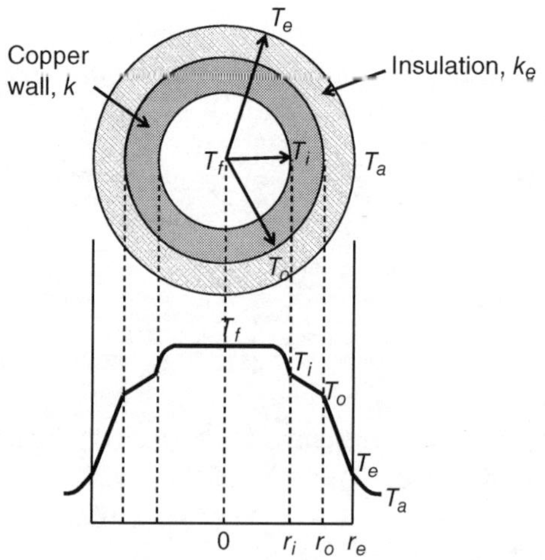

FIGURE 3.6 Cross section of an insulated pipe with internal and external convective heat transfer resistances for Example 3.3. The bottom graph is a rough sketch of the expected temperature profile, which is symmetrical relative to the pipe centerline.

compared to the bare pipe. We solve this problem by first calculating the heat transfer loss for the bare pipe. From Eq. (3.43), we have:

$$R_{\text{overall,bare}} = \frac{1}{2\pi \times 0.015 \times 1 \times 5000} + \frac{\ln(0.02/0.015)}{2\pi \times 1 \times 390} + \frac{1}{2\pi \times 0.02 \times 1 \times 100}$$

$$= 0.0021 + 0.00012 + 0.080 = 0.082 \text{ K/W}$$

We can see that, in this case, the external convective resistance is dominant since it is, by far, the largest of the three terms, accounting for about 98% of the total resistance. The resistance through the pipe wall itself is, instead, quite small and nearly negligible. The rate of heat loss for the bare pipe will be:

$$Q_r = \frac{100 - 0}{0.082} = 1{,}220 \text{ W}$$

If insulation is added, the overall resistance will become equal to the resistance of the bare pipe plus the resistance of the insulating layer. Thus:

$$R_{\text{overall,insulated}} = \frac{1}{2\pi r_i L h_i} + \frac{\ln(r_o/r_i)}{2\pi L k} + \frac{\ln(r_e/r_o)}{2\pi L k_e} + \frac{1}{2\pi r_e L h_e}$$

$$= 0.0021 + 0.00012 + \frac{\ln(r_e/0.02)}{2\pi \times 1 \times 1.2} + \frac{1}{2\pi r_e \times 1 \times 100}$$

$$= 0.0022 + 0.13 \times \ln(r_e/0.02) + 0.0016/r_e$$

In order to reduce the heat loss by 50%, the overall resistance will have to increase two-fold compared to the bare pipe. Hence, we will need $R_{\text{overall,insulated}} = 2 \times 0.082 = 0.164$ K/W. From the equation above, by trial and error, we can find that the value of r_e that gives $R_{\text{overall,insulated}} = 0.164$ K/W is $r_e = 0.055$ m. This means that a 3.5 cm layer of insulation will be needed. The heat loss will become 610 W. Note that we have assumed that the external convective heat transfer coefficient, h_e, is equal to the value given for h_o for the bare pipe regardless of the thickness of insulation. This is an assumption that must be checked for accuracy.

As seen in Example 3.3, adding insulation increases the overall resistance and reduces the rate of heat transfer. But is this always the case? If we look at the expression for $R_{\text{overall,insulated}}$ in Example 3.3, we see that increasing the thickness of insulation (i.e., increasing r_e) will increase the resistance of the insulating layer, which is represented by the term $\ln(r_e/r_o)/2\pi L k_e$ in the expression for the overall resistance. At the same time, however, increasing r_e will also decrease the external convective resistance, which is represented by the term $1/2\pi r_e L h_e$ in the expression for the overall resistance. The latter occurs because increasing r_e will increase the surface area for convective heat transfer. Because of their counter-acting effects, the net result of adding insulation will depend on the balance of these two terms. In general, if we keep the pipe dimensions constant, adding an incremental thickness of insulation will be beneficial if:

$$\frac{dR_{\text{overall,insulated}}}{dr_e} > 0$$

Since the overall resistance is expressed by:

$$R_{\text{overall,insulated}} = \frac{1}{2\pi r_i L h_i} + \frac{\ln(r_o/r_i)}{2\pi L k} + \frac{\ln(r_e/r_o)}{2\pi L k_e} + \frac{1}{2\pi r_e L h_e}$$

adding an incremental thickness of insulation will be beneficial if:

$$\frac{d}{dr_e}\left[\frac{1}{2\pi r_i L h_i}+\frac{\ln(r_o/r_i)}{2\pi L k}+\frac{\ln(r_e/r_o)}{2\pi L k_e}+\frac{1}{2\pi r_e L h_e}\right]>0$$

In this equation, the first two terms do not vary with r_e. Thus, the derivative with respect to r_e will yield:

$$\frac{1}{2\pi L k_e r_e}-\frac{1}{2\pi L h_e r_e^2}>0$$

This inequality will be satisfied if:

$$r_e>\frac{k_e}{h_e}$$

In Example 3.3, we found $r_e = 0.055$ m for a case where k_e/h_e was equal to 0.012 m. Thus, this inequality was definitely satisfied and, as we have seen, adding insulation was beneficial. On the other hand, if we had a thin pipe with, say, $r_o = 0.005$ m, and wanted to use an insulating material with $k_e = 5$ W/m K, the heat loss would actually increase by adding insulation until r_e becomes larger than $5/100 = 0.05$ m, which is 10 times the radius of the pipe. For this reason, very small pipes are often left uninsulated in order to avoid having to use a very thick insulating layer.

3.4 Spherical Walls

In this section, we consider heat conduction through a spherical shell with inner radius r_i and outer radius r_o. Temperatures T_i and T_o are specified at the inner and outer surfaces of the spherical shell, respectively, and both are assumed to be uniformly distributed over their respective surfaces. In this case, heat transfer will occur exclusively in the radial dimension of a spherical coordinate system and T will be a function of r only, or $T = T(r)$. The control volume for our energy balance will thus be a spherical shell with thickness Δr and with its surfaces located at r and $r+\Delta r$. The steady steady energy balance for this volume with no heat generation yields the following result:

$$Q_r\big|_r = Q_r\big|_{r+\Delta r}$$

where $Q_r\big|_r$ and $Q_r\big|_{r+\Delta r}$ are the rates of heat transfer in and out of the control volume. Dividing this equation by Δr and taking the limit $\Delta r \to 0$, we obtain:

$$\lim_{\Delta r\to 0}\left(\frac{Q_r\big|_{r+\Delta r}-Q_r\big|_r}{\Delta r}\right)=0$$

or

$$\frac{dQ_r}{dr}=0 \tag{3.44}$$

This result shows that, just as in the case of the flat wall and of the cylindrical wall, at steady state and with no heat generation, Q_r will be constant across the thickness of the spherical shell. In this case, however, the cross-sectional area A will vary with the square of r according to the following equation:

$$A = 4\pi r^2 \tag{3.45}$$

As a result,

$$q_r = \frac{Q_r}{4\pi r^2} \tag{3.46}$$

Thus, for a spherical shell, the flux will decrease with the square of radial position rather than with r as we saw for a cylindrical shell. In spherical coordinates, Fourier's law is as follows:

$$q_r = -k\frac{dT}{dr} \tag{3.47}$$

Combining Eqs. (3.46) and (3.47) then yields:

$$\frac{Q_r}{4\pi r^2} = -k\frac{dT}{dr} \tag{3.48}$$

Since Q_r is a constant with r, we can separate the variables r and T and integrate as follows between the inner and outer shell surfaces:

$$\frac{Q_r}{2\pi}\int_{r_i}^{r_o}\frac{dr}{r^2} = -\int_{T_i}^{T_o} k\,dT \tag{3.49}$$

As in the case of the slab or cylindrical wall, the integration limits are based on the temperatures T_i and T_o that are specified at the inner and outer surfaces, respectively. Finally, performing the integrals with the assumption that k is constant, we obtain the following expression for the heat transfer rate:

$$Q_r = \frac{4\pi k}{\dfrac{1}{r_i}-\dfrac{1}{r_o}}(T_i - T_o) = \frac{T_i - T_o}{R} \tag{3.50}$$

The quantity $R = (1/r_i - 1/r_o)/4\pi k$ is the heat transfer resistance of the spherical shell. We can also write R as follows:

$$R = \frac{r_o - r_i}{r_o r_i}\frac{1}{4\pi k} = \frac{r_o}{r_i}\frac{\Delta}{4\pi r_o^2 k} \tag{3.51}$$

where $\Delta = r_o - r_i$ is the thickness of the shell. This equation shows that if the shell is very thin, i.e., if $r_o/r_i \sim 1$, we will have $R \sim \Delta/4\pi r_o^2 k$, which is the result obtained for a flat wall with thickness Δ and area $A = 4\pi r_o^2$. For these conditions, we will be able to neglect the curvature and treat the spherical shell as if it were a flat wall.

Analogous expressions can be found for a multilayer spherical shell or when convective resistances at the shell surfaces are considered by adding together the corresponding resistances. For instance, for a spherical shell with internal and external convective heat transfer resistances described by the heat transfer coefficients h_i and h_o, we can write the following three expressions:

$$\begin{aligned} Q_r &= 4\pi r_i^2 h_i (T_f - T_i) \\ Q_r &= \frac{4\pi k}{\dfrac{1}{r_i}-\dfrac{1}{r_o}}(T_i - T_o) \\ Q_r &= 4\pi r_o^2 h_o (T_o - T_a) \end{aligned} \tag{3.52}$$

where T_f and T_a are the bulk temperatures of the fluids inside and outside the spherical shell, respectively. At steady state and with no heat generation, Q_r will be the same in all three expressions. As before, in order to eliminate the unknown temperatures T_i and T_o, we rearrange these expressions as follows:

$$\frac{Q_r}{4\pi r_i^2 h_i} = T_f - T_i$$

$$\frac{\dfrac{Q_r}{4\pi k}}{1/r_i - 1/r_o} = T_i - T_o \tag{3.53}$$

$$\frac{Q_r}{4\pi r_o^2 h_o} = T_o - T_a$$

Adding these expressions together we obtain the following result:

$$Q_r = \frac{T_f - T_a}{R_{\text{overall}}} \tag{3.54}$$

where

$$R_{\text{overall}} = \frac{1}{4\pi r_i^2 h_i} + \frac{1/r_i - 1/r_o}{4\pi k} + \frac{1}{4\pi r_o^2 h_o} \tag{3.55}$$

is the overall heat transfer resistance.

Example 3.4. Heat transfer through a spherical shell. A 60-W light bulb is lit inside a spherical shell made of an opaque solid with unknown thermal conductivity. The inner and outer radii are 10 cm and 15 cm, respectively. By measuring the steady-state temperatures at the inner and outer surfaces of the shell, we can determine the thermal conductivity of the material. In this example problem, the measured temperatures are $T_i = 230°\text{C}$ and $T_o = 50°\text{C}$. Since we know that, at steady state, $Q_r = 60$ W, from Eq. (3.50) we can calculate $R = (T_i - T_o)/Q_r = (230 - 50)/60 = 3.0$ K/W. In turn, using this value of R, we can calculate the unknown thermal conductivity from Eq. (3.51), which gives:

$$k = \frac{r_o}{r_i}\frac{\Delta}{4\pi r_o^2 R} = \frac{0.15}{0.10}\frac{0.05}{4\pi \times (0.15)^2 \times 3.0} = 0.088 \text{ W/m K}$$

3.5 Heat Conduction with Simultaneous Heat Generation

In Example 3.4, we considered a problem where heat generated within the inner space of a solid spherical shell was conducted across the shell. In that problem, no heat was generated within the solid shell itself. We now consider a problem where thermal energy is generated uniformly within the solid where heat conduction takes place.

Example 3.5. Temperature profile is an electrically heated solid rod. Let's consider a cylindrical solid rod heated by the passage of electrical current and suspended in air at a temperature, T_a. We want to find the temperature at the surface of the rod as well as the temperature profile within the rod itself. As shown in Fig. 3.7, the rod has a radius r_o and a length L and has an electrical resistance R_e. A voltage

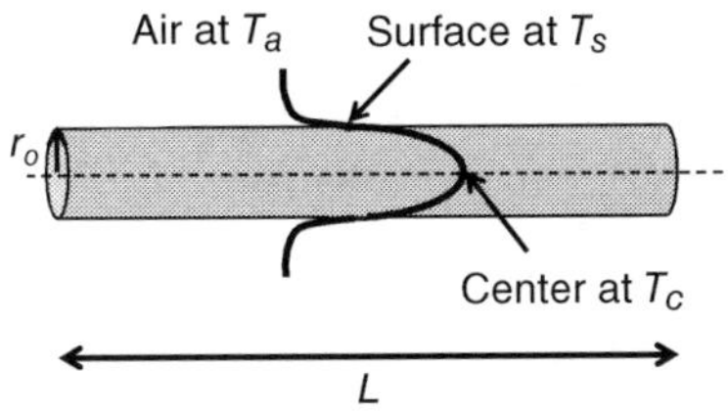

Figure 3.7 Cylindrical rod heated by an electric current for Example 3.5.

difference ΔV is applied across the length of the rod. We first find the power generated by the passage of the electrical current according to the Joule effect, which is:

$$P = \frac{\Delta V^2}{R_e} \tag{3.56}$$

At steady state, the temperature of the rod will be constant in time. For this to happen, the heat generated by the electrical current must be transported away from the rod surface by convective heat transfer. In this case, an overall energy balance around the rod will give:

$$P = 2\pi r_o L h (T_s - T_a) \tag{3.57}$$

where h is the convective heat transfer coefficient at the rod surface. The rod surface temperature is easily obtained from this equation as:

$$T_s = T_a + \frac{P}{2\pi r_o L h} \tag{3.58}$$

Next, we turn our attention to the temperature profile inside the rod. This calculation will require a differential energy balance for a cylindrical shell within the rod itself with its surfaces located at positions r and $r + \Delta r$. This shell will have a thickness Δr and a volume equal to $2\pi r L \times \Delta r$. We need to balance heat conduction in and out of this shell against the heat generated within the shell itself as follows:

$$\text{Energy accumulation} = \text{Energy in} - \text{Energy out} + \text{Energy generation}$$

Energy accumulation will be zero at steady state. Energy in and energy out will be expressed by $Q_r|_r$ and $Q_r|_{r+\Delta r}$, respectively. Finally, energy generation will be expressed in terms of G, which is the power generated per unit volume of rod. The latter is given by:

$$G = \frac{P}{\pi r_o^2 L} \tag{3.59}$$

where $\pi r_o^2 L$ is the volume of the rod. For simplicity, we can assume that G is a constant with r. Balancing these four energy terms, we obtain:

$$0 = Q_r|_r - Q_r|_{r+\Delta r} + 2\pi r L \Delta r G$$

Dividing this equation by Δr and taking the limit $\Delta r \to 0$ gives the following result:

$$\lim_{\Delta r \to 0} \left(\frac{Q_r|_{r+\Delta r} - Q_r|_r}{\Delta r} \right) = 2\pi r L G$$

or

$$\frac{dQ_r}{dr} = 2\pi r L G \tag{3.60}$$

Equation (3.60) shows that, in this case, Q_r is not constant but actually increases with r. Physically, this occurs because more and more heat will need to be conducted as we move away from the center of the rod and toward its surface to compensate for the increased cumulative heat generation. From Fourier's law in cylindrical coordinates, we have:

$$q_r = -k\frac{dT}{dr} \tag{3.61}$$

Thus,

$$Q_r = -2\pi rLk\frac{dT}{dr} \tag{3.62}$$

Combining this result with Eq. (3.60), we obtain the following differential equation:

$$-2\pi L\frac{d}{dr}\left(kr\frac{dT}{dr}\right) = 2\pi rLG \tag{3.63}$$

If k is constant, canceling the common terms, this equation becomes:

$$\frac{d}{dr}\left(r\frac{dT}{dr}\right) = -\frac{G}{k}r \tag{3.64}$$

This equation is subject to the following boundary conditions:

$$\begin{aligned} r = 0 &: \frac{dT}{dr} = 0 \\ r = r_o &: T = T_s \end{aligned} \tag{3.65}$$

The first of these conditions arises because of symmetry and will lead to a maximum temperature at the center of the rod. The second of these conditions sets the surface temperature equal to T_s, which we already know from Eq. (3.58). We can solve Eq. (3.64) by separating the variables r and T and integrating as follows:

$$\int_0^{r\frac{dT}{dr}} d\left(r\frac{dT}{dr}\right) = -\frac{G}{k}\int_0^r r\,dr \tag{3.66}$$

Performing the integrals, we obtain:

$$r\frac{dT}{dr} = -\frac{Gr^2}{2k} \tag{3.67}$$

A second integration gives:

$$\int_{T_c}^{T_s} dT = -\frac{G}{2k}\int_0^{r_o} r\,dr \tag{3.68}$$

or

$$T_c = T_s + \frac{Gr_o^2}{4k} \tag{3.69}$$

where T_c is the centerline temperature. For a sample calculation, let's consider the case of a 10-mm-diameter graphite rod with a length of 0.8 m and an applied voltage of 12 V in air at $T_a = 20°C$ where the convective heat transfer coefficient is $h = 100$ W/m^2 K. For graphite, the thermal conductivity is $k = 2.5$ W/m K and the electrical resistivity is $\rho_e = 4.0 \times 10^{-5}$ Ω m. The electrical resistance, R_e will be proportional to the length of the rod and inversely proportional to its cross-section. Thus, $R_e = \rho_e L/\pi r_o^2 = 4.0 \times 10^{-5} \times 0.8/\pi \times 0.005^2$ = 0.403 Ω. The total power generated will thus be $P = (12)^2/0.403 = 357$ W, while the power generated per unit volume will be

$G = P/\pi r_o^2 L = 357/(\pi \times 0.005^2 \times 0.8) = 5.68 \times 10^6\ \text{W/m}^3$. For these conditions, Eq. (3.58) will give a surface temperature equal to:

$$T_s = 20 + \frac{357}{2\pi \times 0.005 \times 0.8 \times 100} = 162°\text{C}$$

while Eq. (3.69) will give the centerline temperature:

$$T_c = 162 + \frac{5.68 \times 10^6 \times (0.005)^2}{4 \times 2.5} = 176°\text{C}$$

Note that we have neglected radiant heat transfer from the surface of the rod. This is probably reasonable, in this case, because the surface temperature is not so high, but could become very important at higher surface temperatures.

3.6 Problems

P.3.1. Double-pane glass windows, designed to reduces heat losses, consist, as the name implies, of two glass panes separated by a thin gap filled with dry air and sealed. Let's compare the efficiency of a single-pane window (just one pane without an air gap) with that of a double-pane window with the same overall thickness of glass. The single-pane window has a surface area of 1.5 m^2 and the glass thickness is 0.5 cm. The double-pane window has the same area but comprises two 0.25-cm-thick glass panes separated by a 2-mm air gap. The convective heat transfer coefficients inside and outside the room are 10 and 20 W/m^2 K and the inside and outside temperatures are 25°C and 4°C, respectively. Compute the heat losses from the window for the two cases. In each case, also compute the temperature at the inside and outside surfaces of the window. Assume that heat transfer occurs only by conduction in the air gap and neglect radiant heat transfer.

P.3.2. Serpe [Nature, 565 (2019) 438] has reported on the development of "smart windows," where the gap between two glass panes is filled with an aqueous solution containing hydrogel particles made of poly(*N*-isopropylacrylamide) or pNIPAm, which are thermo-responsive. At low temperature (e.g., < 32°C), the particles are swollen and become transparent allowing light to pass through. However, at higher temperatures, the hydrogel particles shrink and scatter the sunlight, preventing radiant heat transfer into the room during a sunny day and reducing air conditioning costs. Suppose that the double pane window described in Problem 3.1 is filled with such a hydrogel-containing solution. By how much will the convective/conductive heat transfer rate change compared to the case where the gap is filled with air? The thermal conductivity of pNIPA hydrogel has been reported to be on the order of 0.40 W/m K.

P.3.3. The walls of a cold-storage room kept at 8°F are insulated with a 4-in. layer of cork. The external air is at 72°F and has a dew point of 60°F. Since cork is porous, air diffuses across the insulating layer and water vapor will start condensing at the point where the temperature is 60°F and freezing where the temperature reaches 32°F. (a) Find the depths within the cork where liquid water and ice are formed, if cork is assumed to consist of wood solid with $k = 0.058$ BTU/h ft °F with air-filled pores with porosity $\varepsilon_p = 0.50$ and if the effects of condensation and freezing in the pores on the thermal conductivity of cork are neglected. Also neglect any convective heat transfer resistance. (b) Repeat part (a) if the effects of condensation and freezing in the pores on the thermal conductivity of cork are considered. Note that condensation and freezing within the pores will alter the thermal conductivity.

P.3.4. The design of an astronaut's glove requires that you select the thickness of insulation needed to keep the skin surface temperature at 25°C for long periods of time when the outer surface of the insulation is at –271°C. The skin and insulation are separated by a 1-mm-thick polymer membrane with $k = 0.35$ W/m K. Blood vessels 3.0 mm below the skin surface keep the temperature there at a constant 37°C. Neglecting heat transfer by radiation: (a) Determine the thickness of insulation required if its thermal conductivity is 0.07 W/m K. (b) For the thickness you have selected, what temperature would be sensed by nerve endings located 1.5 mm from the skin surface?

P.3.5. An insulated cardboard box is to be designed to transport a labile vaccine product that has to be kept at –78°C in areas where low-temperature refrigeration is not widely available. The box is a cube 0.3 m on a side and is filled halfway with dry ice, which keeps the contents cold by slowly subliming. The box will be insulated with a layer of Styrofoam insulation ($k = 0.030$ W/m K). Determine the thickness of insulation needed if the low-temperature has to last for 1 week assuming that the outside temperature is 30°C and that the heat transfer resistance of the insulation is dominant. Assume that the bulk density of the dry ice is 1,200 kg/m^3.

P.3.6. For a class demonstration project, I placed an empty stainless-steel pan (k=15 W/m K) onto a hot plate with its temperature set at 150°C as shown in Fig. P.3.6. After a few minutes, I asked the students whether I could touch the bottom of the pan with my bare finger. A few said "yes, do it!" (I wonder if they did not like the last test…), but most said an emphatic "no." I then removed the pan from the hot plate and filled it with water at room temperature. I also placed a magnetic stirrer in the pan, which kept the water well mixed in the pan. After a few minutes on the hot plate, I asked the students again whether I could touch the bottom of the pan. Few answered this time and there seemed to be some surprised gasps when I stuck my finger in and touched the bottom of the pan without getting burned. I repeated this a few times, but only until the water got hot, of course. The bottom of the pan is 5 mm thick. (a) Explain what happened in this demonstration. (b) Estimate what the temperature at the bottom surface of the pan was when there was only air at 22°C. (c) Estimate what the temperature at the bottom surface of the pan was after a few minutes when it was full of water at 22°C.

P.3.7. A cylindrical exhaust duct has a constant inside temperature of 650°F and is insulated on the exterior with a 4-in. layer of rock wool, which has an outer surface temperature of 95°F.

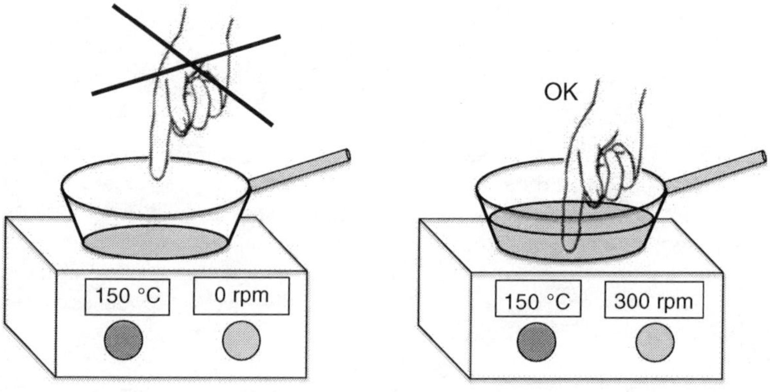

Figure P.3.6

The duct has an inner diameter of 3.5 in. and ½-in. thick walls of a ceramic material with a thermal conductivity of 0.78 BTU/h ft°F. Find the rate of heat loss from per linear foot of duct.

P.3.8. The thermal conductivity of an opaque insulating material is to be determined by lighting a 5-W light bulb inside a hollow sphere made of that material and measuring the surface temperatures. The sphere has inside and outside radii of 0.01 and 0.015 m, respectively, and is suspended in air at 25°C, which is blown over the sphere at a velocity of 5 m/s. The inner and outer surface temperatures of the sphere are measured with thermocouples at steady state and are 90°C and 60°C, respectively. (a) Write the differential energy balance equation and proper boundary conditions needed to describe steady-state heat transfer through the spherical shell. (b) Solve your equation to find an expression for the heat transfer rate. (c) Determine the thermal conductivity of the material. (d) Determine the convective heat transfer coefficient at the outer surface of the sphere. Note: assume that the thermal conductivity of the material is independent of temperature.

P.3.9. Liquid nitrogen loaded into an insulated tank at 1 atm absolute pressure is kept at the boiling point by allowing a small amount of it to gradually evaporate and flow out of the tank. In order to keep a constant temperature, the rate at which nitrogen vapor is vented from the tank must be such that the latent heat of vaporization exactly balances the rate of heat transfer through the tank wall. A thin-walled liquid nitrogen cylinder, 40 cm in diameter and 1.5 m tall, is covered with 4 cm of an insulating material with $k = 0.0020$ W/m K. The external heat transfer coefficient is $h = 20$ W/m^2 K and the external air temperature is 22°C. (a) Determine the steady rate of heat transfer into the tank. (b) Determine the rate at which nitrogen vapor needs to be vented from the tank in kg/s. (c) Determine how long the tank will last if it is initially completely filled with liquid nitrogen.

P.3.10. The proposed design of a fuel element for a nuclear power reactor consists of 2-in.-diameter cylindrical fuel core surrounded by a 0.25-in.-thick aluminum-alloy cladding. The outside surface of the cladding is kept at 200°F. The fuel core gives off heat at the rate of 5.8×10^7 BTU/h ft^3. For safety reasons, the aluminum-alloy temperature cannot exceed 800°F at any point. Is the proposed design satisfactory? Data: $k_{\text{aluminum alloy}} = 118\,(1 - 4.90 \times 10^{-4}\,T)$ where T is in °F and $k_{\text{aluminum alloy}}$ in BTU/h ft°F.

P.3.11. A cube, 1.0-in. long on each side, is constructed of equal-sized, alternating parallel layers of copper and Pyrex glass. There are five layers of copper and five layers of glass. A question has arisen whether this cube would be a better conductor of heat (for a fixed overall ΔT) if it were oriented so that heat was conducted parallel to the alternating layers or normal to them. Find the ratio of the heat conduction rates for the two different orientations.

P.3.12. An electrically heated glove consists of a flat thin-sheet electrode embedded within a 0.01 m layer of an insulating material with a thermal conductivity of 0.02 W/m K, which is directly in contact with the skin. Blood vessels, 0.004 m below the skin surface, keep the temperature at 37°C. The external air temperature is –10°C and the external heat transfer coefficient is 5.0 W/m^2 K. The electrode temperature is 60°C. For this problem, assume that the thermal conductivity of skin is 0.20 W/m K. (a) Sketch the temperature profiles for this problem assuming a flat wall geometry with the thin-sheet electrode located at a distance x from the outer glove surface. Note that the temperature is maximum at the electrode. (b) Write explicit expressions for the heat transfer fluxes to the surrounding air and to the skin in terms of known temperatures only, when the thin-sheet electrode is located at a distance x from the outer glove surface. (c) Determine a distance from the outer glove surface where the thin-sheet electrode should be located if the power generated is 250 W/m^2. (d) For the conditions in part c, determine the temperature at the inner glove surface in contact with the skin.

P.3.13. Consider a slab-shaped graphite electrode that is 3 ft long, 3 in. wide, and 0.5 in. thick. A voltage difference of 24 V is applied across the 3-ft length. Assuming that the temperature is uniform over the 3-ft length and the 3-in. width, plot the temperature profile along the 0.5-in. thickness and determine the temperature at the center of the electrode if the surface temperature of the electrode is 176°F on both sides. Assume that the temperature in the electrode is only a function of the coordinate along the 0.5-in.-thick dimension of the slab. What would the external heat transfer coefficient have to be in order to have the surface temperature of 176°F if the external ambient temperature is 68°F? How would your answer change if the slab is insulated on one side and the surface temperature is 176°F on the un-insulated side? The thermal conductivity of the graphite used is 2.9 BTU/h ft°F and its electrical resistivity is $1.4 \times 10^{-4}\ \Omega \cdot$ ft.

CHAPTER 4

Combined Conductive and Convective Heat Transfer

4.1 Introduction

In this chapter, we consider situations where conductive and convective heat transfer processes occur simultaneously and are combined to determine the overall heat transfer rate. Unlike the problems in the previous chapters, where conduction and convection were combined in series only at system boundaries, here we consider cases where conduction and convection are locally coupled with convective transport occurring in a direction perpendicular to that of conductive transport. The main example covered in this chapter is that of "fins" or, more generally, extended surfaces. Fins are important in a variety of applications usually involving heat transfer from solid surfaces to air or other gases. As seen in Chap. 1 (e.g., see Table 1.2), convective heat transfer coefficients are lowest for cases that involve transport to gases, especially when only free convection takes place. Thus, the convective heat transfer resistance can be quite large in these cases. Blowing the gas over the surface at a high speed helps by increasing the heat transfer coefficient, but there are limits as to what can be achieved in practice. A more practical way to overcome such limitations is to increase the heat transfer rate by increasing the heat transfer area. This can be obtained by attaching fins made of a highly conductive material to the surface that we need to either cool or heat. Figure 4.1 shows some examples of such fins. In all these cases, if the objective is to cool the underlying surface, heat is conducted away from the surface that we want to cool through the base of the fin and is dissipated to the surrounding fluid by convective heat transfer from the fin surfaces. More heat can be transported to the surrounding fluid because of the larger area compared to the area of the underlying bare surface. Of course, multiple fins can be attached to the same surface, thereby multiplying the cooling effect. Analogous considerations can be made if the objective is to transfer heat to the underlying surface from a hotter gas. Fins are used extensively in air-cooled heat exchangers, in various kinds of room heating and air conditioning devices, in air-cooled engines, such as those used in motorcycles, in automobile radiators, and to cool electronic devices. Obviously, fins add weight, complexity, and cost. Thus, a careful design is usually needed.

The main consideration affecting fin design is the fact that, starting at the base of the fin, as heat is gradually convected away to the surrounding fluid, the temperature will

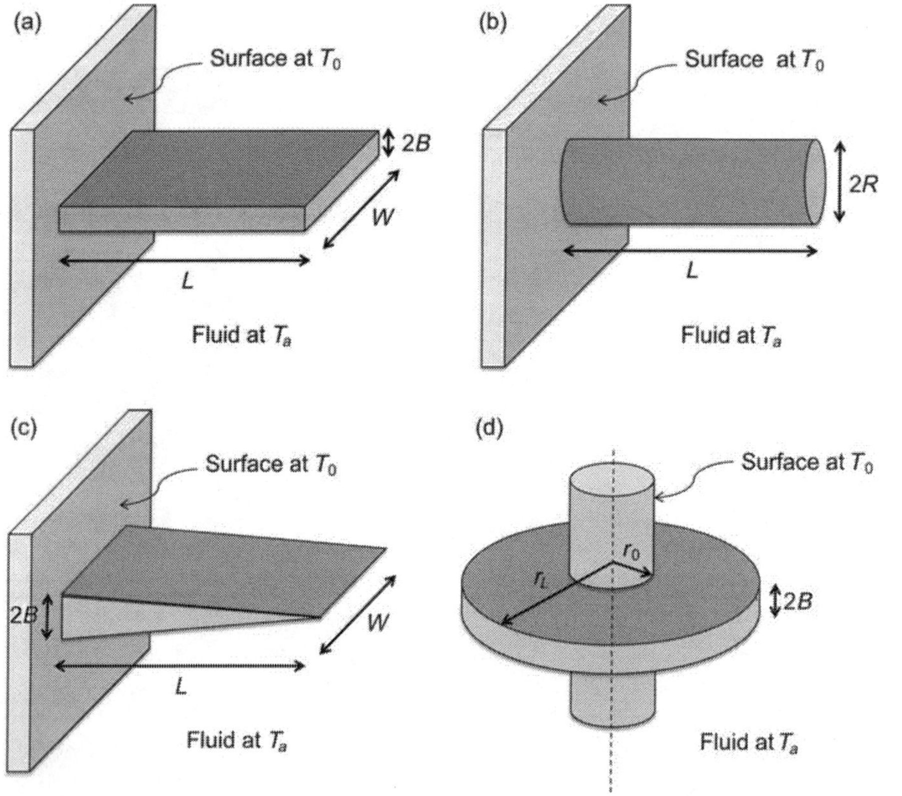

FIGURE 4.1 Examples of designs of rectangular (a), cylindrical (b), triangular (c), and circular fins (d). Note that multiple fins can be attached to the underlying surface at temperature T_0. T_a is the temperature of the surrounding fluid, normally air.

decrease along the fin from the temperature T_0 at its base to lower temperatures as we move away from the surface that we want to cool. In turn, since convective transport depends on the difference between the fin temperature and the fluid temperature, the local convective heat transfer flux will decrease as we move away from the base of the fin to its tip. If the fin is very long, there will be more surface area. However, the temperature will also drop off more from the base value making heat transfer less efficient. Analogous considerations apply to cases where rather than cooling the underlying surface with a cooler gas we want to transfer heat to it from a hotter fluid. Intuitively, in either case, we will want to design fins that are "thin," so they use small amounts of material, that provide a large surface area while being relatively short, and that are made of a highly conductive material, which will minimize the temperature change along the length of the fin. The quantitative analysis useful for such designs is presented in the following sections. It should be noted, however, that similar analyses will apply to many other situations beside fins, where convective heat transfer occurs perpendicular to the direction of conductive transport. A number of such situations are covered by some of the end-of-chapter problems.

4.2 Analysis and Design of Straight Fins

In this section, we consider the analysis and design of straight fins such as those illustrated in Fig. 4.1. Let's consider a fin with an arbitrary shape as illustrated in Fig. 4.2. This generic fin has a cross-section $A(x)$ and a perimeter $p(x)$, both of which can vary along the length of the fin from its base at $x = 0$ to its tip at $x = L$. The cross-section at the base of the fin is $A_0 = A(0)$, while the cross-section at the tip is $A_L = A(L)$. The total fin area over which convective transport to the surrounding air will occur is given by the following expression:

$$S = \int_0^L p(x)dx + A_L \tag{4.1}$$

Note that, in practice, A_L will usually be a small fraction of the total area, S. Thus, the perimeter area, i.e., the first term on the right-hand-side of Eq. (4.1), will generally be dominant.

If the fin is "thin" compared to its length, which will usually be the case in practice, we can assume that the temperature within the fin itself is uniform at each distance x from the base. Thus, the fin temperature will be a function of x only, or $T = T(x)$. In this case, we can obtain the function $T(x)$ from a one-dimensional differential energy balance. We start by writing an energy balance for a section of the fin with sides located at x and at $x + \Delta x$ as shown in the right-hand panel of Fig. 4.2. Heat is conducted in and out of this volume within the fin at rates $Q_x|_x$ and $Q_x|_{x+\Delta x}$, respectively, while it is simultaneously transported out of the fin by convection at the rate $p(x)\Delta x\, h(T - T_a)$, where h is the convective heat transfer coefficient. Note that $p(x)\Delta x$ approximates the surface area of the control volume. Balancing these energy flows, we obtain:

$$Q_x\big|_x = Q_x\big|_{x+\Delta x} + p(x)\Delta x\, h(T - T_a)$$

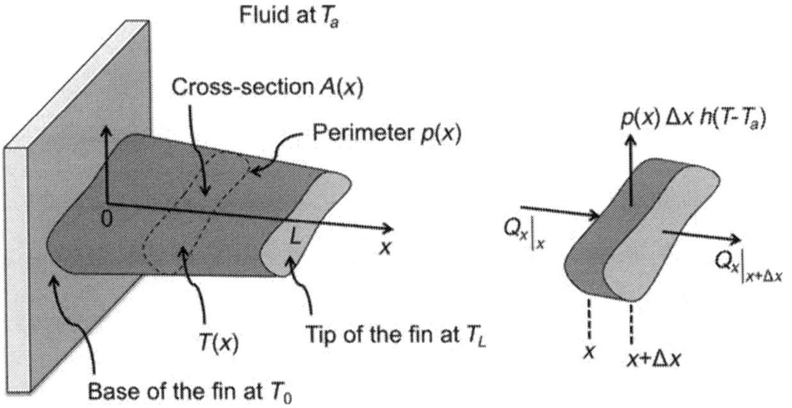

Figure 4.2 Generic straight fin (left) and control volume used to derive the differential energy balance (right).

Dividing this equation by Δx and taking the limit $\Delta x \to 0$, we have:

$$\lim_{\Delta x \to 0}\left(\frac{Q_x\big|_{x+\Delta x} - Q_x\big|_x}{\Delta x}\right) = -p(x)h(T - T_a)$$

or

$$\frac{dQ_x}{dx} = -p(x)h(T - T_a) \tag{4.2}$$

Equation (4.2) shows that the rate of heat conduction through the fin, Q_x, will decrease as we move from the base of the fin and toward its tip because of the heat lost by convection.

In order to go further, we introduce Fourier's law, which gives:

$$q_x = -k\frac{dT}{dx} \tag{4.3}$$

where k is the thermal conductivity of the fin material. Since $Q_x = A(x)q_x$, we will have:

$$Q_x = -A(x)k\frac{dT}{dx} \tag{4.4}$$

Combining Eqs. (4.2) and (4.4) gives the following result:

$$\frac{d}{dx}\left[A(x)k\frac{dT}{dx}\right] = p(x)\, h(T - T_a) \tag{4.5}$$

This is a second-order ordinary differential equation that requires two boundary conditions. In general, these conditions can be written as follows:

$$x = 0 : T = T_0 \tag{4.6}$$

$$x = L : -k\,\frac{dT}{dx} = \, h(T - T_a) \tag{4.7}$$

The first of these conditions obviously states that the fin base temperature will be the same as the temperature of the surface to which the fin is attached. The second condition at $x = L$ is derived from the following energy balance:

$$Q_x\,|_{x=L} = -A_L k\,\frac{dT}{dx}\bigg|_{x=L} = A_L h\left(T\big|_{x=L} - T_a\right) \tag{4.8}$$

This equation expresses that the conductive heat transfer reaching the tip of the fin will be equal to the convective heat transfer away from the tip of the fin to the surrounding fluid. In practice, since most of the heat conducted into the base of the fin will be dissipated by convection along the perimeter surface of the fin, $Q_x\big|_{x=L}$ is typically very small. As a result, we can usually replace the second boundary condition with $q_x\big|_{x=L} = 0$ or

$$x = L : \frac{dT}{dx} = 0 \tag{4.9}$$

This revised boundary condition allows some simplification of the ensuing mathematics and is commonly used. Note that at the base of the fin, the conductive heat transfer rate is given by Fourier's law as follows:

$$Q_x|_{x=0} = -A_0 k \left.\frac{dT}{dx}\right|_{x=0} \tag{4.10}$$

where $(dT/dx)_{x=0}$ is the temperature gradient at the base of the fin. Essentially all of this heat will be dissipated to the surrounding environment by convective transport along the fin perimeter area S. We can calculate this quantity by solving Eq. (4.5) to find $T(x)$ and then calculating dT/dx at $x = 0$.

The solution of Eq. (4.5) subject to boundary conditions Eqs. (4.6) and (4.7) or Eq. (4.9) depends on how A and p vary with x (i.e., on the shape of the fin) and on whether the thermal conductivity of the fin, k, varies with temperature or not. For rectangular or cylindrical fins (such as those shown in Figs. 4.1a and b), A and p will be constant along the length of the fin. For a rectangular fin we will have $A = 2\,BW$ and $p = 2W + 4B$, while for a cylindrical fin we will have $A = \pi R^2$ and $p = 2\pi R$. In either case, if k is constant, Eq. (4.5) will become:

$$\frac{d^2T}{dx^2} = \frac{ph}{Ak}(T - T_a) \tag{4.11}$$

Before solving this equation, it is useful to define the following dimensionless temperature and dimensionless distance variables:

$$\theta = \frac{T - T_a}{T_0 - T_a} \tag{4.12}$$

$$\xi = \frac{x}{L} \tag{4.13}$$

Since $T = T_a + (T_0 - T_a)\theta$ and $x = L\xi$, substituting these variables into Eq. (4.11), we obtain:

$$\frac{(T_0 - T_a)}{L^2}\frac{d^2\theta}{d\xi^2} = \frac{ph}{Ak}(T - T_a) \tag{4.14}$$

or:

$$\frac{d^2\theta}{d\xi^2} = N^2\theta \tag{4.15}$$

This equation is subject to the following conditions easily derived from Eqs. (4.6) and (4.9):

$$\xi = 0: \quad \theta = 1 \tag{4.16}$$

$$\xi = 1: \quad \frac{d\theta}{d\xi} = 0 \tag{4.17}$$

The parameter N that appears in Eq. (4.15) is called the "fin modulus" and is given by the following equation:

$$N = L\sqrt{\frac{ph}{Ak}} \tag{4.18}$$

An advantage of using these dimensionless variables is that we can generalize the solution and use it for different cases. Rendering the equation and boundary conditions dimensionless also gives us an immediate appreciation for the impact of the different physical parameters. Since, in the dimensionless form, N is the only parameter that appears in the equation and in its boundary conditions, it is clear that the different physical properties, L, p, A, and k, will affect the fin behavior only in the combination indicated by Eq. (4.18). Thus, for instance, we can see that doubling the fin length will have the same effect as decreasing the thermal conductivity four-fold or increasing the ratio of perimeter and area four-fold.

In order to find the exact behavior of the fin, we need to solve Eq. (4.15) with the associated boundary conditions. This is a standard ODE and its solution, subject to boundary conditions in Eqs. (4.16) and (4.17), is found in many differential equations textbooks and is given by any of the following three equivalent expressions:

$$\begin{aligned}\theta &= \frac{e^{-N}}{e^{N}+e^{-N}}e^{N\xi} + \frac{e^{N}}{e^{N}+e^{-N}}e^{-N\xi} \\ &= \cosh(N\xi) - \tanh(N)\sinh(N\xi) \\ &= \frac{\cosh[N(1-\xi)]}{\cosh(N)}\end{aligned} \tag{4.19}$$

Setting $\xi = 1$ in Eq. (4.19), we obtain the dimensionless temperature at the tip of the fin:

$$\theta\big|_{\xi=1} = \frac{T_L - T_a}{T_0 - T_a} = \frac{\cosh(0)}{\cosh(N)} = \frac{1}{\cosh(N)} \tag{4.20}$$

In order to calculate the heat transfer rate at the base of the fin, we need the derivative of the temperature profile dT/dx at $x = 0$. We can obtain this derivative from Eq. (4.19) if we consider that

$$\frac{dT}{dx} = (T_0 - T_a)\frac{d\xi}{dx}\frac{d\theta}{d\xi} = \frac{T_0 - T_a}{L}\frac{d\theta}{d\xi} \tag{4.21}$$

In this case, from Eq. (4.19) we have:

$$\frac{d\theta}{d\xi} = \frac{d}{d\xi}\left\{\frac{\cosh[N(1-\xi)]}{\cosh(N)}\right\} = -N\frac{\sinh[N(1-\xi)]}{\cosh(N)} \tag{4.22}$$

Setting $\xi = 0$, we obtain the derivative at the base of the fin:

$$\frac{dT}{dx}\bigg|_{x=0} = \frac{T_0 - T_a}{L}\frac{d\theta}{d\xi}\bigg|_{\xi=0} = \frac{T_0 - T_a}{L}\times\left[-N\frac{\sinh(N)}{\cosh(N)}\right] \tag{4.23}$$

Finally, inserting this result into Eq. (4.10) gives us the following expression for the conductive heat transfer rate at the base of the fin:

$$Q_x\big|_{x=0} = \frac{AkN}{L}\frac{\sinh(N)}{\cosh(N)}(T_0 - T_a) = \frac{AkN}{L}\tanh(N)(T_0 - T_a) \tag{4.24}$$

It is useful at this point to introduce the concept of "fin efficiency" according to the following reasoning. In an ideal, infinitely efficient fin, the temperature within the fin would be uniform and equal to the base temperature T_0. In such an ideal fin, since the driving force for convective heat transfer would be the maximum possible, the rate of heat dissipation would also be the maximum possible and none of the differential mathematics above would be needed to calculate the heat transfer rate. In fact, for the ideal fin, we can calculate the heat transfer rate simply by multiplying the fin perimeter area S, defined by Eq. (4.1), by the convective heat transfer coefficient and by the temperature difference driving force $T_0 - T_a$. For a thin fin with constant p, neglecting the fin tip area, A_L, we will have $S \sim p \times L$. Thus, the rate of heat dissipation in such an ideal fin would be approximated by:

$$Q_{\text{ideal}} = pLh(T_0 - T_a) \tag{4.25}$$

which is obviously a very easy calculation. Next, we define the fin efficiency, η, as the heat that is actually dissipated by a real fin divided by the heat that would be dissipated by an ideal fin with the same geometry, or:

$$\eta = \frac{Q_x\big|_{x=0}}{Q_{\text{ideal}}} \tag{4.26}$$

Combining Eqs. (4.24), (4.25), and (4.26), we arrive at the following expression for η:

$$\begin{aligned} \eta &= \frac{\dfrac{AkN}{L}\tanh(N)(T_0 - T_a)}{PLh(T_0 - T_a)} \\ &= \frac{Ak}{L^2Ph} N \tanh(N) \\ &= \frac{\tanh(N)}{N} \end{aligned} \tag{4.27}$$

This result shows that the fin efficiency will be a function of the fin modulus only.

Figure 4.3 shows a plot of the fin efficiency according to this equation as a function of the fin modulus, N, on logarithmic scales. As seen from Fig. 4.3, when $N < 0.2$, η is predicted to be very close to 1, indicating a highly efficient fin. On the other hand, for values of $N > 1$, η is predicted to be low, indicating that the fin becomes inefficient. For $N \geq 2$, $\tanh(N)$ becomes very close to 1 so that for values of $N > 2$ we will have $\eta \sim 1/N$. On the log scales of Fig. 4.3, this result is represented by a straight line with slope equal to -1. The physical meaning of N can now be understood. From Eq. (4.18), $N^2 = L^2ph/Ak = (L/Ak)/(1/Lph)$. The quantity L/Ak represents the conductive resistance to heat transfer along the length of the fin. Conversely, the quantity $1/Lph$ represents the convective heat transfer resistance that limits heat dissipation to the surrouding fluid. As a result, N^2 represents the ratio of conductive and convective resistances. When N is small, the conductive resistance is small and heat dissipation will be largely determined by the convective resistance resulting in a fin efficiency close to 1. On the other hand, if N is large, the conductive resistance will be large. In this case, the temperature will drop substantially over the length of the fin leading to a poor fin efficiency. Examples 4.1 and 4.2 illustrate some applications of this concept.

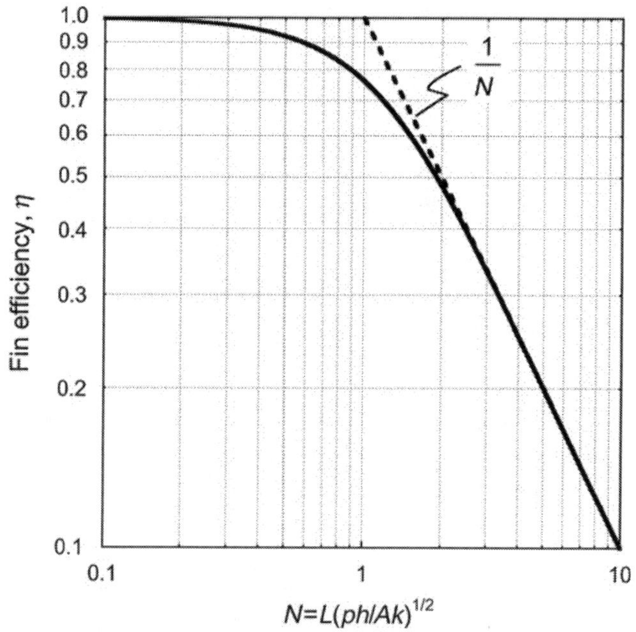

FIGURE 4.3 Thermal efficiency of a straight fin with constant A, p, and k as a function of the fin modulus.

Example 4.1. Thermal efficiency of a rectangular fin. Let's calculate the heat dissipation rate and the efficiency of a rectangular fin made of a steel alloy with $k = 10$ W/m K and having the following characteristics: $L = 3$ cm, $W = 5$ cm, $2B = 0.2$ cm, $h = 50$ W/m^2 K, $T_0 = 300$°C, and $T_a = 25$°C. The easiest way to solve this problem is to start by calculating the heat dissipation rate for an ideal fin with the same dimensions. Since the fin is very thin, we will have $p \sim 2W$. Thus, working in SI units, we have:

$$Q_{\text{ideal}} = pLh(T_0 - T_a) = 2WLh(T_0 - T_a) = 2 \times 0.05 \times 0.03 \times 50 \times 275 = 41.3 \text{ W}$$

Next, we calculate the fin modulus. The fin cross-section is $A = 2BW$. Thus,

$$N = L\sqrt{\frac{ph}{Ak}} = L\sqrt{\frac{2Wh}{2BWk}} = L\sqrt{\frac{h}{Bk}} = 0.03\sqrt{\frac{50}{0.001 \times 10}} = 2.12$$

From Eq. (4.27) or, directly, from Fig. 4.3, we have:

$$\eta = \frac{\tanh(2.12)}{2.12} = \frac{0.972}{2.12} = 0.458$$

Finally, we find the heat dissipated by the actual fin by multiplying Q_{ideal} by the fin efficiency, which gives:

$$Q_x\big|_{x=0} = \eta \times Q_{\text{ideal}} = 0.458 \times 41.2 = 18.9 \text{ W}$$

We can also calculate the temperature at the tip of the fin from Eq. (4.20), which gives:

$$\theta\big|_{\xi=1} = \frac{1}{\cosh(2.12)} = 0.237$$

Thus,

$$T_L = T_a + (T_0 - T_a)\theta\big|_{\xi=1} = 25 + 275 \times 0.237 = 90.1\text{°C}$$

Note that this fin is rather inefficient as indicated by the low value of η and by the corresponding low value of the tip temperature. If the fin were made of a different material, say copper, for which $k = 390$ W/m K, instead of the steel alloy, N would become equal to 0.34. In this case, the fin efficiency would be $\eta = \tanh(0.34)/0.34 = 0.96$, while the heat dissipation rate would be $Q_x|_{x=0} = 0.96 \times 41.3 = 40$ W. In this case, $\theta|_{\xi=1}$ would become equal to 0.945 and the tip temperature would become 285°C. The higher fin efficiency obtained in this case is a result of the much higher conductivity of copper which allows heat to be much more easily transported from the base of the fin toward the tip with a small temperature gradient. In turn, since the fin temperature remains close to the base temperature, a higher convective heat transfer rate is obtained.

4.3 Optimum Design of Straight Fins

The true optimum design of a fin will depend on both technical and economic factors that account for the weight of the fin as well as the material and manufacturing costs along with various constraints that may limit the choice of parameter values. For example, in a given application, the fin length may be restricted by space limitation and its minimum thickness may be limited by mechanical strength. For a rectangular fin of a given width, W, we can do a simple analysis to determine, directly, the fin length that will maximize the heat dissipated for a given weight of material. While this may not be the true economic optimum, it will serve as a reasonable initial estimate. The procedure for this determination follows.

The mass of a rectangular fin of length L, width W, and thickness $2B$ is given by:

$$m = 2BWL\rho_s \tag{4.28}$$

where ρ_s is the density of the material of which the fin is made and $2BWL$ is the volume of the fin. From this equation, the fin length can be written as $L = m/2BW\rho_s$. Inserting this expression for L into the expression for the fin modulus, we obtain:

$$N = L\sqrt{\frac{h}{Bk}} = \frac{m}{2BW\rho_s}\sqrt{\frac{h}{Bk}} = \frac{\gamma}{B^{3/2}} \tag{4.29}$$

where γ is defined as:

$$\gamma = \frac{m\sqrt{h/k}}{2W\rho_s} \tag{4.30}$$

From Eq. (4.29), we can express the fin half-thickness as a function of N as follows:

$$B = \left(\frac{\gamma}{N}\right)^{2/3} \tag{4.31}$$

Next, we express the heat dissipated by the fin as a function of the mass of the fin according to the following relationship:

$$Q_x\big|_{x=0} = 2WLh\frac{\tanh(N)}{N}\Delta T = 2W\left(\frac{m}{2BW\rho_s}\right)h\frac{\tanh(N)}{N}\Delta T \tag{4.32}$$

Canceling the common terms, this equation gives:

$$Q_x\big|_{x=0} = \frac{mh}{B\rho_s}\frac{\tanh(N)}{N}\Delta T \tag{4.33}$$

Substituting Eq. (4.31) for B in Eq. (4.33), we obtain:

$$Q_x\big|_{x=0} = \frac{mh}{\rho_s}\left(\frac{N}{\gamma}\right)^{2/3}\frac{\tanh(N)}{N}\Delta T = c\Delta T\, N^{-1/3}\tanh(N) \tag{4.34}$$

where $c = mh/\rho_s\gamma^{2/3}$ is a constant involving a combination of different parameter values. Since c is a constant, we can find the fin modulus that maximizes the heat dissipated by the fin by setting the derivative $dQ_x|_{x=0}/dN = 0$. From Eq. (4.34), we have:

$$\frac{dQ_x\big|_{x=0}}{dN} = c\Delta T\,\frac{d}{dN}[N^{-1/3}\tanh(N)] = c\Delta T\left[-\frac{1}{3}N^{-4/3}\tanh(N) + \frac{N^{-1/3}}{\cosh^2(N)}\right] \tag{4.35}$$

According to this equation, $dQ_x|_{x=0}/dN$ is equal to zero when the term in brackets is zero, which occurs when $N = 1.419$. This will be the "optimum" value of N that will maximize heat dissipation. The corresponding fin efficiency of the optimized fin will be:

$$\eta_{\text{opt}} = \frac{\tanh(1.419)}{1.419} = 0.627 \tag{4.36}$$

From Fig. 4.3, we can see that the optimum value of N lies in the region where the fin efficiency curve bends. For these conditions, there will be a good balance of fin efficiency and material utilization. Example 4.2 demonstrates the application of these relationships for optimum fin design.

Example 4.2. Calculation of optimum fin length. In this example, we consider the prediction of the optimum length of a 10-cm-wide rectangular fin made of aluminum for the following conditions: $m = 10$ g, $\rho_s = 2{,}700$ kg/m^3, $k = 250$ W/m K, $h = 50$ W/m^2 K. From Eq. (4.30), we have:

$$\gamma = \frac{0.010\sqrt{50/250}}{2\times 0.10\times 2{,}700} = 8.3\times 10^{-6}\ \text{m}^{3/2}$$

From Eq. (4.31), with $N = 1.419$, we obtain:

$$B = \left(\frac{8.3\times 10^{-6}}{1.419}\right)^{2/3} = 3.2\times 10^{-4}\ \text{m}$$

Finally, from Eq. (4.28), we obtain the optimum fin length:

$$L = \frac{0.010}{2\times 3.2\times 10^{-4}\times 0.10\times 2{,}700} = 0.058\ \text{m}$$

As seen from these numbers, the optimum fin will have to be relatively short (5.8 cm) and very thin (0.64 mm), which may or may not be achievable in practice.

4.4 Design of Circular Fins

Circular fins, such as the one shown in Fig. 4.1d, are used extensively especially around pipes. They are easy to construct by fitting punched-out disks or, nearly equivalently, by welding a continuous helical ribbon around a cylindrical metal pipe. The circular geometry works well, for example, when there is a cross-flow of air around the finned pipe. In practice, circular fins are also thin, which means that essentially all of the heat will be dissipated to the surroundings by convective heat transfer through the upper and lower annular surfaces of the fin. As a result, neglecting heat transfer from the outer edge of the fin will typically introduce only a small error. For these conditions, the heat transfer rate for an ideal circular fin, obtained when the fin is at a uniform temperature T_0 equal to the base temperature, is given by the following equation:

$$Q_{\text{ideal}} = 2\pi(r_L^2 - r_0^2)h(T_0 - T_a) \tag{4.37}$$

where we have assumed that convective heat transfer occurs on both faces of the fin and that the convective heat transfer coefficient is the same for both faces. Assuming that the temperature in the fin is uniform along the fin thickness at each radial position, r, the differential energy balance, Eq. (4.5), also applies to the circular film provided that we write it in cylindrical coordinates. We obtain the following equation and boundary conditions:

$$\frac{d}{dr}\left[A(r)k\frac{dT}{dr}\right] = p(r)\,h(T - T_a) \tag{4.38}$$

$$r = r_0: \;\; T = T_0 \tag{4.39}$$

$$r = r_L: \;\; \frac{dT}{dr} = 0 \tag{4.40}$$

where we have assumed, again, that heat transfer from the outer edge of the fin is negligible. The cross-sectional area and perimeter of the circular fin are as follows:

$$A = 2\pi r \times 2B = 4\pi r B \tag{4.41}$$

$$p = 2 \times 2\pi r = 4\pi r \tag{4.42}$$

Assuming that k is constant, by combining Eqs. (4.38), (4.41), and (4.42), we obtain the following result:

$$\frac{1}{r}\frac{d}{dr}\left[r\frac{dT}{dr}\right] = \frac{h}{Bk}(T - T_a) \tag{4.43}$$

We can solve this equation with boundary conditions in Eqs. (4.39) and (4.40) to find the temperature profile $T(r)$. The rate of heat transfer at the base of the fin can then be found from the following equation:

$$Q_r\big|_{r=r_0} = -A(r_0)k\frac{dT}{dr}\bigg|_{r=r_0} = -4\pi r_0 Bk\frac{dT}{dr}\bigg|_{r=r_0} \tag{4.44}$$

In dimensionless form, Eqs. (4.43), (4.39), and (4.40) can be written as follows:

$$\frac{1}{\rho}\frac{d}{d\rho}\left[\rho\frac{d\theta}{d\rho}\right] = N^2\theta \tag{4.45}$$

$$\rho = \frac{r_0}{r_L - r_0}: \;\; \theta = 1 \tag{4.46}$$

$$\rho = \frac{r_L}{r_L - r_0}: \;\; \frac{d\theta}{d\rho} = 0 \tag{4.47}$$

where

$$\theta = \frac{T - T_a}{T_0 - T_a} \tag{4.48}$$

$$\rho = \frac{r}{r_L - r_0} \tag{4.49}$$

The fin modulus that appears in Eq. (4.45) is defined as:

$$N = (r_L - r_0)\sqrt{\frac{h}{Bk}} \tag{4.50}$$

Note that unlike the rectangular fin, the solution will now depend on two dimensionless parameters: the fin modulus, N, and the aspect ratio of the fin r_L/r_0. The solution of Eqs. (4.45), (4.46), and (4.47) is a complex function $\theta = \theta(\rho, N, r_L/r_0)$ involving Bessel functions (see M.E.4.54). The fin efficiency is given by the following expression:

$$\eta = \frac{Q_r|_{r=r_0}}{Q_{\text{ideal}}} \tag{4.51}$$

where:

$$Q_r|_{r=r_0} = -2\pi r_0 \times 2B \times k \frac{dT}{dr}\bigg|_{r=r_0} \tag{4.52}$$

In terms of the dimensionless variables θ and ρ, we can write Eq. (4.52) as follows:

$$Q_r|_{r=r_0} = -4\pi Bk \frac{r_0}{r_L - r_0} \frac{d\theta}{d\rho}\bigg|_{\rho=\frac{r_0}{r_L - r_0}} \tag{4.53}$$

Finally, substituting Eqs. (4.53) and (4.37) into Eq. (4.51) gives the final result:

(M.E.4.54)

$$\eta = -\frac{2}{N^2} \frac{1}{r_L/r_0 - 1} \frac{d\theta}{d\rho}\bigg|_{\rho=\frac{1}{r_L/r_0 - 1}} \tag{4.54}$$

This function is also complicated and is given graphically in Fig. 4.4 for different values of r_L/r_0. Note that the curve with $r_L/r_0 = 1$ corresponds to the limit of a very short circular fin and is the same as that obtained for a rectangular fin with length $L = r_L - r_0$. Example 4.3 provides a sample calculation based on these relationships.

Example 4.3. Efficiency of a circular fin. Let's calculate the efficiency of a circular film made of steel for the following conditions: $r_0 = 1$ cm, $r_L = 4$ cm, $B = 1$ mm, $k = 20$ W/m K, $h = 50$ W/m^2 K, $T_0 = 300°$C, and $T_a = 25°$C. For these conditions, Eqs. (4.37) and (4.50) give, respectively:

$$Q_{\text{ideal}} = 2\pi(0.04^2 - 0.01^2) \times 50 \times (300 - 25) = 130 \text{ W}$$

and

$$N = (0.04 - 0.01)\sqrt{\frac{50}{0.001 \times 20}} = 1.5$$

From Fig. 4.4 or M.E.4.54, with $r_L/r_0 = 4$, we obtain $\eta = 0.42$. Thus, $Q_r|_{r=r_0} = \eta \times Q_{\text{ideal}} = 0.42 \times 130 =$ 54.6 W.

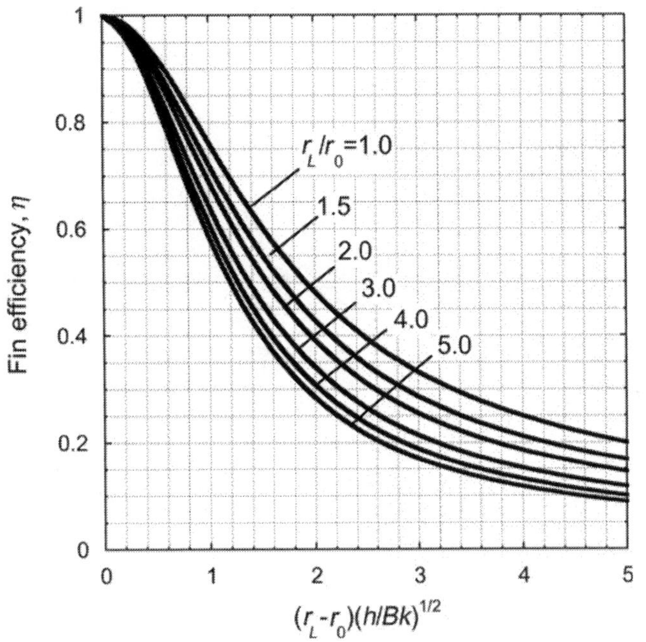

FIGURE 4.4 Thermal efficiency of circular fins (Fig. 4.1d) with different values of the ratio r_L/r_0. Mathematica notebook: (M.F.4.4).

Other fin shapes with variable cross-section and perimeter or cases where the fin thermal conductivity varies with temperature can be handled in a similar manner. In complicated cases, the solution is best found by solving the differential energy balance equation numerically subject to the desired boundary conditions.

4.5 Overall Comparison of Finned and Un-Finned Surfaces

An overall comparison of finned and bare surfaces can be made by accounting for both the added heat transfer area contributed by the fins as well as for the area of the underlying surface that remains bare. With reference to Fig. 4.5, in an un-finned surface, the entire area, A_{total}, which is at temperature T_0, is available for convective heat transfer. Thus, for the un-finned surface, the heat transfer rate will be:

$$Q_{\text{unfinned}} = A_{\text{total}} h (T_0 - T_a) \tag{4.55}$$

On the other hand, for a finned surface, the portion of the underlying surface that is occupied by the bases of the fins, A_{bases}, is no longer available for direct heat transfer to the surrounding area. In this case, for the finned surface, the total rate of heat transfer will be the sum of two terms—one accounting for convective heat transfer from the fins and the other for convective heat transfer from the underlying area that remains bare. The following expression is obtained:

$$\begin{aligned} Q_{\text{finned}} &= A_{\text{fins}} \eta h (T_0 - T_a) + (A_{\text{total}} - A_{\text{bases}})\, h (T_0 - T_a) \\ &= [(A_{\text{fins}} \eta + (A_{\text{total}} - A_{\text{bases}})]\, h (T_0 - T_a) \end{aligned} \tag{4.56}$$

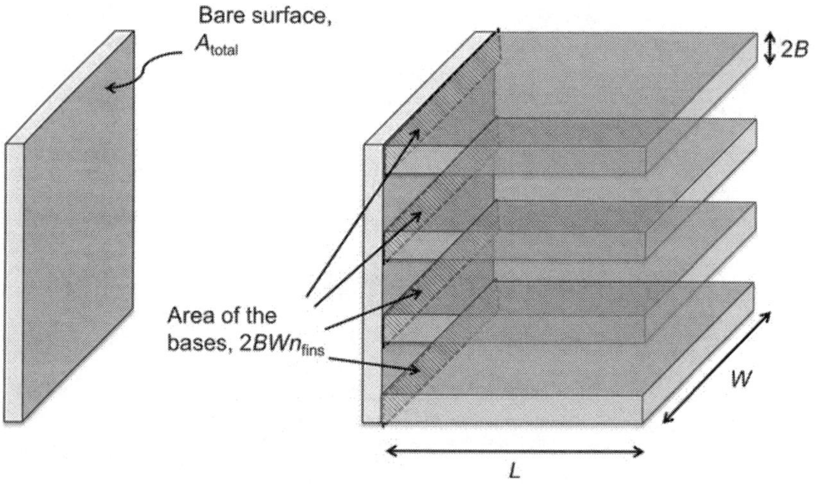

FIGURE 4.5 Comparison of a bare surface (left) and a finned surface with four fins (right). Heat dissipation occurs both from the fin surfaces and from the portion of the underlying surface that remains bare.

where A_{fins} is the total convective heat transfer area of the fins, and $A_{total} - A_{bases}$ is the portion of the underlying surface area that remains bare. The equation shows that calculating the total heat transfer rate requires that we discount the area of the fins, A_{fins}, by multiplying it by the fin efficiency while we discount the area of the underlying surface by subtracting the area occupied by the fin bases from the total area. Example 4.4 illustrates the use of these relationships.

Example 4.4. Overall comparison of un-finned and finned surfaces. Consider a 50 cm long pipe with a 1.0 cm radius that is fitted with 20 circular fins with the characteristics of the fin in Example 4.3. The surface temperature of the pipe is 300°C, while the air temperature is 25°C. For the conditions of Example 4.3, for the bare pipe, the heat transfer area is $A_{total} = 2\pi r_0 L = 2\pi \times 0.01 \times 0.5 = 0.031$ m². Thus, for the un-finned pipe, the heat transfer rate will be:

$$Q_{unfinned} = 0.031 \times 50 \times (300 - 25) = 430 \text{ W}$$

For the finned pipe, the area occupied by the fin bases will be $A_{bases} = 20 \times 2\pi r_0 \times 2B = 20 \times 4\pi \times 0.01 \times 0.001 = 0.0025$ m². The total area of the fins over which convective heat transfer occurs will be $A_{fins} = 20 \times 2 \times \pi \times (r_L^2 - r_0^2) = 20 \times 2\pi(0.04^2 - 0.01^2) = 0.188$ m². The fin efficiency is $\eta = 0.42$ as calculated in Example 4.3. The final result, obtained from Eq. (4.56), is thus:

$$Q_{finned} = [0.188 \times 0.42 + (0.031 - 0.0025)] \times 50 \times 275 = 1{,}480 \text{ W}$$

which is nearly 3.5 times greater than for the bare pipe.

4.6 Problems

P.4.1. Heat is to be provided to a small room with the finned heater sketched in Fig. P.4.1. Circulating hot oil keeps the base of the fins at 140°F. The fins are made of iron with $k = 40$ BTU/h ft°F, and the convective heat transfer coefficient is h = 10 BTU/h ft²°F. The room air is at 65°F and

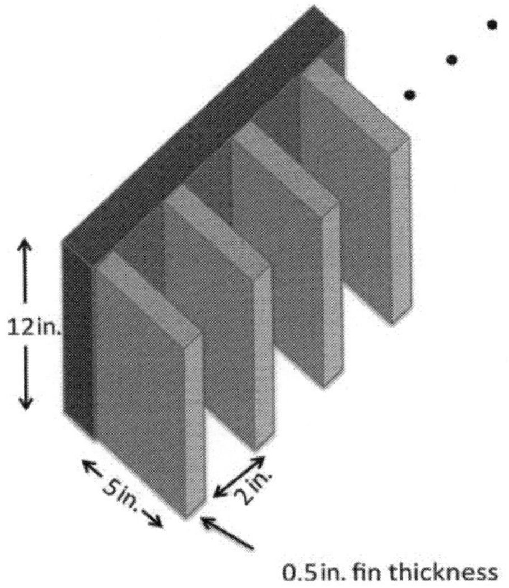

FIGURE P.4.1

can be considered to be well mixed. The room loses heat to the surroundings at the rate of 6,500 BTU/h. Determine the thermal efficiency of the fins in the heater, the fin tip temperature, and the number of fins required.

P.4.2. To increase heat dissipation from a long tube with a 3.0-cm outer diameter, 12 longitudinal rectangular fins made of aluminum (k = 250 W/m K) are attached to the outer surface of the tube projecting radially outward with an even spacing. The fins are 0.1 cm thick and 2.0 cm long. The tube surface temperature is 100°C, and the ambient air temperature is 25°C. The convective heat transfer coefficient at the pipe and fin surfaces is h = 65 W/m^2 K. Determine: (a) The rate of heat loss from the tube per m of tube length if no fins are present. (b) The fin efficiency and the rate of heat dissipation per meter of tube length if the fins are installed. (c) Repeat your calculation in (b) if the fins are made of 316 stainless steel instead of aluminum.

P.4.3. A cylindrical power transistor that generates heat at a constant rate is kept cool by a set of 12 rectangular aluminum fins mounted radially as sketched in Fig. P.4.3. The transistor is

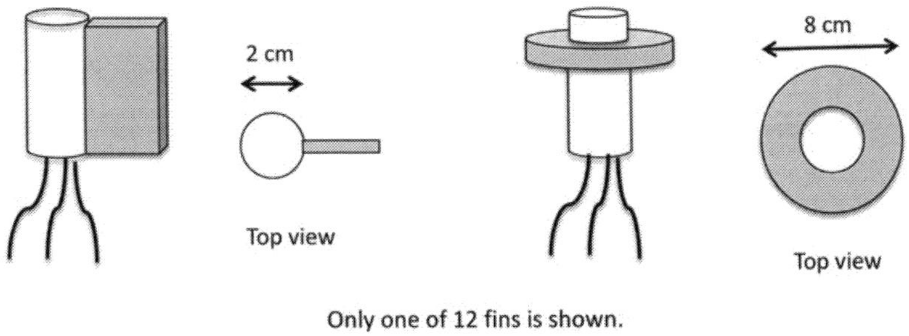

FIGURE P.4.3

1 cm in radius and 5 cm tall. The fins are 1.0 mm thick, 3.0 cm long, and 5.0 cm tall. The ambient temperature is 30°C, and the convective heat transfer coefficient is 50.0 W/m² K. The steady-state temperature at the tip of the fins is measured and is 45°C. Heat transfer from the bare transistor surface can be neglected. Assuming steady state, calculate: (a) The transistor surface temperature for these conditions. (b) The fin efficiency. (c) The heat generated by the transistor. (d) The fin efficiency if the fins above are replaced with 12 circular aluminum fins with the same thickness as the rectangular fins above and an outer radius of 4.0 cm. (e) The transistor surface temperature that would be obtained with 12 circular fins as shown if the thickness is the same.

P.4.4. A 45-W power integrated circuit chip mounted on an insulated circuit board is to be kept at its maximum operating temperature of 70°C using five identical straight rectangular aluminum-alloy fins (k = 150 W/m K), 1 mm thick and 4 cm wide, as sketched below. Air at 20°C is circulated through and around the fins by a fan yielding a convective heat transfer coefficient h = 80 W/m² K. (a) Determine what the transistor temperature would be if no fins were used. (b) Determine the fin length that is required to keep the transistor at 70°C and estimate the corresponding fin efficiency.

P.4.5. A metal pot holding boiling water has a handle made of an un-insulated hollow, stainless steel tube (k = 20 W/m K) with inner and outer diameters of 1.6 cm and 2.0 cm, respectively, and is 25 cm long. The ambient temperature is 22°C, and the convective heat transfer coefficient is h = 10 W/m² K. (a) If the maximum safe temperature is 50°C, at what distance from the pot can we safely grab the handle with bare hands? (b) Repeat your calculations in (a) if the handle is made of an un-insulated solid cylinder of the same material and outside diameter and length as the hollow handle. (c) Repeat the calculations in (b) if the solid handle is covered with a 2-mm-thick layer of an insulating material with k = 0.02 W/m K. For all three cases, assume steady-state conditions.

P.4.6. A large metal pot containing fondue (melted cheese) held at 120°C is to be placed on a wooden table supported on a trivet with four vertical legs. The legs are solid cylindrical rods with a diameter of 0.5 cm and are made of stainless steel. The maximum temperature to which the surface of the wood can be exposed is 60°C. The convective heat transfer coefficient at the rods' surface is estimated to be 8.0 W/m² K. The air temperature is 20°C. Assuming steady state: (a) Write down the differential energy balance and the boundary conditions needed to find the temperature profile in the rods. State your assumptions clearly. (b) Use available solutions of your equation in (a) to determine how long the rods need to be. (c) Determine the rate of heat transfer from the pot to each leg.

P.4.7. A thermocouple well, sketched in Fig. P.4.7, is used to measure the temperature of the air flowing inside a metal duct at a velocity of 10 m/s. The well consists of a cylindrical copper tube

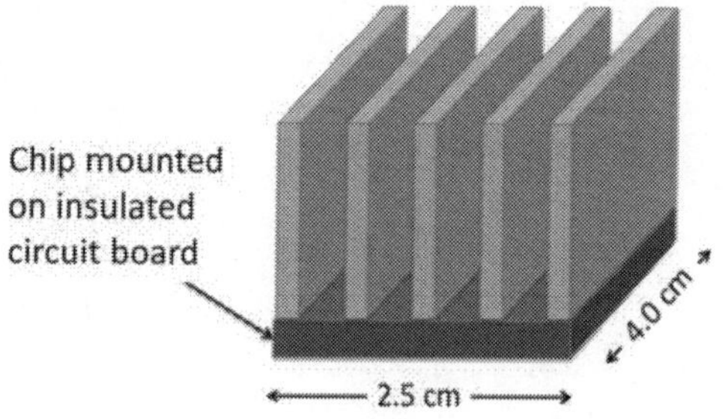

FIGURE P.4.4

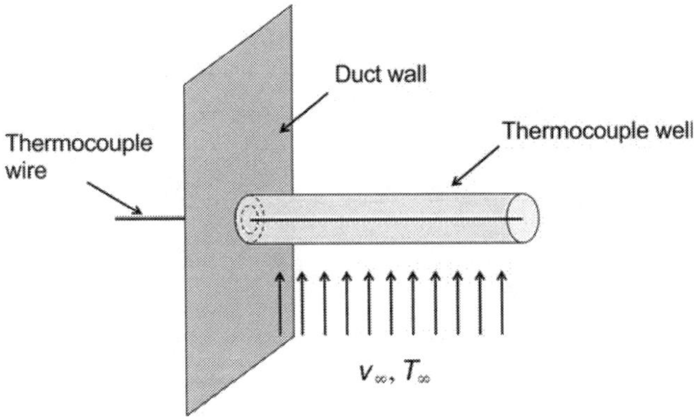

FIGURE P.4.7

(0.01 m outside diameter, 0.001 m wall thickness, and 0.1 m long) which is closed at the end that protrudes into the flowing gas stream and is soldered to the duct wall at the other end. A thin thermocouple wire is placed in the well and senses the temperature at its closed end. The duct wall temperature is 300 K, and the temperature indicated by the thermocouple is 313 K. The heat transfer coefficient describing convective heat transfer from the air to the thermocouple well is estimated to be $h = 50\ \mathrm{W/m^2\ K}$. Assuming steady state, determine the actual temperature of the flowing air. Note that heat conduction from the duct wall will affect the temperature reading. Suggest ways in which we can make the thermocouple readout closer to the actual stream temperature.

P.4.8. The liquid phase reactor shown in Fig. P.4.8 is operated at 250°C and is agitated with a propeller mounted on a 10-cm-diameter steel shaft with a thermal conductivity $k = 50\ \mathrm{W/m\ K}$. The agitator shaft is driven by a belt-pulley system, whose maximum operating temperature is 75°C. A design concern is that the belt must be located on the shaft at a distance sufficiently far away from

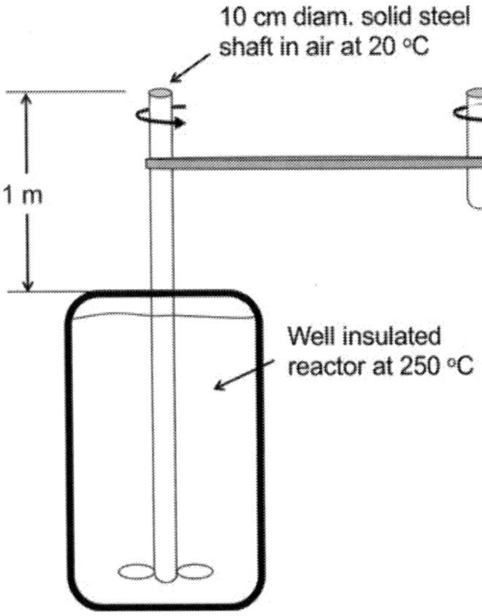

FIGURE P.4.8

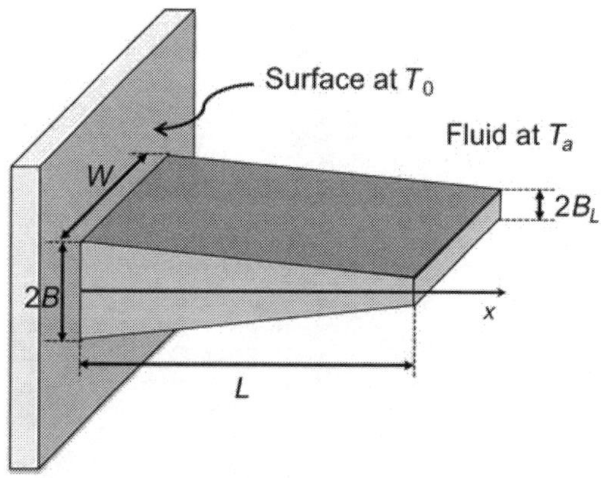

FIGURE P.4.9

the reactor so that this maximum temperature will not be exceeded. The total length of the shaft from the reactor up is 1 m. The air is at 20°C, and the convective heat transfer coefficient is estimated to be 10 W/m² K. Neglect any effects of friction. (a) Write the differential energy balance and boundary conditions needed to address this design concern. (b) Determine the minimum distance from the reactor at which the belt must be placed. (c) Determine the rate of total heat loss from the reactor through the stirrer shaft.

P.4.9. Consider a straight fin with the trapezoidal shape sketched in Fig. P.4.9. Assuming that the temperature in the fin is only a function of the axial coordinate x (i.e., it is uniform at each vertical location): (a) Write the differential balances needed to describe conductive heat transfer in the fin and simultaneous convective heat transfer at the fin surfaces and the boundary conditions assuming that the fin is long and thin. Note that the cross-section A and the perimeter p of the fin vary as a function of x. (b) Solve your equations numerically (note that this is a boundary value problem) to find the temperature profile in the fin and the fin efficiency. (c) Plot the temperature profile in the fin and determine the fin efficiency and the heat transfer rate for a 316 stainless-steel fin with $T_0 = 100°C$, $T_a = 20°C$, $L = 0.05$ m, $B_0 = 0.005$ m, $B_L = 0.001$ m, and $W = 0.1$ m, if the convective heat transfer coefficient is $h = 50$ W/m² K.

P.4.10. An air-cooled utility engine needs to dissipate 2,000 W of heat in order to operate at a steady temperature of 150°C when the ambient temperature is 25°C. The engine block can be represented as a 3-in.-diameter cylinder, 20 cm long. Cooling is achieved by attaching 10 fins to the engine block. The fins are made of a metal with a thermal conductivity $k = 65$ W/m K and a density $\rho_s = 7{,}000$ kg/m³. The convective heat transfer coefficient is estimated to be $h = 100$ W/m² K. Weight is a concern. Thus, it is important to optimize the fin design. (a) Determine the minimum-weight fin design and the fin-tip temperature if 10 rectangular fins spanning the length of the engine block are installed. (b) Determine the minimum-weight fin design and the fin-tip temperature if 10 circular fins are installed along the length of the engine block. For each case plot, the total weight of the fins as a function of their half-thickness B to find the minimum weight design.

CHAPTER 5

Multidimensional and Transient Heat Conduction

5.1 Introduction

In this chapter, we consider heat transfer processes where the temperature within a given domain varies in more than one dimension and/or as a function of time. We also consider the existence of a velocity field within the domain where heat transfer takes place to account for the simultaneous occurrence of conductive and convective processes. The basis for this analysis is the transient differential thermal energy balance coupled with the continuity equation and with Fourier's law. Boundary conditions are established at the boundaries of the domain of interest either by specifying the temperature or by imposing a condition on the heat transfer flux as we have seen in Chap. 3 for steady-state problems. Below, we derive the differential energy balance in rectangular coordinates and provide the results for cylindrical and spherical coordinates. We then consider the application of these equations to transient heat transfer problems.

5.2 General Differential Mass and Energy Balances

5.2.1 Rectangular Coordinates

The derivation of general differential mass and energy balances is based on the definition of a control volume as a cube with sides Δx, Δy, and Δz as shown in Fig. 5.1. The faces of this cube are located at positions x and $x+\Delta x$, y and $y+\Delta y$, and z and $z+\Delta z$. Within the domain of interest, at each point x, y, z, and time t, we will have density ρ and temperature T, which are scalar quantities, and velocity v and heat transfer flux q, which are vectors. Because v and q are vectors, there will be v_x, v_y, and v_z components of the velocity vector and q_x, q_y, and q_z components of the heat transfer flux vector, each, in general, also a function of x, y, z, and t. We also define three additional scalar quantities: the specific internal energy, $\hat{U}$, the specific enthalpy, $\hat{H}$, and the net rate of energy generation per unit volume, G. The latter can be due to an electrical current, chemical reactions, etc. In general, all of these quantities will also be functions of time and will vary along the three spatial coordinates.

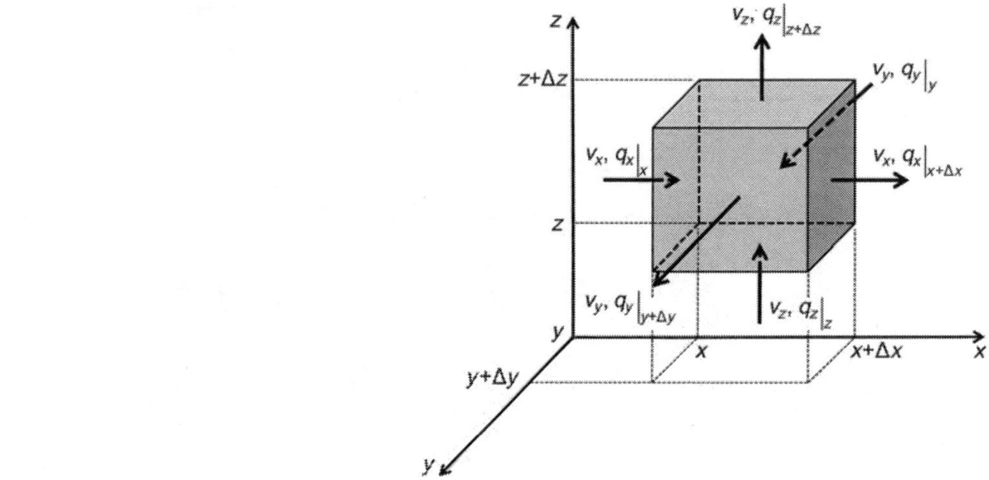

Figure 5.1 Control volume used to derive the general differential mass and energy balances. v and q are the velocity and conductive heat transfer flux vectors whose x, y, and z, components are shown on each face at x and $x+\Delta x$, y and $y+\Delta y$, and z and $z+\Delta z$, respectively.

a. Differential mass balance

We derive the differential mass balance by writing the following for our chosen control volume:

$$\text{Mass accumulation} = \text{Mass input} - \text{Mass output}$$

Accumulation will be the time-derivative of the mass contained in the control volume, while input and outputs are found from the mass flow rates passing through each face of the control volume. The following expressions are obtained for these terms:

Mass accumulation: $\frac{\partial}{\partial t}(\Delta x \Delta y \Delta z \times \rho)$

Mass input: $\Delta y \Delta z \times v_x \rho\big|_x + \Delta x \Delta z \times v_y \rho\big|_y + \Delta x \Delta y \times v_z \rho\big|_z$

Mass output: $\Delta y \Delta z \times v_x \rho\big|_{x+\Delta x} + \Delta x \Delta z \times v_y \rho\big|_{y+\Delta y} + \Delta x \Delta y \times v_z \rho\big|_{z+\Delta z}$

Note that since, for now, we are balancing mass around a volume of finite size, the values ρ, v_x, v_y, and v_z will be averaged over volume and respective surfaces. Balancing these three terms yields the following result:

$$\frac{\partial}{\partial t}(\Delta x \Delta y \Delta z \times \rho) = \left(\Delta y \Delta z v_x \rho\big|_x + \Delta x \Delta z \times v_y \rho\big|_y + \Delta x \Delta y \times v_z \rho\big|_z \right)$$

$$- \left(\Delta y \Delta z \times v_x \rho\big|_{x+\Delta x} + \Delta x \Delta z \times v_y \rho\big|_{y+\Delta y} + \Delta x \Delta y \times v_z \rho\big|_{z+\Delta z} \right)$$

Next, we divide all terms by $\Delta x \Delta y \Delta z$, which gives:

$$\frac{\partial \rho}{\partial t}+\frac{v_x\rho|_{x+\Delta x}-v_x\rho|_x}{\Delta x}+\frac{v_y\rho|_{y+\Delta y}-v_y\rho|_y}{\Delta y}+\frac{v_z\rho|_{z+\Delta z}-v_z\rho|_z}{\Delta z}=0$$

Finally, we take the limit for $\Delta x \to 0$, $\Delta y \to 0$, and $\Delta z \to 0$ to obtain the following partial differential equation:

$$\frac{\partial \rho}{\partial t}+\frac{\partial (v_x\rho)}{\partial x}+\frac{\partial (v_y\rho)}{\partial y}+\frac{\partial (v_z\rho)}{\partial z}=0 \tag{5.1}$$

Since we have made the control volume differential by taking the limit $\Delta x \to 0$, $\Delta y \to 0$, and $\Delta z \to 0$, the density and velocity components appearing in this equation will now be point values.

b. Differential energy balance

The general differential energy balance is derived by starting with the following balance for the control volume:

Energy accumulation = Energy input – Energy output + Energy generation

Accumulation will be the time-derivative of the internal energy contained in the control volume. Input and outputs will be found from the enthalpy flows passing through each face of the control volume plus the conductive heat transfer through each face. Enthalpy and internal energy are related to each other by $\hat{H}=\hat{U}+P\hat{V}$, where P is pressure and $\hat{V}$ is the specific volume (i.e., volume/mass). Thus, $\hat{H}$ will account for both the internal energy carried by the bulk flow in and out of the control volume as well as for the associated flow work. Expressing each of these terms as a function of these variables, we obtain:

Energy accumulation: $\frac{\partial}{\partial t}\left(\Delta x\Delta y\Delta z\times\rho\hat{U}\right)$

Energy input: $\Delta y\Delta z\left(v_x\rho\hat{H}+q_x\right)\Big|_x+\Delta x\Delta z\left(v_y\rho\hat{H}+q_y\right)\Big|_y+\Delta x\Delta y\left(v_z\rho\hat{H}+q_z\right)\Big|_z$

Energy output: $\Delta y\Delta z\left(v_x\rho\hat{H}+q_x\right)\Big|_{x+\Delta x}+\Delta x\Delta z\left(v_y\rho\hat{H}+q_y\right)\Big|_{y+\Delta y}+\Delta x\Delta y\left(v_z\rho\hat{H}+q_z\right)\Big|_{z+\Delta z}$

Energy generation: $\Delta x\Delta y\Delta z\times G$

Balancing these four terms yields the following result:

$$\begin{aligned}\frac{\partial}{\partial t}\left(\Delta x\Delta y\Delta z\times\rho\hat{U}\right)=&\left\{\Delta y\Delta z\left(v_x\rho\hat{H}+q_x\right)\Big|_x+\Delta x\Delta z\left(v_y\rho\hat{H}+q_y\right)\Big|_y+\Delta x\Delta y\left(v_z\rho\hat{H}+q_z\right)\Big|_z\right\}\\&-\left\{\Delta y\Delta z\left(v_x\rho\hat{H}+q_x\right)\Big|_{x+\Delta x}+\Delta x\Delta z\left(v_y\rho\hat{H}+q_y\right)\Big|_{y+\Delta y}+\Delta x\Delta y\left(v_z\rho\hat{H}+q_z\right)\Big|_{z+\Delta z}\right\}\\&+\Delta x\Delta y\Delta z\times G\end{aligned}$$

As before, since, for now, the control volume is finite, the dependent variables appearing in this equation should be regarded as averages over the appropriate volume or surface to which they apply. The next step is to divide all terms by $\Delta x\Delta y\Delta z$ as we did for the mass balance, which gives the following result:

$$\frac{\partial(\rho\hat{U})}{\partial t}+\frac{v_x\rho\hat{H}\big|_{x+\Delta x}-v_x\rho\hat{H}\big|_x}{\Delta x}+\frac{v_y\rho\hat{H}\big|_{y+\Delta y}-v_y\rho\hat{H}\big|_y}{\Delta y}+\frac{v_z\rho\hat{H}\big|_{z+\Delta z}-v_z\rho\hat{H}\big|_z}{\Delta z}$$

$$=-\left[\frac{q_x\big|_{x+\Delta x}-q_x\big|_x}{\Delta x}+\frac{q_y\big|_{y+\Delta y}-q_y\big|_y}{\Delta y}+\frac{q_z\big|_{z+\Delta z}-q_z\big|_z}{\Delta z}\right]+G$$

Finally, taking the limit of this equation for $\Delta x\to 0$, $\Delta y\to 0$, and $\Delta z\to 0$, we obtain:

$$\frac{\partial(\rho\hat{U})}{\partial t}+\frac{\partial(v_x\rho\hat{H})}{\partial x}+\frac{\partial(v_y\rho\hat{H})}{\partial y}+\frac{\partial(v_z\rho\hat{H})}{\partial z}=-\left(\frac{\partial q_x}{\partial x}+\frac{\partial q_y}{\partial y}+\frac{\partial q_z}{\partial z}\right)+G \tag{5.2}$$

The left-hand side of this equation can be simplified by recognizing that the derivatives of the products $\rho\hat{U}$, $v_x\rho\hat{H}$, $v_y\rho\hat{H}$, and $v_z\rho\hat{H}$ can be expressed as follows:

$$\frac{\partial(\rho\hat{U})}{\partial t}=\hat{U}\frac{\partial\rho}{\partial t}+\rho\frac{\partial\hat{U}}{\partial t} \tag{5.3}$$

$$\frac{\partial(v_x\rho\hat{H})}{\partial x}=\hat{H}\frac{\partial(v_x\rho)}{\partial x}+v_x\rho\frac{\partial\hat{H}}{\partial x} \tag{5.4}$$

$$\frac{\partial\left(v_y\rho\hat{H}\right)}{\partial y}=\hat{H}\frac{\partial\left(v_y\rho\right)}{\partial y}+v_y\rho\frac{\partial\hat{H}}{\partial y} \tag{5.5}$$

$$\frac{\partial\left(v_z\rho\hat{H}\right)}{\partial z}=\hat{H}\frac{\partial\left(v_z\rho\right)}{\partial z}+v_z\rho\frac{\partial\hat{H}}{\partial z} \tag{5.6}$$

Further we can recognize that, in practice, $\hat{U}\sim\hat{H}$. This will be an excellent approximation for liquids and solid phases, but will also be a good assumption for gases where the $P\hat{V}$ term is usually small except at very high pressures. Substituting these relationships in Eq. (5.2) and rearranging the terms, we obtain the following result:

$$\hat{H}\left[\frac{\partial\rho}{\partial t}+\frac{\partial(v_x\rho)}{\partial x}+\frac{\partial(v_y\rho)}{\partial y}+\frac{\partial(v_z\rho)}{\partial z}\right]+\rho\left[\frac{\partial\hat{H}}{\partial t}+v_x\frac{\partial\hat{H}}{\partial x}+v_y\frac{\partial\hat{H}}{\partial y}+v_z\frac{\partial\hat{H}}{\partial z}\right]$$
$$=-\left(\frac{\partial q_x}{\partial x}+\frac{\partial q_y}{\partial y}+\frac{\partial q_z}{\partial z}\right)+G \tag{5.7}$$

The first term in brackets on the left-hand side of this equation is equal to zero because of the differential mass balance, Eq. (5.1). Additionally, we can express the enthalpy based on the heat capacity, C_p, according to the following equation:

$$d\hat{H}=C_p dT \tag{5.8}$$

Substituting this expression in Eq. (5.7) gives the following result:

$$\rho C_p\left(\frac{\partial T}{\partial t}+v_x\frac{\partial T}{\partial x}+v_y\frac{\partial T}{\partial y}+v_z\frac{\partial T}{\partial z}\right)=-\left(\frac{\partial q_x}{\partial x}+\frac{\partial q_y}{\partial y}+\frac{\partial q_z}{\partial z}\right)+G \tag{5.9}$$

The final result is obtained by combining this equation with the expressions for the components of the conductive heat transfer flux according to Fourier's law, which are given in Table 3.1. The following result is obtained:

$$\rho C_p\left(\frac{\partial T}{\partial t}+v_x\frac{\partial T}{\partial x}+v_y\frac{\partial T}{\partial y}+v_z\frac{\partial T}{\partial z}\right)=\frac{\partial}{\partial x}\left(k\frac{\partial T}{\partial x}\right)+\frac{\partial}{\partial y}\left(k\frac{\partial T}{\partial y}\right)+\frac{\partial}{\partial z}\left(k\frac{\partial T}{\partial z}\right)+G \tag{5.10}$$

Finally, if we treat the thermal conductivity, k, as constant, we obtain the following:

$$\frac{\partial T}{\partial t}+v_x\frac{\partial T}{\partial x}+v_y\frac{\partial T}{\partial y}+v_z\frac{\partial T}{\partial z}=\alpha\left(\frac{\partial^2 T}{\partial x^2}+\frac{\partial^2 T}{\partial y^2}+\frac{\partial^2 T}{\partial z^2}\right)+\frac{G}{\rho C_p} \tag{5.11}$$

where the quantity

$$\alpha=\frac{k}{\rho C_p} \tag{5.12}$$

is called the thermal diffusivity. Equation (5.11) can be written in a more compact form in vector notation as follows:

$$\frac{DT}{Dt}=\alpha\nabla^2 T+\frac{G}{\rho C_p} \tag{5.13}$$

where:

$$\frac{DT}{Dt}=\frac{\partial T}{\partial t}+v\cdot\nabla T \tag{5.14}$$

is the "substantial" or "material" derivative of the temperature (note that $v\cdot\nabla T$ is the dot product of the velocity vector and the temperature gradient vector) and

$$\nabla^2=\frac{\partial^2}{\partial x^2}+\frac{\partial^2}{\partial y^2}+\frac{\partial^2}{\partial z^2} \tag{5.15}$$

is the Laplacian operator. The thermal diffusivity, α, is a material property with special significance. Since the dimensions of k, ρ, and C_p are, respectively, power/(length × temperature), mass/length^3, and energy/(mass × temperature), the dimensions of α will be area per time and its units will typically be m^2/s or ft^2/h. Interestingly, these are the same dimensions and units of the kinematic viscosity of a fluid, $\nu=\mu/\rho$. We will see that in the energy balance, α plays a role analogous to that played by ν in the momentum balance. We will explore the consequences of this analogy later.

5.2.2 Cylindrical and Spherical Coordinates

The derivation of the general differential mass balance and of the general differential energy balance in cylindrical and spherical coordinates follows the same steps used for the rectangular coordinate system. The derivation is a bit more complex, however, because the control volume will no longer be a cube so that the input and output areas will vary with position in the chosen coordinate system. In cylindrical coordinates, the control

A. Rectangular coordinates: $T = T(t, x, y, z)$
$\frac{\partial T}{\partial t} + v_x \frac{\partial T}{\partial x} + v_y \frac{\partial T}{\partial y} + v_z \frac{\partial T}{\partial z} = \alpha \left(\frac{\partial^2 T}{\partial x^2} + \frac{\partial^2 T}{\partial y^2} + \frac{\partial^2 T}{\partial z^2} \right) + \frac{G}{\rho C_p}$
B. Cylindrical coordinates: $T = T(t, r, z, \theta)$
$\frac{\partial T}{\partial t} + v_r \frac{\partial T}{\partial r} + v_z \frac{\partial T}{\partial z} + \frac{v_\theta}{r} \frac{\partial T}{\partial \theta} = \alpha \left[\frac{1}{r} \frac{\partial}{\partial r} \left(r \frac{\partial T}{\partial r} \right) + \frac{\partial^2 T}{\partial z^2} + \frac{1}{r^2} \frac{\partial^2 T}{\partial \theta^2} \right] + \frac{G}{\rho C_p}$
C. Spherical coordinates: $T = T(t, r, \theta, \phi)$
$\frac{\partial T}{\partial t} + v_r \frac{\partial T}{\partial r} + \frac{v_\theta}{r} \frac{\partial T}{\partial \theta} + \frac{v_\phi}{r \sin\theta} \frac{\partial T}{\partial \phi} = \alpha \left[\frac{1}{r^2} \frac{\partial}{\partial r} \left(r^2 \frac{\partial T}{\partial r} \right) + \frac{1}{r^2 \sin\theta} \frac{\partial}{\partial \theta} \left(\sin\theta \frac{\partial T}{\partial \theta} \right) + \frac{1}{r^2 \sin^2\theta} \frac{\partial^2 T}{\partial \phi^2} \right] + \frac{G}{\rho C_p}$

Table 5.1 General Differential Thermal Energy Balance in Different Coordinate Systems with Constant k and $\hat{U} = \hat{H}$

volume will be a cylindrical shell with sides of length Δr, Δz, and $r\Delta\theta$, where $\Delta\theta$ is the arc-length in radians and r is the radial coordinate. Similarly, in spherical coordinates, the control volume will be a spherical shell with sides of length Δr, $r\Delta\theta$, and $r \sin\theta \times \Delta\phi$ where $\Delta\theta$ and $\Delta\phi$ are the arc-lengths in radians and r is the radial coordinate. The conductive heat transfer flux expressions, given in Table 3.1, are also a bit more complicated than those written for the rectangular coordinate system. Rather than presenting the somewhat lengthy derivation of these balances, the final equations are given in Table 5.1 for the three coordinate systems. In each case, the only restrictive assumptions are that we have taken k to be constant and $\hat{U} = \hat{H}$, which are reasonable for many practical systems.

The equations in Table 5.1 are readily adapted to describe simpler systems. Simplified forms for five common cases are as follows:

1. Steady-state processes: $\frac{\partial T}{\partial t} = 0$
2. Systems with no fluid flow: $v = 0$
3. No y and z temperature gradients in rectangular coordinates:

$$\frac{\partial T}{\partial t} + v_x \frac{\partial T}{\partial x} = \alpha \frac{\partial^2 T}{\partial x^2} + \frac{G}{\rho C_p} \tag{5.16}$$

4. No z and θ temperature gradients in cylindrical coordinates:

$$\frac{\partial T}{\partial t} + v_r \frac{\partial T}{\partial r} = \alpha \frac{1}{r} \frac{\partial}{\partial r} \left(r \frac{\partial T}{\partial r} \right) + \frac{G}{\rho C_p} \tag{5.17}$$

5. No θ and ϕ temperature gradients in spherical coordinates:

$$\frac{\partial T}{\partial t} + v_r \frac{\partial T}{\partial r} = \alpha \frac{1}{r^2} \frac{\partial}{\partial r} \left(r^2 \frac{\partial T}{\partial r} \right) + \frac{G}{\rho C_p} \tag{5.18}$$

5.3 Transient Heat Transfer Processes

We will not cover in any detail heat transfer problems in multiple dimensions. While such situations are obviously common, the solution of the corresponding differential balances is usually complex and typically best handled by numerical simulation. Numerical solutions are often most efficiently obtained by finite element methods that can be easily adapted to complex geometries. Computational packages, such as COMSOL and FLUENT, are recommended for handling these simulations. Since the aim of this book is to address mainly conceptual aspects, we will focus, instead, on one-dimensional transient systems, which, in many cases, can be handled with analytical mathematics. Despite this limitation, many practical systems can be approximated as being one-dimensional so that the analyses presented in this section will have not only conceptual significance but also practical value.

Here we consider cases where a solid object at an initial temperature T_0 is cooled or heated as a result of heat transfer to or from a surrounding fluid at temperature T_a, as sketched in Fig. 5.2. In general, within the object, heat transfer will occur as a result of heat conduction, while in the surrounding fluid heat transfer will occur as a result of convective transport. The temperature at the interface between the object and the surrounding fluid is T_s and will be the same on both sides of the interface. The heat transfer fluxes will also be the same on the two sides of this interface assuming that no heat is generated right at the surface. We can envision three distinct scenarios depending on the relative importance of the internal conductive resistance within the object and the external convective resistance in the fluid:

1. The heat transfer resistance within the object is very small: In this case, the temperature within the object will be the same everywhere and will vary with time only.
2. The heat transfer resistance within the object is very large: In this case, the temperature within the object will vary both in time and with position while the surface temperature T_s will be close to the temperature of the bulk fluid T_a.
3. The internal and external heat transfer resistances are comparable in magnitude: In this case, the temperature within the object will vary with time and position while the surface temperature T_s will be intermediate between the temperature of the object and the temperature of the bulk fluid T_a.

The relative importance of the two heat transfer resistances can be expressed in terms of the Biot number, Bi, which is defined as follows:

$$\mathrm{Bi} = \frac{\text{Internal conductive resistance}}{\text{External convective resistance}}$$

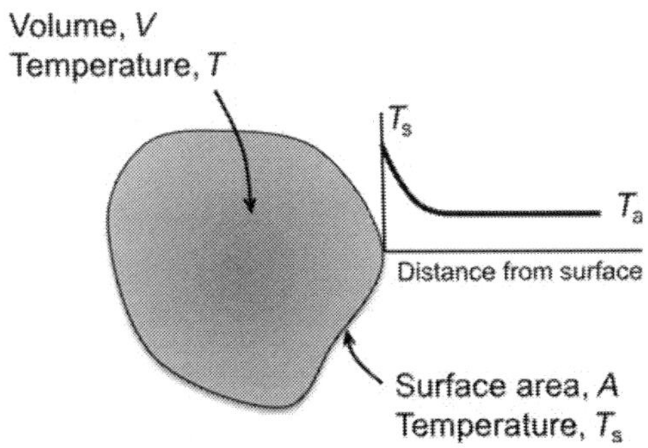

Figure 5.2 Heat transfer between a solid object of volume V and surface area A and a fluid at temperature T_a. T_s is the surface temperature.

In general, if Bi << 1, the external resistance will be dominant and we will be in the first of the three scenarios above. If Bi >> 1, the internal resistance will be dominant and we will be in the second of the three scenarios. Finally, if Bi ~ 1, both resistances will be important and we will be in the third scenario. Note that in an actual process, the two resistances may vary with time. Thus, for now, we are only interested in their order of magnitude, which can be estimated as follows:

$$\text{Internal conductive resistance} \sim \frac{\ell}{k}$$

$$\text{External convective resistance} \sim \frac{1}{h}$$

where ℓ is a characteristic length describing the size of the object, k is the thermal conductivity of the object, and h is the external convective heat transfer coefficient. Accordingly, we have:

$$\text{Bi} = \frac{h\ell}{k} \tag{5.19}$$

If h is small, for instance, because the surrounding fluid is a gas and the flow velocity is small, the external heat transfer resistance will be large and we would predict a small Bi. Conversely, if the object is large in size and/or its thermal conductivity is small, the internal conductive resistance will be large and we would predict a large Bi. Note that, in practice, the actual value of Bi will depend on how ℓ is defined. For example, if the object is a large slab with thickness $2B$, ℓ would normally be defined as the slab half-thickness B. On the other hand, if the object is either a long cylinder or a sphere with radius R, ℓ would normally be defined as the radius R. For irregular shapes, one can take $\ell \sim V/A$ but only for estimation purposes.

Although in the description above we have made reference to the case of heat transfer between a solid object and a fluid, under certain conditions, the same analysis will also apply to cases where the object is actually a fluid, such as a viscous liquid. An example is the case of a liquid droplet suspended in a gaseous stream. If the droplet is well mixed because of internal recirculation, the temperature in the droplet will be uniform and the external resistance will be dominant, which is equivalent to having Bi << 1. On the other hand, if the liquid within the droplet is stagnant because of its high viscosity, heat transfer within the droplet will occur by conduction. In this case, the relative importance of internal and external resistances will depend on the actual value of Bi.

5.3.1 Processes Controlled by the External Convective Resistance

When the external resistance is dominant (i.e., situations where Bi << 1), the temperature within the object will be essentially uniform. The surface temperature will, thus, be the same as the object temperature and will vary with time only; i.e., $T_s = T(t)$. In this case, the following transient macroscopic energy balance applied to the object in question will be sufficient to find the function $T(t)$:

$$\text{Energy accumulation} = \text{Net energy input} + \text{Energy generation}$$

The terms in this balance can be expressed as follows:

Energy accumulation: $V\frac{d(\rho\hat{U})}{dt}$

Net energy input: $Ah(T_a - T_s)$

Energy generation: $V \times G$

Combining these expressions, we obtain the following equation:

$$V\frac{d(\rho\hat{U})}{dt} = Ah(T_a - T_s) + VG \tag{5.20}$$

Note that if $T_a > T_s$, energy will be transferred from the surrounding to the object and the net energy input will be positive. Conversely, if $T_a < T_s$, energy will be transferred from the object to the surrounding and the net energy input will be negative. In either case, since $d\hat{U} \sim d\hat{H} = C_p dT$, assuming that the density of the object remains constant, Eq. (5.20) will become:

$$\rho C_p V\frac{dT}{dt} = Ah(T_a - T) + VG \tag{5.21}$$

where we have taken $T = T_s$. Dividing each term in this equation by $\rho C_p V$, we obtain:

$$\frac{dT}{dt} = \frac{Ah}{\rho C_p V}(T_a - T) + \frac{G}{\rho C_p} \tag{5.22}$$

This ordinary differential equation is subject to the following initial condition:

$$t = 0: \quad T = T_0 \tag{5.23}$$

If we assume that $G = 0$ and that the external fluid temperature, T_a, is constant, we can separate the variables T and t in Eq. (5.22) and integrate as follows:

$$\int_{T_0}^{T} \frac{dT}{T_a - T} = \frac{Ah}{\rho C_p V}\int_0^t dt \tag{5.24}$$

Carrying out the integrations gives the final result:

$$\frac{T - T_a}{T_0 - T_a} = \exp\left(-\frac{Ah}{\rho C_p V}t\right) \tag{5.25}$$

Note that this result is independent of the particular geometry of the object and that only the area-to-volume ratio will matter. In practice, the external temperature T_a can be considered constant if the object is in contact with a large volume of a constant-temperature fluid. In this case, heat transfer to or from the object will not affect the external temperature. If this is not the case, for example, if the object exchanges heat with a finite volume of fluid, an additional overall energy balance will be needed to find both the object temperature and the external temperature as illustrated in Example 5.1.

Example 5.1. Cooling of solid spheres in a finite volume of fluid with the external resistance controlling. Let's consider a case where 1 kg of aluminum spheres with radius R = 10 mm initially at 90°C are immersed in 1 L of water initially at 20°C in a tank. The tank is insulated, and heat losses are negligible. The water is assumed to be perfectly mixed. Let's find the time needed to cool the aluminum spheres to 40°C assuming that the external heat transfer resistance at the sphere-fluid interface is dominant and that $h = 500\ \text{W/m}^2\ \text{K}$. The density of aluminum is $\rho = 2{,}700\ \text{kg/m}^3$ and its heat capacity is $C_p = 900$ J/kg K. The total volume of the spheres is 1/2,700 = 0.00037 m^3. Since, for one sphere, the surface area is $4\pi R^2$ and the volume is $4\pi R^3/3$, the surface-to-volume ratio will be $A/V = 3/R = 3/0.01 = 300\ \text{m}^{-1}$. If we assume that the water temperature remains constant, Eq. (5.25) can be used directly giving:

$$t = -\frac{\rho C_p}{hA/V}\ln\left(\frac{T - T_a}{T_0 - T_a}\right) = \frac{2{,}700\times 900}{500\times 300}\ln\left(\frac{40-20}{90-20}\right) = 20.3\ \text{s}$$

Coversely, if the changing temperature of the water is considered, we need to go back to Eq. (5.22). In this case, the fluid temperature T_a is a variable. Thus, in order to use this equation, we need to write an additional overall energy balance expressing that the heat lost by the spheres will be equal to the heat gained by the water. Mathematically, this balance is expressed as follows:

$$\rho C_p V(T_0 - T) = \rho_f C_{p,f} V_f (T_a - T_{a,0}) \tag{5.26}$$

where ρ_f, $C_{p,f}$, V_f, and $T_{a,0}$ are, respectively, the density, heat capacity, volume, and initial temperature of the external fluid. Solving Eq. (5.26) for T_a, we obtain the following result:

$$T_a = T_{a,0} + \frac{\rho C_p V}{\rho_f C_{p,f} V_f}(T_0 - T) = T_{a,0} + \frac{1}{\Lambda}(T_0 - T) \tag{5.27}$$

where $\Lambda = \rho_f C_{p,f} V_f / \rho C_p V$. The solution of the transient problem will thus be obtained by combining Eqs. (5.22) and (5.27). Inserting Eq. (5.27) for T_a in Eq. (5.22), we obtain:

$$\frac{dT}{dt} = -\frac{Ah}{\rho C_p V}\left\{T - \left[T_{a,0} + \frac{1}{\Lambda}(T_0 - T)\right]\right\} = -\frac{Ah}{\rho C_p V}\left[\left(1 + \frac{1}{\Lambda}\right)T - T_{a,0} - \frac{1}{\Lambda}T_0\right] \tag{5.28}$$

We can now separate the variables T and t and integrate this equation with the initial condition that $T = T_0$ at $t = 0$, which gives:

$$\int_{T_0}^{T} \frac{dT}{(1 + 1/\Lambda)T - T_{a,0} - T_0/\Lambda} = -\frac{Ah}{\rho C_p V}\int_0^t dt \tag{5.29}$$

Performing the integrals, we obtain the final result:

$$\frac{1}{1 + 1/\Lambda}\ln\left[\frac{(1 + 1/\Lambda)T - T_{a,0} - T_0/\Lambda}{T_0 - T_{a,0}}\right] = -\frac{Ah}{\rho C_p V}t \tag{5.30}$$

For our problem, we have:

$$\frac{1}{\Lambda} = \frac{2{,}700\times 900\times 0.00037}{1{,}000\times 4{,}200\times 0.001} = 0.21$$

Equation (5.30) then gives:

$$t = -\frac{\rho C_p}{(1 + 1/\Lambda)hA/V}\ln\left[\frac{(1 + 1/\Lambda)T - T_{a,0} - T_0/\Lambda}{T_0 - T_{a,0}}\right] =$$

$$= -\frac{2{,}700\times 900}{(1 + 0.21)\times 500\times 300}\ln\left[\frac{(1 + 0.21)40 - 20 - 0.21\times 90}{90 - 20}\right] = 26.7\ \text{s}$$

The final temperature of the water will be $T_a = T_{a,0} + (T_0 - T)/\Lambda = 20 + 0.21(90 - 40) = 20.5°\text{C}$. Note that cooling will take a little longer than initially predicted when we neglected the temperature change of the fluid, since, in reality, the water will heat up, thereby reducing the driving force for heat transfer and slowing the overall process down a bit. Finally, let's determine if the assumption that we have made that the external heat transfer resistance is dominant is correct. Based on the thermal conductivity of aluminum, $k = 250$ W/m K (see Table 2.3), we have $\text{Bi} = hR/k = 500 \times 0.01/250 = 0.02$. This value is pretty small, confirming that the external resistance will in fact be dominant.

5.3.2 Processes Controlled by the Internal Conductive Resistance

When the internal resistance is controlling, the internal temperature distribution will depend both on the size of the object and on its particular shape. We consider four different cases where the object is: (a) an infinite slab, (b) a semi-infinite slab with thickness $2B$, (c) a long cylinder with radius R, and (d) a sphere with radius R. These geometries can be reasonable approximations for a variety of practical systems. In all four cases, the object's surface temperature is assumed to be equal to the temperature of the external fluid T_a, which, for simplicity, we will initially assume is constant. We also assume that $G = 0$ so that there will be no heat generation within the object.

a. Solution for an infinite slab

Consider the situation in Fig. 5.3 that depicts a slab that is infinite in all dimensions but bounded by a surface at $x = 0$. The slab is at an initially uniform temperature T_0 and its surface is brought suddenly to a uniform temperature T_s. As a result of conductive transport of heat from the surface into the slab, a temperature profile will develop within the slab penetrating deeper and deeper from the surface as time goes on. For a practical system, "infinite" will mean that the actual slab is thick enough that the temperature profile will not be affected by any boundary on the right-hand side of the surface where we specify $T = T_s$. For a slab of finite size, the analysis will be valid provided that we limit its application to relatively short times. Since there is no fluid motion within the slab and no heat generation, the general differential energy balance (item A in Table 5.1) simplifies to the following:

$$\frac{\partial T}{\partial t} = \alpha \frac{\partial^2 T}{\partial x^2} \tag{5.31}$$

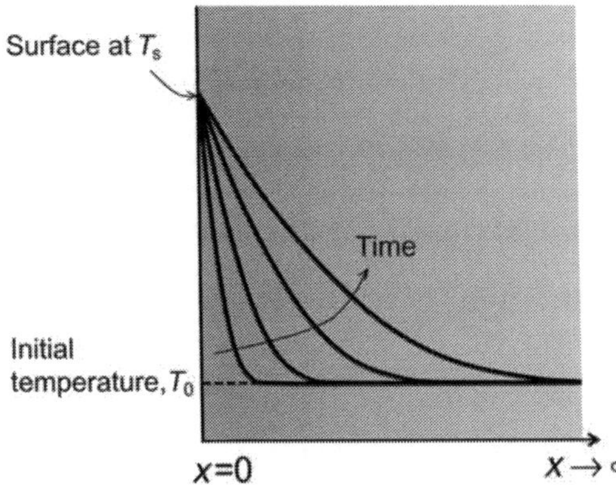

FIGURE 5.3 Evolution of the temperature profile for heat transfer in an infinite slab with an initial temperature T_0 and a constant surface temperature T_s.

This equation is subject to the following initial and boundary conditions:

$$
\begin{aligned}
&t = 0: \ T = T_0 \ \text{ for any } x \\
&x = 0: \ T = T_s \ \text{for any } t > 0 \\
&x \to \infty: \ T = T_0 \ \text{for any } t
\end{aligned}
\tag{5.32}
$$

The first of these conditions specifies that the initial temperature T_0 is uniform, the second imposes a step change in temperature from T_0 to T_s at $x = 0$, and the third specifies that the slab is infinite so that the temperature far away from the boundary at $x = 0$ will remain at the initial value T_0. Next, we introduce the following dimensionless temperature variable:

$$
Y = \frac{T - T_s}{T_0 - T_s} \tag{5.33}
$$

Note that

$$
1 - Y = \frac{T - T_0}{T_s - T_0} \tag{5.34}
$$

Since $T_s - T_0$ is the maximum temperature change that could occur in the slab, $1 - Y$ represents the fraction of this maximum temperature change that is actually attained at given values of t and x. From Eq. (5.33), we have $T = T_s + (T_0 - T_s)Y$. Thus, Eqs. (5.31) and (5.32) become:

$$
\begin{aligned}
&\frac{\partial Y}{\partial t} = \alpha \frac{\partial^2 Y}{\partial x^2} \\
&t = 0: \quad Y = 1 \ \text{ for any } x \\
&x = 0: \quad Y = 0 \ \text{for any } t > 0 \\
&x \to \infty: \ Y = 1 \ \text{for any } t
\end{aligned}
\tag{5.35}
$$

We now introduce the following expression:

$$
\eta = \frac{x}{2\sqrt{\alpha t}} \tag{5.36}
$$

which combines x and t into the single variable η, and assume that Y can be expressed as a function of this variable alone. This is called the Boltzmann transformation and it is possible because of the particular boundary conditions for this problem. Using the chain rule of differentiation, we can find that:

$$
\begin{aligned}
\frac{\partial Y}{\partial t} &= \frac{dY}{d\eta}\frac{\partial \eta}{\partial t} = \frac{dY}{d\eta}\left(-\frac{x}{4t\sqrt{\alpha t}}\right) \\
\frac{\partial^2 Y}{\partial x^2} &= \frac{\partial}{\partial x}\left(\frac{dY}{d\eta}\frac{\partial \eta}{\partial x}\right) = \frac{d^2 Y}{d\eta^2}\left(\frac{\partial \eta}{\partial x}\right)^2 + \frac{dY}{d\eta}\frac{\partial^2 \eta}{\partial x^2} \\
&= \frac{d^2 Y}{d\eta^2}\left(\frac{1}{4\alpha t}\right)
\end{aligned}
\tag{5.37}
$$

Substituting these expressions into Eq. (5.35) gives the following result:

$$\frac{d^2Y}{d\eta^2}+2\eta\frac{dY}{d\eta}=0$$
$$\eta=0: \quad Y=0 \tag{5.38}$$
$$\eta=\infty: \quad Y=1$$

Note that, now, we have only two boundary conditions since the dependence on t and x has been combined into the single variable η. Either $t = 0$ or $x=\infty$ gives $\eta=\infty$, which, in turn, corresponds to $Y=1$. This result is what makes the Boltzmann transformation work. The solution of Eq. (5.38) is found in standard differential equations textbooks and is given by the following:

$$Y=\text{erf}(\eta) \tag{5.39}$$

where erf is the error function. The fraction of the maximum temperature change achieved at a given value of η will thus be:

$$1-Y=1-\text{erf}(\eta)=\text{erfc}(\eta) \tag{5.40}$$

where erfc = 1 − erf is the complementary error function. Figure 5.4 shows this result graphically. We can reconstruct the actual temperature at t and x from this figure, or directly from the erf function, by calculating the corresponding value of η according to Eq. (5.36). The heat transfer flux at the surface is, of course, of interest and can be found from the temperature profile. From Fourier's law, we have:

$$q_x\big|_{x=0}=-k\frac{\partial T}{\partial x}\bigg|_{x=0} \tag{5.41}$$

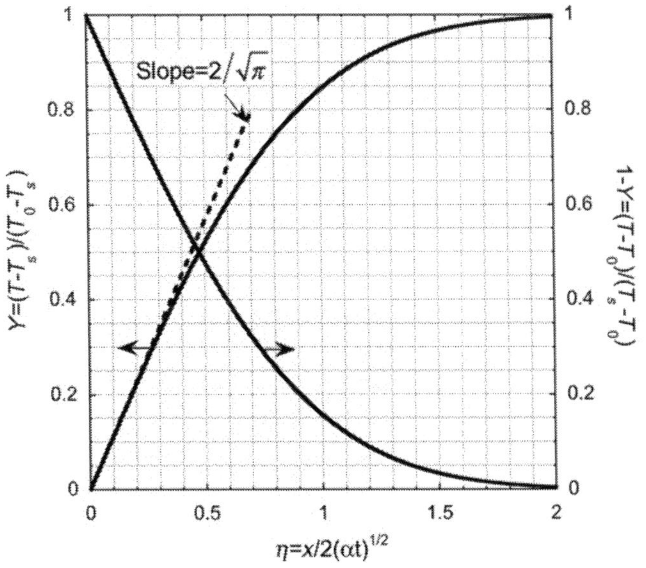

FIGURE 5.4 Error function $Y=\text{erf}(\eta)$ and complementary error function $1-Y=\text{erfc}(\eta)$ describing conductive heat transfer in an infinite slab. T_s and T_0 are the slab surface temperature and the initial temperature, respectively.

In terms of the variables Y and η, this equation can be written as follows:

$$q_x\Big|_{x=0} = -k(T_0 - T_s)\frac{\partial \eta}{\partial x}\frac{dY}{d\eta}\Big|_{x=0} = k(T_0 - T_s)\frac{1}{2\sqrt{\alpha t}}\frac{dY}{d\eta}\Big|_{x=0} \tag{5.42}$$

Finally, since the derivative of $\text{erf}(\eta)$ at $\eta = 0$ is $2/\sqrt{\pi}$ (see Fig. 5.4), we will have:

$$q_x\Big|_{x=0} = -k(T_0 - T_s)\frac{\partial \eta}{\partial x}\frac{dY}{d\eta}\Big|_{x=0} = \frac{k}{\sqrt{\pi \alpha t}}(T_s - T_0) \tag{5.43}$$

In addition to the instantaneous flux, the cumulative amount of heat conducted into the slab between time $t = 0$ and an arbitrary time t is also of interest. Normalized by the cross-sectional area, this quantity can be calculated by integrating $q_x\big|_{x=0}$ over time. The following result is obtained:

$$\int_0^t q_x\Big|_{x=0} dt = \int_0^t \frac{k}{\sqrt{\pi \alpha t}}(T_s - T_0)\,dt = k\sqrt{\frac{4t}{\pi \alpha}}(T_s - T_0) \tag{5.44}$$

A very important observation can be made based on Fig. 5.4. In Fig. 5.4, we see that $Y \sim 1$ for values of $\eta \sim 2$ or larger. Since a value of $Y \sim 1$ corresponds to a temperature that is nearly the same as the initial temperature T_0, $\eta = 2$ gives us the depth of penetration of the temperature profile for a given time t. In dimensional form, the depth of penetration that corresponds to $\eta = 2$ will be:

$$\delta = 4\sqrt{\alpha t} \tag{5.45}$$

We can also use this relationship to find the time required to attain a certain change in temperature at a particular distance from the surface. The actual value will depend on what fractional change we are looking for relative to the maximum possible. For example, the time needed to attain a temperature change equal to 0.5% of the maximum possible $(1 - Y = 0.005)$ at a distance δ will be:

$$t_{0.5\%} \sim \frac{1}{16}\frac{\delta^2}{\alpha} \tag{5.46}$$

Conversely, the time needed to attain a temperature change equal to 50% of the maximum possible, corresponding to $1 - Y = 0.5$ or $\eta \sim 0.5$, at the same distance δ will be

$$t_{50\%} \sim \frac{\delta^2}{\alpha} \tag{5.47}$$

Either result shows that time must increase four-fold if we want to double the depth of penetration of the temperature profile into the slab. These results also show a bit more clearly the meaning of α. In Chap. 3, we saw that for steady-state heat conduction, the only relevant physical property was the thermal conductivity, k, of the material where conduction takes place. However, as seen in this section, transient heat conduction is affected by both the thermal conductivity of the material and its value of ρC_p through their ratio α. This different dependence occurs because, while in a steady-state process all of the heat conducted passes through the slab, in a transient process, a portion of the heat conducted accumulates within the slab itself. This accumulation, measured by the heat capacity per unit volume, ρC_p, ceases, of course, once a steady state is attained. Note that only the ratio of k and ρC_p actually matters as far as the determination of the temperature profile is concerned. Thus, a high α can be obtained either because of a high thermal conductivity or because of a low value of ρC_p. Let's compare, for example, conductive heat transfer through air with conductive heat transfer through liquid water at 1 atm and 298 K. The thermal conductivity of air is much lower than that of water (about 0.026 W/m K vs. about 0.65 W/m K). Thus, we can expect steady-state heat conduction to be much faster through water than through air for otherwise identical conditions. On the other hand, the thermal diffusivity of air is much larger than that of water (about 2.2×10^{-5} m^2/s vs. about 1.5×10^{-7} m^2/s), because ρC_p is much smaller for air. In this case, we can expect that the time needed for the temperature profile to propagate through air will be much shorter than for water for otherwise identical conditions. In other words, a steady state will be achieved much faster for heat transfer through air than through water. Example 5.2 provides an additional illustration of this concept.

Example 5.2. When will you start sensing that your finger has been dipped in hot water? Let's see a simple application of the equations developed in this section. If we assume that the external resistance is negligible, the skin surface temperature will reach 60°C instantly upon immersion in water at 60°C, which is hot enough to cause pain and deep tissue damage. However, you will not feel a thing until a temperature change occurs at the nerve endings, which are located at a certain depth below the skin surface. The thermal conductivity of skin is $k = 0.37$ W/m K, its density is $\rho = 1{,}100$ kg/m^3, and its heat capacity is $C_p = 3390$ J/kg K. These approximate values give a thermal diffusivity $\alpha = 9.9 \times 10^{-8}$ m^2/s. From Eq. (5.46), if we assume that the nerve endings are 0.5 mm below the skin surface ($\delta = 0.0005$ m), we get, $t \sim 0.16$ s. This is a short time, but not zero. Let's now calculate how long it will take to sense a temperature of 50°C when the finger is dipped in water at 60°C. This is when we will actually begin to feel pain. In terms of dimensionless temperature, with $T_0 = 37$°C, this means $Y = (50-60)/(37-60) = 0.43$. Treating again the finger as an infinite slab and neglecting the effects of blood flow, from Fig. 5.4 we see that for $Y = 0.43$ we will have $\eta = 0.4$. The corresponding time will then be:

$$t = \frac{x^2}{4\eta^2\alpha} = \frac{0.0005^2}{4 \times 0.4^2 \times 9.9 \times 10^{-8}} = 3.9 \text{ s}$$

Of course, if the external heat transfer resistance is significant, these times would be longer. For instance, if your finger is suddenly exposed to cold air, where the convective heat transfer coefficient is small (*cf.* Table 1.2), it could take quite a while for you to sense the air temperature.

b. Solution for a semi-infinite slab

Figure 5.5 shows the evolution of the temperature profile expected in a slab of thickness $2B$ starting at an initial uniform temperature T_0 and suddenly exposed to a constant surface temperature T_s on both sides at $x = -B$ and at $x = +B$. The difference between this case and the infinite slab case is that, now, the profile is symmetrical around the centerline of the slab at $x = 0$. For short times, the depth of penetration of the temperature profile will be small and the solution will be essentially the same as that obtained for the infinite slab. However, as time runs along, the temperature profile will penetrate deeper into the slab eventually extending to the slab center. After a long time, the entire slab will approach the surface temperature T_s.

The differential energy balance describing this problem is exactly the same as that written for the infinite slab. The initial condition is also the same. However, the boundary conditions are different. We can write the boundary conditions either at $x = -B$ and at $x = +B$, where we know the surface temperature, or, equivalently at $x = +B$ where $T = T_s$ and at $x = 0$, where the heat transfer flux must be zero. Since, according to Fourier's law, the flux is proportional to the temperature gradient, the temperature profile must be flat at the centerline. Thus, its derivative with respect to x will be zero.

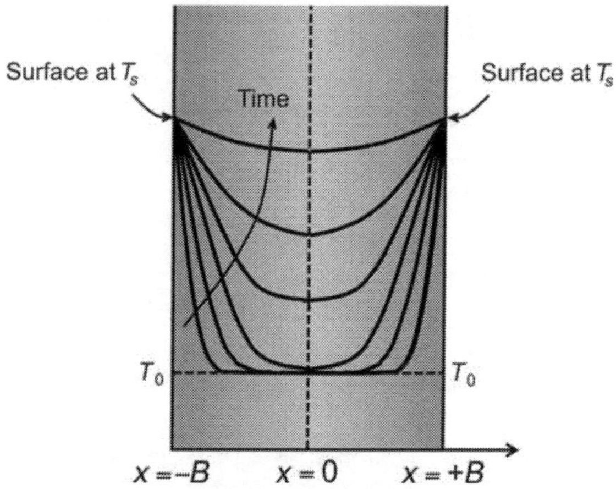

Figure 5.5 Evolution of the temperature profile for heat transfer in a semi-infinite slab of thickness $2B$ with initial temperature T_0 exposed to a constant surface temperature T_s on both sides of the slab.

The differential equation and initial and boundary conditions describing this process are then as follows:

$$
\begin{aligned}
&\frac{\partial T}{\partial t} = \alpha \frac{\partial^2 T}{\partial x^2} \\
&t = 0: \quad T = T_0 \text{ for any } x \\
&x = 0: \quad \frac{\partial T}{\partial x} = 0 \text{ for any } t \\
&x = B: \; T = T_s \text{ for any } t > 0
\end{aligned}
\tag{5.48}
$$

It is useful to write this equation and initial and boundary conditions in dimensionless form so we can get a generalized solution by defining a dimensionless temperature, Y, a dimensionless time, τ, and a dimensionless distance variable, ξ, according to the following relationships:

$$Y = \frac{T - T_s}{T_0 - T_s} \tag{5.49}$$

$$\tau = \frac{\alpha t}{B^2} \tag{5.50}$$

$$\xi = \frac{x}{B} \tag{5.51}$$

Since $T = T_s + (T_0 - T_s)Y, t = B^2\tau/\alpha$, and $x = B\xi$, substituting these variables into Eq. (48), we obtain:

$$
\begin{aligned}
&\frac{\partial Y}{\partial \tau} = \frac{\partial^2 Y}{\partial \xi^2} \\
&\tau = 0: \; Y = 1 \text{ for any } \xi \\
&\xi = 0: \quad \frac{\partial T}{\partial \xi} = 0 \text{ for any } \tau \\
&\xi = 1: \quad Y = 0 \text{ for any } \tau > 0
\end{aligned}
\tag{5.52}
$$

Geometry	Series Solution	Mathematica Notebook
A. Slab	$Y=\frac{T-T_s}{T_0-T_s}=\sum_{n=1}^{\infty}\frac{-2(-1)^n}{\lambda_n}\cos(\lambda_n\xi)\exp(-\lambda_n^2\tau)$ $Y_{avg}=\frac{T_{avg}-T_s}{T_0-T_s}=\sum_{n=1}^{\infty}\frac{2}{\lambda_n^2}\exp(-\lambda_n^2\tau)$ $\lambda_n=\frac{(2n-1)\pi}{2}$	(M.T.5.2-A)
B. Cylinder	$Y=\frac{T-T_s}{T_0-T_s}=2\sum_{n=1}^{\infty}\frac{J_0(\lambda_n\xi)}{\lambda_n J_1(\lambda_n)}\exp(-\lambda_n^2\tau)$ $Y_{avg}=\frac{T_{avg}-T_s}{T_0-T_s}=4\sum_{n=1}^{\infty}\frac{1}{\lambda_n^2}\exp(-\lambda_n^2\tau)$ $\lambda_n\rightarrow$ roots of $J_0(\lambda_n)=0$	(M.T.5.2-B)
C. Sphere	$Y=\frac{T-T_s}{T_0-T_s}=\frac{2}{\xi}\sum_{n=1}^{\infty}\frac{(-1)^{n+1}}{\lambda_n}\sin(n\pi\xi)\exp(-\lambda_n^2\tau)$ $Y_{avg}=\frac{T_{avg}-T_s}{T_0-T_s}=6\sum_{n=1}^{\infty}\frac{1}{\lambda_n^2}\exp(-\lambda_n^2\tau)$ $\lambda_n=\pi n$	(M.T.5.2-C)

TABLE 5.2 Solutions for Transient Heat Conduction in a Slab with Thickness $2B$, in a Long Cylinder, with Radius R, and in a Sphere with Radius R with Negligible External Resistance (Bi = ∞) ($\tau=\alpha t/B^2$ and $\xi=x/B$ for the Slab. $\tau=\alpha t/R^2$ and $\xi=r/R$ for the Cylinder and for the Sphere)

The solution of this equation with associated initial and boundary conditions will be a function $Y(\tau,\ \xi)$. This function, obtained by separation of variables, is given by item A of Table 5.2 in the form of a convergent Fourier series. In general, calculations using such series solutions require the addition of a number of terms sufficient to provide suitable precision. In this case, the number of terms in the series that is required to obtain a desired precision will depend on the value of τ. Many terms will be required for short times, but as few as one term in the series will be sufficient to approximate the exact solution for longer times. Example 5.3 illustrates the effect of the number of terms in the series on the predicted temperature profile.

Example 5.3. Evaluate the dimensionless temperature profiles in a slab for different values of τ. The calculation is easily done using item A in Table 5.2 with a spreadsheet or other computational tool. Figure 5.6 shows the results of these calculations. For $\tau = 0.01$, which corresponds to short times, many terms (about 10) are needed in the series to obtain a precise, nonoscillating profile. On the other hand, for $\tau = 0.15$, even just one term is sufficient to approximate the exact solution. In this case, the solution can be approximated by:

$$Y\sim\frac{4}{\pi}\cos\left(\frac{\pi\xi}{2}\right)\exp\left(-\frac{\pi^2}{4}\tau\right)\tag{5.53}$$

which is obtained from item A in Table 5.2 with $n = 1$. This one-term approximation becomes even better for higher values of τ and is useful for practical calculations. It shows that, for long times, the dimensionless temperature at each point within the slab will vary exponentially with time.

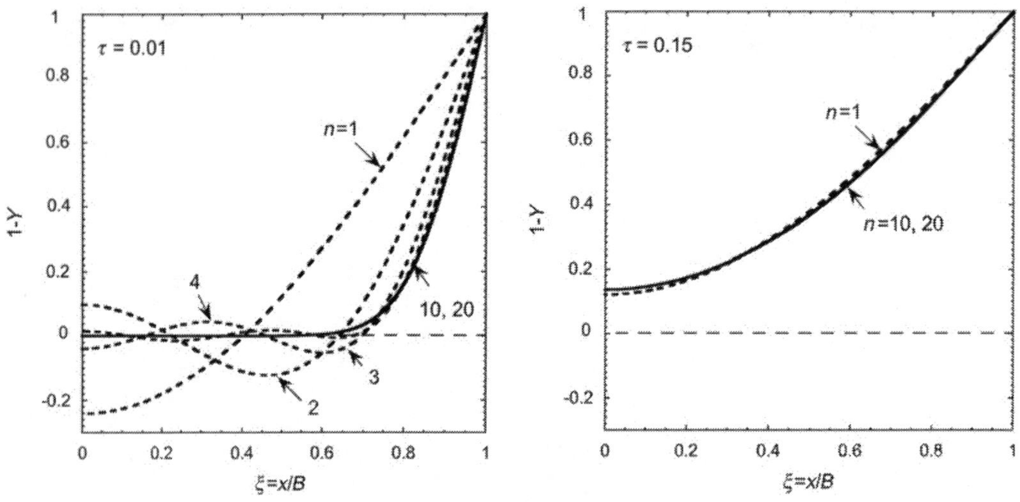

FIGURE 5.6 Dimensionless temperature profile in a slab with thickness $2B$ calculated from the series solution with a different number of terms for a short time $\tau = 0.01$ (left) and a relatively long time $\tau = 0.15$ (right). Note that the figures show $1 - Y = (T - T_0)/(T_s - T_0)$.

The series solution for the slab is useful for different purposes. Beside the case of a slab of thickness $2B$ with both faces held at T_s, the same solution can also be used to describe conductive heat transfer in a slab of thickness B where one side is kept at T_s while the other side is well insulated. For the insulated side, q_x and, thus, $\partial T/\partial x$ will be zero, which is exactly the condition at the centerline of the slab with thickness $2B$. Another useful result obtainable from the series solution is the average temperature in the slab. The latter can be found from the following integral:

$$T_{\text{avg}}(t) = \frac{1}{B}\int_0^B T(t, x)\,dx \tag{5.54}$$

In dimensionless form, Eq. (5.54) becomes:

$$Y_{\text{avg}}(\tau) = \frac{T_{\text{avg}} - T_s}{T_0 - T_s} = \int_0^1 Y(\tau, \xi)\,d\xi \tag{5.55}$$

The corresponding expression, found by inserting the series solution for Y in Eq. (5.55) and integrating term-by-term, is also given in Table 5.2 (item A). For relatively long times, e.g., with $\tau > 0.15$, we can use just the first term of the series solution, Eq. (5.53), in the integral above. In this case, we obtain:

$$\begin{aligned} Y_{\text{avg}} &\sim \int_0^1 \frac{4}{\pi}\cos\left(\frac{\pi\xi}{2}\right)\exp\left(-\frac{\pi^2}{4}\tau\right)d\xi \\ &= \frac{4}{\pi}\exp\left(-\frac{\pi^2}{4}\tau\right)\int_0^1 \cos\left(\frac{\pi\xi}{2}\right)d\xi \\ &= \frac{8}{\pi^2}\exp\left(-\frac{\pi^2}{4}\tau\right) \end{aligned} \tag{5.56}$$

This result shows that for relatively long times the average dimensionless temperature will also vary exponentially with time.

Let's now consider the problem of transient heat conduction through a semi-infinite slab of thickness B where we specify different temperatures on its two sides. In this case, starting with a uniform temperature T_0, we suddenly bring the temperature to T_s on one side of the slab while keeping the other side at the initial temperature T_0. The difference between this case and the previous one where both sides were exposed to the same temperature is that, in this case, at long times, rather than achieving a uniform temperature with a zero flux, we will achieve a steady-state condition with a constant, nonzero heat transfer flux across the slab. Intuitively, we recognize that it will take some time to attain such steady-state conditions and that this time will depend on a combination of thermal conductivity, density, and heat capacity of the material. We expect that, initially, little heat will reach the side of the slab kept at T_0 while most of the heat accumulates in the slab. Over time, more and more heat will pass through the slab until a maximum value is obtained at steady state when there is no longer any accumulation of energy within the slab itself.

The differential energy balance and initial and boundary conditions that describe this problem are as follows:

$$\begin{aligned}
&\frac{\partial T}{\partial t} = \alpha \frac{\partial^2 T}{\partial x^2} \\
&t = 0: \quad T = T_0 \text{ for any } x \\
&x = 0: \quad T = T_s \text{ for any } t > 0 \\
&x = B: \quad T = T_0 \text{ for any } t
\end{aligned} \tag{5.57}$$

Note that the differential energy balance is exactly the same as what we wrote previously. The difference is in the boundary conditions, which results in a different solution. In terms of the dimensionless variables Y, τ, and ξ, Eq. (5.57) becomes:

$$\begin{aligned}
&\frac{\partial Y}{\partial \tau} = \frac{\partial^2 Y}{\partial \xi^2} \\
&\tau = 0: \quad Y = 1 \text{ for any } \xi \\
&\xi = 0: \quad Y = 0 \text{ for any } \tau > 0 \\
&\xi = 1: \quad Y = 1 \text{ for any } \tau
\end{aligned} \tag{5.58}$$

We can solve this mathematical problem more readily by defining a new dimensionless temperature $Y_\xi = Y - \xi$. Inserting this new variable into Eq. (5.58), we obtain:

$$\begin{aligned}
&\frac{\partial Y_\xi}{\partial \tau} = \frac{\partial^2 Y_\xi}{\partial \xi^2} \\
&\tau = 0: \quad Y_\xi = 1 - \xi \text{ for any } \xi \\
&\xi = 0: \quad Y_\xi = 0 \text{ for any } \tau > 0 \\
&\xi = 1: \quad Y_\xi = 0 \text{ for any } \tau
\end{aligned} \tag{5.59}$$

Note that, in terms of Y_ξ, the boundary conditions become homogeneous (i.e., $Y_\xi = 0$ at both $\xi = 0$ and $\xi = 1$), which simplifies obtaining the solution by separation of variables. This solution, expressed in terms of $Y = \xi + Y_\xi$, is as follows:

(M.E.5.60)
$$Y = \xi + \frac{2}{\pi}\sum_{n=1}^{\infty}\frac{\sin(n\pi\xi)}{n}\exp(-n^2\pi^2\tau) \tag{5.60}$$

Beside predicting the temperature profile, this result also allows us to calculate the heat transfer flux at $x = B$. From Fourier's law, we have:

$$q_x\big|_{x=B} = -k\left(\frac{\partial T}{\partial x}\right)_{x=B} \tag{5.61}$$

or, in terms of Y and ξ:

$$q_x\big|_{x=B} = -\frac{k(T_0 - T_s)}{B}\left(\frac{\partial Y}{\partial \xi}\right)_{\xi=B} \tag{5.62}$$

Substituting Eq. (5.60), we obtain:

(M.E.5.63)
$$\begin{aligned} q_x\big|_{x=B} &= \frac{k(T_s - T_0)}{B}\left\{\frac{\partial}{\partial \xi}\left[\xi + \frac{2}{\pi}\sum_{n=1}^{\infty}\frac{\sin(n\pi\xi)}{n}\exp(-n^2\pi^2\tau)\right]\right\}_{\xi=1} \\ &= \frac{k(T_s - T_0)}{B}\left[1 + 2\sum_{n=1}^{\infty}\cos(n\pi)\exp(-n^2\pi^2\tau)\right] \end{aligned} \tag{5.63}$$

Note that, for long times, τ becomes large and the exponential terms in this equation become very small. In this case, the second term within the brackets will become negligible, and the flux will approach the steady-state value:

$$q_{x,SS} = \frac{k(T_s - T_0)}{B} \tag{5.64}$$

This result is, of course, the same flux that we would have obtained directly by applying Fourier's law under steady-state conditions. Figure 5.7 shows the evolution of the dimensionless temperature profile in the slab calculated from Eq. (5.60) and the corresponding heat transfer flux at $x = B$ based on Eq. (5.63) normalized by the steady-state flux, i.e., $q_x\big|_{x=B}/q_{x,SS}$. We can see that for "long" times, the temperature profile becomes linear and the heat transfer flux becomes constant. This happens when steady-state conditions are approached. As seen from Fig. 5.7, this occurs when $\tau \sim 0.5$, which corresponds to a time $t \sim 0.5B^2/\alpha$. As expected, this time is proportional to the square of the thickness of the slab and inversely proportional to the thermal diffusivity of the slab. As seen from these results, k/B determines the steady-state flux, while B^2/α determines the time needed to attain steady state.

c. Solutions for long cylinders and spheres of radius R

The differential energy balances and associated initial and boundary conditions for cylinders and spheres are easily derived from Table 5.1 by considering only the radial dependence of T and no energy generation and are given by the following equations:

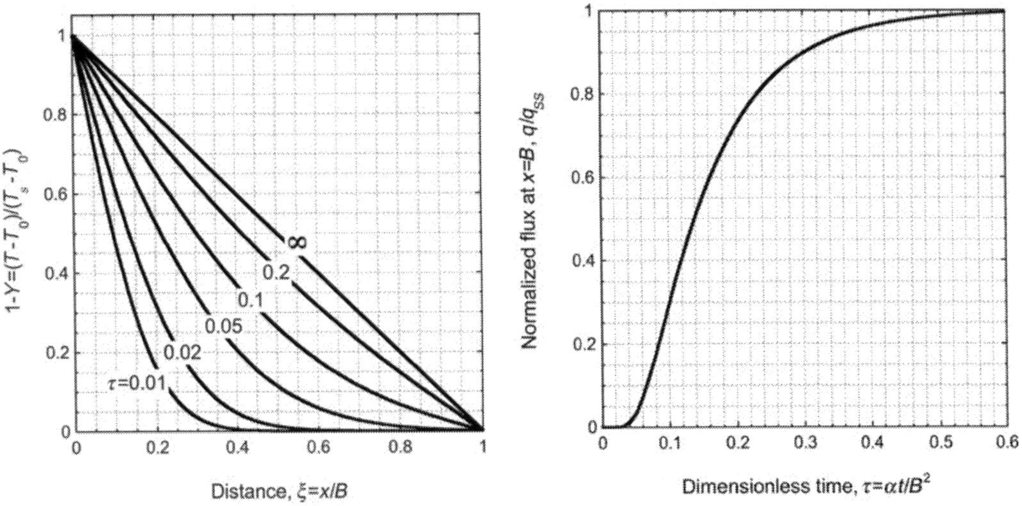

Figure 5.7 Dimensionless temperature profile in a semi-infinite slab initially at T_0 with surface temperatures kept at T_s for $x = 0$ and at T_0 for $x = B$ (left) and normalized heat transfer flux at $x = B$ (right). $q_{x,SS} = k(T_s - T_0)/B$ is the steady-state flux. Mathematica notebook: (M.F.5.7).

Long cylinders

$$\begin{aligned}
&\frac{\partial T}{\partial t} = \frac{\alpha}{r}\frac{\partial}{\partial r}\left(r\frac{\partial T}{\partial r}\right)\\
&t=0: \quad T = T_0 \quad \text{for any } r\\
&r=0: \quad \frac{\partial T}{\partial r} = 0 \text{ for any } t\\
&r=R: \quad T = T_s \text{ for any } t>0
\end{aligned} \tag{5.65}$$

Spheres

$$\begin{aligned}
&\frac{\partial T}{\partial t} = \frac{\alpha}{r^2}\frac{\partial}{\partial r}\left(r^2\frac{\partial T}{\partial r}\right)\\
&t=0: \quad T = T_0 \quad \text{for any } r\\
&r=0: \quad \frac{\partial T}{\partial r} = 0 \text{ for any } t\\
&r=R: \quad T = T_s \text{ for any } t>0
\end{aligned} \tag{5.66}$$

Series solutions are obtained for each of these two cases in terms of the dimensionless variables:

$$Y = \frac{T - T_s}{T_0 - T_s} \tag{5.67}$$

$$\tau = \frac{\alpha t}{R^2} \tag{5.68}$$

$$\xi = \frac{r}{R} \tag{5.69}$$

and are given in Table 5.2, items B and C. Expressions for the average dimensionless temperature Y_{avg} are also given in this table. Note that the average temperatures given by Y_{avg} are cylindrical and spherical averages, respectively. These special averages are defined in such a way that multiplying the average temperature by the product of density, heat capacity, and volume will give the heat content of the object. For a long cylinder, we can derive an expression for the cylindrical-average temperature as follows. Consider first the contribution of a cylindrical shell of thickness dr to the total energy content of the cylinder. The volume of this shell is $2\pi rL \times dr$ where L is the length of the cylinder. The energy contained in this shell is obtained by multiplying this volume by the density, ρ, and heat capacity, C_p, and by the local value of T and is as follows:

$$2\pi rL\rho C_p T(t,r)dr$$

Integrating this quantity across the radius of the cylinder will give the total energy content of the cylinder. We can also obtain the total energy content of the cylinder by multiplying the cylinder volume, $\pi R^2 L$, by ρ and C_p, and by the cylinder average temperature, which gives $\pi R^2 L\rho C_p T_{avg}(t)$. Equating these two quantities, we obtain:

$$\pi R^2 L\rho C_p T_{avg}(t) = \int_0^R 2\pi rL\rho C_p T(t,r)dr \tag{5.70}$$

or, canceling the common terms:

$$T_{avg}(t) = \frac{2}{R^2}\int_0^R rT(t,r)dr \tag{5.71}$$

In terms of the dimensionless temperature variable, this result can be written as follows:

$$Y_{avg}(\tau) = \frac{T_{avg} - T_s}{T_0 - T_s} = 2\int_0^1 \xi Y(\tau,\xi)d\xi \tag{5.72}$$

Analogously, for the spherical average, we consider the energy content of a spherical shell of thickness dr and volume $4\pi r^2 \times dr$, which is given by:

$$4\pi r^2 \rho C_p T(t,r)dr$$

Integrating this quantity across the radius of the sphere will give us the total energy content of the sphere. Similarly, we can obtain the total energy content of the sphere by multiplying the volume of the sphere, $4\pi R^3/3$, by ρ and C_p, and by the sphere average temperature, which gives $(4\pi R^3/3) \times \rho C_p T_{avg}(t)$. The final result is obtained by equating these two expressions for the total energy content which gives:

$$\frac{4\pi R^3}{3}\rho C_p T_{avg}(t) = \int_0^R 4\pi r^2 \rho C_p T(t,r)dr \tag{5.73}$$

or, canceling the common terms:

$$T_{avg}(t) = \frac{3}{R^3}\int_0^R r^2 T(t,r)dr \tag{5.74}$$

Finally, in terms of the dimensionless variable, we will have:

$$Y_{avg}(\tau) = \frac{T_{avg} - T_s}{T_0 - T_s} = 3\int_0^1 \xi^2 Y(\tau,\xi)\,d\xi \tag{5.75}$$

Figure 5.8 compares the dimensionless average temperature Y_{avg} calculated from item A in Table 5.2 for a slab, from item B in Table 5.2 for a long cylinder, and from item C in Table 5.2 for a sphere, all as a function of dimensionless time. As seen in Fig. 5.8, while the shapes of the curves are similar and each shows an exponential decay at high values of τ, quantitative differences exist between the different geometries. For example, for the same radius, heat transfer will be faster in a sphere compared to a cylinder since the surface to volume ratio is higher for the latter.

d. Solutions for multidimensional geometries—Newman's rule

Exact solutions for the problem of transient heat conduction in multidimensional geometries (i.e., for a brick or for a stocky cylinder) are best handled by a direct numerical solution using either finite difference methods or, for more complex shapes, finite element-based methods. However, a useful approximation for geometries that are not too different from the standard slab, long cylinder, and sphere shapes is Newman's rule. According to this rule, the dimensionless temperature Y at a point in a three-dimensional object at a particular time t can be obtained as the product of the values of Y calculated for each dimension using the value of τ based on the length scale associated with that dimension. For example, consider the shapes in Fig. 5.9. For the brick-shaped object, assuming that the initial temperature is uniform and equal to T_0 and that

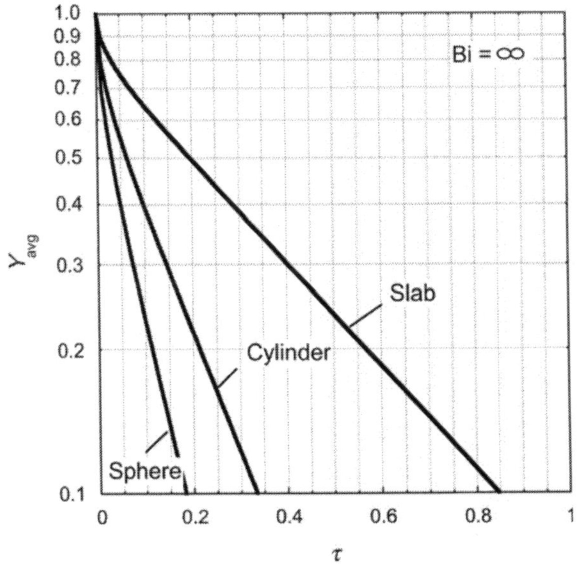

FIGURE 5.8 Dimensionless average temperature vs. dimensionless time in a slab of thickness $2B$, in a long cylinder with radius R, and in a sphere with radius R. $\tau = \alpha t/B^2$ for the slab and $\tau = \alpha t/R^2$ for the cylinder and the sphere.

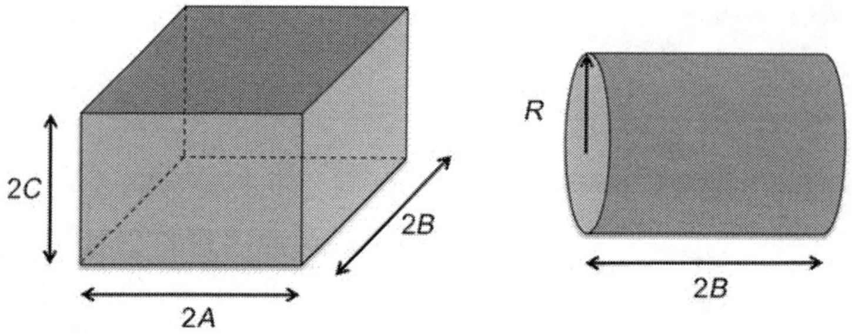

Figure 5.9 Brick and stocky cylinder geometries.

all faces are held at the same temperature T_s, the dimensionless temperature at any point with coordinates x, y, and z within the brick can be written as follows:

$$Y = Y_A(\tau_A, \xi_A) \times Y_B(\tau_B, \xi_B) \times Y_C(\tau_C, \xi_C) \tag{5.76}$$

where the dimensionless time and distance variables are:

$$\begin{aligned} \tau_A &= \frac{\alpha t}{A^2}, \quad \tau_B = \frac{\alpha t}{B^2}, \quad \tau_C = \frac{\alpha t}{C^2} \\ \xi_A &= \frac{x}{A}, \quad \xi_B = \frac{y}{B}, \quad \xi_C = \frac{z}{C} \end{aligned} \tag{5.77}$$

Similarly, for the stocky cylinder, we will have:

$$Y = Y_B(\tau_B, \xi_B) \times Y_R(\tau_R, \xi_R) \tag{5.78}$$

where

$$\begin{aligned} \tau_B &= \frac{\alpha t}{B^2}, \quad \tau_R = \frac{\alpha t}{R^2} \\ \xi_B &= \frac{y}{B}, \quad \xi_R = \frac{r}{R} \end{aligned} \tag{5.79}$$

Example 5.4 illustrates the application of these approximate relationships.

Example 5.4. Temperature at the center of a brick and at the center of a stocky cylinder. Let's apply Newman's rule to calculate the dimensionless temperature at the center of a cube 2 cm on a side with $\alpha = 1.5 \times 10^{-8}$ m^2/s for $t = 60$ min (1,200 s). We first calculate:

$$\tau_A = \tau_B = \tau_C = \frac{1.5 \times 10^{-8} \times 1{,}200}{0.01^2} = 0.18$$

From the series solution for a slab, item A in Table 5.2, or from the one-term approximation, Eq. (5.53), which is appropriate in this case since $\tau > 0.15$, at the center of the brick, where $\xi_A = \xi_B = \xi_C = 0$, we will find $Y_A = Y_B = Y_C = 0.81$. Thus, $Y = 0.81 \times 0.81 \times 0.81 = 0.53$. Similarly, for a cylinder that is 2 cm long and has a 1 cm radius with the same α at $t = 60$ min, we will have:

$$\tau_B = \frac{1.5 \times 10^{-8} \times 1{,}200}{0.01^2} = 0.18$$

$$\tau_R = \frac{1.5 \times 10^{-8} \times 1{,}200}{0.01^2} = 0.18$$

At the center ($\xi_B = \xi_R = 0$), the solution for the slab gives again $Y_B = 0.81$, while the series solution for the cylinder (one term is again sufficient) gives $Y_B = 0.56$. Thus, we will have $Y = Y_B \times Y_R = 0.81 \times 0.56 = 0.45$. Exact, numerically computed values of Y for these geometries can be found, for example, in Schneider, Conduction Heat Transfer, Addison-Wesley, 1955, p. 249. For these conditions, the exact values are $Y = 0.48$ for the cube and $Y = 0.40$ for the stocky cylinder. The values obtained from the simple Newman's rule are only about 10% off from the exact values for the cube and only about 13% off for the stocky cylinder.

e. Solution for spheres in a finite volume of fluid

The case of heat transfer to or from spheres at an initial temperature T_0 immersed in a finite volume of a fluid at an initial temperature $T_{f,0}$ is important in a number of practical situations. We have dealt with this problem already in Example 5.1 for a case where the external heat transfer resistance was dominant. We now consider the same situation for a case where the internal resistance is dominant (i.e., when the external resistance is negligible). With reference to Fig. 5.10, if the volume of fluid is finite, its temperature, T_f, will vary because of heat transfer to or from the spheres. In the ensuing development, we assume that the fluid is well mixed. The differential energy balance for each sphere is the same as Eq. (5.66). However, the surface temperature, which, in this case, is equal to the fluid temperature, T_f, will vary with time. Conductive transport inside the spheres will be described mathematically by the following equation and boundary conditions:

$$\begin{aligned} &\frac{\partial T}{\partial t} = \frac{\alpha}{r^2}\frac{\partial}{\partial r}\left(r^2 \frac{\partial T}{\partial r}\right) \\ &t = 0: \quad T = T_0 \quad \text{for any } r \\ &r = 0: \quad \frac{\partial T}{\partial r} = 0 \text{ for any } t \\ &r = R: \quad T = T_f(t) \text{ for any } t > 0 \end{aligned} \tag{5.80}$$

Note at the surface of the spheres, $r = R$, we have specified that the temperature will be equal to the fluid temperature, T_f. Since the latter varies with time, we need another

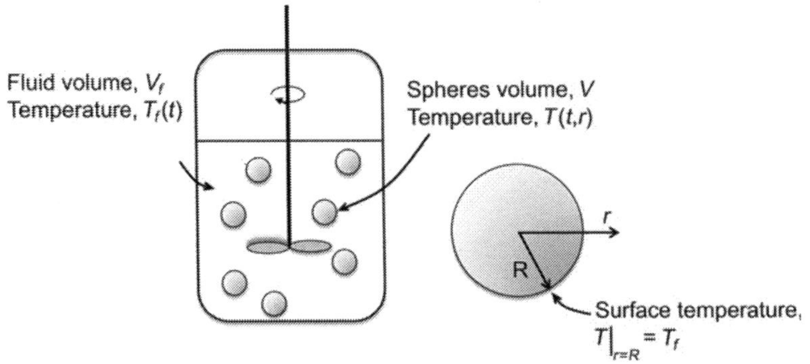

FIGURE 5.10 Heat transfer to spheres in a finite volume of fluid.

equation to solve the problem. This equation can be found by writing a transient macroscopic energy balance as follows:

$$\rho_f C_{p,f} V_f \frac{dT_f}{dt} = 4\pi R^2 \times N_{\text{spheres}} \times \left(-k \frac{\partial T}{\partial r}\bigg|_{r=R} \right) \tag{5.81}$$

$$t = 0, \quad T_f = T_{f,0}$$

In this equation, ρ_f and $C_{p,f}$ are the density and heat capacity of the fluid, k is the thermal conductivity of the material of which the spheres are made, N_{spheres} is the number of spheres, and the quantity $-k\,dT/dr|_{r=R}$ is the heat transfer flux at the sphere surface. Thus, this equation equates the rate of accumulation of energy in the fluid to the rate of heat transfer from the spheres. Note that we can calculate the number of spheres from the total volume of spheres, V, as $N_{\text{spheres}} = V/(4\pi R^3/3)$. The analytical solution of the system of Eqs. (5.80) and (5.81), in dimensionless form, is as follows:

(M.E.5.82)

$$Y_f = \frac{\Lambda}{1+\Lambda} + 6\Lambda \sum_{n=1}^{\infty} \frac{\exp(-b_n^2 \tau)}{\Lambda^2 b_n^2 + 9(1+\Lambda)} \tag{5.82}$$

where:

$$Y_f = \frac{T_f - T_0}{T_{f,0} - T_0}, \quad \Lambda = \frac{\rho_f C_{p,f} V_f}{\rho C_p V}, \quad \tau = \frac{\alpha t}{R^2} \tag{5.83}$$

The b_n values in Eq. (5.82) are obtained as the roots of the following equation for each n:

$$\tan(b_n) = \frac{3(b_n)}{3 + \Lambda (b_n)^2} \tag{5.84}$$

Figure 5.11 provides a graph of the dimensionless solution calculated from Eq. (5.82) for different values of the parameter Λ. Note that Y_f gives the dimensionless fluid temperature. The average temperature of the spheres can be found from the overall energy balance, which is still given by Eq. (5.26) by taking $T = T_{\text{avg}}$. This balance applies regardless of which resistance is dominant. Large values of Λ will correspond to cases where there is a large amount of fluid compared to the volume of spheres. In these cases, the fluid temperature is predicted to remain close to its initial value. Example 5.5 illustrates the application of these relationships.

Example 5.5. Heat transfer to or from spheres in a finite volume of fluid. Let's calculate the fluid temperature 10 min after we have added 100 kg of 10-cm-diameter solid glass spheres initially at 90°C to 100 L of water initially at 20°C in a well-insulated tank. The relevant properties of glass are $\rho = 2{,}800 \text{ kg/m}^3$, $C_p = 800 \text{ J/kg K}$, and $k = 0.8 \text{ W/m K}$, while the relevant properties of water are $\rho_f = 1{,}000 \text{ kg/m}^3$ and $C_{p,f} = 4{,}200 \text{ J/kg K}$. For these conditions, we will have:

$$\alpha = \frac{k}{\rho C_p} = \frac{0.8}{2{,}800 \times 800} = 3.6 \times 10^{-7} \text{ m}^2/\text{s}$$

$$\tau = \frac{\alpha t}{R^2} = \frac{3.6 \times 10^{-7} \times (10 \times 60)}{(0.05)^2} = 0.086$$

$$\Lambda = \frac{\rho_f C_{p,f} V_f}{\rho C_p V} = \frac{1{,}000 \times 4{,}200 \times 0.1}{2{,}800 \times 800 \times (100/2{,}800)} = 5.2$$

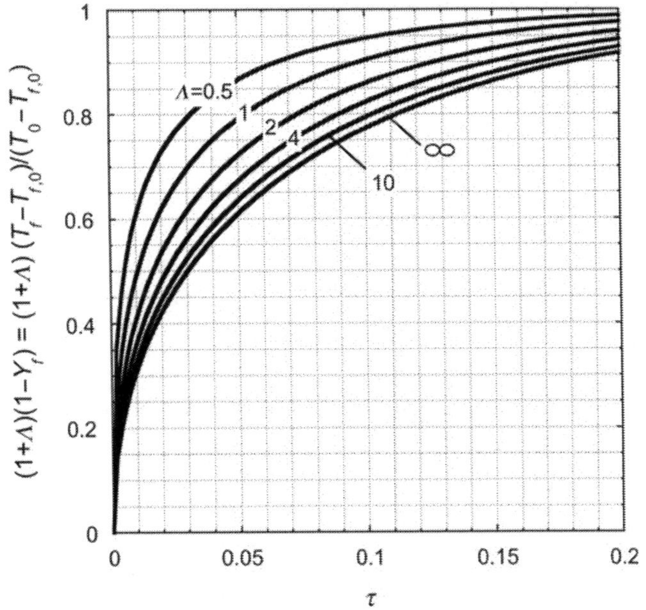

Figure 5.11 Dimensionless fluid temperature for heat transfer to spheres in a finite volume of a well-mixed fluid based on Eq. (5.82). Mathematica notebook: (M.F.5.11).

Reading from Fig. 5.11 or, more precisely, from M.E.5.82, we obtain:

$$(1+\Lambda)(1-Y_f)=0.78$$

or

$$(1+5.2)\left(1-\frac{T_f-90}{20-90}\right)=0.78$$

from which we get a fluid temperature $T_f = 29°\text{C}$.

5.3.3 Processes Where Both the Internal and External Resistances Are Important

Heat transfer processes where both internal and external resistances are important are characterized by intermediate values of the Biot number, generally between 0.1 and 10. Let's consider first a semi-infinite slab. The differential energy balance used to describe the internal conductive heat transfer process will be exactly the same as that used for cases where the external resistance is negligible and is again:

$$\frac{\partial T}{\partial t}=\alpha\frac{\partial^2 T}{\partial x^2} \tag{5.85}$$

The initial condition and the boundary condition at the center of the slab will also be the same as before and are:

$$\begin{aligned} t=0:&\quad T=T_0 \\ x=0:&\quad \frac{\partial T}{\partial x}=0 \end{aligned} \tag{5.86}$$

However, the boundary condition at the surface of the object will be different. In this case, as shown schematically in Fig. 5.12, the surface temperature T_s will not be constant but will vary between the temperature T_a of the external bulk fluid and the object's temperature. Let's first look at the physical picture considering, for example, a cold slab, initially at T_0, suddenly immersed in a hot fluid. Immediately after immersing the slab in the fluid, the slab's surface temperature will be close to T_0 and heat will be transferred by convection from the bulk fluid to the surface driven by the temperature difference $T_a - T_0$. The temperature gradient in the slab will initially be very steep. As a consequence, conductive transport will tend to be fast. Over time, the temperature profile will penetrate deeper into the slab resulting in a smaller temperature gradient and, thus, slower conductive transport. In turn, the surface temperature will gradually increase. Although there will be a smaller driving force since the surface temperature will be higher than T_0, convective transport will still keep up with internal conduction keeping the flux on the two sides of the interface equal to each other. Eventually the surface temperature will approach T_a when thermal equilibrium is attained. Note that, throughout this process, at the centerline (located at $x = 0$), the temperature gradient will remain equal to zero. The same process will also occur, of course, within a slab of thickness B that is insulated at $x = 0$.

Now that we have established a physical basis, we can specify the boundary condition at the surface by equating the conductive heat transfer flux on the slab side to the convective heat transfer flux on the fluid side of the interface, which gives the following:

$$-k\left.\frac{\partial T}{\partial x}\right|_{x=B} = h\left(T\big|_{x=B} - T_a\right) \tag{5.87}$$

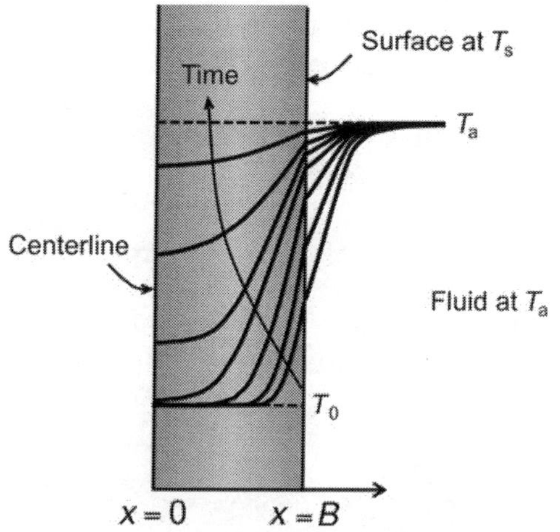

Figure 5.12 Evolution of the temperature profile for heat transfer in a slab of thickness $2B$ with initial temperature T_0 in contact with a fluid at constant temperature T_a on both surfaces when both internal and external heat transfer resistances are important. Only one-half of the slab is shown. Thus, $x = 0$ corresponds to the centerline of the slab.

where k is the thermal conductivity of the object, h is the convective heat transfer coefficient, and $T|_{x=B}$ is the time-varying surface temperature T_s. Next, let's define the following dimensionless variables for the slab:

$$Y = \frac{T - T_a}{T_0 - T_a} \tag{5.88}$$

$$\tau = \frac{\alpha t}{B^2} \tag{5.89}$$

$$\xi = \frac{x}{B} \tag{5.90}$$

τ and ξ are the same as before. However, Y is now based on T_a rather than T_s. Since $T = T_a + (T_0 - T_a)Y$, inserting these variables in Eqs. (5.85), (5.86), and (5.87), we obtain:

$$(T_0 - T_a)\frac{\alpha}{B^2}\frac{\partial Y}{\partial \tau} = (T_0 - T_a)\frac{\alpha}{B^2}\frac{\partial^2 Y}{\partial \xi^2} \tag{5.91}$$

$$\tau = 0: \quad T_a + (T_0 - T_a)Y = T_0 \tag{5.92}$$

$$\xi = 0: \quad \frac{(T_0 - T_a)}{B}\frac{\partial Y}{\partial \xi} = 0 \tag{5.93}$$

$$\xi = 1: \quad -k\frac{(T_0 - T_a)}{B}\left.\frac{\partial T}{\partial \xi}\right|_{\xi=1} = h\left[T_a + (T_0 - T_a)Y|_{\xi=1} - T_a\right] \tag{5.94}$$

or, canceling common terms:

$$\frac{\partial Y}{\partial \tau} = \frac{\partial^2 Y}{\partial \xi^2} \tag{5.95}$$

$$\tau = 0: \quad Y = 1 \tag{5.96}$$

$$\xi = 0: \quad \frac{\partial Y}{\partial \xi} = 0 \tag{5.97}$$

$$\xi = 1: \quad \left.\frac{\partial Y}{\partial \xi}\right|_{\xi=1} = -\text{Bi} \times Y|_{\xi=1} \tag{5.98}$$

where the Biot number, Bi, is given by:

$$\text{Bi} = \frac{hB}{k} \tag{5.99}$$

We can verify that Eq. (5.87) is also applicable at the surface of a cylinder or at the surface of a sphere if written in terms of the radial coordinate. Its dimensionless form, Eq. (5.98), is also the same provided that, for a long cylinder or a sphere, we define Bi and ξ as follows:

$$\text{Bi} = \frac{hR}{k} \tag{5.100}$$

$$\xi = \frac{r}{R} \tag{5.101}$$

Geometry	Series Solution	Mathematica Notebook
A. Slab	$Y=\frac{T-T_a}{T_0-T_a}=\sum_{n=1}^{\infty}\frac{4\sin(\lambda_n)}{2\lambda_n+\sin(2\lambda_n)}\cos(\lambda_n\xi)\exp(-\lambda_n^2\tau)$ $Y_{avg}=\frac{T_{avg}-T_a}{T_0-T_a}=\sum_{n=1}^{\infty}\frac{4\sin^2(\lambda_n)}{2\lambda_n^2+\lambda_n\sin(2\lambda_n)}\exp(-\lambda_n^2\tau)$ $\lambda_n\rightarrow$ roots of $\lambda_n\tan(\lambda_n)-\mathrm{Bi}=0$	(M.T.5.3-A)
B. Cylinder	$Y=\frac{T-T_a}{T_0-T_a}=2\mathrm{Bi}\sum_{n=1}^{\infty}\frac{J_0(\lambda_n\xi)}{(\lambda_n^2+Bi^2)J_0(\lambda_n)}\exp(-\lambda_n^2\tau)$ $Y_{avg}=\frac{T_{avg}-T_a}{T_0-T_a}=4\mathrm{Bi}\sum_{n=1}^{\infty}\frac{J_1(\lambda_n)}{\lambda_n(\lambda_n^2+Bi^2)J_0(\lambda_n)}\exp(-\lambda_n^2\tau)$ $\lambda_n\rightarrow$ roots of $\lambda_nJ_1(\lambda_n)-J_0(\lambda_n)\mathrm{Bi}=0$	(M.T.5.3-B)
C. Sphere	$Y=\frac{T-T_a}{T_0-T_a}=\frac{2\mathrm{Bi}}{\xi}\sum_{n=1}^{\infty}\frac{\left[\lambda_n^2+(\mathrm{Bi}-1)^2\right]\sin(\lambda_n)}{\lambda_n^2\left[\lambda_n^2+\mathrm{Bi}(\mathrm{Bi}-1)\right]}\sin(\lambda_n\xi)\exp(-\lambda_n^2\tau)$ $Y_{avg}=\frac{T_{avg}-T_a}{T_0-T_a}=6\mathrm{Bi}\sum_{n=1}^{\infty}\frac{\left[\lambda_n^2+(\mathrm{Bi}-1)^2\right]\left[\sin^2(\lambda_n)-\frac{\lambda_n}{2}\sin(2\lambda_n)\right]}{\lambda_n^4\left[\lambda_n^2+\mathrm{Bi}(\mathrm{Bi}-1)\right]}\exp(-\lambda_n^2\tau)$ $\lambda_n\rightarrow$ roots of $\lambda_n\cot(\lambda_n)+\mathrm{Bi}-1=0$	(M.T.5.3-C)

TABLE 5.3 Solutions for Transient Heat Conduction in a Slab with Thickness $2B$, in a Long Cylinder, with Radius R, and in a Sphere with Radius R Considering the External Resistance (Bi < 10) ($\tau=\alpha t/B^2$ and $\xi=x/B$ for the Slab. $\tau=\alpha t/R^2$ and $\xi=r/R$ for the Cylinder and the Sphere)

The dimensionless temperature profiles obtained as the solution of Eqs. (5.95)–(5.98) for a semi-infinite slab or of the corresponding equations for a long cylinder and for a sphere will be functions of τ, Bi, and ξ. These functions are given in Table 5.3 along with the corresponding slab, cylindrical, and spherical averages, where Y_{avg} is defined as:

$$Y_{avg}=\frac{T_{avg}-T_a}{T_0-T_a} \tag{5.102}$$

Figures 5.13 and 5.14 show plots of these functions for different values of Bi in a semi-log format commonly known as Gurney–Lurie charts. Figure 5.13 shows the value of Y for the three geometries. In each case, for each value of Bi, the solid upper line in the graph is calculated at $\xi=0$ and is, thus, the centerline temperature. The dashed bottom line is calculated at $\xi=1$ and, is thus, the surface temperature. For Bi $=\infty$, there is no dashed line since, in this case, there is no external resistance and, thus, $T_s=T_a$ giving $Y=0$ at the surface. Thus, for Bi $=\infty$, the dashed line will coincide with the vertical axis. Note that, as Bi decreases, the difference between the solid and dashed lines becomes progressively smaller. This occurs because, as Bi decreases, the external resistance becomes increasingly important making the difference between the centerline and surface temperatures of the object smaller and smaller.

Figure 5.14 shows the corresponding average dimensionless temperature values calculated from the equations in Table 5.3 for different values of Bi. If plotted along with

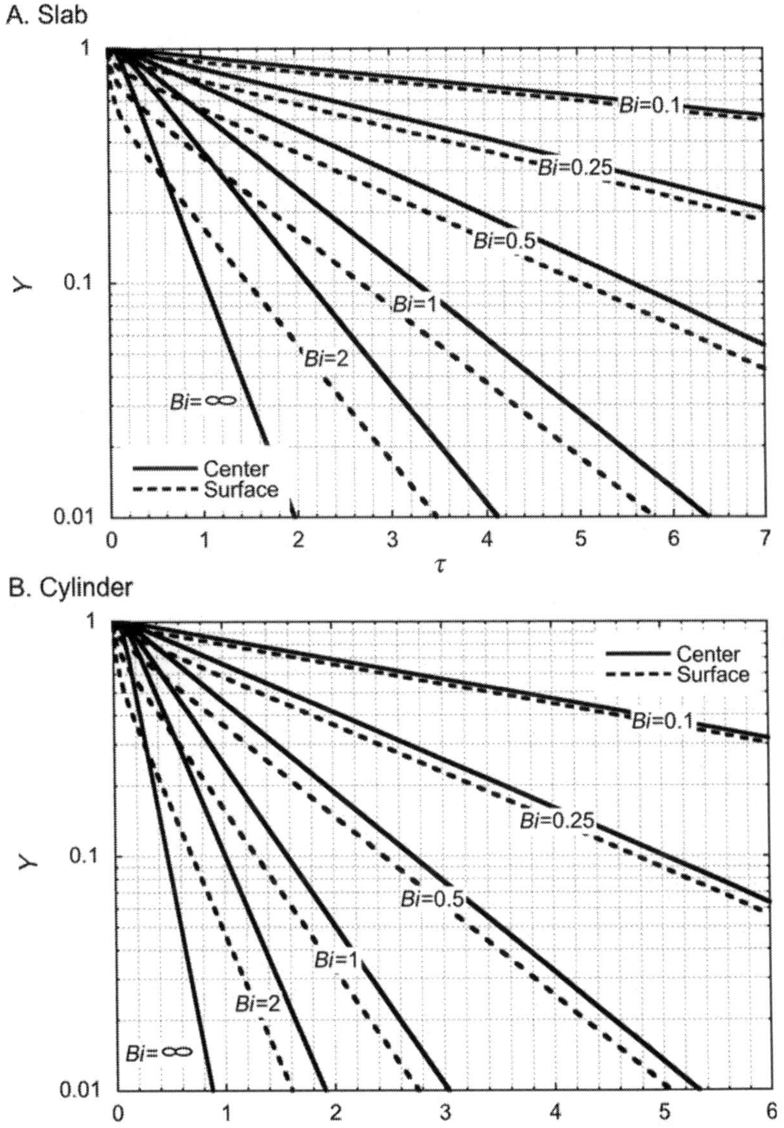

FIGURE 5.13 Dimensionless temperature, $Y = (T - T_a)/(T_0 - T_a)$, vs. dimensionless time τ at different values of the Biot number in: (A) a slab with thickness $2B$; (B) a long cylinder, with radius R; and (C) a sphere with radius R considering the external resistance (Bi < 10). $\tau = \alpha t/B^2$ and $\xi = x/B$ for the slab and $\tau = at/R^2$ and $\xi = r/R$ for the cylinder and for the sphere. Mathematica notebooks: (M.F.5.13-A), (M.F.5.13-B), and (M.F.5.13-C).

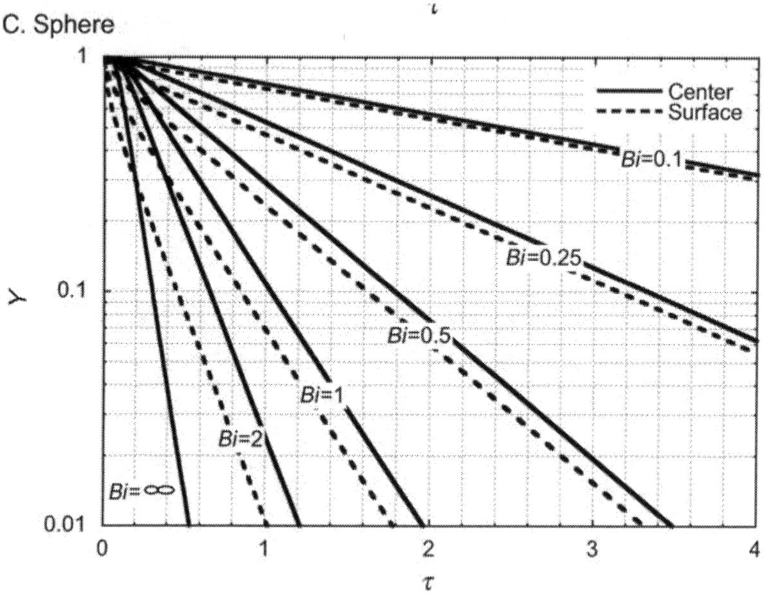

FIGURE 5.13 *(Continued)*

the Y-value curves in Fig. 5.13, these Y_{avg} curves would fall in between the upper and lower Y-curves. The Gurney–Lurie charts can be used conveniently for quick calculations of temperatures at specified points and of the average temperature by interpolating, as necessary, between the few curves plotted. For more precise calculations, the series solutions can be used directly remembering that the first term is sufficient in each case when τ is relatively large (e.g., for $\tau > 0.15$). Example 5.6 illustrates an application of these relationships.

Example 5.6. Transient heat transfer in the kitchen. Let's use the equations developed in this chapter to compare two different ways of cooking a carrot. The first one is to put the carrot on a rack in a convection oven set at 375°F. The second is called "*sous vide*" cooking (French for "under vacuum"). This style of cooking, whose ability to produce healthier and juicier food and with less mess than conventional cooking has been extolled, involves vacuum-sealing the carrot in a thin plastic baggie and submerging it in a water bath agitated and held at a constant temperature by constant recirculation of the water through an external heater. It has been suggested that using this method food can be cooked in a comparable time while avoiding the temperature extremes that can occur in a conventional oven. Consider a long carrot, 1.5 in. diameter. The carrot is initially at 40°F, and has a density of 40 lb/ft^3, a heat capacity of 0.91 BTU/lb °F, and a thermal conductivity of 0.32 BTU/h ft °F. In order to attain the desired texture, the carrot must reach a minimum temperature of 180°F. Let's first find the time required to cook the carrot in a convention oven assuming that the convective heat transfer coefficient is 5.0 BTU/h ft^2 °F. We first calculate Y. We want to get a centerline temperature T = 180°F. Thus, $Y = (180-375)/(40-375) = 0.58$. Next, we calculate the Biot number, $\text{Bi} = hR/k = 5\times(0.75/12)/0.32 = 0.98$. This number is close to 1 indicating that both external and internal resistances will be important. From Fig. 5.13B or, more precisely using M.T.5.3-B, for $Y = 0.58$ and $\text{Bi} = 0.98$, we find $\tau = \alpha t/R^2 = 0.46$. Since $\alpha = k/\rho C_p = 0.32/(40\times 0.91) = 0.0088$ ft^2/h, we will have $t = 0.46\times(0.75/12)^2/0.0088 = 0.20$ h. We can find the temperature at the surface of the carrot from the dashed line in Fig. 5.13B. At the same value of $\tau = 0.46$ with $\xi = 1$, we find $Y = 0.38$, which gives $T = 375 + 0.38\times(40-375) = 248$°F. Let's now

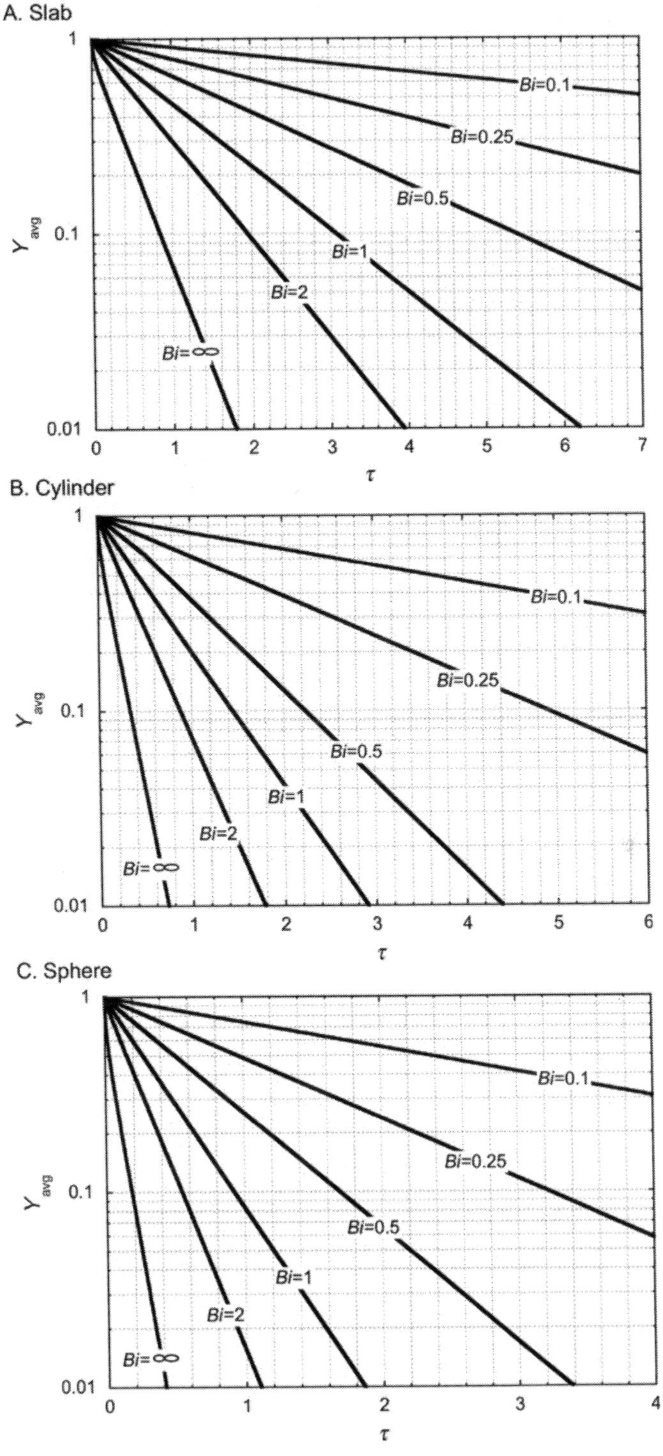

FIGURE 5.14 Dimensionless average temperature, $Y = (T_{avg} - T_a)/(T_0 - T_a)$, vs. dimensionless time τ at different values of the Biot number in: (A) a slab with thickness $2B$; (B) a long cylinder, with radius R; and (C) a sphere with radius R considering the external resistance (Bi < 10). $\tau = \alpha t/B^2$ and $\xi = x/B$ for the slab and $\tau = \alpha t/R^2$ and $\xi = r/R$ for the cylinder and the sphere.

look at the *sous vide* cooking method assuming that the water bath is kept at a constant temperature of 185°F and that the heat transfer coefficient is 500 BTU/h ft^2 °F. A higher value of h is expected compared to that used for the convection oven because we now have heat transfer from liquid water instead of air. We can neglect the heat transfer resistance of the thin plastic layer of the baggie. We now have $Y = (180-185)/(40-185) = 0.034$ and $\text{Bi} = 500 \times (0.75/12)/0.32 = 98$. This large value of Bi indicates that the external resistance will be negligible. From Fig. 5.13B at Bi = ∞ (or from M.T.5.2-B), we find $\tau = 0.66$ or $t = 0.66 \times (0.75/12)^2/0.0088 = 0.29$ h. Since the external resistance is negligible for this case, the surface temperature will be 185°F ($Y = 0$). So, in this analysis, *sous vide* will take a little longer, but it will indeed avoid the high surface temperature of the conventional oven, which may lead to poor texture and loss of nutrients. Note that for the conventional oven we have neglected radiant heat transfer, which, in practice, could speed up the cooking time.

5.4 Problems

P.5.1. We want to bond together two slabs of a solid material, each with a thickness of 1.5 cm, to form a laminate by using a thin layer of a thermosetting glue, which fuses and forms a bond between the two plates when the temperature is brought to 150°C. To accomplish this, the two plates are pressed between two metal plates maintained at a constant temperature of 200°C. (a) Determine how long will the two materials have to be pressed together between the metal plates if they are initially at 20°C using the full series solution. (b) Compare your answer in (a) with that obtained from Fig. 5.12 and with that obtained using just the first term in the series solution. Data: The thermal diffusivity of the solid material is 1.2×10^{-3} cm^2/s.

P.5.2. A stainless-steel plate, 1.0 m^2 in area and 5.0 cm thick, is to be attached to a glass substrate with a thermosetting glue. For this purpose, a thin layer of the glue is applied to the glass and the stainless-steel plate is then laid on top of it. The glue sets when the temperature reaches 120°C. The initial temperature is 20°C. Two cases are considered. In the first case (I), a metal plate kept at 150°C is firmly pressed against the stainless-steel plate. In the second case (II), air at 200°C is blown over the stainless-steel plate resulting in a heat transfer coefficient $h = 300$ W/m^2 K. Assuming that the glass substrate serves as a perfect insulator: (a) Write down the differential equations and initial and boundary conditions needed to describe each of the two cases making sure that you define all of the variables appearing in your equations. (b) Determine the time required to set the glue in case I as well as the amount of heat transferred to the stainless-steel plate when the glue setting temperature is reached. (c) Determine the time required to set the glue in case II as well as the temperature at the stainless steel air interface when the glue temperature reaches 150°C.

P.5.3. A long, cast-iron cylinder 0.2 m in diameter and initially at 95°C is cooled by contacting it with a stream of liquid water at a constant temperature of 25°C. The convective heat transfer coefficient is $h = 1{,}200$ W/m^2 K. (a) Determine the time needed for the cylinder to reach 55°C if the external resistance is assumed to be dominant. (b) Determine the time needed for the average temperature of the cylinder to reach 55°C if the internal resistance is assumed to be dominant. (c) Determine the time needed for the average temperature of the cylinder to reach 55°C if both internal resistance and external resistances are considered.

P.5.4. It is March 7th, the first day of Spring break, and you have not made it quite to beach destination. Perhaps you could have, except for that sheriff with the radar trap—but that is behind you as you shower in your motel room not too far from your final destination. In the morning you and your seven friends will be off bright and early in your compact car and you'll all be at the beach for a late afternoon swim and the beginning of a well-deserved Spring break—and so to bed.

You are startlingly awakened shortly before midnight by a loud knocking on the door. Good grief! It's the same sheriff you saw earlier in the day! "Sorry for what happened north of here today, folks, but I need your help. There has been a murder; my coroner has gone fishing, but I noticed a copy of your Heat & Mass Transfer book on the rear window deck of your vehicle and I have a thermometer. There are several suspects and your help is needed to tell me which one is the likely culprit," It seems that the victim, one John Lurie, was known as a dealer in hot cars, and may have been ready to confess to the police the roles of his accomplices, Gus Fourier and Bill Gurney. John's body was found on the beach, where the waves of the outgoing tide were gently washing over it. Gus has an iron-tight alibi for the day of the murder—he was with a group of gambling friends from dinner (6 PM) until interviewed by the sheriff an hour ago. Bill was seen shopping with a girlfriend from early morning until 6 PM, and the sheriff has been unable to locate him since. You accompany the sheriff to the beach, where you determine that John Lurie's centerline temperature is 86.5°F and the water temperature is 80°F. He looks as if he might be 5 ft, 8 or 9 in. tall and weigh 200 lb. Which of the two suspects should the sheriff apprehend? [Problem originally posed by A.B. Metzner, Univ. of Delaware].

P.5.5. A solid teflon cylinder 0.05 m in diameter and 1 m in length initially at a uniform temperature of 100°C is cooled by blowing air at 20°C over it. The Biot number is Bi = 2.0. (a) Determine the time needed for the centerline of the cylinder to reach 40°C. (b) Make a table and plot the temperature profile in the cylinder as a function of radial position from the center to its outer surface when the centerline temperature reaches 40°C. (c) Based on the results in (b), determine the average temperature in the cylinder using the appropriate "cylindrical average" by integrating the tabulated values in part (b). (d) Compare your result in (c) with that obtained directly from Fig. 5.12 or item B in Table 5.3.

P.5.6. Pyrex-glass spheres 2 cm in diameter and initially at 100°C are cooled by a stream of air at 30°C. The convective heat transfer coefficient is estimated to be 60 W/m^2 K. (a) Determine the controlling heat transfer resistance. (b) Determine the time needed for the center of the spheres to reach 50°C if the external resistance is assumed to be dominant. (c) Determine the time needed for the center of the spheres to reach 50°C if both resistances are considered. (d) Repeat your calculations in (c) if instead of an air stream we use a stream of liquid water at 30°C which gives a heat transfer coefficient of 1,500 W/m^2 K.

P.5.7. A long concrete beam with a rectangular cross-section measuring 8 × 12 in. is exposed to a temperature of 1,500°F in a fire. Using Newman's rule, determine the time of exposure needed for the temperature at the center of the beam to reach 700°F if the beam is at an initial temperature of 70°F. The thermal diffusivity of the concrete beam is 0.018 ft^2/h.

P.5.8. A long cylindrical fiber of a plastic material with thermal conductivity $k = 0.13$ W/m K, density $\rho = 1{,}400$ kg/m^3, and heat capacity $C_p = 1{,}050$ J/kg K is continuously extruded and cooled by pulling it through a chamber where air at 10°C flows perpendicular to the fiber at a speed of 10 m/s as sketched in Fig. P.5.8. The fiber, having a diameter of 3.0 mm, enters the chamber at a uniform temperature of 100°C, and is pulled through the chamber at the speed of 0.5 m/s. The convective heat transfer coefficient at the fiber surface is estimated to be $h = 100$ W/m^2 K. (a) Write down the differential energy balance and the proper initial and boundary conditions needed to describe the fiber cooling process. Include only the relevant terms assuming that the fiber is very long compared to its radius. There is no need to solve the equation. (b) Using available solutions, calculate the length of the chamber needed to attain a centerline fiber temperature of 30°C. Note that time = length/velocity. (c) Calculate the surface temperature of the fiber when it leaves the chamber.

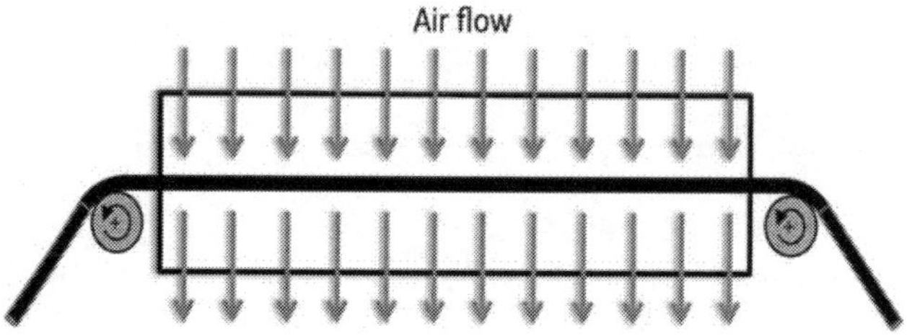

FIGURE P.5.8

P.5.9. A soft drink in a cup can be cooled without dilution using chilled plastic spheres that are 1.0 cm in diameter instead of ice. Assume that 300 mL of the spheres initially at –20°C are added to 300 mL of the soft drink initially at 25°C in a well-insulated cup. The relevant properties of the plastic material are thermal conductivity $k = 0.20$ W/m K, density $\rho = 1{,}400$ kg/m^3, and heat capacity $C_p = 1{,}800$ J/kg K. Neglecting the heat transfer resistance outside the spheres and assuming that the soft drink is well mixed, determine: (a) The lowest temperature to which the soft drink can be cooled for these conditions. (b) The time required to cool the soft drink to within 10% of the lowest possible temperature.

P.5.10. It has been claimed that "smart," hydrogel-filled double-pane windows can be used to shift energy usage in buildings to off-peak times by delaying the heat transfer as a result of the heat capacity of the hydrogel. Let's consider a double-pane window with a 5.0 mm gap filled with a polymer-based hydrogel having a thermal conductivity of 0.30 W/m K, and density of 1,050 kg/m^3, and a heat capacity of 4,000 J/kg K. The room temperature is kept at 25°C while the external temperature varies between 25°C at night to 35°C in the morning. (a) Assuming that the heat transfer resistance through the hydrogel is dominant, predict the heat transfer flux into the room as a function of time if, starting at 25°C, the external temperature suddenly reaches 35°C. Determine the cumulative heat transfer flux into the room over a 1-h period and compare your answer with the cumulative flux that would be obtained during the same time if energy accumulation in the hydrogel were neglected.

CHAPTER 6

Convective Heat Transfer

6.1 Introduction

As discussed in Chap. 1, we use the term "convective heat transfer" to describe the transport at the interface between two phases, usually a fluid and a solid phase, that are in relative motion with each other. Convective heat transfer, as defined here, is not to be confused with the bulk transport of energy by a flowing fluid. The latter can occur without a temperature gradient. On the other hand, convective heat transfer is driven by a temperature gradient at the interface between fluid and solid phases. When a fluid flows over a solid surface, a velocity gradient will exist at the surface with the velocity varying from zero at the solid surface to a bulk- or free-stream value away from the surface. The combination of conductive heat transfer in the flowing fluid and that resulting from the fluid motion near the interface is what we call convective heat transfer. As already noted, the convective heat transfer flux at a fluid–solid interface can be described by the following equation:

$$q_y\Big|_{y=0} = h(T_s - T_a) \tag{6.1}$$

where h is the convective heat transfer coefficient, typically measured in units of W/m^2 K or BTU/h ft^2°F, T_s is the surface temperature, and T_a is the bulk- or free-stream temperature of the fluid far away from the surface. Note that, in this notation, y is a coordinate perpendicular to the surface with $y = 0$ denoting the location of the surface itself. Examples of convective heat transfer are the exchange of heat between a fluid flowing over a flat surface, the exchange of heat between a fluid flowing in a pipe and the pipe wall, and the exchange of heat between a fluid flowing over a solid body and its surface. These are examples of forced convection, which are characterized by the fact that the fluid flow is generated mechanically (i.e., by a pump or a blower) and not dependent on the heat transfer process itself. Free convection, on the other hand, occurs when the flow originates from the heat transfer process itself. An example is the heat transfer between a hot surface and a cold gas. In this case, flow is generated by the density gradient created by the heat transfer process and the associated temperature gradient. Other special cases are the condensation of a vapor over a cold surface or the vaporization of a liquid over a hot surface. These more complicated processes involving changes of phase can result from a combination of forced and free convection effects.

The film model provides a simple representation useful to conceptualize convective heat transfer under many conditions. In this model, we assume that the convective heat transfer resistance is represented by the resistance offered by a stagnant film of fluid with thickness Δ adjacent to the solid surface. If we assume that only heat conduction takes place across this hypothetical stagnant film, we obtain the following result [*cf.* Eq. (3.4)]:

$$q_y\big|_{y=0} = \frac{k}{\Delta}(T_s - T_a) \tag{6.2}$$

where k is the thermal conductivity of the fluid. Comparing Eqs. (6.1) and (6.2), we obtain the following expression for the convective heat transfer coefficient:

$$h = \frac{k}{\Delta} \tag{6.3}$$

This equation is useful but only to some extent. For example, it does show that h is generally higher for a fluid with higher k and, thus, explains, at least partially, why the h-values for gases are much smaller than those observed for liquids, since gases have much lower values of k. On the other hand, in general, we do not know Δ. Based on Eq. (6.3), we can indeed say that, for example, faster flow over a surface will result in a higher h because the equivalent Δ will be smaller, but in fact, there really isn't a stagnant film in most practical systems and Eq. (6.3) is just an abstract model. Thus, by itself, this equation does not have predictive ability.

As seen in the previous chapters (e.g., see Table 1.2), convective heat transfer coefficients vary widely in magnitude depending on whether the fluid is a gas or a liquid, the specific physical properties of the fluid, the flow velocity, and the geometry of the surface. Although, heat transfer coefficients can be measured empirically, for example by measuring a heat transfer rate or a flux with a known temperature difference, the prediction and correlation of heat transfer coefficients is important for practical applications, especially given the enormous range over which h varies.

There are two distinct approaches that are useful for this purpose:

1. For laminar flow, h can be predicted either analytically, for simple geometries, or using computational methods (e.g., computational fluid dynamics or CFD) for complex systems.
2. For other cases, h can be predicted using generalized semi-empirical correlations of heat transfer data based on analogies between momentum transfer and heat transfer at fluid-solid interfaces (and, as we will see later, analogies between heat and mass transfer).

Purely empirical approaches are also useful, but they are limited to specific fluids or specific conditions and are difficult to generalize. In this chapter, we consider first analytical solutions for systems involving laminar flow of Newtonian fluids, which have both practical and conceptual value. We then explore methods for the prediction of h in different systems based on analogies. Rather than attempting to provide a comprehensive treatment of all available correlations, we will focus mainly on conceptual aspects and illustrative examples. Section 5 of *Perry's Chemical Engineers' Handbook*, 9th edition, provides a useful reference with extensive coverage of convective heat transfer for many practical cases.

6.2 Forced Convection in Laminar Flow

6.2.1 Laminar Flow Over a Flat Plate

The simplest case that we can consider is the process of steady-state heat transfer at the interface between a fluid and a wide flat plate. The situation is the same as that introduced in Sec. 1.3 and is illustrated in more detail in Fig. 6.1. As seen in Fig. 6.1, the fluid approaches the leading edge of the plate with velocity v_∞ and temperature T_∞, while the surface is kept at a uniform temperature T_s. This situation can be treated as a two-dimensional problem if the plate is wide and both velocity and temperature can be described in terms of the x and y coordinates only. Since the fluid velocity is zero at the surface and the surface temperature is T_s, velocity and temperature profiles will develop starting at the leading edge of the plate. The velocity will go from a value of zero at the surface to the value v_∞ far away from it. The temperature will go from the value T_s at the surface to the value T_∞ far from it. We refer to the distance from the surface where the velocity v approaches v_∞ as the momentum transport boundary layer, δ, and to the distance from the surface where T approaches T_∞ as the thermal boundary layer, δ_T. In general, δ and δ_T will be different from each other but they will both vary with distance from the edge of the plate (i.e., $\delta = \delta(x)$ and $\delta_T = \delta_T(x)$), both becoming larger as we move further downstream along the x-axis.

The equations needed to describe transport in this system are the differential mass balance (or continuity equation), the differential momentum balance, and the differential thermal energy balance. The general differential mass balance and the general differential energy balance were derived in Chap. 5 (see Table 5.1). The differential momentum balance can be found in many standard books and handbooks. It is given, for example, by equation 6.26 of *Perry's Chemical Engineers' Handbook*, 9th edition, for

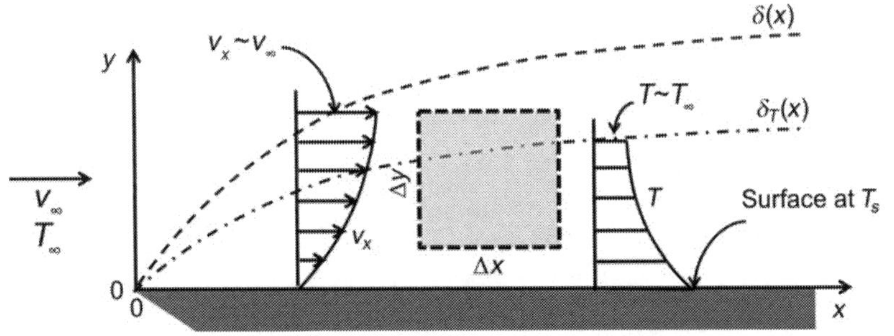

Figure 6.1 Convective heat transfer for flow past the edge of a wide flat plate. The volume used to derive differential momentum and energy balances has dimensions Δx and Δy as shown. $\delta(x)$ is the thickness of the momentum transport boundary layer, while $\delta_T(x)$ is the thickness of the thermal boundary layer.

the general case, and by equation 6.27 also in *Perry's Chemical Engineers' Handbook* for an incompressible Newtonian fluid with constant viscosity. The latter is called the Navier-Stokes equation. In this analysis, we limit our attention to an incompressible Newtonian fluid with constant density, viscosity, and thermal conductivity. We also assume that there is no energy generation within the fluid. While these assumptions are obviously restrictive, the ensuing analysis still has both conceptual and practical value for many systems.

For our system, in rectangular coordinates, with only x and y dependence of v_x, v_y, and T, the differential mass, momentum, and energy balances are as follows:

$$\frac{\partial v_x}{\partial x}+\frac{\partial v_y}{\partial y}=0 \tag{6.4}$$

$$v_x\frac{\partial v_x}{\partial x}+v_y\frac{\partial v_x}{\partial y}=\nu\left(\frac{\partial^2 v_x}{\partial x^2}+\frac{\partial^2 v_x}{\partial y^2}\right) \tag{6.5}$$

$$v_x\frac{\partial T}{\partial x}+v_y\frac{\partial T}{\partial y}=\alpha\left(\frac{\partial^2 T}{\partial x^2}+\frac{\partial^2 T}{\partial y^2}\right) \tag{6.6}$$

where $\nu=\mu/\rho$ is the kinematic viscosity and $\alpha=k/\rho C_p$ is the fluid's thermal diffusivity. Note that for the momentum balance, we consider only the x-component, Eq. (6.5), which is sufficient here for illustrative purposes. It is important to recognize, however, that while the main component of the velocity vector is v_x, because of Eq. (6.4), the velocity will also have a finite y-component, v_y, everywhere except right at the surface where both v_x and v_y are zero. Without much loss of generality, we can neglect the first term in parenthesis on the right-handside of both Eqs. (6.5) and (6.6). In Eq. (6.5), the term $\partial^2 v_x/\partial x^2$ arises from the contribution of viscous transport to total momentum transport in the x-direction. This contribution will typically negligible, since momentum transport in this direction will be dominated by bulk flow and, thus, v_x. Similarly, in Eq. (6.6), the term $\partial^2 T/\partial x^2$ arises from the contribution of conductive transport to total thermal energy transport in the x-direction. Again, since flow is mainly along x, energy transport in this direction will be dominated by the bulk flow and, thus, v_x. Not so, of course, for either momentum or heat transfer in the y-direction along which viscous momentum transport and conductive heat transfer will both be important. The simplified forms of Eqs. (6.5) and (6.6) are as follows:

$$v_x\frac{\partial v_x}{\partial x}+v_y\frac{\partial v_x}{\partial y}=\nu\frac{\partial^2 v_x}{\partial y^2} \tag{6.7}$$

$$v_x\frac{\partial T}{\partial x}+v_y\frac{\partial T}{\partial y}=\alpha\frac{\partial^2 T}{\partial y^2} \tag{6.8}$$

These equations are subject to the following boundary conditions:

$$\begin{aligned}
&x=0: v_x=v_\infty, T=T_\infty\\
&y=0: v_x=0, T=T_s\\
&y\to\infty: v_x=v_\infty, T=T_\infty
\end{aligned} \tag{6.9}$$

The similarity of these equations and boundary conditions for momentum and heat transfer is already striking, but we can make them even more similar to each other by introducing the following dimensionless variables:

$$u_x = \frac{v_x}{v_\infty}; \quad u_y = \frac{v_y}{v_\infty} \tag{6.10}$$

$$\theta = \frac{T - T_s}{T_\infty - T_s} \tag{6.11}$$

u_x and u_y are the dimensionless components of the velocity vector and θ is a dimensionless temperature. Inserting these variables in Eqs. (6.7) to (6.9), we obtain:

$$u_x \frac{\partial u_x}{\partial x} + u_y \frac{\partial u_x}{\partial y} = \frac{\nu}{v_\infty} \frac{\partial^2 u_x}{\partial y^2} \tag{6.12}$$

$$u_x \frac{\partial \theta}{\partial x} + u_y \frac{\partial \theta}{\partial y} = \frac{\alpha}{v_\infty} \frac{\partial^2 \theta}{\partial y^2} \tag{6.13}$$

$$\begin{aligned} x &= 0 : u_x = \theta = 1 \\ y &= 0 : u_x = \theta = 0 \\ y &\to \infty : u_x = \theta = 1 \end{aligned} \tag{6.14}$$

Note that, in this form, the boundary conditions for Eqs. (6.12) and (6.13) are exactly the same. Moreover, these two equations too become exactly the same for a fluid for which $\nu = \alpha$. In this case, since the two equations are the same and they have the same boundary conditions, we will also have the same solution for the momentum and the heat transfer problems. In other words, the solution of Eq. (6.12), which is a function $u_x(x, y)$, will also be the solution of Eq. (6.13), which is the function $\theta(x, y)$ that we are looking for.

The case of $\nu = \alpha$ is, in reality, practically important. For simplicity, we express the ratio of ν and α in terms of the following quantity called the Prandtl number:

$$\text{Pr} = \frac{\nu}{\alpha} = \frac{\mu C_p}{k} \tag{6.15}$$

When $\text{Pr} = 1$, we have $\nu = \alpha$ and momentum and heat transfer problems will have the same solution. We will see that for gases at ambient conditions, Pr is usually close to 1. Conversely, for liquids, Pr is usually closer to 10 or larger.

The solution of Eq. (6.12) subject to boundary conditions (6.14) is given by the Blasius theory and is shown in Fig. 6.2. This theoretical result expresses the dimensionless velocity u_x as a unique function of the following variable:

$$\eta = \frac{y}{2}\left(\frac{v_\infty}{\nu x}\right)^{1/2} \tag{6.16}$$

which combines x and y into a single dimensionless quantity. For a given x, η is proportional to the distance from the surface along the y-axis. As seen in Fig. 6.2, $u_x \sim 1$ when

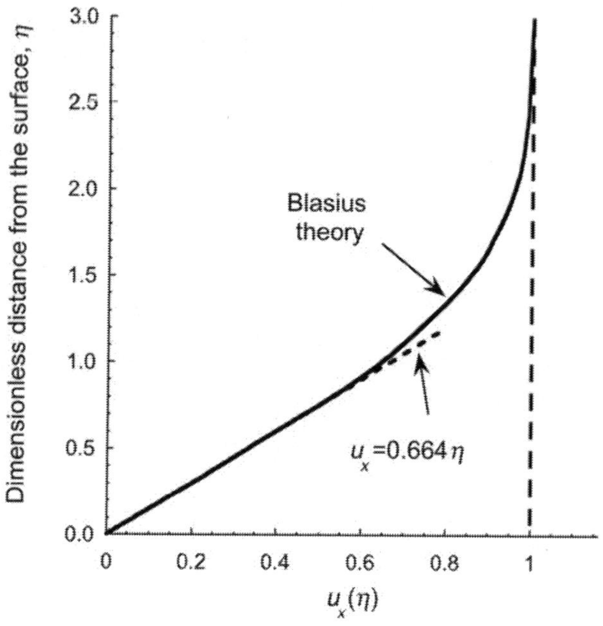

FIGURE 6.2 Dimensionless velocity profile, $u_x = v_x/v_\infty$, according to Blasius theory for laminar flow past the edge of a flat plate.

$\eta \sim 2.5$. Thus, this value of η can be used to define the thickness of the momentum transport boundary layer.

Dimensionally, setting $\eta \sim 2.5$ in Eq. (6.16), we obtain:

$$2.5 = \frac{y}{2}\left(\frac{v_\infty}{\nu x}\right)^{1/2} \tag{6.17}$$

or

$$\delta = 5\left(\frac{\nu x}{v_\infty}\right)^{1/2} \tag{6.18}$$

If $\text{Pr} = 1$, we will have $u_x = \theta$ and thus $\delta_T = \delta$.

Let's now find the local heat transfer coefficient, h_x, at a particular point at a distance x from the leading edge of the plate according to the following definition:

$$q_y\Big|_{y=0} = h_x(T_s - T_\infty) \tag{6.19}$$

Since the fluid velocity is zero at the surface, we can also write the flux at the surface according to Fourier's law, which gives:

$$q_y\Big|_{y=0} = -k\frac{\partial T}{\partial y}\Big|_{y=0} \tag{6.20}$$

Intuitively, we expect that h_x will be larger near the edge of the plate where the temperature gradient in the y direction is steep and heat transfer is locally fast, and smaller downstream where the temperature gradient in the y direction is shallower and, thus, heat transfer is locally slower. Quantitatively, since Eqs. (6.19) and (6.20) must give the same result, we can find the relationship between h_x and the temperature gradient by equating these two expressions for q_y, which gives:

$$h_x = -\frac{k}{T_s - T_\infty}\left(\frac{\partial T}{\partial y}\right)_{y=0} \tag{6.21}$$

In terms of dimensionless temperature, this result can be expressed as follows:

$$h_x = k\left(\frac{\partial \theta}{\partial y}\right)_{y=0} \tag{6.22}$$

If $\Pr = 1$, we will have $\theta = u_x(\eta)$. In this case, using the chain rule of differentiation, we obtain:

$$h_x = k\frac{\partial \eta}{\partial y}\left(\frac{d\theta}{d\eta}\right)_{y=0} = \frac{k}{2}\left(\frac{v_\infty}{\nu x}\right)^{1/2}\left(\frac{d\theta}{d\eta}\right)_{\eta=0} = \frac{k}{2}\left(\frac{v_\infty}{\nu x}\right)^{1/2}\left(\frac{du_x}{d\eta}\right)_{\eta=0} \tag{6.23}$$

From Fig. 6.2, we can see that the function $u_x(\eta)$ is linear at $\eta = 0$ with a slope $(du_x/d\eta) = 0.664$. Substituting this value for the derivative in Eq. (6.23), we finally obtain:

$$h_x = 0.332k\left(\frac{v_\infty}{\nu x}\right)^{1/2} \tag{6.24}$$

from which we can predict the local heat transfer coefficient. We are also often interested in predicting the average heat transfer coefficient over a distance $x = L$. This quantity can be found by integrating along the distance x as follows:

$$h = \frac{1}{L}\int_0^L h_x(x)dx = \frac{1}{L}\int_0^L 0.332k\left(\frac{v_\infty}{\nu x}\right)^{1/2} dx = 0.664k\left(\frac{v_\infty}{\nu L}\right)^{1/2} \tag{6.25}$$

This is useful because it allows us to calculate the heat transfer rate for the entire plate by simply multiplying h by the temperature difference between the surface and the bulk stream and by the surface area of the plate. Accordingly, the heat transfer rate over a distance L from the leading edge of a plate of width W will be:

$$Q_y\Big|_{y=0} = 0.664WLk\left(\frac{v_\infty}{\nu L}\right)^{1/2}(T_s - T_\infty) \tag{6.26}$$

It is also useful to express these quantities in terms of dimensionless variables, which, on the one hand, will allow us to express the results in a more compact notation while, on the other, will allow us to generalize the conclusions. The local Reynolds number at a distance x from the leading edge is defined as follows:

$$\mathrm{Re}_x = \frac{\rho v_\infty x}{\mu} = \frac{v_\infty x}{\nu} \tag{6.27}$$

Accordingly, Eqs. (6.18) and (6.24) become:

$$\delta = \delta_T = 5x(\mathrm{Re}_x)^{-1/2} \tag{6.28}$$

$$\mathrm{Nu}_x = \frac{h_x x}{k} = 0.332(\mathrm{Re}_x)^{1/2} \tag{6.29}$$

The quantity $\mathrm{Nu}_x = h_x x/k$ is called the local Nusselt number. Similarly, averaging over the distance L, with the Reynolds number defined as:

$$\mathrm{Re} = \frac{\rho v_\infty L}{\mu} = \frac{v_\infty L}{\nu} \tag{6.30}$$

we obtain:

$$\mathrm{Nu} = \frac{hL}{k} = 0.664(\mathrm{Re})^{1/2} \tag{6.31}$$

where $\mathrm{Nu} = hL/k$ is the average Nusselt number.

The final question is what happens for cases where the Prandtl number is not equal to 1. One can, of course, substitute the Blasius solution in the differential energy balance, Eq. (6.13), and solve the resulting equation numerically, which is obviously complicated. In practice, however, the following empirical modification of Eqs. (6.29) and (6.31) is usually sufficient. If we assume that:

$$\frac{\delta_T}{\delta} = (\mathrm{Pr})^{-1/3} \tag{6.32}$$

we obtain:

$$\mathrm{Nu}_x = 0.332(\mathrm{Re}_x)^{1/2}(\mathrm{Pr})^{1/3} \tag{6.33}$$

and

$$\mathrm{Nu} = 0.664(\mathrm{Re})^{1/2}(\mathrm{Pr})^{1/3} \tag{6.34}$$

These expressions are consistent with experimental results and of sufficient accuracy for many practical applications. Note that, as mentioned above, for gases we have $\mathrm{Pr} \sim 1$, indicating that the momentum and thermal boundary layers will have essentially the same thickness. On the other hand, for liquids, since $\mathrm{Pr} > 1$, the thermal boundary layer will usually be thinner than the momentum boundary layer at the same Reynolds number.

Example 6.1. Heat transfer coefficient for laminar flow over a flat plate. Let's use the equations introduced in this section to predict the heat transfer coefficient for flow of air at 1.0 m/s over a 0.5 m long plate at 1 atm and 25°C. For these conditions, from Appendix A, the mass density of air is $\rho = 1.18$ kg/m^3 and its viscosity is $\mu = 1.84 \times 10^{-5}$ Pa s. Thus, the kinematic viscosity is $\nu = \mu/\rho = 1.56 \times 10^{-5}$ m^2/s. The thermal conductivity of air for these conditions is $k = 0.026$ W/m K and its heat capacity is 1,000 J/kg K. Thus, the thermal diffusivity will be $\alpha = k/\rho C_p = 0.026/(1.18 \times 1{,}000) = 2.2 \times 10^{-5}$ m^2/s. As a result, we find $\mathrm{Pr} = \nu/\alpha = 0.71$. The Reynolds number at $L = 0.5$ m is $\mathrm{Re} = 1.0 \times 0.5/1.56 \times 10^{-5} = 3.2 \times 10^4$. For flow over a flat plate, laminar flow conditions will exist for $\mathrm{Re} < 10^6$. Thus, we will have:

$$\delta = 5 \times 0.5(3.2 \times 10^4)^{-1/2} = 0.014 \text{ m}$$

$$\delta_T = 0.014(0.71)^{-1/3} = 0.016 \text{ m}$$

$$\text{Nu} = 0.664(3.2\times10^4)^{1/2}(0.71)^{1/3} = 106$$

$$h = \frac{k}{L}\text{Nu} = \frac{0.026}{0.5}\times 106 = 5.51\ \text{W/m}^2\ \text{s}$$

Let's compare these values with the values that we would obtain for liquid water at 25°C. In this case, we will have $\rho = 990\ \text{kg/m}^3$ and $\mu = 9.2\times10^{-4}$ Pa s. Thus, $\nu = 9.2\times10^{-4}/990 = 9.3\times10^{-7}\ \text{m}^2/\text{s}$. The thermal conductivity of water at 25°C is $k = 0.606$ W/m K and its heat capacity is 4,200 J/kg K. Thus, the thermal diffusivity will be $\alpha = 0.606/(990\times 4{,}200) = 1.45\times10^{-7}\ \text{m}^2/\text{s}$. As a result, we find $\text{Pr} = 9.3\times10^{-7}/1.45\times10^{-7} = 6.4$. The Reynolds number at $L = 0.5$ m will be $\text{Re} = 1.0\times0.5/9.2\times10^{-7} = 5.4\times10^5$. With these values we obtain:

$$\delta = 5\times0.5(5.4\times10^5)^{-1/2} = 0.0034\ \text{m}$$

$$\delta_T = 0.0034(6.4)^{-1/3} = 0.0018\ \text{m}$$

$$\text{Nu} = 0.664(5.4\times10^5)^{1/2}(6.4)^{1/3} = 906$$

$$h = \frac{k}{L}\text{Nu} = \frac{0.606}{0.5}\times 906 = 1{,}090\ \text{W/m}^2\ \text{s}$$

As expected, for flow of water, the boundary layers are much thinner compared to those predicted for air, the Nusselt number is much higher, and, most importantly, the heat transfer coefficient is much higher compared to the value obtained for flow of air. This result comes about because of the combination of a higher thermal conductivity of liquid water compared to air and of a thinner thermal boundary layer. Note that in all of these calculations we have used the physical properties of the fluid. The heat transfer coefficient is independent of what material the plate is made of provided that the boundary condition $T = T_s$ is satisfied at its surface. An important question is what temperature should the physical properties be evaluated at for such calculations. We can see that the dependence of Nu on Pr is very weak (only 1/3 power). Even the dependence on Re is weak (1/2 power). As a result, usually we do not really need to use extremely precise values of the physical properties for the estimation of Pr and Re. On the other hand, the heat transfer coefficient becomes proportional to k for given Re and Pr. As shown in Chap. 2, for liquids, the dependence of k on temperature is usually small. For gases, on the other hand, this dependence is linear but the slope is not very large. As a result, practical calculations of h from these equations can often be done with adequate accuracy using properties evaluated at a temperature averaged between the surface and the bulk temperature values.

6.2.2 Laminar Flow in a Circular Pipe

In this section, we consider convective heat transfer between a fluid flowing in a pipe and the pipe wall as sketched in Fig. 6.3. We assume that the fluid is Newtonian and that its physical properties (density, viscosity, and thermal conductivity) remain constant. We also

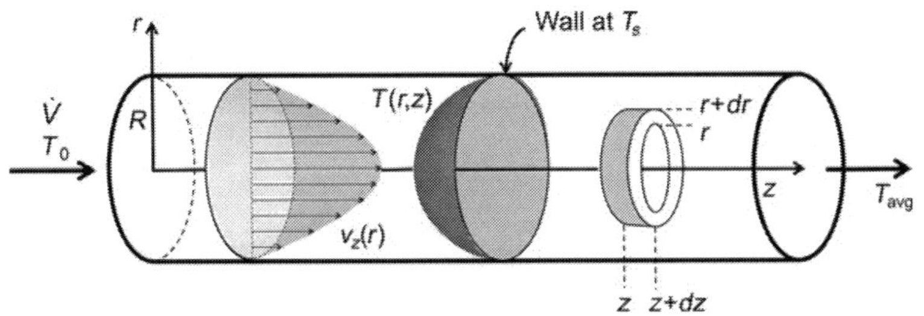

FIGURE 6.3 Velocity and temperature profiles for fully developed laminar flow in a circular pipe.

assume that we have a fully developed velocity profile. Obviously, these assumptions are restrictive, but they are still found to be adequate for many practical systems.

Let's consider first the fluid mechanics in this system. Under these conditions, the velocity profile will be parabolic and, thus, described by the following equation as a function of the radial coordinate r:

$$v_z(r) = v_{z,\max}\left[1-\left(\frac{r}{R}\right)^2\right] \tag{6.35}$$

where $v_{z,\max}$ is the maximum velocity found at the pipe center. The velocity will be zero at the wall, which is a result of the no-slip boundary condition. The average velocity, v_{avg}, is defined in such a way that:

$$\dot{V} = \pi R^2 v_{avg} \tag{6.36}$$

where $\dot{V}$ is the volumetric flow rate. We can find v_{avg} by integrating the velocity profile across the radius. With reference to the differential volume with widths dr and dz sketched in Fig. 6.3, the quantity $2\pi r v_z(r)dr$ represents the differential flow rate passing through this volume. Integrating this quantity across the radius of the pipe will give the total volumetric flow rate. Thus, from Eq. (6.36) we obtain:

$$v_{avg} = \frac{2}{R^2}\int_0^R r v_z(r)\,dr \tag{6.37}$$

Combining Eqs. (6.35) and (6.37) gives the following result:

$$v_{avg} = \frac{2}{R^2}\int_0^R r v_{z,\max}\left[1-\left(\frac{r}{R}\right)^2\right]dr = \frac{1}{2}v_{z,\max} \tag{6.38}$$

This equation shows that the velocity in the center of the pipe will be double the value of the average velocity. Whether the flow will be laminar or not will depend on the value of the Reynolds number, which, for flow in a circular pipe with inner diameter $D = 2R$, is defined as:

$$\text{Re} = \frac{\rho v_{avg} D}{\mu} = \frac{v_{avg} D}{\nu} \tag{6.39}$$

Laminar flow will require that $\text{Re} < 2{,}100$.

Next, we turn our attention to the problem of heat transfer between the wall, which we assume is kept at a uniform temperature T_s and the fluid that enters the pipe at a uniform temperature T_0. As sketched in Fig. 6.3, as a result of heat transfer between the wall and the fluid, the fluid temperature, T, will change as we move down the pipe becoming a function of both r and z. Just like we did for the average velocity, it is useful to define an average temperature, T_{avg}, for the fluid at each axial distance z from the pipe entrance. We define T_{avg} in such a way that the quantity $\dot{V}\rho C_p T_{avg}(z)$ will be the thermal energy relative to T_0 carried by the flowing stream at a particular value of z. With reference again to the differential volume with sides dr and dz in Fig. 6.3, the quantity $2\pi r v_z(r)dr \times \rho C_p T(r,z)$ represents the differential amount of thermal energy carried by the fluid through this volume. In a manner analogous to the calculation of the average velocity, integrating this quantity over the pipe radius will give us the total thermal energy carried by the fluid at a given z-location. Thus,

$$T_{\text{avg}}(z) = \frac{2\pi}{\dot{V}} \int_0^R r v_z(r) T(r,z)\,dr = \frac{2}{v_{z,\text{avg}} R^2} \int_0^R r v_z(r) T(r,z)\,dr \tag{6.40}$$

Combining this equation with Eq. (6.35) gives the final result:

$$T_{\text{avg}}(z) = \frac{4}{R^2} \int_0^R r \left[1 - \left(\frac{r}{R}\right)^2\right] T(r,z)\,dr \tag{6.41}$$

T_{avg} is called the "mixing-cup average" temperature because it is the temperature that we would measure if the fluid leaving the pipe at a distance z from the entrance flowed into a cup or tank becoming completely mixed. The outflow from this mixing cup will be at a temperature equal to T_{avg}. Typically, this is the temperature that really matters when, for example, we want to determine the length of a heat exchanger tube.

Finally, we consider the differential energy balance for the fluid in the pipe, which, together with appropriate boundary conditions, will allow us to actually find T_{avg}. We can derive this balance with reference to the differential control volume with sides dr and dz sketched in Fig. 6.3. Alternatively, we can start from Table 5.1. From item B in Table 5.1, assuming steady-state, that there is no heat generation, and that the temperature is axisymmetric (i.e., $\partial T/\partial\theta = 0$), in cylindrical coordinates we will have:

$$v_z(r)\frac{\partial T}{\partial z} = \alpha\left[\frac{1}{r}\frac{\partial}{\partial r}\left(r\frac{\partial T}{\partial r}\right) + \frac{\partial^2 T}{\partial z^2}\right] \tag{6.42}$$

As we did for the case of flow over a flat plate, we can neglect the term $\partial^2 T/\partial z^2$ on the right-hand side of this equation. This term arises from the conductive heat transfer in the z-direction, which is usually small compared to the energy transport associated with the bulk flow of the fluid. Accordingly, we obtain the following simplified equation and boundary conditions:

$$v_{z,\text{max}}\left[1 - \left(\frac{r}{R}\right)^2\right]\frac{\partial T}{\partial z} = \frac{\alpha}{r}\frac{\partial}{\partial r}\left(r\frac{\partial T}{\partial r}\right) \tag{6.43}$$

$$\begin{aligned} z = 0 &: T = T_0 \\ r = R &: T = T_s \\ r = 0 &: \frac{\partial T}{\partial r} = 0 \end{aligned} \tag{6.44}$$

The first of these boundary conditions sets that the entrance temperature is uniform at T_0. The second sets the fluid interfacial temperature equal to the tube wall temperature T_s. Finally, the last of these boundary conditions specifies that the temperature profile will be symmetrical with respect to r or, equivalently, that the heat transfer flux will be zero at the pipe's centerline. Solving Eq. (6.43) along with boundary conditions in Eq. (6.44) will provide the temperature profile in the fluid as a function $T(r,z)$. Using this function we will then be able to calculate $T_{avg}(z)$ from Eq. (6.41) and use it, in turn, to calculate the local convective heat transfer coefficient, h_z. The latter is defined by the following equation:

$$q_r\big|_{r=R} = h_z(T_{avg} - T_s) \tag{6.45}$$

In a manner analogous to the case of heat transfer for flow past the edge of a flat plate, we can also write the heat transfer flux in terms of the radial temperature gradient according to Fourier's law:

$$q_r\big|_{r=R} = -k\frac{\partial T}{\partial r}\bigg|_{r=R} \tag{6.46}$$

Equating these two expressions for q_r at the pipe wall, we obtain the heat transfer coefficient:

$$h_z = -\frac{k}{(T_{avg} - T_s)}\frac{\partial T}{\partial r}\bigg|_{r=R} \tag{6.47}$$

Note that Eq. (6.45) expresses the flux at the pipe wall in terms of a driving force that is equal to the difference between the pipe wall temperature and the mixing-cup average temperature of the fluid at each z-location. Obviously, h_z will vary along z from the entrance of the pipe down its length. As was the case for flow past the edge of a flat plate, in this case too we expect that h_z will be larger near the pipe entrance, where the radial temperature gradient at the pipe wall is steeper, and smaller downstream, where the radial temperature gradient at the pipe wall becomes shallower.

The solution of Eq. (6.43) with boundary conditions in Eq. (6.44) is complicated, but was obtained a long time ago by Graetz in 1882 as a series solution. In dimensionless form, the solution for the temperature profile is as follows:

(M.E.6.48)
$$Y = \sum_{n=1}^{\infty} A_n \exp(-\lambda_n^2\zeta)\exp(-\lambda_n^2\rho/2)M_n[a,b,c] \tag{6.48}$$

where

$$Y = \frac{T - T_s}{T_0 - T_s} \tag{6.49}$$

$$\zeta = \frac{z}{R}\frac{1}{\text{RePr}} \tag{6.50}$$

$$\rho = \frac{r}{R} \tag{6.51}$$

$M[a,b,c]$ is the confluent hypergeometric function (also known as the Kummer function) of the three arguments:

$$a = \frac{1}{2} - \frac{\lambda_n}{4}$$
$$b = 1 \tag{6.52}$$
$$c = \lambda_n\rho^2$$

This function is expressed by an infinite series (e.g., see Abramowitz and Stegun, *Handbook of Mathematical Functions*, Dover, New York, 1965) and is also available as a special function in mathematical packages such as `Hypergeometric1F1[a,b,c]` in Mathematica Version 12.1 or higher (Wolfram Research, Inc., Champaign, IL, 2020). The values of λ_n are obtained as the roots of the following equation:

$$M_n\left[\frac{1}{2}-\frac{\lambda_n}{4},1,\lambda_n^2\right]=0 \tag{6.53}$$

while the values of A_n are calculated as follows:

$$A_n=\frac{\int_0^1 \exp(-\lambda_n\rho^2/2)M_n\left[\frac{1}{2}-\frac{\lambda_n}{4},1,\lambda_n\rho^2\right]\rho(1-\rho^2)d\rho}{\int_0^1\left\{\exp(-\lambda_n\rho^2/2)M_n\left[\frac{1}{2}-\frac{\lambda_n}{4},1,\lambda_n\rho^2\right]\right\}^2\rho(1-\rho^2)d\rho} \tag{6.54}$$

The first 10 values of λ_n and A_n are:

n	1	2	3	4	5	6	7	8	9	10
λ_n	2.704	6.679	10.67	14.67	18.67	22.67	26.67	30.67	34.67	38.67
A_n	1.476	−0.806	0.589	−0.476	0.405	−0.356	0.319	−0.291	0.268	−0.249

The mixing-cup average temperature and the heat transfer coefficients are calculated by inserting Eq. (6.48) into Eqs. (6.41) and (6.47), respectively, which yields:

(M.E.6.55)

$$\begin{aligned}Y_{\text{avg}}&=\frac{T_{\text{avg}}-T_s}{T_0-T_s}=4\int_0^1\rho(1-\rho^2)Y(\rho,\zeta)d\rho\\&=4\sum_{n=1}^{\infty}A_n\exp(-\lambda_n^2\zeta)\int_0^1\rho(1-\rho^2)\exp(-\lambda_n^2\rho/2)M_n\left(\frac{1}{2}-\frac{\lambda_n}{4},1,\lambda_n^2\rho\right)d\rho\end{aligned} \tag{6.55}$$

and

(M.E.6.56)

$$\text{Nu}_z=\frac{h_zD}{k}=\frac{\sum_{n=1}^{\infty}A_n\exp(-\lambda_n^2\zeta)\exp(-\lambda_n/2)\left.\frac{dM_n}{d\rho}\right|_{\rho=1}}{2\sum_{n=1}^{\infty}A_n\exp(-\lambda_n^2\zeta)\frac{\exp(-\lambda_n/2)}{\lambda_n^2}\left.\frac{dM_n}{d\rho}\right|_{\rho=1}} \tag{6.56}$$

Finally, we can calculate the mass transfer coefficient averaged over a pipe of a length L as:

$$h=\frac{1}{L}\int_0^L h_z(z)dz \tag{6.57}$$

or, in terms of the average Nusselt number as:

(M.E.6.58)

$$\text{Nu}=\frac{1}{Z}\int_0^1 Nu_z(\zeta)d\zeta \tag{6.58}$$

where

$$Z=\frac{L}{R}\frac{1}{\text{RePr}} \tag{6.59}$$

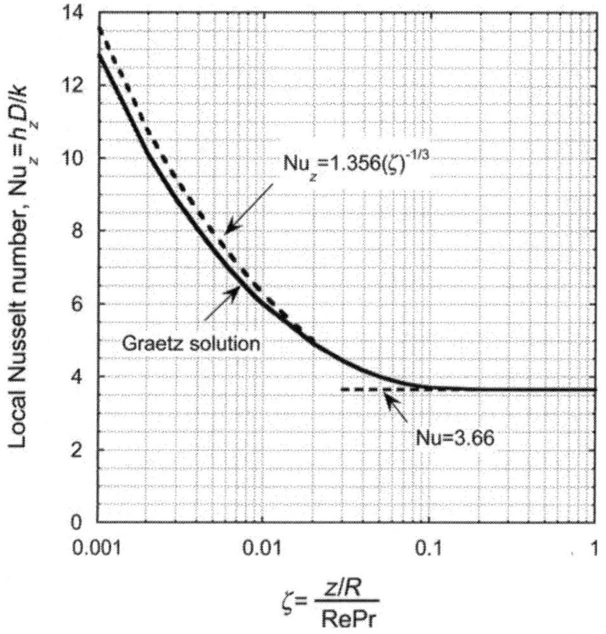

FIGURE 6.4 Local Nusselt number vs. dimensionless position along the pipe length for fully developed laminar flow in a circular pipe with uniform wall temperature.

Figures 6.4 and 6.5 show the local and average Nusselt number along the dimensionless length of the pipe based on Eqs. (6.56) and (6.58), respectively, using 20 terms in the series solution. As seen in Fig. 6.4, the local Nusselt number decreases along the length of the pipe, which means that the heat transfer flux will become smaller for the same $T_s - T_{avg}$. This occurs because as we move down the pipe, the temperature profile will penetrate deeper toward the center of the pipe. As a result, the temperature gradient at the pipe wall, $(\partial T/\partial r)_{r=R}$, will become shallower. As seen in Fig. 6.4, for long pipes, when $z/(\text{RePr}\,R) > 0.1$, the Nusselt number approaches the constant value $\text{Nu}_z = 3.66$. The average Nusselt number, shown in Fig. 6.5, follows the same trend also becoming approximately equal to 3.66 when $L/(\text{RePr}\,R) > 0.1$.

The calculation of the Graetz series solution is not straightforward when $z/(\text{Re Pr}\,R)$ is small since the series becomes slowly convergent and many terms are needed. On the other hand, for $z/(\text{RePr}\,R) > 0.1$ one term is sufficient giving:

$$Y_{avg} \sim 0.82\exp(-2.704^2\zeta) = 0.82\exp\left(-7.31\frac{z/R}{\text{RePr}}\right) \tag{6.60}$$

and

$$\text{Nu}_z \sim \text{Nu} \sim 3.66 \tag{6.61}$$

For smaller values of $z/(\text{RePr}\,R)$, Middleman, *An Introduction to Mass and Heat Transfer*, Wiley, 1998, recommends:

$$\begin{aligned} \text{Nu}_z &= 1.356(\zeta)^{-1/3} \quad \text{for} \quad \zeta < 0.02 \\ \text{Nu} &= 2.033(\text{Z})^{-1/3} \quad \text{for} \quad \text{Z} < 0.06 \end{aligned} \tag{6.62}$$

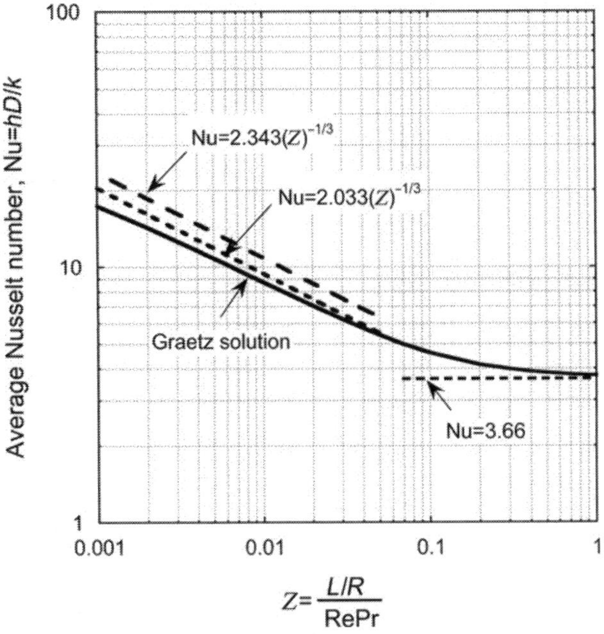

FIGURE 6.5 Nusselt number averaged over the length of the pipe vs. dimensionless pipe length for fully developed laminar flow in a circular pipe with uniform wall temperature.

Sieder and Tate [Ind. Eng. Chem. 28 (1936) 1429] recommend the following equation modified based on experimental data:

$$\mathrm{Nu} = 1.86\left(\frac{L/D}{\mathrm{Re\,Pr}}\right)^{-1/3}\left(\frac{\mu}{\mu_s}\right)^{0.14} = 2.343(\mathrm{Z})^{-1/3}\left(\frac{\mu}{\mu_s}\right)^{0.14} \quad \text{for } \mathrm{Z} < 0.2 \tag{6.63}$$

This equation has the same form as those recommended by Middleman, but a different numerical coefficient. The term μ/μ_s in this equation is the ratio of the viscosity of the fluid at the average temperature and its viscosity at the temperature of the surface. Since this term is to the 0.14 power, the effect of this ratio is generally small. As seen in Figs. 6.4 and 6.5, these approximate relationships are close to the theoretical Graetz limit and are adequate for practical estimates.

The prediction of convective heat transfer in pipes with laminar flow is also possible for other conditions at the pipe surface. For example, instead of assuming that the wall is at a constant temperature, one could assume that the heat transfer flux at the wall is constant. Analytical solutions are available also for such cases. However, they do not deviate a lot from the solution above with a constant wall temperature and are generally within about 20% of this solution. For example, with a constant heat flux at the wall, at large values of Z (i.e., in a long tube) one finds Nu ~ 4.36, which is about 20% higher than the value of 3.66 found above.

Example 6.2. Pipe length required to cool a fluid in laminar flow. Let's consider the calculation of the length of pipe needed to cool the clarified supernatant of a cell culture process used to produce a monoclonal antibody from its initial temperature of 37°C to 8°C flowing at a rate of 0.5 L/min in a 5 mm ID tube.

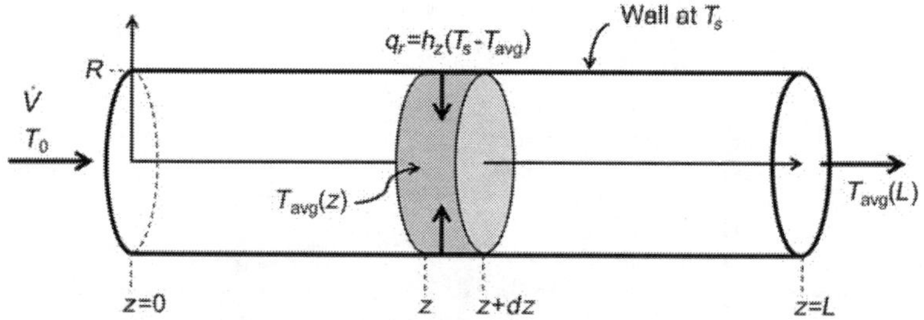

FIGURE 6.6 Heat transfer between a fluid flowing entering a circular pipe at a uniform temperature T_0 and the pipe wall kept at a uniform temperature T_s.

The tube is immersed in an ice bath that keeps its surface temperature at 0°C. The relevant physical properties are $\rho = 1{,}000\ \text{kg/m}^3$, $\nu = 5.0\times 10^{-6}\ \text{m}^2/\text{s}$, $k = 0.6\ \text{W/m K}$, $C_p = 4{,}000\ \text{J/kg K}$. To solve this problem, let's consider the tube sketched in Fig. 6.6. An energy balance for the differential volume with thickness dz shown in Fig. 6.6 gives the following result:

$$\dot{V}\rho C_p dT_{avg} = -\pi D h_z(z)(T_{avg} - T_s)dz$$

where $h_z(z)$ is the local heat transfer coefficient. Separating the variables T_{avg} and z in this equation and integrating, we obtain:

$$\int_{T_0}^{T_{avg}(L)} \frac{dT_{avg}}{T_{avg} - T_s} = -\frac{\pi D}{\dot{V}\rho C_p}\int_0^L h_z(z)dz \tag{6.64}$$

or, carrying out the integrals:

$$\frac{T_{avg}(L) - T_s}{T_0 - T_s} = \exp\left(-\frac{\pi D h L}{\dot{V}\rho C_p}\right) \tag{6.65}$$

where $T_{avg}(L)$ is the mixing-cup average temperature at the pipe exit ($z = L$) and h is the heat transfer coefficient averaged over the pipe length L. Note that this result is obtained regardless of whether h_z is constant or not or whether the flow is laminar or turbulent. Since $\pi DhL/\dot{V}\rho C_p = 2L\text{Nu}/(\text{Re}\,\text{Pr}\,R)$, in terms of our dimensionless variables, this equation can also be written as follows:

$$Y_{avg}(Z) = \exp\left(-\frac{2L/R}{\text{RePr}}\text{Nu}\right) = \exp(-2Z\,\text{Nu}) \tag{6.66}$$

For our system, we have:

$$Y_{avg}(L) = \frac{T_{avg}(L) - T_s}{T_0 - T_s} = \frac{8-0}{37-0} = 0.216$$

Thus:

$$Z\text{Nu} = -\frac{1}{2}\ln\,[Y_{avg}(Z)] = -\frac{1}{2}\ln(0.216) = 0.766$$

or:

$$\text{Nu} = \frac{0.766}{Z}$$

This equation is a straight line on Fig. 6.7. Its intercept with the theoretical line describing Nu vs. Z will be our solution. For this case, the intercept is at $Z = 0.18$ with Nu = 4.2. For our

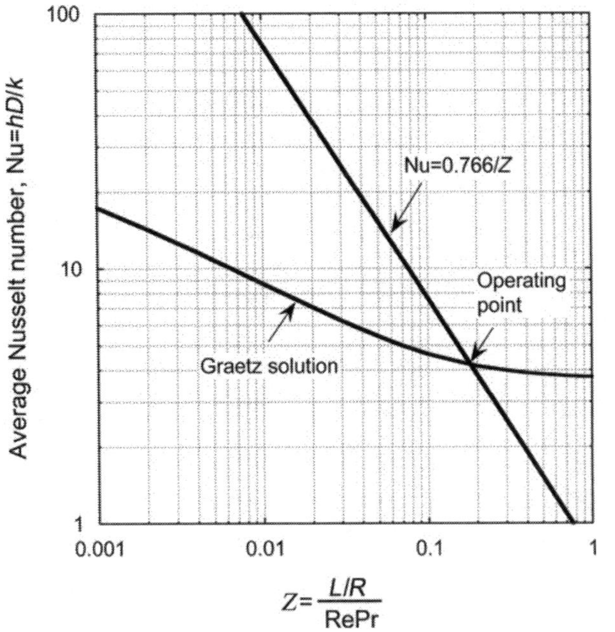

FIGURE 6.7 Graphical solution of Example 6.2.

system, we have $\text{Re} = v_{\text{avg}}D/\nu = 4\dot{V}/\pi D\nu = 4\times(0.0005/60)/(\pi\times 0.005\times 5\times 10^{-6}) = 424$, $\text{Pr} = \nu/\alpha = 33$. Thus, $L = 0.18R\times\text{RePr} = 0.18\times 0.0025\times 424\times 33 = 6.3$ m. Note that we could have also calculated L directly from Eq. (6.55) or, approximately, from Eq. (6.60). The latter gives:

$$Z = -\frac{1}{7.31}\ln(0.216/0.82) = 0.182$$

which is the same result obtained above. This happens because, in this case, the tube is long and Nu is close to the limiting value of 3.66. The more general construction above will work, however, even if the tube is short.

6.3 Forced Convection in Turbulent Flow—Analogies Between Momentum and Heat Transfer

Turbulent flow conditions are likely to prevail in many practical situations. For example, since turbulence enhances transport at fluid-solid interfaces, turbulent flow conditions will often be found in heat exchangers. A theoretical prediction of convective heat transfer under turbulent flow conditions is, however, not straightforward. The momentum transport problem alone is often intractable because of the chaotic nature of turbulent flow. As a result, in practice, predicting heat transfer under turbulent flow is handled through generalized correlations of empirical data. Although purely empirical approaches can be used, using "analogies" between momentum and heat transfer has the advantage that measurements for a particular system can be used, through proper scaling, to make predictions for other systems. In Chap. 13, we will see that such approaches can be extended through analogies between momentum, heat, and mass transfer leading to generalized correlations that connect the three phenomena together despite their vastly different underlying molecular basis.

A well-known feature of turbulent transport at fluid–solid interfaces is that both momentum transport and thermal boundary layers (and, we will see later, mass transport boundary layers) are generally thin. This means that, in turbulent flow, the velocity will reach its free stream value at a very short distance away from the interface. Likewise, the temperature will reach its free stream value also at a very short distance from the interface.

A useful analogy between momentum and mass transfer can be derived with reference to Fig. 6.8, which depicts, schematically, momentum and heat transfer at a generic fluid–solid interface. This could be a flat plate or a pipe wall. Since inertial forces are dominant in turbulent flow, we can describe the shear stress at the wall based on the following equation:

$$\tau_w = \frac{1}{2}\rho v_\infty^2 f \tag{6.67}$$

where $\rho v_\infty^2/2$ is the fluid-specific kinetic energy and f is the friction factor. The latter is generally a function of the Reynolds number. Analogously, for the heat transfer flux at the same interface, we can write:

$$q = h(T_s - T_\infty) \tag{6.68}$$

Let's now assume that momentum and heat transfer within the respective boundary layers can be represented by the following equations:

$$\tau_w = \rho E_M \frac{dv_x}{dy} \tag{6.69}$$

$$q = -\rho C_p E_H \frac{dT}{dy} \tag{6.70}$$

These equations have the same form as Newton's law for viscous momentum transport and as Fourier's law for conductive heat transfer. However, instead of viscosity and thermal conductivity they make use of the eddy viscosity E_M and of the eddy thermal diffusivity E_H, respectively. These quantities have dimensions of length square over time, just like the kinematic viscosity ν and the thermal diffusivity α.

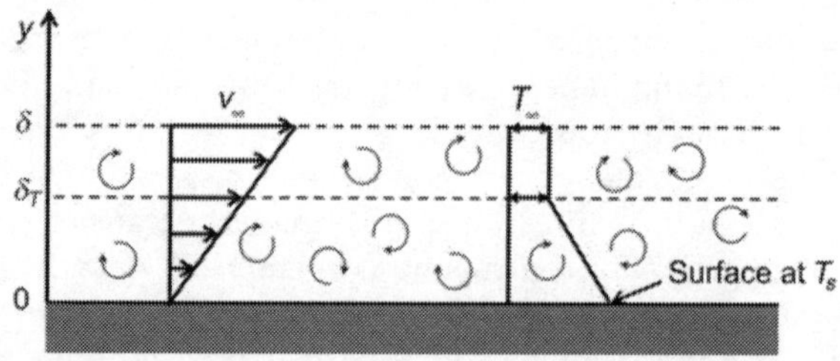

FIGURE 6.8 Heat transfer between a surface and a fluid in turbulent flow. δ and δ_T are the momentum transfer and thermal boundary layer thicknesses, respectively.

However, while ν and α are molecular properties of the fluid, E_M and E_H represent transport associated with turbulent eddies and are, thus, dependent on the local level of turbulence. Combining Eqs. (6.67) to (6.70), we obtain:

$$\frac{1}{2}\rho v_\infty^2 f = \rho E_M \frac{dv_x}{dy} \tag{6.71}$$

and

$$h(T_s - T_\infty) = -\rho C_p E_H \frac{dT}{dy} \tag{6.72}$$

In the next step, we divide these two equations by each other to obtain:

$$\frac{E_M}{\frac{1}{2}v_\infty^2 f} dv_x = -\frac{\rho C_p E_H}{h(T_s - T_\infty)} dT \tag{6.73}$$

Having eliminated the differential dy, we can now integrate this equation from the interface where $v_x = 0$ and $T = T_s$ to the edge of each boundary layers where $v_x = v_\infty$ and $T = T_\infty$ which gives:

$$\frac{E_M}{\frac{1}{2}v_\infty^2 f} \int_0^{v_\infty} dv_x = -\frac{\rho C_p E_H}{h(T_s - T_\infty)} \int_{T_s}^{T_\infty} dT \tag{6.74}$$

The final result is as follows:

$$\frac{E_M}{\frac{1}{2}v_\infty f} = \frac{\rho C_p E_H}{h} \tag{6.75}$$

With some slight rearrangement, this result can also be written as:

$$\frac{f}{2} = \frac{h}{\rho C_p v_\infty} \frac{E_M}{E_H} = \mathrm{St_H} \frac{E_M}{E_H} \tag{6.76}$$

where

$$\mathrm{St_H} = \frac{h}{\rho C_p v_\infty} \tag{6.77}$$

is called the Stanton number for heat transfer. Note that Eq. (6.76) sets forth a quantitative relationship between the friction factor, f, which is a purely fluid mechanical concept, and the Stanton number, which describes convective heat transfer. Generally, we can easily measure f or we can estimate it as a function of the Reynolds number. For example, for turbulent flow in a smooth pipe we can use the Fanning friction factor correlation to predict f as a function of Re (e.g., see Section 6.1.4 in *Perry's Chemical Engineers' Handbook*, 9th edition). We can then use Eq. (6.75) or (6.76) to predict the heat transfer coefficient in the same pipe. This prediction will depend on what assumption is made for the relationship between E_M and E_H. One assumption, based on empirical observations for gases at ambient conditions, is to take $E_M = E_H$. This assumption leads to the so-called Reynolds analogy:

$$\frac{f}{2} = \mathrm{St_H} \tag{6.78}$$

Another assumption, based on empirical observations for a broad range of fluids, is to take $E_M/E_H = \mathrm{Pr}^{2/3}$. This assumption leads to the so-called Chilton–Colburn analogy:

$$\frac{f}{2} = \mathrm{St_H Pr}^{2/3} \tag{6.79}$$

With reference, for example, to flow in a pipe with diameter D, if the volumetric flow rate is $\dot{V}$, v_∞ will be equal to the average velocity in the pipe, which is given by $v_{avg} = \dot{V}/(\pi D^2/4)$. In this case, $\mathrm{Re} = \rho v_{avg} D/\mu = v_{avg}D/\nu$ and $\mathrm{St_H} = (hD/k)/(v_{avg}D/\nu)(\nu/\alpha) = \mathrm{Nu}/(\mathrm{RePr})$. As a result, Eq. (6.79) can also be written as follows:

$$\frac{f}{2} = \frac{\mathrm{Nu}}{\mathrm{RePr}^{1/3}} = \mathrm{J_H} \tag{6.80}$$

where $\mathrm{J_H} = \mathrm{Nu}/\mathrm{RePr}^{1/3}$ is the Chilton–Colburn factor. Obviously, for fluids where Pr~1, such as gases at low pressure, the Reynolds and Chilton–Colburn results coincide. For other systems, the Chilton–Colburn analogy is usually found to be more accurate.

Example 6.3. Heat transfer coefficient from a pressure drop measurement. Let's say that we have measured a pressure drop $\Delta P = 0.05$ atm for a 2.5-cm diameter, 3.0-m-long smooth heat exchanger pipe with water at 20°C flowing at a volumetric flow rate $\dot{V} = 1.0 \times 10^{-3}$ m³/s. We can obtain the heat transfer coefficient from the Chilton–Colburn analogy. The average velocity in the pipe is $v_{avg} = 4\dot{V}/\pi D^2 = 4 \times 1 \times 10^{-3}/\pi(0.025)^2 = 2.04$ m/s. Since for water at 20°C $\nu = 1 \times 10^{-6}$ m²/s, the Reynolds number will be $\mathrm{Re} = v_{avg}D/\nu = 2.04 \times 0.025/1.0 \times 10^{-6} = 51{,}000$. The flow is clearly turbulent since Re >> 2,100. Based on a force balance, we can relate the measured pressure drop to the shear stress at wall through the following equation:

$$\pi D L \tau_w = \Delta P \pi D^2/4$$

Using this expression in combination with Eq. (6.67), we obtain:

$$\frac{f}{2} = \frac{\Delta P}{4\rho v_{avg}^2}\frac{D}{L}$$

Thus, for our system:

$$\frac{f}{2} = \frac{0.05 \times 101{,}325}{4 \times 1000 \times 2.04^2}\frac{0.025}{3.0} = 0.0025$$

For water at 20°C, we have $k = 0.6$ W/m K, $\alpha = k/\rho C_p = 0.6/1{,}000 \times 4{,}200 = 1.4 \times 10^{-7}$ m²/s, and $\mathrm{Pr} = \nu/\alpha = 1.0 \times 10^{-6}/1.4 \times 10^{-7} = 7.1$. With these values, we finally obtain:

$$\mathrm{Nu} = \frac{f}{2}\mathrm{RePr}^{1/3} = 0.0025 \times 51{,}000 \times 7.1^{1/3} = 246$$

and

$$h = \frac{k}{D}\mathrm{Nu} = \frac{0.6}{0.025}246 = 5{,}890 \text{ W/m}^2\text{ K}$$

The Chilton–Colburn analogy ($f/2 = \mathrm{J_H}$) works, in general, for flow over flat surfaces, including flat plates. The example above demonstrates the application of the analogy to pipe flow, which has been validated for many experimental systems since the publication of the original article by Colburn [Trans. Am. Inst. Chem. Engr. 29 (1933) 174]. Although the pipe wall is not really flat, the flat surface approximation still works well because the boundary layers are very thin in turbulent flow, so that the wall curvature has practically no effect. We must recognize, however, that the equivalence between $f/2$ and $\mathrm{J_H}$

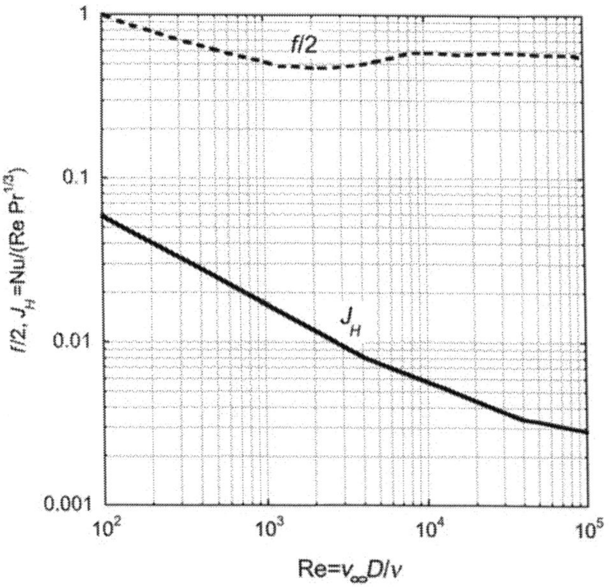

Figure 6.9 Friction factor and Chilton–Colburn factor for momentum and heat transfer for turbulent flow over a long cylinder perpendicular to the cylinder's axis. *D*, used in Nu and Re, is the cylinder diameter.

is not always valid. Consider, for example, Fig. 6.9, which shows the lines describing the experimental results for J_H, obtained from heat transfer measurements for flow over a long cylinder perpendicular to the cylinder's axis along with the corresponding friction factor data. As seen in Fig. 6.9, $f/2$ and J_H are very different from each other in this case. This occurs because, for flow over a cylinder, $f/2$ is affected both by friction at the cylinder surface and by form drag, with the latter becoming increasingly dominant at higher Reynolds numbers. On the other hand, J_H is affected only by what happens in the boundary layer near the cylinder surface. This is also true, of course, for flow over objects with different geometries including banks of tubes and spheres. We will see later, that while the analogy between momentum and heat transfer is limited to flow over essentially flat surfaces, analogies between heat and mass transfer in turbulent flow are generally valid regardless of geometry.

6.4 Correlations for Heat Transfer Coefficients

Generalized correlations useful to estimate heat transfer coefficients are available for many different cases including forced convection, natural convection, heat transfer in agitated tanks, and heat transfer with change of phase (condensation and boiling). Because of the generalized nature of these correlations, we cannot expect extreme precision. Errors are often ±20% and can be larger for more complex cases where the fluid mechanics is not so well defined (e.g., agitated tanks) or where multiple mechanisms can contribute simultaneously to the overall result (e.g., changes of phase). Thus, *h*-values computed from these correlations should be used for estimation purposes only. More precise values will require system-specific correlations or direct measurements.

6.4.1 Forced Convection

An up-to-date collection of correlations to predict heat transfer coefficients for forced convection can be found in Section 5 of *Perry's Chemical Engineers' Handbook*, 9th edition. In addition to tabulated equations, this Section 5 also provides access to spreadsheets to facilitate the calculation of h for (a) laminar pipe flow, (b) turbulent pipe flow, (c) turbulent flow through an annulus, (d) turbulent flow through a noncircular duct, (e) flow normal to a horizontal cylinder, and (f) flow past a flat plate. In these spreadsheets, one can select from any of 74 liquids, and the workbook will calculate the liquid properties required.

In general, correlations for heat transfer coefficients under forced convection are available in the form:

$$\text{Nu} = \text{Nu}(\text{Re}, \text{Pr}) \tag{6.81}$$

typically in accordance with the Chilton–Colburn analogy. In general, if the Nusselt number is known for a certain fluid at a given Re, estimates for a different fluid at the same Re can be made by scaling Nu with $\text{Pr}^{1/3}$. Because of the 1/3 power, the effect of Pr on Nu is generally modest. Thus, in general, we can get reasonable estimates of Nu even without extremely precise values of Pr. Table 6.1 provides a summary of representative equations applicable to many practical systems. Note that, in these equations, Nu and Re are defined based on an appropriate, system-dependent length scale. For example, for flow in pipes, the length scale will be the pipe's inner diameter, for flow past the edge of a flat plate, the length scale will be the length of the plate, while for flow over cylinders or spheres, the length scale will be their diameter.

6.4.2 Natural Convection

Natural convection is fundamentally different from forced convection. In this case, flow is not driven by a pump, fan, or blower. Rather, it is generated by a density gradient established in the fluid as a result of a temperature difference near the solid–fluid interface. Natural convection is most significant when the fluid is a gas, like air, since the density gradient and associated buoyancy effects will be greater. In general, natural convection is slow compared to forced convection and will often be the controlling resistance when multiple heat transfer resistances in series exist. Use of fins or other extended surfaces will thus be warranted in these situations when higher heat transfer rates are desired as discussed in Chap. 4.

Correlations for heat transfer coefficients under natural convections are usually expressed by relationships of the type:

$$\text{Nu} = \text{Nu}(\text{Gr}, \text{Pr}) \tag{6.82}$$

where Gr is the Grashof number, which is defined as follows:

$$\text{Gr} = \frac{g\hat{\beta}\rho^2}{\mu^2} L^3 \Delta T \tag{6.83}$$

where

$$\hat{\beta} = -\frac{1}{\rho}\frac{d\rho}{dT} \tag{6.84}$$

is the fluid's thermal expansion coefficient, L is a characteristic length, and $\Delta T = T_s - T_\infty$ is the difference between surface and ambient temperatures. Values of the group

Case	Reynolds Number	Nusselt Number	Equation	Notes and References
A. Flow over a wide flat plate with length L (M.T.6.1-A)	$\mathrm{Re} = \dfrac{\rho v_\infty L}{\mu}$	$\mathrm{Nu} = \dfrac{hL}{k}$	Laminar flow $\mathrm{Nu} = 0.664\,\mathrm{Re}^{1/2}\mathrm{Pr}^{1/3} \quad \text{for } \mathrm{Re} < 5\times 10^5$	See Eq. (6.34)
			Turbulent flow $\mathrm{Nu} = 0.664\,\mathrm{Re}_{cr}^{1/2}\,\mathrm{Pr}^{1/3} + 0.036\,\mathrm{Re}^{0.8}\,\mathrm{Pr}^{0.43}\left[1-\left(\dfrac{\mathrm{Re}_{cr}}{\mathrm{Re}}\right)^{0.8}\right]$ for $5\times 10^5 < \mathrm{Re} < 3\times 10^7$ with $\mathrm{Re}_{cr} = 5\times 10^5$	Mills (1999)
B. Flow in a circular pipe with diameter D (M.T.6.1-B)	$\mathrm{Re} = \dfrac{\rho v_{avg} D}{\mu}$	$\mathrm{Nu} = \dfrac{hD}{k}$	Laminar flow $\mathrm{Nu} = 1.86\left(\dfrac{L/D}{\mathrm{RePr}}\right)^{-1/3}\left(\dfrac{\mu}{\mu_s}\right)^{0.14}$ for $\mathrm{Re} < 2{,}000,\ \dfrac{L/D}{\mathrm{RePr}} < 0.1$ $\mathrm{Nu} = 3.66$ for $\mathrm{Re} < 2{,}000,\ \dfrac{L/D}{\mathrm{RePr}} > 0.1$	Sieder and Tate (1936) μ =viscosity at bulk temp. μ_s =viscosity at surface temperature
			Turbulent flow $\mathrm{Nu} = \dfrac{(f/2)(\mathrm{Re}-1{,}000)\mathrm{Pr}}{1.0+12.7\sqrt{f/2}\,(\mathrm{Pr}^{2/3}-1)}\left(\dfrac{\mu}{\mu_s}\right)^{0.11}$ with $f/2 = 0.125[0.79\ln(\mathrm{Re}) - 1.64]^{-2}$ for $\mathrm{Re} > 2{,}000$	Gnielinski (1976)
C. Flow around an isolated cylinder with diameter D (M.T.6.1-C)	$\mathrm{Re} = \dfrac{\rho v_\infty D}{\mu}$	$\mathrm{Nu} = \dfrac{hD}{k}$	$\mathrm{Nu} = 0.3 + \dfrac{0.62\,\mathrm{Re}^{1/2}\,\mathrm{Pr}^{1/3}}{[1+(0.4/\mathrm{Pr})^{2/3}]^{1/4}}\left[1+\left(\dfrac{\mathrm{Re}}{282{,}000}\right)^{5/8}\right]^{4/5}$ for $0.1 < \mathrm{Re} < 10^5$	Churchill and Bernstein (1977)

TABLE 6.1 Selected Correlations for Average Heat Transfer Coefficients Under Forced Convection

Case	Reynolds Number	Nusselt Number	Equation	Notes and References
D. Flow perpendicular to banks of tubes with diameter D with velocity v_{max} at the minimum cross-section	$Re = \frac{\rho v_{max} D}{\mu}$	$Nu = \frac{hD}{k}$	$Nu = b(Re)^n Pr^{1/3}$	Grimison (1937) Values of b and n available for different tube arrangements in McAdams, *Heat Transmission*, 3rd ed., McGraw-Hill (1954)
E. Flow around an isolated sphere with diameter D (M.T.6.1-E)	$Re = \frac{\rho v_\infty D}{\mu}$	$Nu = \frac{hD}{k}$	$Nu = 2.0 + 0.60 Re^{1/2} Pr^{1/3}$	Ranz and Marshall (1952)
F. Flow in a packed bed of spheres with diameter D and void fraction ε (M.T.6.1-F)	$Re = \frac{\rho u D}{\mu}$	$Nu = \frac{hD}{k}$	$Nu = \frac{1.09}{\varepsilon} Re^{1/3} Pr^{1/3}$ for $0.0015 < Re < 55$ $Nu = \frac{0.25}{\varepsilon} Re^{0.69} Pr^{1/3}$ for $55 < Re < 1050$	Wilson and Geankoplis (1966). Based on mass transfer data. u is the superficial velocity defined as the volumetric flow rate divided by the bed cross section.

References: Mills, *Heat Transfer*, Prentice Hall, 2nd edition (1999), p. 315.
Sieder and Tate, Ind. Eng. Chem. 28 (1936) 1429.
Gnielinski, Int. Chem. Eng. 16 (1976) 359.
Churchill and Bernstein, J. Heat. Transf. 99 (1977) 300.
Grimison, Trans. ASME 59 (1937) 583.
Ranz and Marshall, Chem. Eng. Progr. 48 (1952) 141.
Wilson and Geankoplis, Ind. Eng. Chem. Fundam. 5 (1966) 9.

TABLE 6.1 Selected Correlations for Average Heat Transfer Coefficients under Forced Convection (*Continued*)

$g\hat{\beta}\rho^2/\mu^2$ are often found tabulated for different gases and, for air, can also be found in Appendix A. The product:

$$\mathrm{Ra} = \mathrm{Gr}\,\mathrm{Pr} = \frac{g\hat{\beta}\rho L^3 \Delta T}{\alpha\mu} \tag{6.85}$$

is called the Rayleigh number. For an ideal gas, the mass density is $\rho = M_r P/\mathcal{R}T$ where M_r is the gas molecular mass. In this case, $\hat{\beta} = 1/T$. For air, Pr is close to 1. Thus, Ra ~ Gr. This number plays the role of a Reynolds number. For instance, for natural convection over a vertical heated wall, increasing L, reducing the kinematic viscosity $\nu = \mu/\rho$, or increasing $\Delta T/T$ will all increase the natural convection effect increasing the heat transfer coefficient. Table 6.2 provides some of the recommended correlations for a few different geometries. In general, fluid properties for use with these equations should be evaluated at the average temperature $(T_s + T_\infty)/2$. Note that, in all cases, the equations in this table provide an average value of the Nusselt number and, thus, h averaged over the surface. Section 5 of *Perry's Chemical Engineers' Handbook*, 9th edition, provides a more complete list including correlations for natural convection in confined spaces such as between two parallel plates as well as a spreadsheet to carry out the calculations. Example 6.4 provides an illustration on the use of these relationships.

Example 6.4. Heat transfer coefficient for natural convection along a vertical wall. Let's estimate the heat transfer coefficient for a 2-m-tall wall held at 50°C in contact with air at 20°C. At the average temperature of 35°C, for air we have $k = 0.027$ W/m K, $\nu = \mu/\rho = 1.6 \times 10^{-5}$ m²/s, $\hat{\beta} = 1/308 = 0.0032$ K⁻¹, and $\mathrm{Pr} = 0.7$. Thus, we can calculate:

$$\mathrm{Gr} = \frac{g\hat{\beta}\rho^2}{\mu^2} H^3 \Delta T = \frac{9.8 \times 0.0032}{(1.6 \times 10^{-5})^2} 2^3 (50 - 20) = 2.9 \times 10^{10}$$

From the equation of Churchill and Chu in Table 6.2 (item C), we obtain:

$$\mathrm{Nu} = \left\{ 0.825 + \frac{0.387(2.9 \times 10^{10} \times 0.7)^{1/6}}{[1 + (0.492/0.7)^{9/16}]^{8/27}} \right\}^2 = 315$$

Finally, we have $h = k\mathrm{Nu}/H = 0.027 \times 315/2 = 4.3$ W/m² K. This low value is typical of what can be expected for natural convection in air with a relatively small temperature difference.

6.4.3 Agitated Tanks

Heat transfer in agitated tanks is important in a number of applications, including chemical reactors, fermentation processes, and many other continuous and batch operations. There are two different considerations. The first is heat transfer to/from the tank wall, which can be jacketed or fitted with integrated coils used to heat of cool the liquid in the tank. The other is heat transfer between the liquid and particles that are suspended in the tank. In either case, there is not an easily discernible velocity that can be used to express a Reynolds number. One could use the velocity of the tip of the impeller or other agitator device, but other factors besides agitator speed will come into play. In general, heat transfer data for agitated tanks are thus often correlated in terms of the power dissipated by the agitator. The theory of isotropic turbulence, attributed to

Case	Grashof Number	Nusselt Number	Equation	Notes and References
A. Long horizontal flat surface with width W facing upward when hot or downward when cold (M.T.6.2-A)	$\mathrm{Gr} = \dfrac{g\hat{\beta}\rho^2}{\mu^2} W^3 \Delta T$	$\mathrm{Nu} = \dfrac{hW}{k}$	Laminar: $\mathrm{Nu} = 0.54(\mathrm{Gr\,Pr})^{1/4}$ with $10^5 < \mathrm{Gr\,Pr} < 10^7$	Lloyd and Moran (1974)
			Turbulent: $\mathrm{Nu} = 0.15(\mathrm{Gr\,Pr})^{1/3}$ with $10^7 < \mathrm{Gr\,Pr} < 10^9$	Lloyd and Moran (1974)
B. Long horizontal flat surface with width W facing downward when hot or facing upward when cold (M.T.6.2-B)	$\mathrm{Gr} = \dfrac{g\hat{\beta}\rho^2}{\mu^2} W^3 \Delta T$	$\mathrm{Nu} = \dfrac{hW}{k}$	Laminar or turbulent $\mathrm{Nu} = 0.27(\mathrm{Gr\,Pr})^{1/4}$ with $10^5 < \mathrm{Gr\,Pr} < 10^{10}$	Lloyd and Moran (1974)
C. Wide vertical surface with height H (M.T.6.2-C)	$\mathrm{Gr} = \dfrac{g\hat{\beta}\rho^2}{\mu^2} H^3 \Delta T$	$\mathrm{Nu} = \dfrac{hH}{k}$	Laminar or turbulent $\mathrm{Nu} = \left\{0.825 + \dfrac{0.387(\mathrm{Gr\,Pr})^{1/6}}{[1+(0.492/\mathrm{Pr})^{9/16}]^{8/27}}\right\}^2$	Churchill and Chu (1975a) Also applicable to vertical cylinders
D. Long horizontal cylinder with diameter D (M.T.6.2-D)	$\mathrm{Gr} = \dfrac{g\hat{\beta}\rho^2}{\mu^2} D^3 \Delta T$	$\mathrm{Nu} = \dfrac{hD}{k}$	Laminar or turbulent $\mathrm{Nu} = \left\{0.60 + \dfrac{0.387(\mathrm{Gr\,Pr})^{1/6}}{[1+(0.559/\mathrm{Pr})^{9/16}]^{8/27}}\right\}^2$	Churchill and Chu (1975b)

References: Lloyd and Moran, ASME Paper 74-WA/HT-66 (1974).
Churchill and Chu, Int. J. Heat Mass Transfer, 18 (1975a) 1323.
Churchill and Chu, Int. J. Heat Mass Transfer, 18 (1975b) 1049.

TABLE 6.2 Selected Correlations for Heat Transfer Coefficients under Natural Convection

Kolmogorov, indicates that the impeller power is dissipated within the agitated liquid as a result of interactions of the smallest turbulent eddies. A consequence of this assumption is that the power dissipated is independent of the particular tank geometry and only dependent on the power supplied to the impeller. Since these turbulent eddies are also responsible for heat transfer at the solid–fluid interfaces, it makes sense to correlate heat transfer coefficients in terms of the power dissipated per unit mass of fluid. The latter is defined as $\bar{\varepsilon} = P/m$, where P is the agitator power and m is the mass of liquid in the tank. Note, however, that the impeller power does depend on tank geometry, impeller speed, and fluid properties. This dependence is expressed by so-called power number correlations where the power number:

$$N_P = \frac{P}{\rho \dot{N}^3 D_i^5} \tag{6.86}$$

is a function of a Reynolds number defined as:

$$\mathrm{Re} = \frac{\dot{N} D_i^2}{\nu} \tag{6.87}$$

In these equations, $\dot{N}$ is the impeller speed in revolutions per second and D_i is the impeller diameter. Section 18 in *Perry's Chemical Engineers' Handbook*, 9th edition, provides various examples of such correlations. Since P and, thus, $\bar{\varepsilon}$, are directly related to Re as defined above, heat transfer data can be correlated directly in terms of Re.

Jackets and Coils

For heat transfer between the fluid and a jacketed wall, *Perry's Handbook* recommends the following equation:

$$\mathrm{Nu} = \frac{hD_T}{k} = a\,\mathrm{Re}^b \mathrm{Pr}^{1/3} \left(\frac{\mu}{\mu_s}\right)^{m'} \tag{6.88}$$

where D_T is the diameter of the tank and μ/μ_s is the ratio of fluid viscosities evaluated at the bulk and surface temperatures. a, b, and m' are constants that depend on the agitator type according to Table 18-1 in *Perry's Chemical Engineers' Handbook*, 9th edition. Similar correlations are available for helical coils in agitated tanks.

Suspended Particles

Heat transfer between particles of diameter D suspended in a liquid in an agitated tank with power input per unit mass $\bar{\varepsilon}$ can be estimated from the following correlation:

$$\mathrm{Nu} = \frac{hD}{k} = 2 + a\left(\frac{\bar{\varepsilon}^{1/3} D^{4/3}}{\nu}\right)^b \mathrm{Pr}^{1/3} \tag{6.89}$$

where a and b are constants. Levins and Glastonbury [Trans. Inst. Chem. Eng., 50 (1972) 132] give $a = 0.5$ and $b = 0.62$, while Armenante and Kirwan [Chem. Eng. Sci., 44 (1989) 2781] give $a = 0.52$ and $b = 0.52$. Although both of these references used mass transfer data, their correlations also describe, approximately, heat transfer data as seen, for example, in Fig. 6.10. The quantity $\bar{\varepsilon}^{1/3} D^{4/3}/\nu$ appearing in this equation plays the role

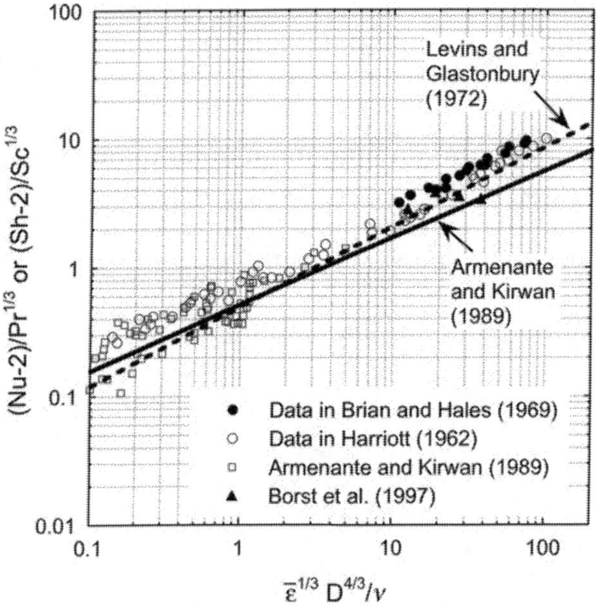

Figure 6.10 Convective heat transfer and convective mass transfer for particles suspended in an agitated tank. The data of Brian and Hales (1969) are for heat transfer, while other data are for mass transfer. For heat transfer between the particles and the fluid Nu = hD/k where D is the particle diameter. For mass transfer, Sh and Sc are defined in Chap. 13.

References: Harriott, AIChE J. 8 (1962) 93;
Brian and Hales, AIChE J. 15 (1969) 419;
Levins and Glastonbury, Trans. Inst. Chem. Eng., 50 (1972) 132;
Armenante and Kirwan, Chem. Eng. Sci., 44 (1989) 2781;
Borst et al., React. Funct. Pol. 32 (1997) 25.

of a Reynolds number for the particles. Note that Nu, in this case, is defined based on the diameter of the particles and not the diameter of the tank. In order to use this correlation, one must first estimate $\bar{\varepsilon}$ from the appropriate power number correlation. Example 6.5 illustrates this calculation for a sample case.

Example 6.5. Heat transfer in an agitated tank. Let's use the equations presented in this section to predict the heat transfer coefficient in a jacketed baffled tank agitated with a disk-type six-blade impeller for the following conditions:

Tank properties	Fluid properties
Tank diameter $D_T = 2$ m	Density $\rho = 1{,}000$ kg/m^3
Impeller diameter $D_i = 0.66$ m	Viscosity $\mu = 0.005$ Pa s
Liquid height $H = 2$ m	Thermal conductivity $k = 0.5$ W/m K
Agitation rate $\dot{N} = 1$ rps (60 rpm)	Heat capacity $C_p = 4{,}200$ J/kg K

Let's first consider heat transfer between the fluid and the tank wall. For this type of impeller, Table 18-1 in *Perry's Chemical Engineers' Handbook*, 9th edition, gives $a = 0.54$, $b = 2/3$, and $m' = 0.14$ for use

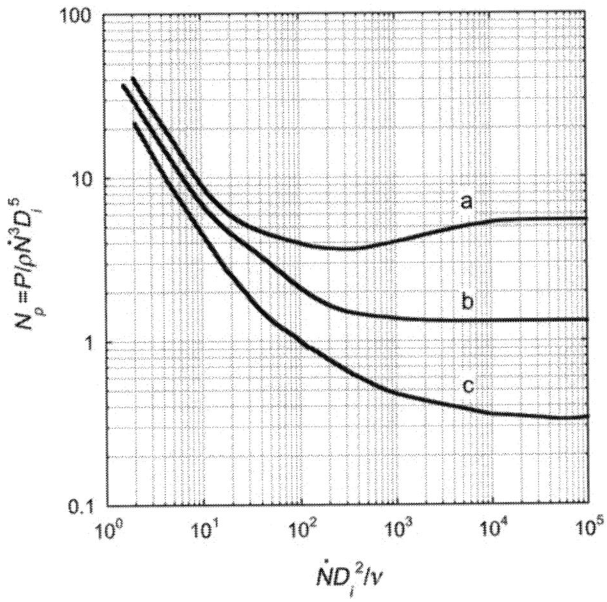

Figure 6.11 Power number correlation for baffled tanks with different agitator types. Data from Fig. 18-1 in *Perry's Chemical Engineers' Handbook*, 9th edition. Curve a: six-blade, disk-style turbine in a tank with four baffles. Curve b: six-blade, 45° pitched-blade turbine in a tank with four baffles. Curve c: marine-style propeller in a tank with four baffles.

in Eq. (6.88) with $40 < \text{Re} < 300{,}000$. In our case, the Reynolds number is $\text{Re} = \dot{N}D_i^2/v = 1 \times 0.66^2/5 \times 10^{-6} = 87{,}100$ and $\text{Pr} = 42$. Neglecting the viscosity correction, Eq. (6.88) then gives $\text{Nu} = 0.54 \times (87{,}100)^{2/3} \times (42)^{1/3} = 3{,}700$. Thus, $h = k\text{Nu}/D_T = 0.5 \times 3{,}700/2 = 930\ \text{W/m}^2\ \text{K}$.

Next, let's consider heat transfer between the fluid and 0.5-mm particles that we assume are fully suspended in the tank. We need to first estimate $\bar{\varepsilon}$ for use in Eq. (6.89). Curve a in Fig. 6.11 based on data in Fig. 18-1 of *Perry's Chemical Engineers' Handbook*, 9th edition, gives the power number correlation for this type of tank/impeller combination (curve 1 in the original figure). At $\text{Re} = 87{,}100$, we read $N_P \sim 6$. Accordingly, the power input will be $P = \rho\dot{N}^3D_i^5N_P = 1{,}000 \times (1.0)^3 \times (0.66)^5 \times 6 = 750\ \text{W}$. The mass of fluid in the tank is $m = \pi D_T^2 H\rho/4 = \pi \times 2^2 \times 2 \times 1{,}000/4 = 6{,}280\ \text{kg}$. Thus, the power input per unit mass is estimated to be $\bar{\varepsilon} = 750/6{,}280 = 0.12\ \text{W/kg}$. For the 0.5-mm particles, we have $\bar{\varepsilon}^{1/3}D^{4/3}/v = (0.12)^{1/3}(0.0005)^{4/3}/5 \times 10^{-6} = 3.9$. Thus, from Eq. (6.89), we obtain $\text{Nu} = 2 + 0.5(3.9)^{0.62}(42)^{1/3} = 6.0$ and $h = k\text{Nu}/D = 0.5 \times 6.0/0.0005 = 6{,}000\ \text{W/m}^2\ \text{K}$.

6.5 Heat Transfer with Change of Phase

6.5.1 Condensation

Condensation of vapors on cold surfaces can occur following two different mechanisms illustrated in Fig. 6.12: drop-wise condensation and film-wise condensation. Drop-wise condensation occurs on an initially dry surface with individual liquid droplets being formed directly on the cold surface. The rate of drop-wise condensation is usually extremely high. However, as more and more drops condense, a continuous film of condensate will form and will run down the condensing surface. In this case, the rate of

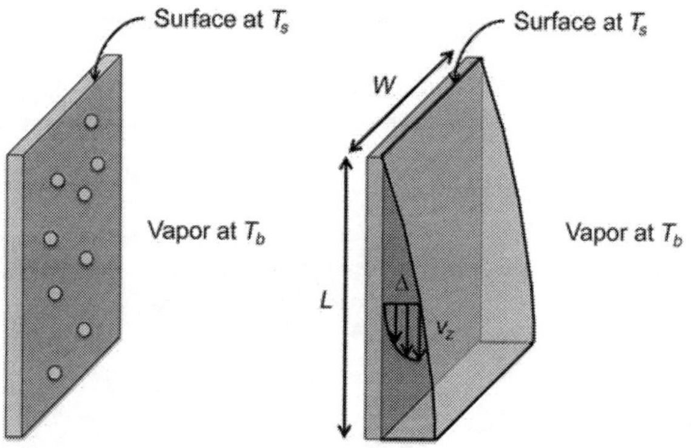

FIGURE 6.12 Drop-wise (left) and film-wise (right) condensation of a saturated vapor at T_b over a vertical surface at T_s.

condensation will become limited by the conductive heat transfer resistance across the condensate film. Although physically very different, we can visualize drop-wise and film-wise condensation if we think about rain on a dusty car windshield. At the onset, we see individual drops forming on the windshield, which then become a continuous film as the drops coalesce and the dust is washed off. Most industrial equipment operates under film-wise condensation which is the usual basis for design.

With reference to Fig. 6.12, a simple model for film-type condensation can be derived by assuming that the heat transfer coefficient for a film of condensate of thickness Δ on a surface is given by the following equation:

$$h = \frac{k}{\Delta} \tag{6.90}$$

where k is the thermal conductivity of the condensate. Note that the condensate flows down by gravity. If we assume that this flow is laminar, the velocity profile as a function of distance from the vapor–condensate interface, x, will be given by:

$$v_z = v_{max}\left[1-\left(\frac{x}{\Delta}\right)^2\right] \tag{6.91}$$

where $v_{max} = 3v_{z,avg}/2$ is the velocity at the vapor–condensate interface and $v_{z,avg}$ is the average velocity in the condensate film. The latter is given by the following equation:

$$v_{z,avg} = \frac{m_c/\rho}{W\Delta} = \frac{\Gamma}{\rho\Delta} \tag{6.92}$$

where m_c is the rate of condensation in mass units (e.g., kg/s), ρ is the density of the condensate (e.g., kg/m^3), and $\Gamma = m_c/W$ is the rate of condensation per unit width of the surface (e.g., kg/s m). Next, we can obtain an expression for $v_{z,avg}$ by performing a force balance on the film. The gravity force pulling the film down will be $WL\Delta\rho g$, while the shear force at the wall will be $WL\mu(dv_z/dx)_{x=\Delta} = WL\mu \times 3v_{z,avg}/\Delta$, where μ is the viscosity of the condensate. Equating these two forces, we obtain:

$$v_{z,avg} = \frac{g\rho\Delta^2}{3\mu} \tag{6.93}$$

Combined with Eq. (6.92), this equation gives the following result for the thickness of the condensate film:

$$\Delta = \left(\frac{3\mu\Gamma}{g\rho^2}\right)^{1/3} \tag{6.94}$$

Finally, from Eq. (6.90) we obtain:

$$h = k\left(\frac{g\rho^2}{3\mu\Gamma}\right)^{1/3} \tag{6.95}$$

or:

$$\text{Nu} = \frac{hL}{k} = c\left(\frac{gL^3\rho^2}{\mu\Gamma}\right)^{1/3} \tag{6.96}$$

where c is a numerical constant. This equation provides the foundation for the correlation of condensing heat transfer coefficients. Since, in practice, the condensate film is thin, even though we have worked out the model for a flat surface, the same relationship can be expected to hold also for curved surfaces. The constant c will be determined empirically to account for the fact that the condensate film thickness will actually vary over the length of the wall as more and more condensate accumulates.

For condensation on the outside of a vertical tube of diameter D and length L, the following equation is recommended:

$$\text{Nu} = \frac{hL}{k} = 0.925\left(\frac{gL^3\rho^2}{\mu\Gamma}\right)^{1/3} \tag{6.97}$$

where $\Gamma = m_c/\pi D$ is the mass rate of condensation per unit perimeter. For condensation over a single horizontal tube of diameter D and length L, the following equation is recommended instead:

$$\text{Nu} = \frac{hD}{k} = 0.76\left(\frac{gD^3\rho^2}{\mu\Gamma}\right)^{1/3} \tag{6.98}$$

where $\Gamma = m_c/2L$. This equation can also be extended to the case of condensation over a bank of n horizontal tubes by taking:

$$h_n = \frac{h_1}{n^{1/4}} \tag{6.99}$$

where h_n is the average of the bank of tubes and h_1 is the condensing coefficient for the top tube. Example 6.6 illustrates the application of some of these relationships.

Example 6.6. Condensation of a vapor over a vertical tube. Let's use the equations presented in this section to estimate the rate of condensation of saturated steam at 100°C on the outside of a 1-in. diameter, 1-m-long tube kept at 30°C. Based on an energy balance, the rate of condensation is related to the rate of heat transfer by the following equation:

$$m_c = \frac{Q}{\Delta\hat{H}_v} = \frac{Ah(T_b - T_s)}{\Delta\hat{H}_v} \tag{6.100}$$

where Q is the heat transfer rate, $\Delta\hat{H}_v$ is the latent heat of vaporization, and $A = \pi DL$ is the heat transfer area. Solving for h, Eq. (6.100) yields:

$$h = \frac{\Delta\hat{H}_v}{T_b - T_s}\frac{m_c}{\pi DL} \tag{6.101}$$

For a vertical tube, combining Eqs. (6.101) and (6.97), we obtain:

$$\frac{\Delta\hat{H}_v}{T_b - T_s}\frac{m_c}{\pi DL} = 0.925\frac{k}{L}\left(\frac{gL^3\rho^2}{\mu m_c/\pi D}\right)^{1/3} \tag{6.102}$$

Finally, solving this equation for m_c we obtain:

$$m_c = 0.943\left(\frac{gL^3\rho^2k^3\Delta T^3\pi^4D^4}{\Delta\hat{H}_v^{\,3}\mu}\right)^{1/4} \tag{6.103}$$

where $\Delta T = T_b - T_s$. For liquid water at 100°C, we have $\rho = 960$ kg/m^3, $k = 0.68$ W/m K, $\mu = 0.00028$ Pa s, and $\Delta\hat{H}_v = 2.25\times10^6$ J/kg. Using these parameters, Eq. (6.103) yields:

$$m_c = 0.943\left(\frac{9.8\times1.0^3\times960^2\times0.68^3\times70^3\times\pi^4\times0.0254^4}{(2.25\times10^6)^3\times0.00028}\right)^{1/4} = 0.0099\ \text{kg/s}$$

Finally, from Eq. (6.101) we obtain:

$$h = \frac{2.25\times10^6}{100-30}\frac{0.0099}{\pi\times0.0254\times1.0} = 3{,}990\ \text{W/m}^2\ \text{K}$$

which is a relatively large value as expected for a condensing vapor. From Eq. (6.94), we can verify that the condensate film is in fact very thin.

6.5.2 Boiling

Vaporization of liquids can occur in a variety of ways including film boiling, which is the vaporization of a thin film of liquid flowing over a hot surface, and pool boiling, which occurs when a large amount of liquid is in contact with the hot surface. Pool boiling is usually dominant in industrial equipment, such as kettle-type reboilers used in distillation applications, and is the focus of this section. Section 5 in *Perry's Chemical Engineers' Handbook*, 9th edition, provides references for other situations. In all cases, the heat transfer rate and the mass rate of vaporization of a saturated liquid, m_v, are related to each other by the following equation:

$$m_v = \frac{qA}{\Delta\hat{H}_v} \tag{6.104}$$

where q is the heat transfer flux, A is the heat transfer area, and $\Delta\hat{H}_v$ is the latent heat of vaporization. Pool boiling occurs through different mechanisms depending mainly on the difference, ΔT, between the surface temperature T_s and the liquid boiling temperature T_b according to the different regimes illustrated qualitatively in Fig. 6.13. In regime I, when the ΔT is small, the heat transfer flux is dominated by natural convection effects associated with buoyancy of the liquid at the surface, which can become slightly superheated. In this case, the change of phase occurs mainly at the liquid–vapor interface, away from the heating surface. Increasing the ΔT brings us into regime II where the flux is dominated by nucleate boiling. In this case, bubbles are formed at the hot surface and rise through the liquid generating large convective flows. In regime III, where the ΔT is even higher, the flux becomes dominated by film boiling. In this case, a layer of vapor forms right at the hot surface resulting in lower fluxes than in regime II because of the relatively large heat transfer resistance offered by this layer of vapor. Note that in this

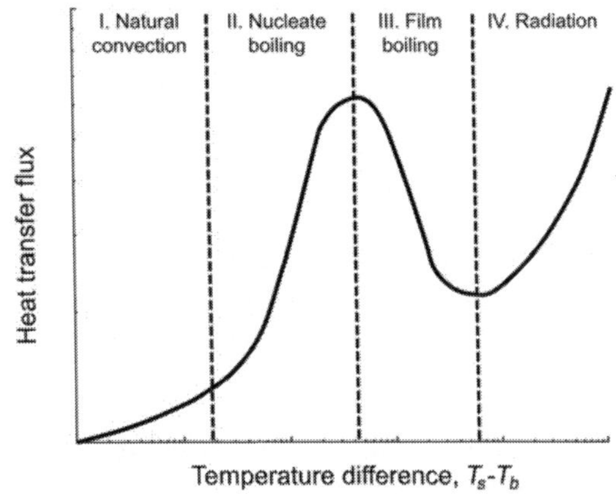

FIGURE 6.13 Qualitative trends of the heat transfer flux for pool boiling of a saturated liquid as a function of the difference between the surface temperature and the boiling temperature.

regime, increasing ΔT actually reduces the heat transfer flux because of the larger thickness of the vapor film formed at the surface. At even higher ΔT values, we enter regime IV where radiant heat transfer becomes important or even dominant and the flux increases again as a function of ΔT. As seen from Fig. 6.13, except for radiant heat transfer, the heat transfer flux is maximum at the transition between regimes II and III. Industrial equipment is typically designed to operate in regime II at a certain percentage of the maximum heat flux in order to have a stable and robust operation.

Two different methods are presented here to predict boiling rates in the nucleate boiling regime (regime II). The first is based on the approaches of Kutateladse and Zuber [Trans. ASME, 80 (1958) 711] and Rohsenow [Trans. ASME, 74 (1952) 969]. According to Kutateladse and Zuber, the maximum heat flux for nucleate boiling can be estimated from the following equation:

$$q_{\max} = 0.15\Delta\hat{H}_v[g\sigma\rho_V^2(\rho_L - \rho_V)]^{1/4} \tag{6.105}$$

where σ is the surface tension of the liquid and ρ_V and ρ_L are the vapor and liquid densities, respectively. According to Rohsenow, the actual flux is given by the following equation:

$$q = \mu_L\Delta\hat{H}_v\left[\frac{g(\rho_L - \rho_V)}{\sigma}\right]^{1/2}\left[\frac{C_{p,L}(T_s - T_b)}{C_{sf}\Delta\hat{H}_v \mathrm{Pr}_L^n}\right]^3 \tag{6.106}$$

where μ_L is the liquid viscosity, $C_{p,L}$ is the liquid heat capacity, and Pr_L is the Prandtl number of the liquid. n is equal to 1 for water and to 1.7 for other liquids. C_{sf} is a numerical constant that varies over an ample range (0.0027–0.015) as a function of the material of which the heating surface is made and of the properties of the liquid being vaporized. Values of C_{sf} for water on various metal surfaces vary between 0.006 for nickel or brass and 0.015 for copper or platinum.

The second method is based on the approach of Mostinskii [Teplenergetika, 4 (1963) 66], who provides the following equations:

$$q_{\max} = aP_c\left(\frac{P}{P_c}\right)^{0.35}\left(1 - \frac{P}{P_c}\right)^{0.9} \tag{6.107}$$

with $a = 0.368$ for use with SI units and:

$$q=\left\{b(T_s-T_b)(P_c)^{0.69}\left[1.8\left(\frac{P}{P_c}\right)^{0.17}+4\left(\frac{P}{P_c}\right)^{1.2}+10\left(\frac{P}{P_c}\right)^{10}\right]\right\}^{1/0.3} \quad (6.108)$$

with $b = 3.60 \times 10^{-5}$ also for use with SI units. In these equations, P is the absolute pressure in the system and P_c is the critical pressure of the liquid being vaporized. This method does not contain material-dependent properties and all of the physical properties of the fluid end up being represented by P_c. Example 6.7 compares the two methods. Section 15 in Rohsenow et al., *Handbook of Heat Transfer*, McGraw-Hill, 1998, should be consulted for additional details.

Example 6.7. Calculation of boiling rate. Let's estimate the boiling rate of water at 1 atm absolute pressure ($T_b = 100$°C) with a heating surface temperature of 110°C. At 100°C, we have the following physical properties:

$$\rho_V = PM_r/RT = 101{,}325\times(18/1000)/8.314\times 373 = 0.59 \text{ kg/m}^3$$

$$\rho_L = 990 \text{ kg/m}^3$$

$$\Delta\hat{H}_v = 2.2\times 10^6 \text{ J/kg}$$

$$C_{p,L} = 4{,}200 \text{ J/kg K}$$

$$\sigma = 0.059 \text{ N/m}$$

$$\mu_L = 0.0003 \text{ Pa s}$$

$$k_L = 0.70 \text{ W/m K}$$

With these values, based on Eq. (6.105) the maximum flux is predicted to be:

$$q_{\max} = 0.15\times 2.2\times 10^6[9.8\times 0.059\times 0.59^2(990-0.59)]^{1/4} = 1.3\times 10^6 \text{ W/m}^2$$

The actual flux will then be:

$$q = 0.0003\times 2.2\times 10^6\left[\frac{9.8(990-0.59)}{0.059}\right]^{1/2}\left[\frac{4{,}200(110-100)}{C_{sf}\times 2.2\times 10^6\times 1.8^1}\right]^3 = \frac{0.319}{C_{sf}^3} \text{ W/m}^2$$

The numerical value of q will depend on C_{sf}. Now let's do the calculation based on the Mostinskii method, which is simpler because it only requires P_c. For water, we have $P_c = 2.21\times 10^7$ Pa. From Eqs. (6.107) and (6.108) with $P/P_c = 1.01\times 10^5/2.21\times 10^7 = 0.0046$ we have, respectively:

$$q_{\max} = 0.368\times 2.21\times 10^7(0.0046)^{0.35}(1-0.0046)^{0.9} = 1.24\times 10^6 \text{ W/m}^2$$

and

$$q = \{3.60\times 10^{-5}(110-100)(2.21\times 10^7)^{0.69}\times[1.8(0.0046)^{0.17}+4(0.0046)^{1.2}+10(0.0046)^{10}]\}^{1/0.3}$$
$$= 8.9\times 10^4 \text{ W/m}^2$$

which gives $q/q_{\max} = 0.073$. The value obtained from Eq. (6.106) is the same as this value if we select $C_{sf} = 0.015$, which is in the range of the experimental values reported by Rohsenow for different surface materials. The corresponding value of the heat transfer coefficient is $h = q/\Delta T = 8{,}940$ W/m^2 K.

6.6 Problems

P.6.1. Air at 200°F and 1 atm absolute pressure is to be cooled to 70°F by passing it through a 2-in. ID copper pipe submerged in an ice bath. The air flow rate is 70 lb/h. Neglecting the heat transfer resistance through the ice bath and through the copper pipe, determine: (a) The heat transfer coefficient inside the pipe. (b) The required length of the pipe. (c) The total rate of heat transfer and the rate at which ice will melt in the ice bath.

P.6.2. We want to cool a solution containing 100 mg/mL of a monoclonal antibody from an initial temperature of 25°C to 10°C by having the solution flow through a coil of stainless-steel tubing immersed in a water bath, which keeps the tube wall at 4°C. The inner diameter of the tube is 1.0 cm. The solution flows at the rate of 10 kg/h and has a viscosity of 20 mPa s at 20°C [Liu et al., J. Pharm. Sci. 94 (2005) 1928]. Assuming that the thermal conductivity of the solution is 0.40 W/m K, that the heat capacity is 4,000 J/kg K, and that the density is 1,030 kg/m^3, determine the length of tubing required.

P.6.3. Quartz-glass spheres 2 cm in diameter and initially at 70°C are cooled by a large toluene stream at 20°C flowing over the spheres at a velocity of 0.01 m/s. Determine: (a) The heat transfer coefficient at the solid/toluene interface. (b) The time required to lower the temperature at center of the spheres to 25°C.

P.6.4. A 10-volt (DC) electrical potential difference is applied to a 1 m long × 1 mm diameter nichrome (80% Ni, 20% Cr) wire. The wire is suspended horizontally in air at 20°C which flows across the wire at a velocity of 5 m/s. Determine the steady-state surface temperature of the wire.

P.6.5. Two options are being considered to cool a can of a soft drink from an initial temperature of 35°C to a maximum temperature of 15°C as quickly as possible. The can is 5 cm in diameter and 20 cm long. Option A: place the can in a refrigerator where air at 4°C is blown over the can perpendicular to its axis of at a velocity of 0.1 m/s. Option B: place the can under a stream water at 10°C flowing perpendicular to its axis at a velocity of 0.5 m/s and covering the entire surface of the can. The soft drink inside the can is assumed to be stagnant in both cases and may be treated as a solid with the physical properties listed below. For both options A and B: (a) Estimate as accurately as possible the convective heat transfer coefficient. (b) Calculate the Biot number and state which resistance is dominant. (c) Estimate the time required to achieve an average temperature of 15°C. (d) Estimate the surface temperature of the can when its average temperature reaches 15°C. Properties of the soft drink: thermal conductivity = 0.5 W/m K, density = 950 kg/m^3, heat capacity = 3,900 J/kg K.

P.6.6. A cylinder of a radioactive solid, 1.5 cm in radius and 2 m long, is encased in a cylindrical shell of pyrex glass, 3.0 cm in outside radius. The radioactive core generates heat at the steady rate of 400 W. The cylindrical assembly is suspended in air at 27°C, which flows crosswise at a speed of 1.0 m/s. The thermal conductivity of the radioactive solid is very high, so that the temperature in the radioactive core is uniform. Assuming steady state: (a) Make a sketch of the temperature profiles for this problem. (b) Write the equation(s) and boundary conditions needed to describe steady-state heat transfer from the radioactive core to the flowing nitrogen. (c) Solve your equation(s) to find an explicit expression for the heat transfer rate (W) from the radioactive core to the flowing nitrogen as a function of the overall ΔT. (d) Estimate the external convective heat transfer coefficient. (e) Determine the temperature of the radioactive core for the conditions described above. (f) Determine the temperature of the outer surface of the glass shell for the conditions described above. (g) Determine what the thermal conductivity of the glass shell material would have to be if it were desired to limit the difference between the temperature of the radioactive core and that of the outer surface of the glass shell to 10°C assuming that all other conditions stay the same.

P.6.7. Plastic spheres, 0.02 m in diameter and initially at 120°C, are cooled by allowing them to drop into a large pool of water at 20°C. The spheres have a density of 1050 kg/m^3, a heat capacity of 900 J/kg K, and a thermal conductivity of 2.8 W/m K. (a) Estimate as accurately as possible the convective heat transfer coefficient. (b) Predict the distance that the spheres would have to fall to reach a temperature of 40°C if the external resistance is assumed to be dominant. (c) Predict the distance that the spheres would have to fall to reach a center temperature of 40°C if the external resistance is neglected. (d) Predict the distance that the spheres would have to fall to reach a center temperature of 40°C if both resistances are considered. (e) For the conditions of part (d), determine the surface temperature of the spheres when the center temperature reaches 40°C.

P.6.8. A square slab of pure silicon to be used to manufacture microelectronics components is to be cooled in air at 25°C from an initial temperature of 200°C to 30°C. The slab is 0.5 m on a side and 5 mm thick and is hung vertically in calm air, suspended by thin metal wires. Neglect radiant heat transfer. (a) Determine the time required to cool the slab to an average temperature of 30°C if the external heat transfer resistance is neglected. (b) Determine the time required to cool the slab to 30°C if natural convection is considered. [Shanks et al., Phys. Rev. 130 (1963) 1743, provide the relevant thermophysical properties of silicon].

P.6.9. Redo problem P.6.4 assuming that there is no forced convection, i.e., all of the heat transfer is by natural convection.

CHAPTER 7

Thermal Design of Heat Exchangers

7.1 Introduction

Heat exchangers are used extensively in many different types of applications including chemical processing, refrigeration, heating and air conditioning, and cooling engines and other heat producing equipment. In the context of chemical processing, heat exchangers are used mainly to heat or cool process streams using different utilities depending on temperature or to exchange heat between a hot process stream and a different colder process stream. For example, one might use a heat exchanger to pre-heat the feed to a chemical reactor used for an exothermic reaction by extracting heat from the reactor effluent. Similarly, one might use the heat available at the condenser of a distillation column to pre-heat the feed to a different column or as the energy source for the reboiler of another column that is operated at a lower temperature.

Utilities normally available as heat sources in chemical process plants include steam, normally available at different pressures and temperatures between 100°C and 200°C, heat transfer fluids, such as Dowtherm A, that, because of their thermochemical stability, can be used at temperatures up to 300°C, and molten salts useful up to 600°C. In the case of steam, condensing saturated steam is usually more efficient since the latent heat of vaporization, which is released upon condensation, is large. Moreover, the heat transfer coefficients for condensing water are large, which help promote fast heat transfer. Low-pressure steam is typically at pressures below 75 psig (T < 121°C), medium-pressure steam is at 75–150 psig (121°C < T < 186°C), and high-pressure steam is at 300 psig (T ~ 220°C). The choice of steam pressure will, thus, depend on the temperature to which we need to heat our process stream. Heat transfer fluids are characterized by being chemically stable at high temperatures. They exchange sensible heat, usually without a change of phase. Molten salts are less commonly used (exceptions are some high-temperature reactors), but their advantage is that higher temperatures can be reached. In general, heating or vaporizing process streams above a few hundred degrees is best accomplished using direct heating in a furnace where energy is supplied by burning a fuel or, in some cases, electricity.

The typical cooling utility in a chemical plant is cooling water. Cooling water is usually treated to prevent corrosion and reduce biofouling and is circulated throughout the plant with a supply temperature in the 15°C to 30°C range. The actual supply temperature will typically depend on environmental conditions. The maximum return temperature of the cooling water is typically around 45°C, since higher temperatures will often lead to precipitation of inorganic salts and consequent fouling of heat exchanger surfaces which can significantly reduce the heat exchanger efficiency. The return, warmed-up cooling water goes to cooling towers where it is contacted with ambient air. In these towers, evaporation of a small fraction of the inlet water causes cooling back down to the supply temperature (see Chap. 15). Plants operated in hot, very humid environments will often use alternative cooling utilities including air-cooled heat exchangers or river, lake, or even sea water in ordinary heat exchangers. Air-cooled heat exchangers are more environmentally friendly. However, since the heat transfer coefficients for air are small, finned surfaces and blowers will often be required to attain reasonable heat transfer rates. The radiator in a gasoline or diesel-powered automobile engine is an example of an air-cooled heat exchanger which takes advantage of the air flow caused by the automobile motion along with finned surfaces. In this case, air flow is supplemented by fans when the automobile is stopped. Using water from natural sources as a coolant is not without problems, including growth of algae and mollusks in pipes and heat exchanger surfaces. Additionally, at large scale, returning the warmed-up water to such natural bodies of water can be environmentally damaging. Cooling to temperatures lower than about 15°C or so will usually require refrigerant fluids that have been chilled in a refrigeration system. Temperatures down to −40°C can be attained with refrigerants, such as ethylene glycol-water mixtures. Lower temperatures will require cryogenic fluids including liquid nitrogen at extremely low temperatures.

Different types of standard heat exchangers are available. Section 11 of *Perry's Chemical Engineers' Handbook*, 9th edition, provides many examples including nomenclature and construction details standardized by the Tubular Exchangers Manufacturers Association (TEMA). Figures 7.1 and 7.2 illustrate two basic types of commonly used heat exchangers: the double-pipe heat exchanger and the shell and tube heat exchanger. In both cases, T_C denotes the temperature of the colder stream, which is heated in the heat exchanger, while T_H denotes the temperature of the hotter stream, which is cooled in the heat exchanger. The double-pipe heat exchanger has the simplest construction

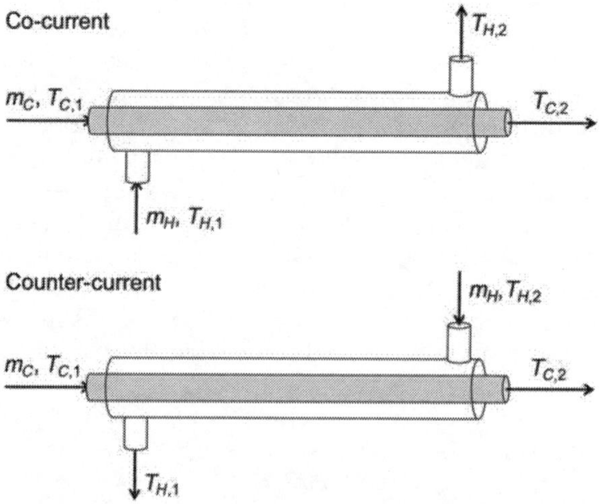

FIGURE 7.1 Double-pipe heat exchanger operated in co-current (top) and counter-current (bottom) mode. m_C and m_H are the mass flow rates of the cold and hot stream, respectively.

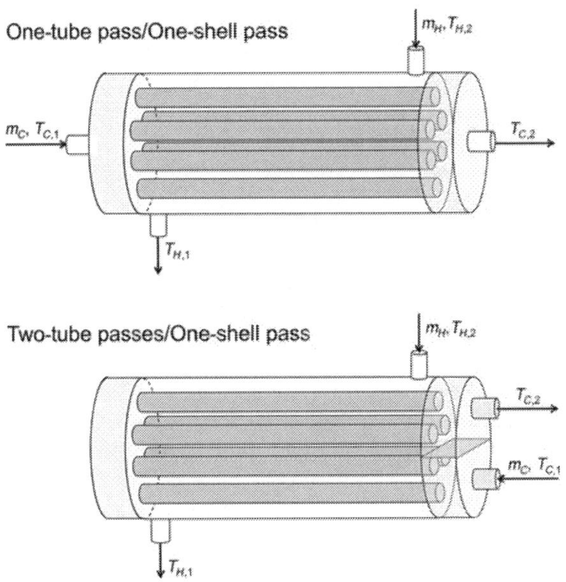

FIGURE 7.2 Examples of shell and tube heat exchangers with a one-tube pass/one-shell pass design (top) and with a two-tube passes/one-shell pass design (bottom).

consisting of one tube inside another. One stream (a process stream or a utility stream) flows inside the inner pipe, while the other flows in the annular space between the inner pipe and the outer pipe. Heat transfer between the two streams occurs across the wall of the inner pipe. The heat transfer rate attained in this arrangement will be affected, in general, by the convective heat transfer resistances on both sides of the inner pipe and by the conductive heat transfer resistance across the inner pipe wall. Heat losses or gains to/from the surrounding will also occur across the wall of the outer pipe, but these can be mitigated by insulation and are often neglected in the initial design of practical heat exchangers especially if the scale is large.

Double-pipe heat exchangers can be operated either co-currently or counter-currently as shown in Fig. 7.1. While simple in construction and easy to clean and maintain, double-pipe heat exchangers can provide only a limited heat transfer area. Increasing the pipe length beyond a few meters becomes impractical unless the double-pipe is coiled or arranged in multiple connected sections in a serpentine configuration. The shell and tube heat exchanger is preferred, however, when large heat transfer areas are needed. Figure 7.2 shows two examples of shell and tube heat exchangers: one with a one-tube pass/one-shell pass design and the other with two-tube passes/one-shell pass design. In both cases, the heat exchanger consists of a bundle of parallel tubes contained inside a shell and connected to inlet and outlet ports through headers at both ends of the tube bundle. In the first case, with one-tube pass, one stream flows inside the inner tubes at one end and exists from the tubes at the other of the heat exchanger, while the other stream flows in the shell. Ideally, this type of heat exchanger is equivalent to multiple double-pipe heat exchangers operated in parallel and can be either counter-current (as shown in Fig. 7.2) or co-current. Construction is simpler than having multiple double pipe heat exchangers or one with very long tubes. However, the heat transfer efficiency can be lower compared to that of a true double-pipe type if the shell is relatively large since flow in the shell may not be ideal. In this case, vertical baffles will often be placed at regular spacings within the shell to help avoiding by-passing of the shell-side fluid. While better than without baffles, the serpentine flow path will only approximate a true counter-current operation. The second example shown in Fig. 7.2 is a shell and tube heat exchanger with two-tube passes and one-shell pass. This design includes a horizontal divider in the header on the right-hand side of the tube bundle, which separates the tubes in two different sections—one above and one below the divider. As shown in this figure, as an example, the cold stream at $T_{C,1}$ enters the bottom section of tubes below the divider, flows from right to left into the header on the left-hand side, and then flows from left to right through the top section of tubes above the divider. Note that, in this arrangement, the hot stream flow is co-current to that of the cold stream in the bottom section of tubes and counter-current to that of the cold stream in the top section. The advantage of this arrangement is that the tube length can be reduced compared to that of a single-pass heat exchanger while still providing a large heat transfer area. At the same time, the heat transfer efficiency can become lower than that of a single-pass system since the flow will be partly counter-current and partly co-current. Additional types of heat exchangers, including shell and tube heat exchangers with multiple number of tube and shell passes, cross-current, and compact plate and frame heat exchangers, are described in Section 11 of *Perry's Chemical Engineers' Handbook*, 9th edition.

Regardless of the type of heat exchangers, given a certain heating or cooling task, thermal design will require a determination of the following quantities:

1. The total heat transfer rate or heat transfer duty, Q_t, which is the total heat exchanged between two process streams or between a process stream and a utility stream
2. The value of any unspecified stream temperature
3. The required heat transfer area, A

These quantities are determined by combining energy balances with a rate equation that describes the rate of heat transfer between the two streams. The latter will depend on the stream temperatures and on the resistances to heat transfer. Once the area is determined, the actual size of the heat exchanger will depend on its physical arrangement and on specific construction details. In practical situations, several decisions will need to be made including:

a. Selecting whether co-current or counter-current will be used
b. Selecting which stream or utility will flow inside the pipes or in the shell
c. Selecting the heat exchanger type depending on physical space, weight, cleanability, maintenance, and minimum fluid velocity constraints
d. Selecting materials of construction depending on temperature, pressure, corrosiveness, and fouling characteristics of the streams

In general, as we will see in detail below, counter-current operation is more efficient and usually preferred. Co-current may be preferable, however, for the rapid cooling or heating of a process stream. If either stream is condensing or boiling at constant temperature, then there will be no difference between co-current and counter-current from the thermal efficiency viewpoint. In this case, other factors, such as physical proximity to external piping, will determine the preferred arrangement. Selecting whether a stream will flow inside the tubes or outside will depend on multiple factors including pressure (usually higher-pressure fluids will be more conveniently accommodated inside the pipes), corrosiveness (usually corrosive fluids will also be better handled inside the pipes), and pressure drop constraints (usually gases and viscous fluids will be on the shell-side where the pressure drop is often lower). The stream temperature can also play a role in this selection. For example, for safety reasons, we may not want to have a very hot stream on the shell-side of an un-insulated heat exchanger that plant operators might inadvertently come in physical contact with. Materials of constructions will, of course, depend on fluid properties. Stainless steel, titanium, or glass-lined tubes might be needed in some cases when handling corrosive fluids. While exhibiting better corrosion resistance, these materials (especially glass) will also often have lower thermal conductivity than carbon steel or copper and can be more expensive.

Other practical considerations involve the physical orientation of the heat exchanger (horizontal or vertical), the direction of flows in and out of the heat exchanger, and its location in the plant. Heat exchangers without change of phase can normally be either horizontal or vertical, with the decision often depending on space constraints and access for maintenance. For instance, it may be easier to remove the tube bundle from the shell for cleaning or maintenance if the heat exchanger is horizontal. The direction of flow matters, however, if changes of phase take place. For instance, flow of cold liquid streams that contain dissolved gases (including cooling water, which is usually air-saturated) will normally be from the bottom of the heat exchanger up. This will facilitate removal of air or other gases that are liberated as the liquid is warmed up in the heat exchanger and their solubility in the liquid is reduced. Accumulation of such gases in the heat exchanger reduces the effective heat transfer area. Thus, venting through a valve located at the top of the heat exchanger will often be necessary. Condensing streams, on the other hand, will typically flow from the top of the heat exchanger down in order to allow proper flow of the condensate. In these cases, a "steam trap," located at the bottom of the heat exchanger, will regulate the outflow allowing only condensate to leave the heat exchanger. Condensers operated below atmospheric pressure, such as those used in low-pressure distillation columns, will often be elevated above ground level. "Barometric condensers" of this type allow the condensate to flow out of a low-pressure heat exchanger by having the liquid flow out through a long vertical tube discharging to atmospheric pressure. Reboilers used in connection with distillation columns are often horizontal. For example, "Kettle" type reboilers comprise a horizontal tube bundle, which provides heat within a large domed shell. Pool boiling (*cf.* Section 6.5.2) will occur over the tube bundle, which is submerged in the liquid being evaporated, with a large dome allowing space for liquid droplets to disengage from the vapor. Vertical "thermosyphon" reboilers take advantage of natural convection effects during boiling to provide fluid flow.

7.2 Thermal Design of Double-Pipe Heat Exchangers

7.2.1 Systems with No Change of Phase

Let's consider first the thermal design of a co-current heat exchanger as sketched in a simplified form in Fig. 7.3. In this case, both streams enter at the "1" end of the heat exchanger with temperatures $T_{C,1}$ and $T_{H,1}$ and leave at the "2" end of the heat exchanger

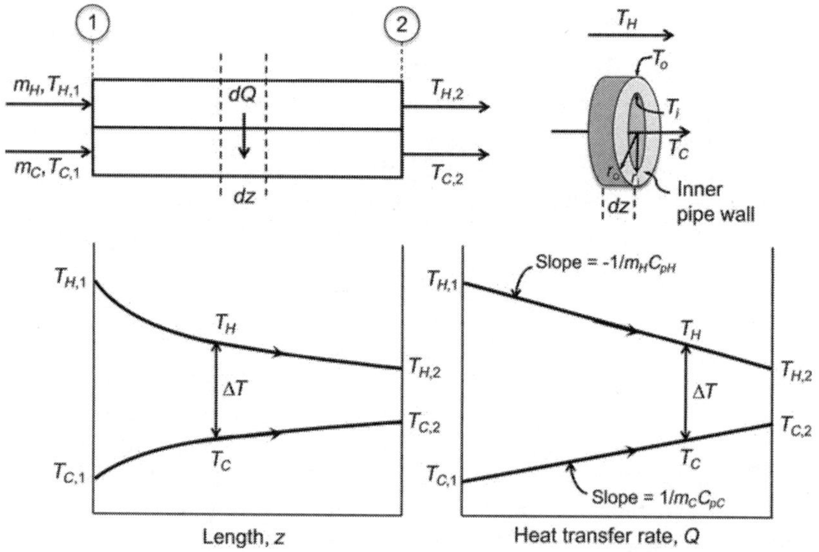

FIGURE 7.3 Temperature profiles for a double-pipe co-current heat exchanger.

with corresponding temperatures $T_{C,2}$ and $T_{H,2}$. The expected temperature profiles along the length of the heat exchanger are also sketched in Fig. 7.3. Note that these profiles are non-linear when plotted along the length of the heat exchanger. Heat transfer will be from the hot stream to the cold stream. Thus, as seen in this figure, the hot stream temperature decreases from the 1-end of the heat exchanger to the 2-end while the cold stream temperature increases in the same direction. Since heat flows from high temperature to low temperature, we must have both $T_{H,1} > T_{C,1}$ and $T_{H,2} > T_{C,2}$ in order to have heat transfer from the hot stream to the cold stream. Thus, in this case, the absolute minimum temperature to which we will be able cool the hot stream will be the outlet temperature of the cold stream. In practice, however, we will need a substantial difference between $T_{H,2}$ and $T_{C,2}$ in order to have a substantial driving force and, thus, reasonably high rates of heat transfer.

We will use the terms $C_{p,C}$ and $C_{p,H}$ to indicate the mean heat capacities of the cold stream and of the hot stream, respectively. Their values will be averaged over the temperature changes that the streams experience in the heat exchanger. Neglecting heat losses or gains to or from the surroundings, separate energy balances for the cold stream and for the hot stream give the following relationships:

$$Q_t = m_H C_{p,H}(T_{H,\text{in}} - T_{H,\text{out}}) = m_H C_{p,H}(T_{H,1} - T_{H,2}) \tag{7.1}$$

$$Q_t = m_C C_{p,C}(T_{C,\text{out}} - T_{C,\text{in}}) = m_C C_{p,C}(T_{C,2} - T_{C,1}) \tag{7.2}$$

where Q_t is the total rate of heat exchange between the two streams. By rearranging Eqs. (7.1) and (7.2) we obtain:

$$\frac{1}{m_H C_{p,H}} = \frac{T_{H,1} - T_{H,2}}{Q_t} \tag{7.3}$$

$$\frac{1}{m_C C_{p,C}} = \frac{T_{C,2} - T_{C,1}}{Q_t} \tag{7.4}$$

Finally, adding Eqs. (7.3) and (7.4) together gives the following result:

$$\frac{1}{m_H C_{p,H}} + \frac{1}{m_C C_{p,C}} = \frac{T_{H,1} - T_{H,2} + T_{C,2} - T_{C,1}}{Q_t} = \frac{\Delta T_1 - \Delta T_2}{Q_t} \tag{7.5}$$

where $\Delta T_1 = T_{H,1} - T_{C,1}$ and $\Delta T_2 = T_{H,2} - T_{C,2}$ are the temperature differences between hot and cold streams at the 1-end of the heat exchanger and at the 2-end of the heat exchanger, respectively. Next, since our goal is to determine the actual temperature profile, we will consider separate differential energy balances for each stream as they flow through a differential section of the heat exchanger with thickness dz as sketched in Fig. 7.3 (top left). These energy balances relate the differential amount of heat exchanged in this section to the differential temperature changes experienced by each stream according to the following equations:

$$dQ = -m_H C_{p,H} dT_H \tag{7.6}$$

$$dQ = m_C C_{p,C} dT_C \tag{7.7}$$

where dQ is the differential amount of heat exchanged between the two streams in this section. Note that we have $-dT_H$ in Eq. (7.6) and $+dT_C$ in Eq. (7.7) since the hot stream temperature decreases while the cold stream temperature increases as heat is exchanged. Rearranging Eqs. (7.3) and (7.4), just as we did for the macroscopic balances, gives the following result:

$$\frac{dT_H}{dQ} = -\frac{1}{m_H C_{p,H}} \tag{7.8}$$

$$\frac{dT_C}{dQ} = +\frac{1}{m_C C_{p,C}} \tag{7.9}$$

These equations show that if we can treat $C_{p,C}$ and $C_{p,H}$ as constants, which is a reasonable assumption in many cases, the slopes dT_H/dQ and dT_C/dQ will also be constants. Thus, plotting T_H and T_C vs. the cumulative rate of heat transfer along the length of the heat exchanger, Q, will give straight lines as sketched in Fig. 7.3 (bottom right). As we will see shortly, the linearity of this plot will help us understand the behavior of heat exchangers. Subtracting Eq. (7.8) from Eq. (7.9) and considering Eq. (7.5), we obtain the following result:

$$\frac{dT_H}{dQ} - \frac{dT_C}{dQ} = -\left(\frac{1}{m_H C_{p,H}} + \frac{1}{m_C C_{p,C}}\right) = -\frac{\Delta T_1 - \Delta T_2}{Q_t} \tag{7.10}$$

This equation is the differential version of Eq. (7.5) and can also be written as follows:

$$\frac{d\Delta T}{dQ} = -\frac{\Delta T_1 - \Delta T_2}{Q_t} \tag{7.11}$$

where $\Delta T = T_H - T_C$ is the local temperature difference between hot and cold streams. Note that in this equation, Q_t is the total rate of heat transfer across the full length of the heat exchanger while dQ is the differential rate of heat transfer over a length dz of the heat exchanger.

In order to proceed further and integrate Eq. (7.11), we need an independent relationship between dQ and ΔT accounting for the resistance to heat transfer. In general, there will be three resistances associated with: (1) convective transport in the fluid outside the pipe, (2) conductive heat transfer through the pipe wall, and (3) convective transport in the fluid inside the pipe. As shown in Sec. 3.3 and with reference to Fig. 7.3 (top right), the following relationships will describe these three processes:

$$dQ = 2\pi r_o h_o (T_H - T_o) dz \tag{7.12}$$

$$dQ = \frac{2\pi k}{\ln(r_o/r_i)} (T_o - T_i) dz \tag{7.13}$$

$$dQ = 2\pi r_i h_i (T_i - T_C) dz \tag{7.14}$$

where r_o and r_i are the outside and inside radii of the inner tube, respectively, h_o and h_i are the convective heat transfer coefficients outside and inside the inner tube, respectively, and k is the thermal conductivity of the tube wall itself. Since these three processes are in series, we can combine the three equations above to eliminate the unknown temperatures T_o and T_i and obtain an expression for the overall resistance. To do this, we rearrange Eqs. (7.12) to (7.14) as follows:

$$\frac{dQ}{2\pi r_o h_o dz} = T_H - T_o \tag{7.15}$$

$$\frac{dQ}{\dfrac{2\pi k}{\ln(r_o/r_i)} dz} = T_o - T_i \tag{7.16}$$

$$\frac{dQ}{2\pi r_i h_i dz} = T_i - T_C \tag{7.17}$$

Adding these three equations, we obtain the following result:

$$\left[\frac{1}{2\pi r_o h_o} + \frac{\ln(r_o/r_i)}{2\pi k} + \frac{1}{2\pi r_i h_i}\right] \frac{dQ}{dz} = T_H - T_C \tag{7.18}$$

where dQ/dz is now a function only of the overall temperature difference between the hot and the cold streams. The quantity $2\pi r_o dz$ is equal to the differential heat transfer area dA_o measured on the outside of the inner tube. Multiplying Eq. (7.18) by $2\pi r_o$ gives the result:

$$\frac{dQ}{dA_o} = U_o (T_H - T_C) = U_o \Delta T \tag{7.19}$$

where:

$$U_o = \left[\frac{1}{h_o} + \frac{r_o \ln(r_o/r_i)}{k} + \frac{1}{h_i}\frac{r_o}{r_i}\right]^{-1} \tag{7.20}$$

is the overall heat transfer coefficient. Note that we use the subscript "o" to indicate that U_o is based on the outside area of the inner pipe. We can now combine Eqs. (7.11) and (7.19) by replacing dQ with the quantity $U_o \Delta T dA_o$ according to Eq. (7.19) which gives:

$$\frac{d\Delta T}{\Delta T} = -\frac{\Delta T_1 - \Delta T_2}{Q_t} U_o dA_o \tag{7.21}$$

The very last step is to integrate this equation from one end of the heat exchanger, where $\Delta T = \Delta T_1$, to the other, where $\Delta T = \Delta T_2$, as follows:

$$\int_{\Delta T_1}^{\Delta T_2} \frac{d\Delta T}{\Delta T} = -\frac{\Delta T_1 - \Delta T_2}{Q_t} \int_0^{A_o} U_o dA_o \tag{7.22}$$

If U_o is treated as a constant, performing the integrals we obtain:

$$\ln \frac{\Delta T_1}{\Delta T_2} = \frac{\Delta T_1 - \Delta T_2}{Q_t} U_o A_o \tag{7.23}$$

This equation is most often written in the following easy-to-remember form:

$$Q_t = U_o A_o \Delta T_{lm} \tag{7.24}$$

where:

$$\Delta T_{lm} = \frac{\Delta T_1 - \Delta T_2}{\ln \dfrac{\Delta T_1}{\Delta T_2}} \tag{7.25}$$

is the "logarithmic mean" temperature difference. We can see that if, for a given ΔT_1, ΔT_2 approaches zero (i.e., if the outlet temperatures of the hot and cold stream become equal to each other), ΔT_{lm} also approaches zero. In this case, the heat transfer area $A_o = Q_t / U_o \Delta T_{lm}$ would become infinite. Thus, it is critical that, in practice, we maintain a finite ΔT between the two streams in order to have a reasonable area.

Let's now consider the case of a counter-current double-pipe heat exchanger as sketched in Fig. 7.4. Note that, compared to Fig. 7.3, we have switched the flow direction of the hot stream but have left the temperature symbols unchanged. Thus, in this case, $T_{C,1}$ and $T_{C,2}$ are still the inlet and outlet temperatures of the cold streams but now $T_{H,2}$ is the inlet temperature of the hot stream while $T_{H,1}$ is its outlet temperature. As seen in Fig. 7.4 (bottom left), in this case, both the hot stream temperature and the cold stream temperature increase from left to right. As before, T_H and T_C are also non-linear functions when plotted versus the length of the heat exchanger. In general, however, compared to the co-current case, $\Delta T = T_H - T_C$ will typically vary less extensively as we move from one end of the heat exchanger to the other. The energy balances for the counter-current configuration are as follows:

$$Q_t = m_H C_{p,H}(T_{H,\text{in}} - T_{H,\text{out}}) = -m_H C_{p,H}(T_{H,1} - T_{H,2}) \tag{7.26}$$

$$Q_t = m_C C_{p,C}(T_{C,\text{out}} - T_{C,\text{in}}) = m_C C_{p,C}(T_{C,2} - T_{C,1}) \tag{7.27}$$

These equations are the same as those that we wrote for the co-current case except that, for the counter-current case, a minus sign precedes m_H on the right-hand side of

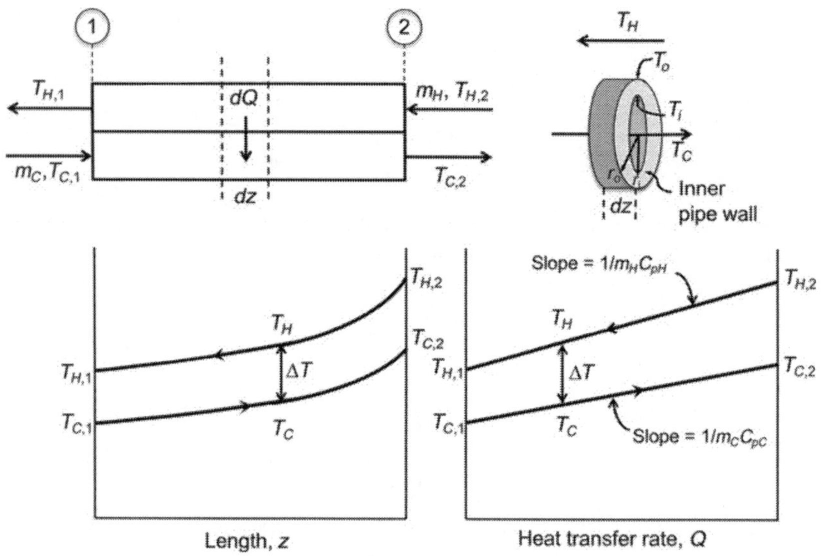

FIGURE 7.4 Temperature profiles for a double-pipe counter-current heat exchanger.

Eq. (7.26) since the direction of flow of this stream is reversed. Rearranging Eqs. (7.26) and (7.27), as we did for the co-current case, we have:

$$\frac{1}{m_H C_{p,H}} = -\frac{T_{H,1} - T_{H,2}}{Q_t} \tag{7.28}$$

$$\frac{1}{m_C C_{p,C}} = \frac{T_{C,2} - T_{C,1}}{Q_t} \tag{7.29}$$

Subtracting the second of these equations from the first, we obtain the following result:

$$\frac{1}{m_H C_{p,H}} - \frac{1}{m_C C_{p,C}} = \frac{T_{H,2} - T_{H,1} - T_{C,2} + T_{C,1}}{Q_t} = -\frac{\Delta T_1 - \Delta T_2}{Q_t} \tag{7.30}$$

Note that the only differences between this equation and the corresponding Eq. (7.5) for the co-current case are the minus signs.

The next step is to consider the differential energy balances. With reference to the heat exchanger section with thickness dz shown in Fig. 7.4 (top left), we can write:

$$dQ = m_H C_{p,H} dT_H \tag{7.31}$$

$$dQ = m_C C_{p,C} dT_C \tag{7.32}$$

or:

$$\frac{dT_H}{dQ} = \frac{1}{m_H C_{p,H}} \tag{7.33}$$

$$\frac{dT_C}{dQ} = \frac{1}{m_C C_{p,C}} \tag{7.34}$$

Note again that the only difference between these equations and Eqs. (7.8) and (7.9) for the co-current case is that now both dT_H/dQ and dT_C/dQ are positive. Thus, just like for the co-current case, the temperature profiles T_H and T_C will be linear when plotted vs. Q provided that we can treat the heat capacities as constants. This time, subtracting Eq. (7.34) from Eq. (7.33) yields:

$$\frac{dT_H}{dQ} - \frac{dT_C}{dQ} = \frac{1}{m_H C_{p,H}} - \frac{1}{m_C C_{p,C}} = -\frac{\Delta T_1 - \Delta T_2}{Q_t} \tag{7.35}$$

or, since $T_H - T_C = \Delta T$,

$$\frac{d\Delta T}{dQ} = -\frac{\Delta T_1 - \Delta T_2}{Q_t} \tag{7.36}$$

Note that Eq. (7.19) is also applicable to the counter-current case since it was derived by considering only the local temperatures T_H and T_C. As a result, combining Eqs. (7.19) and (7.36), we will also obtain:

$$Q_t = U_o A_o \Delta T_{lm} \tag{7.37}$$

where, again,

$$\Delta T_{lm} = \frac{\Delta T_1 - \Delta T_2}{\ln \dfrac{\Delta T_1}{\Delta T_2}} \tag{7.38}$$

This expression for Q_t is exactly the same as that found for the co-current case. The difference, of course, lies in the meaning of ΔT_1 and ΔT_2. For the counter-current case, ΔT_1 is the difference between the outlet temperature of the hot stream and the inlet temperature of the cold streams while ΔT_2 is the difference between the inlet temperature of the hot stream and the outlet temperature of the cold stream. A special situation occurs in a counter-current heat exchanger when $m_H C_{p,H} = m_C C_{p,C}$. In this case, the slopes of the T vs. Q lines are the same. As a result, the ΔT between streams is constant and, thus, $\Delta T_{lm} = \Delta T_1 = \Delta T_2$. Example 7.1 illustrates the application of these equations comparing co-current and counter-current configurations.

Example 7.1. Thermal design of co-current and counter-current double-pipe heat exchangers. Let's use the equations developed in this section to compare co-current and counter-current double-pipe heat exchanger configurations. In this example, let's say that we want to recover energy from a hot process stream at 110°C with $m_H C_{p,H} = 500$ W/K by exchanging heat in a double-pipe heat exchanger with another process stream with $m_C C_{p,C} = 700$ W/K that needs to be heated from its initial temperature of 30°C. We will assume that the overall heat transfer coefficient for the proposed double-pipe heat exchanger is $U_o = 400$ W/m^2 K. We will consider two different versions of this problem based on different specifications. In the first case, we specify that the outlet temperature of the hot stream is to be 70°C. In the second case, we will let the outlet temperature of the hot stream be a variable.

For the first of these two cases, setting $T_{H,\text{out}} = 70$°C, we can use Eq. (7.1) directly to find the heat exchanger duty (i.e., the rate at which heat needs to be exchanged) as follows:

$$Q_t = m_H C_{p,H}(T_{H,\text{in}} - T_{H,\text{out}}) = 500(110 - 70) = 20{,}000 \text{ W}$$

We can then use Eq. (7.2) to find the outlet temperature of the cold stream as:

$$T_{C,\text{out}} = T_{C,\text{in}} + \frac{Q_t}{m_C C_{p,C}} = 30 + \frac{20,000}{700} = 58.6°\text{C}.$$

These two results will be the same regardless of configuration. For a co-current configuration, we will have $T_{H,1} = 110$, $T_{C,1} = 30$, $T_{H,2} = 70$, and $T_{C,2} = 58.6$ all in °C. Thus, the ΔT logarithmic mean will be:

$$\Delta T_{lm} = \frac{(110-30)-(70-58.6)}{\ln\left(\dfrac{110-30}{70-58.6}\right)} = 35.2°\text{C}$$

As a result, we obtain:

$$A_o = \frac{Q_t}{U_o \Delta T_{lm}} = \frac{20,000}{400 \times 35.2} = 1.42\ \text{m}^2$$

The T vs. Q diagram for this case is shown in Fig. 7.5a. As expected, the minimum temperature difference between the two streams occurs at the exit end of the heat exchanger (the 2-end, in this example) and is $\Delta T_{\min} = 70 - 58.6 = 11.4°\text{C}$.

For a counter-current configuration, we will have $T_{H,1} = 70$, $T_{C,1} = 30$, $T_{H,2} = 110$, and $T_{C,2} = 58.6$ all in °C. The corresponding T vs. Q diagram is given in Fig. 7.5b. We can see that, in this case, the minimum temperature difference occurs at the "cold end" of the heat exchanger (the 1-end, in this example) where $\Delta T_{\min} = 70 - 30 = 40°\text{C}$. This minimum ΔT is larger than the minimum found for the co-current case. We now have:

$$\Delta T_{lm} = \frac{(70-30)-(110-58.6)}{\ln\left(\dfrac{70-30}{110-58.6}\right)} = 45.5°\text{C}$$

Thus,

$$A_o = \frac{Q_t}{U_o \Delta T_{lm}} = \frac{20,000}{400 \times 45.5} = 1.10\ \text{m}^2$$

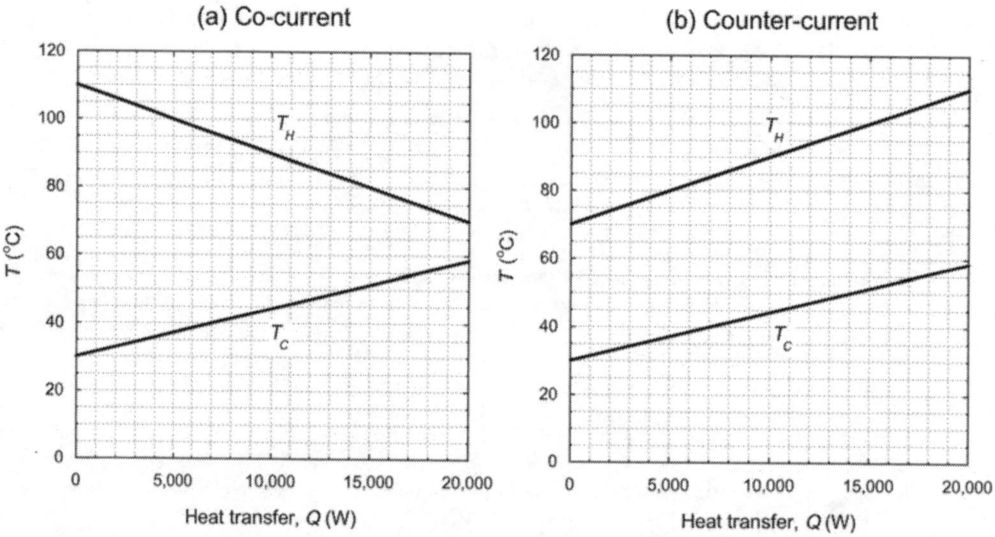

FIGURE 7.5 *T-Q* plots for the double-pipe heat exchanger in Example 7.1 with (a) co-current and (b) counter-current configurations.

We can see that, while the rate of heat transfer and, thus, the energy recovered is the same for both configurations, the area for the counter-current configuration is smaller than that for the co-current configuration (by about 23%). This occurs because, for the counter-current configuration, the ΔT_{lm} is larger. In both cases, we can calculate the length of the heat exchanger based on the diameter of the inner tube. Of course, the co-current configuration will be longer.

Let's now consider a second case where we relax the specification that $T_{H,out}$ must be equal to 70°C. We will let that vary, instead, by selecting different values of the minimum ΔT. In this case, the amount of energy recovered, Q_t, and the heat transfer area required, A_o, will depend on what value we choose for ΔT_{min}. We call the end of the heat exchanger where $\Delta T = \Delta T_{min}$ the "pinched" end of the heat exchanger. In our problem, a smaller ΔT_{min} will lead to greater energy recovery, but also to a larger heat transfer area since the driving force for heat transfer will be smaller. In more practical, economic terms, a higher ΔT_{min} will mean that we save more money by recovering more energy, but at the expense of a higher investment cost which will be required for the larger area and, hence, for the larger heat exchanger needed.

Let's consider only the counter-current configuration, since we have already determined that it is more efficient. For this case, the pinch will be at the cold end of the heat exchanger (i.e., the 1-end in this case). Thus, we will have $T_{H,1} = T_{C,1} + \Delta T_{min}$. In combination with Eqs. (7.26) and (7.27), this relationship leads to the following result:

$$
\begin{aligned}
T_{C,2} &= \frac{(m_C C_{p,C} - m_H C_{p,H})T_{C,1} + m_H C_{p,H}(T_{H,2} - \Delta T_{min})}{m_C C_{p,C}} \\
&= \frac{(700-500)30 + 500(110 - \Delta T_{min})}{700} = \frac{61{,}000 - \Delta T_{min}}{700}
\end{aligned}
\tag{7.39}
$$

For a given ΔT_{min}, we can then calculate Q_t, ΔT_{lm}, and A_o from the following relationships:

$$
\begin{aligned}
Q_t &= m_H C_{p,H}(T_{H,2} - T_{H,1}) = m_H C_{p,H}[T_{H,2} - (T_{C,1} + \Delta T_{min})] \\
&= 500(80 - \Delta T_{min})
\end{aligned}
\tag{7.40}
$$

$$
\begin{aligned}
\Delta T_{lm} &= \frac{(T_{H,1} - T_{C,1}) - (T_{H,2} - T_{C,2})}{\ln \dfrac{T_{H,1} - T_{C,1}}{T_{H,2} - T_{C,2}}} = \frac{(T_{C,1} + \Delta T_{min} - T_{C,1}) - (T_{H,2} - T_{C,2})}{\ln \dfrac{T_{C,1} + \Delta T_{min} - T_{C,1}}{T_{H,2} - T_{C,2}}} \\
&= \frac{\Delta T_{min} - (110 - T_{C,2})}{\ln \dfrac{\Delta T_{min}}{110 - T_{C,2}}}
\end{aligned}
\tag{7.41}
$$

$$
A_o = \frac{Q_t}{U_o \Delta T_{lm}} = \frac{Q_t}{400 \Delta T_{lm}}
\tag{7.42}
$$

The corresponding values of Q_t and A_o for this case are plotted in Fig. 7.6 as a function of ΔT_{min}. As expected, increasing ΔT_{min} will reduce A_o, which is beneficial, but will also reduce the heat recovered Q_t, which is obviously counter-productive. In order to select an optimum ΔT_{min}, we need to assign a monetary value to the energy recovered, say \$0.01/kWh, and an investment cost for the heat exchanger of, say, \$2,500 per square meter of surface area. The net annual revenue will then be the difference between the value of the energy recovered per year and the annualized cost of the heat exchanger. For our case, let's assume that we operate for 8,000 h per year and that we amortize the heat exchanger cost over 10 years; i.e., the investment cost is charged at the rate of \$250 per year per square meter of heat exchanger area. Using these values, we obtain the following expression for the net annual revenue:

$$
\text{Net annual revenue (\$/year)} = 8{,}000 \times Q_t \times \frac{0.01}{1{,}000} - 250 \times A_o
\tag{7.43}
$$

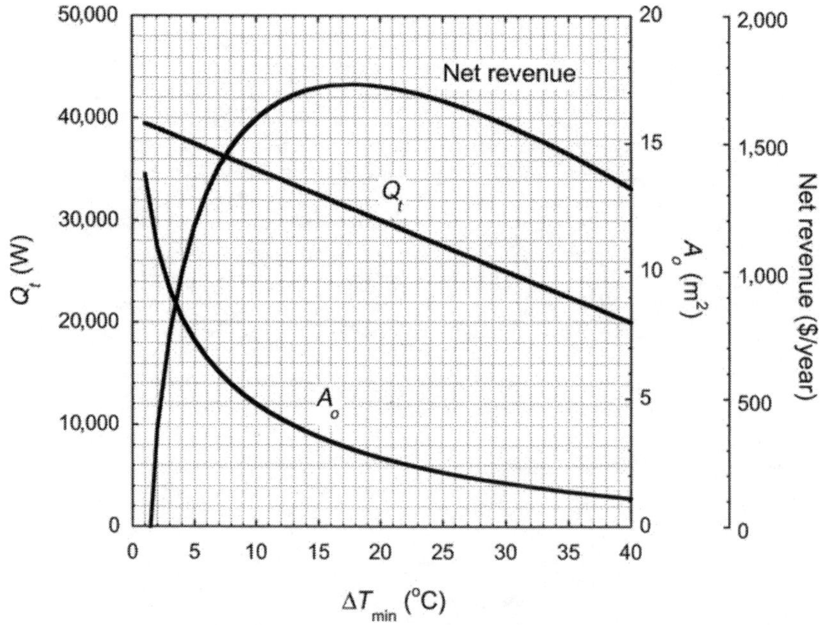

FIGURE 7.6 Optimization of ΔT_{min} for the double-pipe, counter-current heat exchanger in Example 7.1.

with Q_t in units of W and A_o in units of m^2. This function is also plotted in Fig. 7.6. We can see that the net revenue will be maximum with ΔT_{min} ~16°C, which we can consider as the economic optimum. In general, the exact value of the optimum ΔT_{min} will depend, of course, on the specific parameters chosen and on the cost basis. In practice, however, ΔT_{min} values that fall in the range of 10°C to 20°C are actually commonly found for many real systems.

7.2.2 Systems with Changes of Phase

The equations presented in Sec. 7.2.1 are also applicable to cases where one or both streams undergo phase changes provided that the energy balances are re-written to account for the latent heat of vaporization. For a condenser with a saturated vapor feed, the energy balances become:

$$Q_t = m_H \Delta \hat{H}_v \tag{7.44}$$

$$Q_t = m_C C_{p,C}(T_{C,\text{out}} - T_{C,\text{in}}) \tag{7.45}$$

while for a reboiler with a saturated liquid feed, we have:

$$Q_t = m_H C_{p,H}(T_{H,\text{in}} - T_{H,\text{out}}) \tag{7.46}$$

$$Q_t = m_C \Delta \hat{H}_v \tag{7.47}$$

In a sense, we can think of a condenser as a heat exchanger designed to cool by one degree a stream with a heat capacity $C_p = \Delta \hat{H}_v$ or of a reboiler as a heat exchanger designed to heat by one degree a stream with a heat capacity $C_p = \Delta \hat{H}_v$.

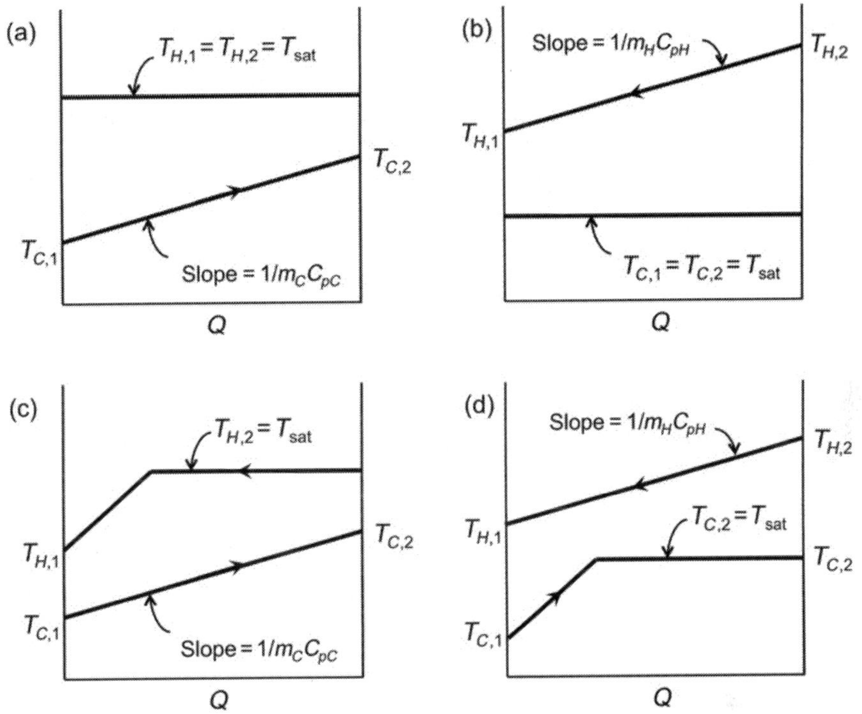

Figure 7.7 *T-Q* diagrams for double-pipe heat exchangers with (a) isotherm condensation; (b) isothermal vaporization; (c) condensation with sub-cooling; and (d) vaporization of a subcooled liquid.

Figures 7.7a and 7.7b show T-Q plots for a hypothetical isothermal condenser and a hypothetical isothermal reboiler, respectively. In both cases, these plots assume that the change of phase occurs at constant temperature. This will happen for the isobaric change of phase of a pure component. In other cases, where the pressure changes significantly over the length of the heat exchanger or for mixtures, the temperature will still change. For isothermal condensation or vaporization, Eq. (7.24) is still valid but now ΔT_{lm} assumes either of the following two simplified forms. For a condenser, we will have:

$$\Delta T_{lm} = \frac{\Delta T_1 - \Delta T_2}{\ln \dfrac{\Delta T_1}{\Delta T_2}} = \frac{(T_{sat} - T_{C,1}) - (T_{sat} - T_{C,2})}{\ln \dfrac{T_{sat} - T_{C,1}}{T_{sat} - T_{C,2}}} = \frac{T_{C,2} - T_{C,1}}{\ln \dfrac{T_{sat} - T_{C,1}}{T_{sat} - T_{C,2}}} \tag{7.48}$$

while for a reboiler, we will have

$$\Delta T_{lm} = \frac{\Delta T_1 - \Delta T_2}{\ln \dfrac{\Delta T_1}{\Delta T_2}} = \frac{(T_{H,1} - T_{sat}) - (T_{H,2} - T_{sat})}{\ln \dfrac{T_{sat} - T_{C,1}}{T_{sat} - T_{C,2}}} = \frac{T_{H,1} - T_{H,2}}{\ln \dfrac{T_{H,1} - T_{sat}}{T_{H,2} - T_{sat}}} \tag{7.49}$$

where T_{sat} is the saturation temperature of the vapor being condensed or of the liquid being vaporized. In either case, the heat exchanger configuration, i.e., whether co- or counter-current, will not affect ΔT_{lm}. As a result, co-current and counter-current configurations will have the same thermal efficiency.

We can also consider situations where a saturated vapor is condensed and sub-cooled or where a subcooled liquid is heated and vaporized in the same heat exchanger. The corresponding *T-Q* diagrams for these cases are shown in Fig. 7.7c and 7.7d. In these cases, the *T-Q* curves are segmented. For example, for a condenser with sub-cooling (Fig. 7.7c), the slope of the T_H profile will be zero in the isothermal condensing region of the heat exchanger and $1/m_H C_{p,H}$ in the sub-cooling region where $C_{p,H}$ is the condensate heat capacity. Conversely, for a vaporizer with a subcooled liquid feed (Fig. 7.7d), the T_C profile will have a slope $1/m_C C_{p,C}$, where $C_{p,C}$ is the heat capacity of the subcooled liquid, in the heating region of the heat exchanger but will be flat in the region where isothermal vaporization occurs.

Example 7.2. Thermal design of a condenser with sub-cooling. Consider the design of a double-pipe heat exchanger designed to condense 0.1 kg/s of saturated toluene vapor at 1 atm absolute pressure and to cool the condensate to 40°C. This will be accomplished in one heat exchanger using cooling water entering at 20°C and leaving at 45°C. Overall heat transfer coefficients are U_o = 1,000 W/m² K in the condensing region of the heat exchanger and U_o = 500 W/m² K in the sub-cooling region. This scenario is illustrated qualitatively in Fig. 7.7c. For design purposes, we can imagine that the heat exchanger is split into two hypothetical exchangers in series as shown in Fig. 7.8.

In this figure, T'_C is the intermediate temperature of the cooling water. The energy balances for the two hypothetical heat exchangers are as follows:

$$Q_1 = m_H C_{p,H}(T_{sat} - T_{H,1}) = m_C C_{p,C}(T'_C - T_{C,1}) \tag{7.50}$$

$$Q_2 = m_H \Delta\hat{H}_v = m_C C_{p,C}(T_{C,2} - T'_C) \tag{7.51}$$

where Q_1 and Q_2 are, respectively, the total heat transfer rate in the sub-cooling and in the condensing regions of the actual heat exchanger. For toluene at 1 atm, we have T_{sat} = 110.6°C and $\Delta\hat{H}_v = 4.2 \times 10^5$ J/kg. For liquid toluene C_p ~ 1,700 J/kg K and for liquid water C_p ~ 4,200 J/kg K. Thus, from Eqs. (7.50) and (7.51) we have:

$$Q_1 = 0.1 \times 1,700(110.6 - 40) = m_C \times 4,200(T'_C - 20)$$

$$Q_2 = 0.1 \times 4.2 \times 10^5 = m_C \times 4,200(45 - T'_C)$$

Solving this set of equations, we obtain:

$$Q_1 = 12,000 \text{ W}$$

$$Q_2 = 42,000 \text{ W}$$

$$T'_C = 25.6°\text{C}$$

$$m_C = 0.51 \text{ kg/s}$$

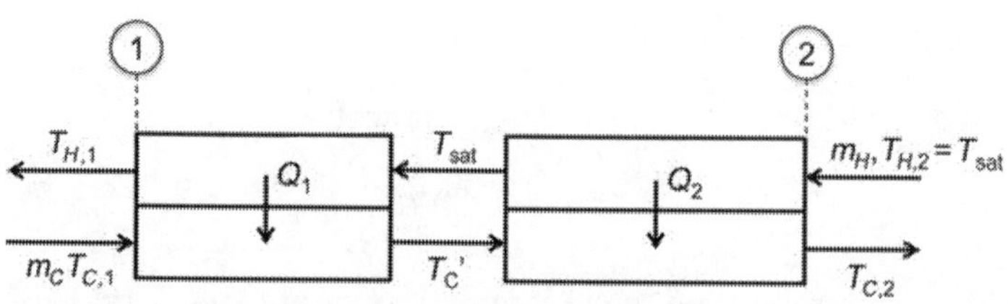

FIGURE 7.8 Hypothetical arrangement of counter-current heat exchangers for Example 7.2.

The heat exchange areas are then calculated based on Eq. (7.37) as follows:

$$A_{o,1} = \frac{Q_1}{U_{o,1}\Delta T_{lm,1}} = \frac{12{,}000}{500[(40-20)-(110.6-25.6)]/\ln[(40-20)/(110.6-25.6)]} = 0.53 \text{ m}^2$$

$$A_{o,2} = \frac{Q_1}{U_{o,2}\Delta T_{lm,2}} = \frac{42{,}000}{1{,}000(45-25.6)/\ln[(110.6-25.6)/(110.6-45)]} = 0.56 \text{ m}^2$$

7.3 Thermal Design of Shell and Tube Heat Exchangers

The thermal design of shell and tube heat exchangers with one-tube pass and one-shell pass is based on the same equations as those developed for double-pipe heat exchangers. From the computational viewpoint, the only difference is that the heat transfer area will be distributed among all the tubes that are contained within the shell. In practical applications, the total number of tubes, N_{total}, will vary from a few to as many as 1,000 or even more, depending on the magnitude of the heat transfer area needed according to the following equation:

$$A_o = 2\pi r_o L N_{\text{total}} \tag{7.52}$$

where r_o is the outside radius of the pipes and L is the pipe length. There are, of course, practical limits to the length of the pipes. When the required length is excessive, a solution is to use multiple passes as sketched, for example, in Fig. 7.2. As already noted, a drawback of multiple passes is, however, that the heat transfer efficiency is reduced compared to that of a pure, counter-current configuration. For example, if we consider two-tube passes and one-shell pass, flow of the tube-side fluid will be split among the two different passes with the direction of flow reversed from one section to the other. As a result, flow in the tubes relative to flow in the shell will be co-current in one pass and counter-current in the other. Thus, the efficiency of the heat exchanger will be intermediate between those of pure co-current and pure counter-current heat exchangers. In practical calculations, this effect is accounted for by introducing a correction factor, F_T, according to the following equation:

$$Q_t = U_o A_o \Delta T_{lm} F_T(S,R) \tag{7.53}$$

where:

$$\Delta T_{lm} = \frac{(T_{H,\text{out}} - T_{C,\text{in}}) - (T_{H,\text{in}} - T_{C,\text{out}})}{\ln\left(\dfrac{T_{H,\text{out}} - T_{C,\text{in}}}{T_{H,\text{in}} - T_{C,\text{out}}}\right)} \tag{7.54}$$

is the logarithmic-mean temperature difference that would be calculated for a true counter-current configuration (i.e., for a heat exchanger with one tube-pass and one-shell pass). The correction factor F_T is a function of the parameters S and R that are defined by the following equations:

$$S = \frac{T_{\text{tube,out}} - T_{\text{tube,in}}}{T_{\text{shell,in}} - T_{\text{tube,in}}} \tag{7.55}$$

$$R = \frac{T_{\text{shell,in}} - T_{\text{shell,out}}}{T_{\text{tube,out}} - T_{\text{tube,in}}} = \frac{m_{\text{tube}} C_{p,\text{tube}}}{m_{\text{shell}} C_{p,\text{shell}}} \tag{7.56}$$

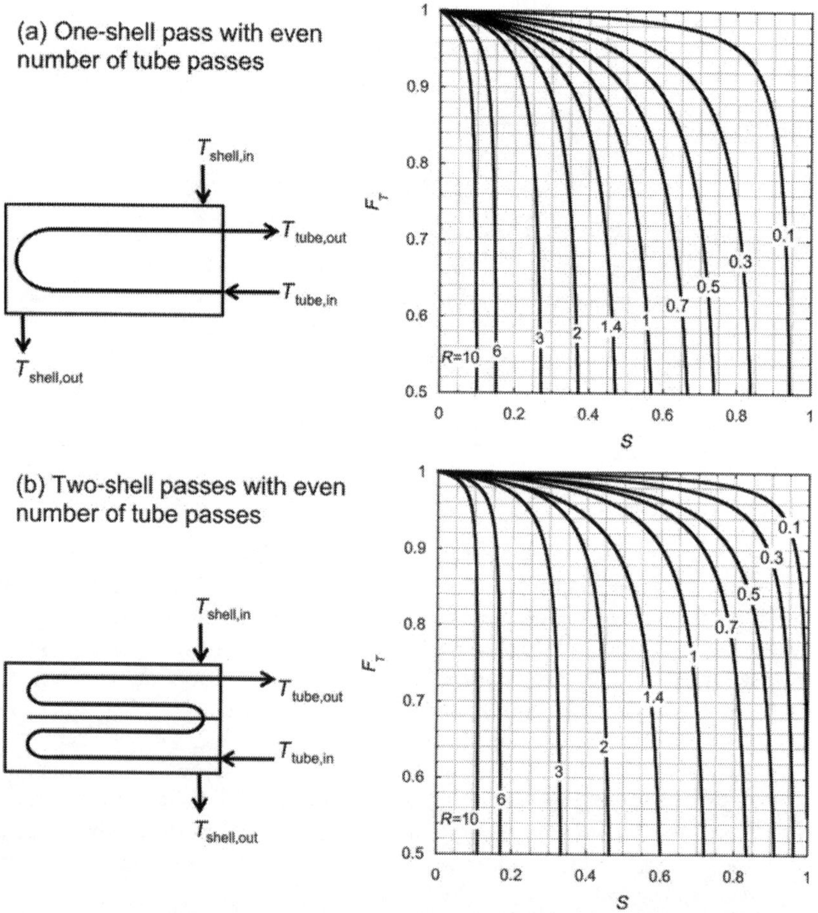

Figure 7.9 Correction factor for shell and tube heat exchangers with multiple passes: (a) one-shell pass and two or more even number of tube passes and (b) two-shell passes and four or more even number of tube passes. $S = (T_{tube,out} - T_{tube,in})/(T_{shell,in} - T_{tube,in})$, $R = (T_{shell,in} - T_{shell,out})/(T_{tube,out} - T_{tube,in})$. Mathematica notebooks: (M.F.7.9a) and (M.F.7.9b).

where T_{tube} and T_{shell} refer to the tube-side and shell-side streams, respectively. The function $F_T(S,R)$ is available in graphical form (e.g., see Fig. 11-4 in *Perry's Chemical Engineers' Handbook*, 9th edition) for many different heat exchanger configurations. Figure 7.9 gives examples for heat exchangers with one-shell pass and two or more even number of tube passes and for heat exchangers with two-shell passes and four or more even number of tube passes. As seen from these figure, small values of R result in values of F_T close to 1. This makes sense because when R is small, the temperature change of the shell-side fluid is small. In this case, the T-Q line will be nearly flat which means that co-current and counter-current performance will be nearly the same. As a result, having multiple-tube passes will be similar to the result obtained with just one pass. As seen from these figures, for a given R, there is a value of S beyond which F_T drops dramatically, which, in turn, would make the heat transfer area very large for a desired heat transfer duty. Obviously, these are conditions that should be avoided. In practice, conditions will be chosen so that F_T is greater than 0.85, before the drop-off. Mathematica notebooks M.F.7.9a and M.F.7.9b provide analytical expressions for F_T in the two cases shown in Fig. 7.9. Example 7.3 illustrates the application of these relationships.

Example 7.3. Thermal design of a shell and tube heat exchanger. Let's design a shell and tube heat exchanger to heat a stream with $m_C C_{p,C} = 5{,}000$ BTU/h°F from 90°F to 210°F flowing in the tubes using as the heat source a stream with $m_H C_{p,H} = 10{,}000$ BTU/h°F entering the shell at 260°F. The heat exchanger will contain tubes with a 0.5 in. outside diameter. In order to maintain a reasonable fluid velocity inside the tubes, $N_t = 30$ tubes per pass will be used. The overall heat transfer coefficient is $U_o = 150$ BTU/h ft²°F. Regardless of configuration, the energy balances give the following results:

$$Q_t = m_C C_{p,C}(T_{C,\text{out}} - T_{C,\text{in}}) = 5{,}000(210-90) = 600{,}000 \text{ BTU/h}$$

$$T_{H,\text{out}} = T_{H,\text{in}} - \frac{Q_t}{m_H C_{p,H}} = 260 - \frac{600{,}000}{10{,}000} = 200°\text{F}$$

The logarithmic mean temperature difference for a counter-current heat exchanger is calculated from Eq. (7.38), which gives:

$$\Delta T_{lm} = \frac{(200-90)-(260-210)}{\ln\left(\dfrac{200-90}{260-210}\right)} = 76.1°\text{F}$$

For a heat exchanger with one-tube pass and one-shell pass, the correction factor F_T is unity. Thus, the heat exchanger area will be:

$$A_o = \frac{Q_t}{U_o \Delta T_{lm}} = \frac{600{,}000}{150 \times 76.1} = 52.6 \text{ ft}^2$$

The required tube length will thus be:

$$L = \frac{A_o}{2\pi r_o N_t} = \frac{52.6}{2\pi(0.25/12)\times 30} = 13.4 \text{ ft}$$

Suppose that we need to have shorter tubes. We can have two shorter heat exchangers in series or we can have one heat exchanger with two-tube passes. In the latter case, we will need to calculate S and R and get F_T. For S and R, we obtain the following values:

$$S = \frac{T_{\text{tube,out}} - T_{\text{tube,in}}}{T_{\text{shell,in}} - T_{\text{tube,in}}} = \frac{210-90}{260-90} = 0.71$$

$$R = \frac{T_{\text{shell,in}} - T_{\text{shell,out}}}{T_{\text{tube,out}} - T_{\text{tube,in}}} = \frac{m_{\text{tube}} C_{p,\text{tube}}}{m_{\text{shell}} C_{p,\text{shell}}} = \frac{5{,}000}{10{,}000} = 0.5$$

From Fig. 7.9a or, more precisely, from M.F.7.9a, we get $F_T = 0.7$. In this case, the heat transfer area will be 52.6/0.7 = 70.1 ft², while, with 30 tubes per pass and two-tube passes, the tube length will be:

$$L = \frac{A_o}{2\pi r_o N_t N_{\text{tube passes}}} = \frac{70.1}{2\pi(0.25/12)\times 30 \times 2} = 8.92 \text{ ft}$$

Since F_T is relatively small, a better design and an even shorter tube length could be obtained if we had four-tube passes and two-shell passes. S and R would remain the same. However, in this case, from Fig. 7.9b or from M.F.7.9b, we would get $F_T \sim 0.95$. Thus, A_o would become 52.6/0.95 = 55.4 ft² while the tube length would become:

$$L = \frac{55.4}{2\pi(0.25/12) \times 30 \times 4} = 3.52 \text{ ft}$$

7.4 Heat Transfer Coefficients in Heat Exchangers

Predicting heat transfer coefficient for one-phase flow in double-pipe heat exchangers is relatively straightforward following the methods presented in Sec. 6.4. The Gnielinski equation (Table 6.1, item B) is recommend to predict h_i for the usually turbulent flow conditions inside the tubes. For h_o, with turbulent flow in the annular space between the inner pipe with outer diameter D_o and the outer pipe with inner diameter D_e, Gnielinski [Heat Trans. Eng. 36 (2015) 9] provides the following correlation valid for long tubes:

$$\text{Nu} = \frac{(f/2)(\text{Re} - 1{,}000)\,\text{Pr}}{1.0 + 12.7\sqrt{f/2}\,(\text{Pr}^{2/3} - 1)} \left(\frac{\mu}{\mu_s}\right)^{0.11} F_{ann} \tag{7.57}$$

$$\text{with } f/2 = 0.125[0.78\ln(\text{Re}^*) - 1.5]^{-2} \quad \text{and Re} > 4{,}000$$

This equation has the same form as that used for flow inside a circular pipe. However, the dimensionless groups that appear in this equation are defined differently according to the following relationships:

$$\text{Nu} = \frac{h_o D_H}{k} \tag{7.58}$$

$$\text{Re} = \frac{v_{\text{avg}} D_H}{\nu} \tag{7.59}$$

$$\text{Re}^* = \text{Re}\frac{(1+a)^2 \ln a + (1-a^2)}{(1-a)^2 \ln a} \tag{7.60}$$

$$F_{\text{ann}} = 0.75a^{-0.17} \tag{7.61}$$

where v_{avg} is the average fluid velocity in the annular space, $D_H = D_e - D_o$, and $a = D_o/D_e$.

In the case of shell and tube heat exchangers, the same correlations for pipe flow also apply to the tube-side coefficient. Note, however, that, in this case, the flow is split between the number of tubes present in each pass and the Reynolds number should be calculated based on the average velocity for an individual tube. The shell-side heat transfer coefficients in shell and tube heat exchangers are often more difficult to predict because flow in the shell is usually more complex. Baffles are typically used to direct the shell-side flow creating complex flow patterns. In this case, correlations specific for shell and tube heat exchangers should be used and examples of which can be found in Section 11 of *Perry's Chemical Engineers' Handbook*, 9th edition.

Typical heat transfer coefficient values are often used for preliminary design calculations. Table 11-3 in Perry's Handbook provides typical value of overall heat transfer coefficients found in many practical systems. For illustrative purposes,

System	h (W/m² K)
Steam condensation (film type)	2,500–20,000
Boiling water (pool boiling)	1,800–3,000
Boiling organics (pool boiling)	1,000–1,500
Condensing organic vapors	1,000–1,600
Cooling water	1,000–1500
Heat transfer oils	1,000–1,500
Light organic liquids	800–1,000
Heavy organic liquids	600–900
Oils	100–600
Air and low-pressure gases (forced convection)	20–100

*Data from Sinnott and Towler, *Chemical Engineering Design*, Elsevier, 2019, 6th edition.

Table 7.1 Typical Values of Individual Heat Transfer Coefficients for Preliminary Design of Shell and Tube Heat-Exchangers*

Table 7.1 provides a selection of such typical values. We can see that, as expected, the highest values are observed for condensation and boiling while the lowest values are obtained for viscous oils and heavy organic liquids and for air and other gases at low pressures.

Fouling of heat transfer surfaces is a common problem affecting many practical heat exchanger applications. The term fouling is used to denote the accumulation of materials at the fluid-solid interface that increase the heat transfer resistance and, thus, reduce the overall heat transfer coefficient. Fouling can occur through different mechanisms including biofilm formation, corrosion of the metal surface, decomposition of reactive or otherwise unstable fluids, and deposition of precipitates. The latter is common when using water as a coolant since the solubility of calcium and magnesium sulfate salts decreases as the temperature increases. As a result, precipitation can occur as the coolant temperature increases along the heat exchangers. The effects of fouling can become so large that periodic cleaning is required. This can be accomplished with chemical washes and/or by mechanical means. In order to account for reduced heat transfer rates that will occur as a result of fouling, heat exchangers are typically designed with a safety factor that accounts for the fouling resistance. In this case, if U_o is the overall heat transfer coefficient for the unfouled heat exchanger, a corrected heat transfer coefficient accounting for the fouling resistance, $U_{o,d}$, is defined as follows:

$$U_{o,d} = \left(\frac{1}{U_o} + \frac{1}{h_{d,i}} \frac{r_o}{r_i} + \frac{1}{h_{d,o}} \right)^{-1} \tag{7.62}$$

where $h_{d,i}$ and $h_{d,o}$ are fouling coefficient for the tube-side and the shell-side fluid, respectively. The reciprocals of their values represent the additional heat transfer resistance caused by fouling. Table 7.2 provides some typical values of h_d encountered with different fluids. A smaller value of h_d indicates greater fouling resistance.

System	h_d (W/m² K)
Steam	10,000–2,000
Overhead vapors in crude oil distillation	5,000
Naphta and light distillates	5,000
Cooling water	4,000
Lake or well water	3,000
Crude oil at velocity >1 m/s	2,500
Crude oil at velocity <1 m/s	1,500

*Data from Bennett and Myers, *Momentum, Heat and Mass Transfer*, McGraw-Hill, 3rd edition, 1982, and Towler and Sinnott, *Chemical Engineering Design*, Elsevier, 2008.

TABLE 7.2 Typical Values of Fouling Coefficients for Preliminary Design of Shell and Tube Heat-Exchangers*

Example 7.4. Estimation of heat transfer coefficients for a shell and tube heat exchanger. Let's use what we learned in this section to estimate the overall heat transfer coefficient for a shell and tube heat exchanger designed to heat 1 kg/s of liquid butanol from 40°C to 80°C on the tube side using saturated steam at 75 psig on the shell side. The heat exchanger will have a total of fifty 1/2-in. nominal size, Schedule 40 stainless steel (Type 308) tubes with two-tube passes and one-shell pass. To solve this problem, we first estimate h_i for the tube-side fluid. From Table 2-146 in *Perry's Chemical Engineers' Handbook*, 9th edition, at the average temperature of 60°C, the relevant properties of liquid butanol are $\rho = 780$ kg/m³, $\mu = 0.00093$ Pa s, $k = 0.16$ W/m K, and $C_p = 2{,}600$ J/kg K. Thus, $\mathrm{Pr} = \mu C_p/k = 15$. From Perry's Table 10-22, for 1/2-in.-diameter "Schedule 40" pipes, actual inner and outer diameters are $D_i = 0.0158$ m and $D_o = 0.0213$ m, respectively. Since there are two tube passes, there are 25 tubes per pass. Thus, the flow rate per tube will be $m_{\text{tube}} = 1/25 = 0.04$ kg/s. In turn, the Reynolds number in each tube will be:

$$\mathrm{Re} = \frac{\rho v_{\text{avg}} D_i}{\mu} = \frac{4 m_{\text{tube}}}{\pi \mu D_i} = \frac{4 \times 0.04}{\pi \times 0.00093 \times 0.0158} = 3{,}500$$

Since Re > 2,100, the flow will be turbulent. Using the Gnielisnki correlations (item B in Table 6.1), we obtain:

$$\frac{f}{2} = 0.125[0.79\ln(\mathrm{Re}) - 1.64]^{-2} = 0.125[0.79\ln(3{,}500) - 1.64]^{-2} = 0.0054$$

and:

$$\mathrm{Nu} = \frac{(f/2)(\mathrm{Re} - 1{,}000)\,\mathrm{Pr}}{1.0 + 12.7\sqrt{f/2}\,(\mathrm{Pr}^{2/3} - 1)}\left(\frac{\mu}{\mu_s}\right)^{0.11} \sim \frac{(0.0054)(3{,}500 - 1{,}000) \times 15}{1.0 + 12.7\sqrt{0.0054}\,(15^{2/3} - 1)} = 35.3$$

Thus:

$$h_i = \frac{\mathrm{Nu} \times k}{D_i} = \frac{35.3 \times 0.16}{0.0158} = 357 \text{ W/m}^2\text{ K}$$

Next, we calculate the heat transfer resistance across the pipe wall. From Perry's Table 2-150, the thermal conductivity of 308 stainless steel is about 8.8 BTU/h ft °F or 15.2 W/m K. Thus, the conductive resistance across the pipe wall will be:

$$\frac{r_o \ln(r_o/r_i)}{k} = \frac{0.0213}{2}\frac{\ln(0.0213/0.0158)}{15.2} = 0.00021 \text{ m}^2\text{ K/W}$$

Finally, from Table 7.1, we can estimate that h_o will be at least 2,500 W/m² K. Since this value is so much larger than h_i, we likely do not need a more precise estimate of h_o. If we take $h_o = 2{,}500$ W/m² K, based on Eq. (7.20), the overall heat transfer coefficient will be:

$$U_o = \left[\frac{1}{h_o} + \frac{r_o \ln(r_o/r_i)}{k} + \frac{1}{h_i}\frac{r_o}{r_i}\right]^{-1} = \left(\frac{1}{2{,}500} + 0.00021 + \frac{1}{357}\frac{0.0213}{0.0158}\right)^{-1}$$

$$= (0.0004 + 0.00021 + 0.0038)^{-1} = 228 \text{ W/m}^2\text{K}$$

We can see that the tube-side convective resistance will be largely dominant. We can now account for fouling using fouling coefficients $h_{d,i} = 5{,}000$ and $h_{d,o} = 2{,}000$ W/m² K according to Table 7.2. Using these values, the overall heat transfer coefficient for the fouled heat exchanger will be:

$$U_{o,d} = \left(\frac{1}{228} + \frac{1}{5{,}000}\frac{2.13}{1.58} + \frac{1}{2{,}000}\right)^{-1} = 194 \text{ W/m}^2\text{ K}$$

7.5 Problems

P.7.1. A double-pipe heat exchanger is to be designed to cool 5.5 gal/min of a hot oil from 250°F to 120°F using 11 gal/min of cooling water available at 70°F. The exchanger is to consist of 12-ft-long sections of ¾-in.-diameter copper tubing placed inside 1½-in.-diameter, well-insulated tubing, with water flowing in the annular space between the two tubes. How many sections of copper tubing are required if the flow of the two streams is (a) counter-current and (b) co-current? The overall heat transfer coefficient is $U_o = 110$ BTU/h ft² °F, the heat capacity of the oil is 0.55 BTU/lb °F, and the density of the oil is 52 lb/ft³. Assume that there is no heat transfer to the surrounding.

P.7.2. Assume that the oil in problem P.7.1 is an organic liquid coming from a chemical reactor. It is necessary to "quench" this hot oil as quickly as possible to stop the reaction and prevent its degradation into a useless by-product. (a) Would you use a counter-current or a co-current configuration in your double-pipe heat exchanger? Plot T vs. Q to explain your answer. (b) Suggest a type of heat exchanger that might be better than a double-pipe for this quenching operation.

P.7.3. Dolphins and other marine mammals are able to maintain a surprisingly high body temperature even though they are swimming in ocean water at 4°C. Since the dolphin tail has a large surface area, much of the heat loss occurs there. (a) Develop a model to describe heat transfer from the tail of the dolphin to the ocean water assuming that the tail is represented as a two-tube pass heat exchanger. Blood enters the tail at 40°C with a flow rate of 0.3 kg/s and returns to the body of the dolphin at a lower temperature. The overall heat transfer coefficient in the tail is $U_o = 420$ W/m² K, and the heat transfer area is $A_o = 3.0$ m². Determine the temperature of the blood returning to the main part of the body of the dolphin for these conditions. (b) Develop a model to describe heat transfer from the tail of the dolphin to the water assuming that the tail is represented again as a two-tube pass heat exchanger, but that prior to returning to the main part of the body the blood passes through an internal counter-current heat exchanger where heat transfer occurs between the warm blood going to the tail and the cooler blood returning from the tail. The overall heat transfer coefficient in this internal heat exchanger is $U_o = 630$ W/m² K and its heat transfer area is $A_o = 2.0$ m². Determine the temperature of the blood returning to the main part of the body of the dolphin under these conditions. Also determine the fraction of heat loss that is saved by this arrangement compared to that in part (a). For this problem, assume that the properties of blood are approximated by those of water. It is possible that the ears of elephants and other large mammals work in a similar way with an internal heat exchanger by-passed when cooling is needed and turned on to conserve energy. [Problem originally proposed by Prof. O. Levenspiel, Oregon State Univ.].

P.7.4. A double-pipe heat exchanger consists of 20 ft of 2-in. "Schedule 40" steel pipe (see Table 10-22 in *Perry's Chemical Engineers' Handbook*, 9th edition), jacketed with 3-in. "Schedule 40" steel pipe. Hot water at the rate of 3,000 lb/h enters the inside pipe under pressure at 300 °F, and 2,230 lb/h of cold water enters the jacket at 60 °F. Calculate the temperatures of the streams leaving the heat exchanger and the total heat transfer rate for (a) counter-current and (b) co-current configurations. The convective heat transfer coefficients inside and outside the tubes are $h_i = 1{,}200$ and $h_o = 1{,}175$ BTU/h ft^2 °F, respectively.

P.7.5. A shell and tube heat exchanger being used to heat 0.80 kg/s of oil from 27°C to 150°C is made of one hundred 1-in.-diameter 16-BWG-gage copper pipes (see Table 11-12 in *Perry's Chemical Engineers' Handbook*, 9th edition), which are 4 m long. The oil is heated by saturated steam condensing on the shell side at 10 bar absolute pressure. The heat capacity of the oil is 3,350 J/kg K and the steam-side heat transfer coefficient is 4,550 W/m^2 K. Using these data, calculate the oil-side heat transfer coefficient h_i.

P.7.6. Acetone vapor at 150°C and 1 atm absolute pressure enters the shell side of a vertical counter-current shell and tube heat exchanger with one-tube pass and one-shell pass at the rate of 0.1 kg/s and leaves as a saturated liquid at essentially the same pressure. Cooling is provided by cooling water with a maximum exit temperature of 50°C. Determine: (a) The recommended direction of flow of acetone and cooling water in the heat exchanger (up or down). (b) The flow rate and inlet temperature of the cooling water if it is desired to have a $\Delta T_{min} = 10$°C. (HINT: plot T_H and T_C vs. Q). (c) The surface area of the heat exchanger if the overall heat transfer coefficients are $U_o = 1{,}000$ W/m^2 K in the condensing section of the heat exchanger and $U_o = 100$ W/m^2 K in the non-condensing section.

P.7.7. A shell and tube heat exchanger with two-tube passes consists of a total of 3,000 straight copper tubes inside a single shell. The tubes have an outer diameter of 2.5 cm and an inner diameter of 2.0 cm. Cooling water entering at 20°C flows on the tube side at the rate of 1,000 kg/s. Condensation of a vapor occurs on the shell side at constant temperature of 56°C with a heat transfer coefficient $h_o = 11{,}000$ W/m^2 K. The heat transfer duty of the condenser is 6.7×10^7 W. Determine the temperature of the cooling water leaving the condenser and the length of the tubes.

P.7.8. A liquid A is heated from 20°C to 60°C flowing on the tube side of a shell and tube heat exchanger at a flow rate of 44 kg/s. Another liquid B flowing on the shell side is cooled from 100°C to a to-be-determined temperature. The heat exchanger has one-shell pass and four-tube passes. The shell inner diameter is 30 in. and contains 1,024 copper tubes ¾-in. diameter 14 BWG, each 3.5 m long. The shell side heat transfer coefficient is 900 W/m^2 K. Using the data below, calculate the outlet temperature of the shell-side liquid and its flow rate in kg/h.

Data:

Heat capacity of liquid A = 1,990 J/kg K

Heat capacity of liquid B = 1,700 J/kg K

Viscosity of liquid A = 2.9 mPa s

Thermal conductivity of liquid A = 0.137 W/m K

Density of liquid A = 830 kg/m^3

The above properties can be assumed to remain constant with temperature.

P.7.9. A shell and tube heat exchanger with one-tube pass and one-shell pass has been designed to cool a hot oil on the shell side from 70°C to 40°C using cooling water that on the tube side entering at 20°C and leaving at 40°C. The product mC_p for the oil is 1,500 W/K. The heat exchanger has 100 tubes made of a metal alloy with $k = 5.0$ W/m K, each 2.0 m long, 0.02 m in outer diameter, and 0.015 m in inner diameter. The tube-side convective heat transfer coefficient is $h_i = 1{,}000$ W/m^2 K. An automatic control system keeps the outlet temperature of the oil at the desired 40°C by adjusting the flow rate of the cooling water. When the heat exchanger is first put in service, it works exactly according to the design specifications. However, after a few months in service, an engineer notes that, while the oil is being cooled to the correct temperature, the outlet temperature of the cooling water has dropped to 35°C. Fouling is suspected as the culprit. (a) Determine the heat exchanger duty, the cooling water flow rate, and the overall heat transfer coefficient for the clean heat exchanger. (b) Determine the cooling water flow rate and the fouling coefficient for the fouled heat exchanger.

P.7.10. A shell and tube heat exchanger is to be designed to preheat the feed to a reactor with $m_C C_{p,C} = 800$ W/K from 20°C to 90°C using waste saturated steam at 1 atm absolute pressure. The overall heat transfer coefficient is $U_o = 300$ W/m² K. (a) Determine the heat transfer area required. (b) The heat exchanger is built according to the specifications in part (a). However, when it is put in operation, it is found that it is gradually fouled as a function time on stream with a fouling coefficient given by:

$$\frac{1}{h_d} = a \times t$$

where a = 0.0004 m² K/W h and t is the time on stream in hours. Cleaning the heat exchanger restores the original conditions but takes 4 h. It is proposed to operate the existing heat exchanger on a cycle during which the heat exchanger spends a number of hours on stream followed by a 4-h cleaning period. Develop an expression to calculate the cumulative amount of heat exchanged during the time on stream and use this expression to find the cycle time that maximizes the amount of heat exchanged during 1 week (168 h). Note that the number of cycles per week will be $168/(t + 4)$.

PART II

Mass Transfer

CHAPTER 8

Introduction to Mass Transfer

8.1 Introduction

We use the term "mass transfer" to describe a process where there is preferential transport of a component in a mixture driven by a mole fraction or by a mass fraction gradient. Mass transfer can occur as a result of diffusion and/or convection. Just like heat conduction, diffusion can occur in a stagnant fluid or even in a solid. On the other hand, convection occurs at a fluid–solid interface or at the boundary between two immiscible phases, when there is relative motion between the two. Mass transfer problems are often more complex than heat transfer problems since they involve mixtures and varying composition, which, in turn, can affect the physical properties in a very substantial way. Nevertheless, despite the fact that from a molecular viewpoint, heat conduction and diffusion are generally very different, mathematical equivalences between the description of certain mass transfer problems and analogous heat transfer problems exist in many important cases. Such analogies will facilitate understanding and analyzing mass transfer processes. This will not always be the case, however, particularly when mass transfer occurs in nondilute mixtures where the mole fraction can be large or when coupling of diffusion fluxes occurs. In such cases, there might be little, if any, mathematical equivalence between heat and mass transport.

It is important to distinguish between transport of molecules due to bulk flow and mass transfer. For example, flow of a mixture in, say, a pipe without a composition gradient is not what we call mass transfer since the molar flow of each component in the mixture will just be proportional to its mole fraction; hence, there will be no preferential transport. Many of the readers of this book have already heard about Fick's law and may be accustomed to the notion that diffusion is driven by a concentration gradient. While this is true in many cases, a concentration gradient alone is insufficient to drive mass transfer. For instance, consider the case of cold air sitting atop a heated horizontal surface. Because of the temperature gradient, the air density will be lower at the hot surface, where the temperature is higher, which, in turn, will drive natural convection. While there will be a concentration gradient of both oxygen and nitrogen because of the density gradient, there will be no preferential transport because of the absence of a mole or mass fraction gradient.

A final, very important general consideration regards what will happen at the interface between two phases, such as a solid and a fluid, a liquid and a gas, or two immiscible liquids. For both heat transfer and mass transfer problems we usually assume that equilibrium will prevail at these interfaces. For heat transfer, this means that the temperature will be the same on both sides of the interface. Thus, although the slope will change, the temperature profile will be continuous as we move from one side of the interface to the other. Conversely, for mass transfer problems, equilibrium at the interface between two phases will not mean that the composition will be the same. Thus, not only will the slopes of the composition profiles be different, but the composition will also be discontinuous at the interface. From a thermodynamic viewpoint, we could avoid this discontinuity by defining the composition in terms of chemical potential (or fugacity) instead of the more familiar mole or mass fraction scales. The chemical potential will indeed be a continuous function across the interface, just like temperature. However, formulating transport laws in terms of chemical potential will often introduce other complexities so that, in practice, we use mole or mass fractions to define the composition and account for composition discontinuity at the interface by imposing that phase equilibrium is satisfied at these phase boundaries.

Mass transfer is important in many practical applications. One example is the evaporation of a liquid like water or, say, benzene, into a gas like air. In these cases, the rate of evaporation will often be determined by mass transfer on the gas-side of the interface. Water evaporation is important in drying, and humidification operations. Evaporation of organic liquids is important when we want to remove solvents from a solid product, such as a rubbery polymer, or when we want to estimate vapor emissions from storage tanks to the atmosphere, which has, of course, very important environmental implications. Mass transfer is important in catalysis, for example when the catalyst is contained within porous pellets. In this case, reactants and products must diffuse in and out of the pores for the reaction to take place. If the reaction kinetics is fast enough, the overall rate could actually become completely limited by mass transfer. Mass transfer is also important in drug delivery and in processes used to manufacture biotechnology products. Consider, for example, the design of a patch for the transdermal delivery of a drug. The objective is to get gradual and sustained delivery of the drug (e.g., for motion sickness) over long periods of time which can be obtained if the material of which the patch is made slows down the rate at which the drug reaches the skin. If the patch is too thin, diffusion will be too fast and the drug will be delivered too quickly. Conversely, if the patch is too thick, the delivery rate will be too slow for the drug to have a meaningful effect. Obviously, in this example, in addition to geometry, material properties of the patch and the chemical characteristics of the drug will influence the design. A final example is the recovery and purification of biological drugs, such as monoclonal antibodies (mAbs). mAbs are large proteins that have many current and future therapeutic uses for the treatment of previously untreatable ailments such as cancer, autoimmune diseases, and serious infections. They are produced in cell culture bioreactors and, in most cases, they are purified via their highly selective binding to a protein called "protein A" that is used as a ligand. To make this efficient as a proecss, the "protein A" ligand is immobilized within porous particles. Obviously, the mAb must get to the ligand in order to bind and be separated from the cell culture milieu and this process will usually be controlled by the rate of mass transfer of the mAb in the particle pores. Since mAbs are very large, they diffuse very slowly. This means that the binding process might take a very long time, which may not be economical. Making the particles too big or the pores too small may render the time needed so long to as to become impractical, say days or even weeks. Conversely, making the particles too small may result in very fast binding. However, other aspects associated with using very tiny particles might make the process impractical, for example, by requiring inordinately high operating pressures if the particles are used in a packed bed or column.

The following example provides a view of mass transfer that illustrates some of these concepts by drawing comparisons with analogous heat transfer processes.

Example 8.1. Absorber design problem. Let's consider the task of designing an absorber to remove ammonia (identified as component A) from an air stream flowing at the rate of 1.0 mol/s by absorbing it into a stream of liquid water at 1 atm and 30°C. Since ammonia is fairly soluble in water, as long as we use enough water, such a process will be able to remove most of the ammonia from the air. The ammonia-rich water leaving the absorber can then be sent to a vaporizer where the ammonia can be boiled off at a higher temperature and recovered. The remaining water can then be cooled and recirculated back to the absorber. Here we will focus only on the absorber part of such a process. There are obviously similarities between this mass transfer problem and the heat exchanger problem discussed in Example 1.1. In fact, we can call the absorber a "mass exchanger" that transfers ammonia from one stream to the other just like a heat exchanger transfers heat from one stream to the other. The absorber problem also has two different "dimensions." The first dimension is about the amount of ammonia that needs to be exchanged and the flow of water that we will need to accomplish this task. The second is about the size of the unit that we will need to achieve the desired rate of exchange. We can solve for the first dimension, at least in part, using material balances and phase-equilibrium relationships. However, just like we needed to learn about heat transfer concepts to determine the size of the heat exchanger in Example 1.1, we will need to learn mass transfer concepts before we can address the second dimension of the gas-absorber design task.

We begin again with a simple process flow diagram (PFD) showing all labeled input and output streams as in Fig. 8.1.

The quantity $M_{t,A}$, measured, for example, in units of mol/s, represents the total rate of mass transfer of ammonia from the air stream to the liquid water. Note that, unlike the heat exchanger in Example 1.1 where the two streams were only in thermal contact, in this case there is no physical separation between gas and liquid phases. Thus, mass transfer of ammonia will occur across the gas–liquid interface. The material balances for this process are as follows. For the total moles we have:

$$G_{M,\text{in}} + L_{M,\text{in}} = G_{M,\text{out}} + L_{M,\text{out}} \tag{8.1}$$

while for the moles of ammonia we have:

$$y_{A,\text{in}} G_{M,\text{in}} + x_{A,\text{in}} L_{M,\text{in}} = y_{A,\text{out}} G_{M,\text{out}} + x_{A,\text{out}} L_{M,\text{out}} \tag{8.2}$$

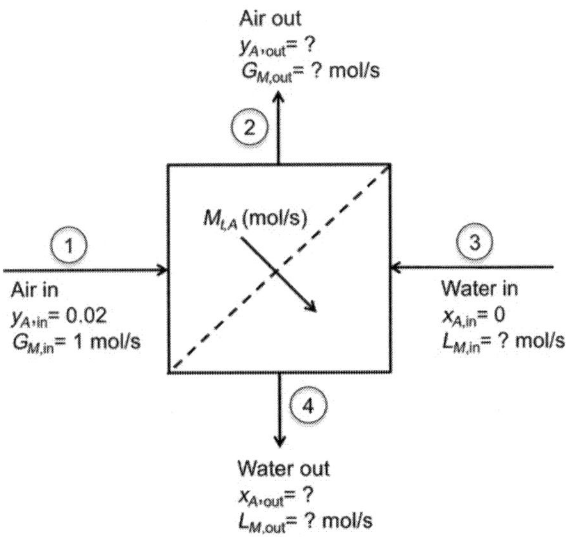

FIGURE 8.1 Process flow diagram for Example 8.1.

In these equations, G_M and L_M are the molar flow rates of the gas and liquid streams, respectively, while y_A and x_A are the corresponding mole fractions of ammonia. We will simplify the problem by considering only dilute conditions; i.e., conditions where y_A and x_A are both much less than one. In this case, we can assume that inlet and outlet gas flow rates and inlet and outlet liquid flow rates will be the same and equal to G_M and L_M, respectively. Only one material balance will be needed in this case and can be written as follows:

$$y_{A,\text{in}}G_M + x_{A,\text{in}}L_M = y_{A,\text{out}}G_M + x_{A,\text{out}}L_M \tag{8.3}$$

With some rearrangement, this equation becomes:

$$L_M = G_M \frac{y_{A,\text{in}} - y_{A,\text{out}}}{x_{A,\text{out}} - x_{A,\text{in}}} \tag{8.4}$$

or

$$y_{A,\text{in}} = \frac{L_M}{G_M} x_{A,\text{out}} - \frac{L_M}{G_M} x_{A,\text{in}} + y_{A,\text{out}} \tag{8.5}$$

Let's assume that $x_{A,\text{in}} = 0$ (i.e., we feed pure water with no ammonia to the absorber), that $y_{A,\text{in}} = 0.02$, as an example, and that we want to absorb 90% of the ammonia fed. We can easily calculate the rate at which we need to absorb ammonia as $M_{t,A} = 0.9 \times y_{A,\text{in}}G_M = 0.9 \times 0.02 \times 1.0 = 0.018$ mol/s. The outlet ammonia mole fraction will then be $y_{A,\text{out}} = 0.02 \times (1 - 0.90) = 0.002$. However, we cannot yet find L_M since $x_{A,\text{out}}$ is also unknown in Eq. (8.4) or (8.5). In order to find $x_{A,\text{out}}$, we need to consider the actual arrangement of the absorber.

Let's consider first a hypothetical co-current arrangement of the absorber as shown in Fig. 8.2. In this case, the gas and liquid streams flow in the same direction and are separated at the outlet. For this arrangement to work, ammonia must transfer from the gas to the liquid everywhere along the absorber. In turn, this means that the mole fraction of ammonia in the outlet gas cannot fall below the value that would be in equilibrium with the mole fraction of ammonia in the outlet liquid. For the dilute conditions of this problem, Henry's law will likely apply. In this case, the relationship between the equilibrium value of y_A in the gas phase and the value of x_A in the liquid phase will be linear and is described by the following equation:

$$y_A = Kx_A \tag{8.6}$$

where $K = H/P$. In this expression for K, H is the Henry's law constant (in units of pressure) and P is the absolute pressure. For the ammonia–water system at 30°C and 1 atm, we have $K = 1.2$. Combining Eqs. (8.6) and (8.4) yields:

$$L_M = G_M \frac{y_{A,\text{in}} - y_{A,\text{out}}}{\dfrac{y_{A,\text{out}}}{K} - x_{A,\text{in}}} = 1.0 \times \frac{0.02 - 0.002}{\dfrac{0.002}{1.2} - 0} = 10.8 \text{ mol/s}$$

This is the minimum liquid flow rate that we will have to use in order to remove 90% of the ammonia fed. We can also do this calculation graphically, as shown in Fig. 8.2. On a graph of y_A vs. x_A, we plot both the equilibrium line, Eq. (8.6), and a line representing Eq. (8.5). Note that if we treat $y_{A,\text{out}}$ and $x_{A,\text{out}}$ as the variables y_A and x_A, this equation is also a straight line on this graph, which we call the operating line, given by:

$$y_A = -\frac{L_M}{G_M} x_A + \frac{L_M}{G_M} x_{A,\text{in}} - y_{A,\text{in}} \tag{8.7}$$

This line passes through the point with coordinates $x_{A,\text{in}}, y_{A,\text{in}}$ and has slope $-L_M/G_M$. If equilibrium is established, this line will cross the equilibrium line precisely at $y_A = y_{A,\text{out}}$. The intersection is shown in Fig. 8.2 and, of course, gives the same result as the analytical solution above for the minimum value of L_M. Note that, in practice, we will need a flow rate higher than the minimum of 10.8 mol/s in order to stay away from equilibrium at the outlet of the absorber. Intuitively, the closer we get to equilibrium, the slower the rate of absorption of ammonia will become. Ultimately, equilibrium between the outlet

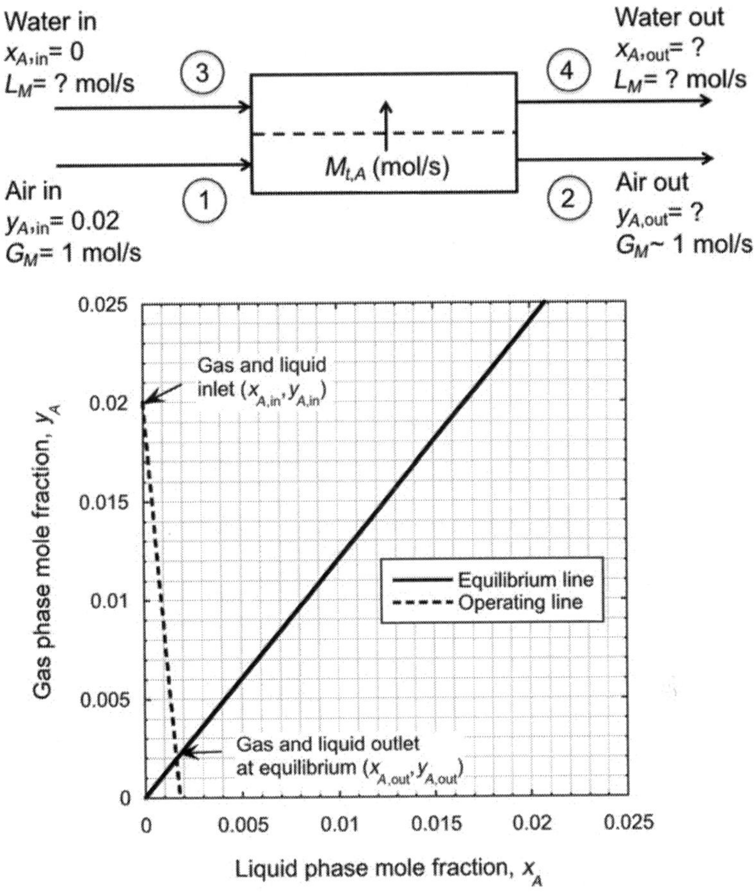

FIGURE 8.2 Absorption of ammonia from air in water in a co-current gas–liquid contactor.

gas and liquid stream will lead to an infinitely long absorber. Thus, the actual L_M will have to be larger than this minimum value in order to be able to operate with an absorber of finite length.

Let's now consider the counter-current arrangement of the absorber shown in Fig. 8.3. In this case, the gas flows in at one end (left) while the liquid flows in at the opposite end (right) of the absorber. The material balance is exactly the same as before. However, in this case, since there is no ammonia in the entering water ($x_{A,in} = 0$), equilibrium can only be approached at the gas-inlet end of the absorber, where $y_A = y_{A,in}$ and $x_A = x_{A,out}$. In this case, Eq. (8.4) gives:

$$L_M = G_M \frac{y_{A,in} - y_{A,out}}{\dfrac{y_{A,in}}{K} - x_{A,in}} = 1.0 \times \frac{0.02 - 0.002}{\dfrac{0.02}{1.2} - 0} = 1.08 \text{ mol/s}$$

This minimum value of L_M is about 10 times smaller than the minimum found for the co-current case. This occurs because, with the counter-current arrangement, we no longer have an equilibrium limitation between the two streams leaving the absorber. Instead, the equilibrium limitation is now between the inlet gas and the outlet liquid, which, in turn, leads to a more favorable result. Ultimately, the overall conclusion is the same that we ended up drawing when comparing co- and counter-current heat exchanger configurations in Example 1.1. In many cases, as shown here, counter-current

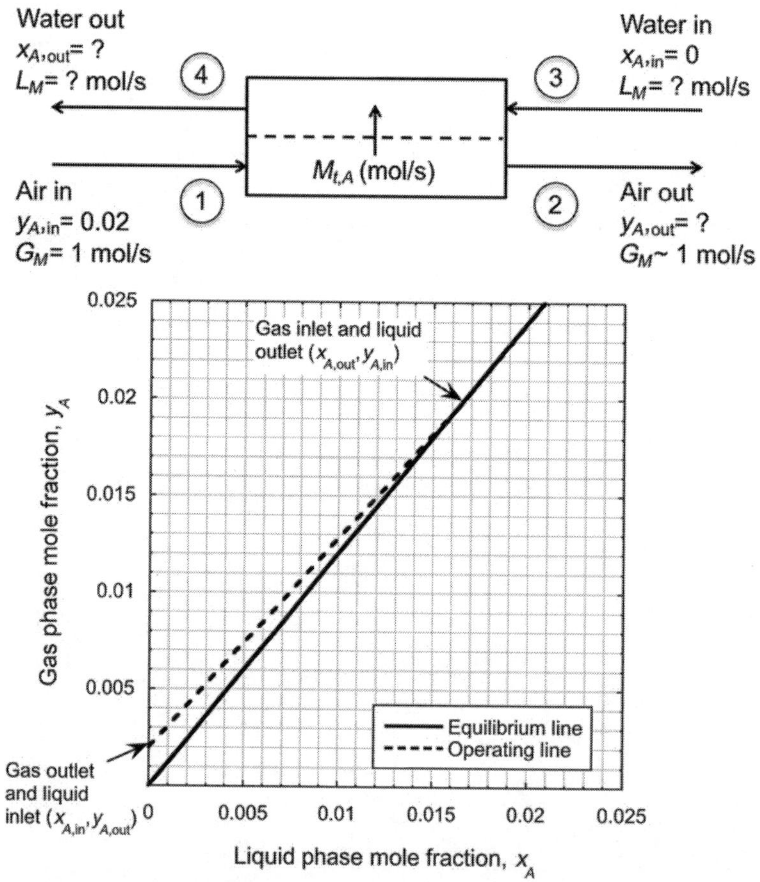

FIGURE 8.3 Absorption of ammonia from air in water in a counter-current gas–liquid contactor.

will be better. Figure 8.3 shows the corresponding graphical approach to determine L_M. In this case, treating again $x_{A,\text{out}}$ and $y_{A,\text{in}}$ as the x_A and y_A variables, Eq. (8.7) yields:

$$y_A = \frac{L_M}{G_M} x_A - \frac{L_M}{G_M} x_{A,\text{in}} - y_{A,\text{out}} \tag{8.8}$$

This line passes through a point with coordinates $x_{A,\text{in}}$, $y_{A,\text{out}}$ and has a positive slope L_M/G_M. Note that, of course, in the counter-current case too, we will still need to use a liquid flow rate higher than the minimum calculated above in order to maintain a reasonable driving force for absorption and, thus, reasonable rates of mass transfer.

At this point, we have exhausted what we can do for the solution of this problem just based on material balances and phase equilibrium relationships. In order to determine the physical size of the absorber in either co- or counter-current configuration, we will need to describe how the rate of absorption of ammonia is related to the composition y_A and x_A at each point along the length of the absorber. Thus, while the total rate of mass transfer of A is $M_{t,A}$, there will be a local rate that varies from point to point in the absorber. Conceptually, this is analogous to what happens in a heat exchanger (see Chap. 7). The absorption rate will be controlled by the rate at which ammonia can transfer from the bulk gas to the gas–liquid interface and from the gas–liquid interface into the bulk liquid. We will learn that this is a convective mass transfer process.

8.2 Mass Transfer Rate and Mass Transfer Flux

As discussed in Example 8.1, we define the molar rate of mass transfer of a component i as M_i typically in units of mol/s. In analogy to the heat transfer flux, we can also define a molar mass transfer flux, N_i, typically in units of mol/m^2 s which we obtain by dividing M_i by the area, S, across which mass transfer takes place as follows:

$$N_i = \frac{M_i}{S} \tag{8.9}$$

Note that the area S is perpendicular to the flux. For instance, in the case of Example 8.1, the area S would be the gas–liquid interfacial area. Obviously, since $M_i = S \times N_i$, one of the ways in which we can increase the rate of mass transfer is to increase the interfacial area. In the case of the absorbers discussed in Example 8.1 that is exactly the approach taken by designing actual systems that provide large gas–liquid interfacial areas.

Units of mass can also be used to describe mass transfer problems. In this case, m_i will be the rate of mass transfer in units of kg/s and n_i will be the corresponding mass transfer flux in units of kg/m^2 s defined as follows:

$$n_i = \frac{m_i}{S} \tag{8.10}$$

Note that both N_i and n_i are vectors. Thus, for example, in rectangular coordinates, we can write:

$$N_i = N_{i,x}\vec{i} + N_{i,y}\vec{j} + N_{i,z}\vec{k} \tag{8.11}$$

where $N_{i,x}$, $N_{i,y}$, and $N_{i,z}$ are the x, y, and z components of the molar flux and $\vec{i}$, $\vec{j}$, and $\vec{k}$ are the corresponding unit vectors.

8.3 Mass Transfer Mechanisms

Just as we saw for heat transfer, there are distinct mechanisms through which mass transfer can occur:

1. Diffusive mass transfer and
2. Convective mass transfer.

These mechanisms are analogous to conductive and convective heat transfer. There is, however, no obvious mass transfer equivalent of radiant heat transfer. *Diffusive mass transfer* or, simply, *diffusion*, can occur in the absence of bulk fluid motion and is driven by a spatial composition gradient according to Fick's law. This is covered in Chaps. 9 to 12. *Convective mass transfer* is associated with fluid motion at the interface between two phases, such as the interface between a moving fluid and a stationary solid or the interface between two fluids in relative motion and is covered in Chap. 13. The relative importance of diffusive and convective mass transfer varies widely depending on the physical system and, in particular, on exactly what happens at the interface between phases when transport occurs in multiphase systems.

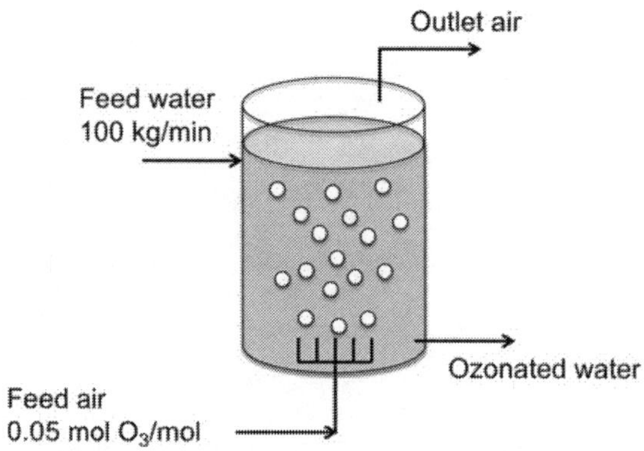

FIGURE P.8.1

8.4 Problems

P.8.1. Ozonization of water can be used to control odor and for disinfection. In the simple ozonizer sketched in Fig. P.8.1, air containing 5.0 mol % O_3 is bubbled through water at 20°C flowing at the rate of 100 kg/min. The absolute pressure in the ozonizer is 1.2 atm. The solubility of pure O_3 gas in water at 1 atm and 20°C is 0.57 g/L. (a) What is the maximum concentration of O_3 in water that can be obtained for these conditions? Express your result in g/L and as a mole fraction. (b) In the actual operation, the mole fraction of O_3 in the ozonated water is 4.5×10^{-6} while the mole fraction of O_3 in the air leaving the ozonizer is 0.02. Determine the molar (mol/min) and volumetric (m^3/min) flow rate of the feed air and the rate of mass transfer of O_3. (c) Assume that the air bubbles in the ozonizer occupy a volume of 0.3 m^3 and that they are, on overage, 2 mm in diameter. Determine the mass transfer flux of O_3 at the bubble–water interface.

P.8.2. Sampling by the local Department of Environmental Quality (DEQ) of the water in a small creek has revealed a steady concentration of a highly toxic chemical of 0.015 g/m^3. At some distance upstream from the sampling point, the DEQ personnel have spotted a cylindrical concrete beam lying in the creek and it is suspected that the contaminating chemical has been leaching out of the beam into the water. The creek flows at a velocity of 0.2 m/s and its cross-section is 10 m^2. The concrete beam is 0.3 m in diameter and 1.8 m long. Assuming that the DEQ suspicion is correct, determine (a) the rate of mass transfer of the chemical and (b) the mass transfer flux of the chemical at the concrete–water interface.

CHAPTER 9
Fick's Law

9.1 Introduction

This chapter introduces Fick's law, which describes diffusional mass transfer resulting from a spatial composition gradient. In general, diffusion occurs as a result of the random walk of atoms or molecules and is dependent on temperature and on the physical properties of the diffusing species and of the medium where diffusion takes place. This dependence on physical properties is expressed by the diffusion coefficient or diffusivity, for which we use the symbol $\mathcal{D}$. For a mixture of two components A and B, $\mathcal{D}_{A,B}$ represents the binary diffusion coefficient or binary diffusivity. Before we can properly introduce Fick's law, we need to precisely define composition and material velocity scales, which are introduced in the following section.

9.2 Composition and Velocity Scales

9.2.1 Composition Scales

The following quantities are defined to express the composition of a mixture on a mass basis:

$$\rho = \frac{\text{Total mass}}{\text{Volume}} = \frac{m_{\text{total}}}{V} \tag{9.1}$$

$$\rho_i = \frac{\text{Mass of component } i}{\text{Volume}} = \frac{m_i}{V} \tag{9.2}$$

$$w_i = \frac{\text{Mass of component } i}{\text{Total mass}} = \frac{m_i}{m_{\text{total}}} \tag{9.3}$$

ρ is the mass density (or just "density") of the mixture, typically expressed in units of kg/m^3 or lb/ft^3. ρ_i is the mass concentration of component i, typically in units of kg of i/m^3 or lb of i/ft^3. Finally, w_i is the mass fraction of component i as kg of component i per kg of mixture or lb of i per lb of mixture. In these definitions, the quantities m and V can be replaced, respectively, by the mass and volumetric flow rates of a stream. In either case, the following relationships are easily derived:

$$\rho = \sum_{i=1}^{n_c} \rho_i \tag{9.4}$$

$$w_i = \frac{\rho_i}{\rho} \tag{9.5}$$

where n_c is the number of components in the mixture. The following analogous quantities can be defined on a molar basis:

$$C = \frac{\text{Total moles}}{\text{Volume}} = \frac{M_{\text{total}}}{V} \tag{9.6}$$

$$C_i = \frac{\text{Moles of component } i}{\text{Volume}} = \frac{M_i}{V} \tag{9.7}$$

$$y_i = \frac{\text{Moles of component } i}{\text{Total moles}} = \frac{M_i}{M_{\text{total}}} \tag{9.8}$$

C is the molar density of the mixture, typically expressed in units of mol/m^3 or lbmol/ft^3. C_i is the molar concentration of component i, typically in units of mol of i/m^3 or lbmol of i/ft^3. Finally, y_i, which we use for the gas phase or, alternatively, x_i, which we will use for the liquid phase, is the mole fraction of component i. Either will be expressed as mol of component i per mol of mixture of lbmol of i per lbmol of mixture. As before, the M and V quantities in these definitions can be replaced by molar and volumetric flow rates, respectively. The following relationships are also easily derived:

$$C = \sum_{i=1}^{n_c} C_i \tag{9.9}$$

$$y_i = \frac{C_i}{C} \tag{9.10}$$

where n_c is again the number of components in the mixture.

The reasons why we introduce both mass and molar bases to define composition is that mass and molar densities vary differently with composition depending on the particular phase for which they are used. For example, in a gas at low pressure, when the ideal gas law is valid, the molar density will be:

$$C = \frac{M_{\text{total}}}{V} = \frac{P}{\mathcal{R}T} \tag{9.11}$$

which is completely independent of composition. The mass density of a gas mixture, on the other hand, will generally vary substantially with composition. In a liquid mixture, the situation is usually the opposite. In this case, while the molar density C can change dramatically with composition, the mass density ρ will often remain fairly constant. Consider, for example, a liquid mixture of water (M_r = 18 g/mol) and ethanol (M_r = 46 g/mol) at 20°C. For pure water, ρ = 1,000 kg/m^3, while for a mixture containing 50% ethanol by mass, ρ = 910 kg/m^3, which is only about 8.7% smaller. On the other hand, the molar density of pure water will be C = 1,000/18 = 55.5 kmol/m^3, while the molar density of the 50% mixture will be $C = 0.5 \times 910/18 + 0.5 \times 910/46 = 35.2$ kmol/m^3, which is about 36% smaller. Obviously, for such a mixture, C will vary much more with composition compared to ρ. Since being able to assume either constant mass density or constant molar density will help simplify the analysis of mass transfer processes, we will tend to use the molar basis for gas mixtures and the mass basis for liquid mixtures. Fortunately, it is easy to convert from one basis to the other using the following relationships:

$$C_i = \frac{\rho_i}{M_{r,i}} \tag{9.12}$$

$$\rho_i = C_i M_{r,i} \tag{9.13}$$

$$C = \sum_{i=1}^{n_c} \frac{\rho_i}{M_{r,i}} \tag{9.14}$$

$$\rho = \sum_{i=1}^{n_c} C_i M_{r,i} \tag{9.15}$$

$$y_i = \frac{\rho_i / M_{r,i}}{\sum_{i=1}^{n_c} \rho_i / M_{r,i}} = \frac{w_i / M_{r,i}}{\sum_{i=1}^{n_c} w_i / M_{r,i}} \tag{9.16}$$

$$w_i = \frac{C_i M_{r,i}}{\sum_{i=1}^{n_c} C_i M_{r,i}} = \frac{y_i M_{r,i}}{\sum_{i=1}^{n_c} y_i M_{r,i}} \tag{9.17}$$

9.2.2 Mass Transfer Fluxes

Having defined mass and molar-based composition variables, we can now define the corresponding fluxes and their relationship to these variables. Let's consider the amount of material transported per unit of time through a cross-sectional area S perpendicular to the direction of transport. We first consider the mass basis by defining mass fluxes according to the following relationships:

$$n = \frac{\text{Total mass/Time}}{\text{Area}} = \frac{\text{Total mass}}{\text{Volume}} \times \frac{\text{Distance}}{\text{Time}} = \rho \times v \tag{9.18}$$

$$n_i = \frac{\text{Mass of component } i\text{/Time}}{\text{Area}} = \frac{\text{Mass of component } i}{\text{Volume}} \times \frac{\text{Distance}}{\text{Time}} = \rho_i \times v_i \tag{9.19}$$

In these equations, n is the total mass flux, typically expressed in units of kg/m^2 s, and n_i is the mass flux of component i, typically in units of kg of i/m^2 s. v is the velocity of the mixture, typically in m/s, and v_i is the velocity of component i, also typically in m/s. The mixture and component velocities will be related to each other by the following equation:

$$n = \rho v = \sum_{i=1}^{n_c} n_i = \sum_{i=1}^{n_c} \rho_i v_i \tag{9.20}$$

or

$$v = \sum_{i=1}^{n_c} w_i v_i \tag{9.21}$$

We see that v is just the mass-average of the component velocities, v_i. Let's now define the component velocity v_i' relative to the mass-average velocity as follows:

$$v_i' = v_i - v \tag{9.22}$$

In order to understand why we introduce this definition, we can think of a very simple and often quoted analogy to fish in a river. The fish are swimming around and we can measure their velocity v_i while standing on a bridge. Alternatively, we can measure their velocity v_i' while drifting downriver on a canoe. In the latter case, v_i' will be zero if the fish also drift along (dead?), it will be positive if they swim downstream, or it will be negative if they swim upstream. Based on v_i' we can now define a mass transfer flux j_i relative to the mass average velocity as follows:

$$j_i = \rho_i v_i' \tag{9.23}$$

Combining Eqs. (9.22) and (9.23) yields the following relationship:

$$j_i = \rho_i(v_i - v) = \rho_i v_i - \rho_i v = n_i - \frac{\rho_i}{\rho} n = n_i - w_i n \tag{9.24}$$

or

$$n_i = w_i n + j_i = w_i \sum_{i=1}^{n_c} n_i + j_i \tag{9.25}$$

In the fish-river analogy, j_i would measure the mass of fish going by the drifting canoe, while n_i would measure the mass of fish going by the bridge. We should note that Eq. (9.25) applies regardless of whether or not there is a bulk flow driven by gravity (as in the fish-river analogy) or by a pump or by some other mechanical device. Even if we had diffusion alone, we could still encounter situations where there is a finite total flux n even though no force is applied to drive a bulk flow. In other words, the total flux n can be a result of the diffusion process itself. An important property of the relative fluxes is that their sum is zero. In fact, from Eq. (9.25), we have:

$$\sum_{i=1}^{n_c} n_i = n \sum_{i=1}^{n_c} w_i + \sum_{i=1}^{n_c} j_i = n + \sum_{i=1}^{n_c} j_i \tag{9.26}$$

which gives

$$\sum_{i=1}^{n_c} j_i = 0 \tag{9.27}$$

Let's now look at the definition of fluxes and relative velocity when we use a molar basis. In this case we will have:

$$N = \frac{\text{Total moles/Time}}{\text{Area}} = \frac{\text{Total moles}}{\text{Volume}} \times \frac{\text{Distance}}{\text{Time}} = C \times v^* \tag{9.28}$$

and

$$N_i = \frac{\text{Moles of component } i\text{/Time}}{\text{Area}} = \frac{\text{Moles of component } i}{\text{Volume}} \times \frac{\text{Distance}}{\text{Time}} = C_i \times v_i \tag{9.29}$$

N is the total molar flux, typically in units of mol/m^2 s, and N_i is the molar flux of component i, typically in units of mol of i/m^2 s. v_i is again the velocity of component i, also typically in m/s, but v^* is now a molar-average velocity, which is different from v. We can find v^* by considering that the total molar flux, N, can be written as follows:

$$N = Cv^* = \sum_{i=1}^{n_c} N_i = \sum_{i=1}^{n_c} C_i v_i \tag{9.30}$$

which gives

$$v^* = \sum_{i=1}^{n_c} y_i v_i \tag{9.31}$$

We see that v^* is an average of the component velocities, v_i, weighted by their individual mole fractions. As before, it makes sense to define a new component velocity v_i'' relative to the molar average velocity as follows:

$$v_i'' = v_i - v^* \tag{9.32}$$

and a molar flux J_i relative to the molar average velocity as:

$$J_i = C_i v_i'' \tag{9.33}$$

As in the mass-basis case, Eqs. (9.32) and (9.33) yield the following relationship:

$$J_i = C_i(v_i - v^*) = C_i v_i - C_i v^* = N_i - \frac{C_i}{C} N = N_i - y_i N \tag{9.34}$$

or

$$N_i = y_i N + J_i = y_i \sum_{i=1}^{n_c} N_i + J_i \tag{9.35}$$

Note that in this case too we will have:

$$\sum_{i=1}^{n_c} J_i = 0 \tag{9.36}$$

Going back to the fish-river analogy, the difference between mass and molar basis is that for the molar basis we are counting the number (i.e., the moles) of fish going either by the drifting canoe (J_i) or by the bridge (N_i) while for the mass basis we are measuring the mass of fish.

9.3 Fick's Law for Binary Mixtures

Fick's law describes the flux of a component relative to the appropriate average velocity as a function of the spatial gradient in its mass fraction or mole fraction. This law applies to the flux that results from random motion—thus, we write it, properly, in terms of either j_i or J_i. Here we focus on binary mixtures of two components A and B, which is amenable to a more rigorous treatment while capturing many practical applications. Multicomponent diffusion is substantially more complex, and a brief introduction is provided in Chap. 15. On a mass basis, Fick's law for a binary mixture is expressed as follows:

$$j_A = -\rho \mathcal{D}_{A,B} \nabla w_A \tag{9.37}$$

$$j_B = -\rho \mathcal{D}_{A,B} \nabla w_B \tag{9.38}$$

where $\mathcal{D}_{A,B}$ is the binary diffusion coefficient or binary diffusivity. The negative sign in Eqs. (9.37) and (9.38) accounts for the fact that the flux will be positive in the direction of a decreasing mass fraction. Thus, a negative ∇w_i will result in a positive j_i. The diffusion coefficient $\mathcal{D}_{A,B}$ has dimensions of area/time and is typically measured in units of m^2/s of ft^2/h. For ideal gases, $\mathcal{D}_{A,B}$ is a function of temperature and pressure and of the properties of the mixture components but is independent of composition. For liquids, $\mathcal{D}_{A,B}$ is a function of temperature and of the mixture viscosity and will often vary substantially as a function of composition.

It is not difficult to verify that, for a given binary mixture, there will be only one diffusivity; i.e., we will have $\mathcal{D}_{A,B} = \mathcal{D}_{B,A}$. In fact, adding Eqs. (9.37) and (9.38) together gives:

$$j_A + j_B = -\rho(\mathcal{D}_{A,B}\nabla w_A + \mathcal{D}_{B,A}\nabla w_B) \tag{9.39}$$

which, based on Eq. (9.27) is zero. For a binary mixture, we have $w_A + w_B = 1$. Thus, $\nabla w_B = -\nabla w_A$. Substituting this result in Eq. (9.39), we obtain:

$$0 = \mathcal{D}_{A,B}\nabla w_A - \mathcal{D}_{B,A}\nabla w_A \tag{9.40}$$

which shows that, indeed, $\mathcal{D}_{A,B} = \mathcal{D}_{B,A}$.

Based on Eq. (9.25), in terms of the absolute fluxes relative to fixed coordinates, Fick's law can be written as follows:

$$n_A = w_A(n_A + n_B) - \rho\mathcal{D}_{A,B}\nabla w_A \tag{9.41}$$

$$n_B = w_B(n_A + n_B) - \rho\mathcal{D}_{A,B}\nabla w_B \tag{9.42}$$

We can see from these equations that the flux of a component i is actually the sum of two terms. For component A, the first term, $w_A(n_A + n_B)$, describes transport that results from the combined motion of A and B, while the second term, $-\rho\mathcal{D}_{A,B}\nabla w_A$, describes transport due to random walk of A. If we consider diffusion of A in a very dilute mixture with B in large excess such that, say, $w_A < 0.01$, Eq. (9.41) will give:

$$n_A \sim -\rho\mathcal{D}_{A,B}\nabla w_A \tag{9.43}$$

In such a dilute mixture, the density will be essentially the same as the density of pure B and will thus be a constant independent of the exact composition. As a result, we will have:

$$n_A = -\mathcal{D}_{A,B}\nabla \rho w_A = -\mathcal{D}_{A,B}\nabla \rho_A \tag{9.44}$$

which shows that the mass flux of A will be proportional to its mass concentration gradient.

On a molar basis, Fick's law can be written as follows:

$$J_A = -C\mathcal{D}_{A,B}\nabla y_A \tag{9.45}$$

$$J_B = -C\mathcal{D}_{A,B}\nabla y_B \tag{9.46}$$

where C is the molar density of the mixture. Accordingly, based on Eq. (9.35), the fluxes relative to fixed coordinates will be:

$$N_A = y_A(N_A + N_B) - C\mathcal{D}_{A,B}\nabla y_A \tag{9.47}$$

$$N_B = y_B(N_A + N_B) - C\mathcal{D}_{A,B}\nabla y_B \tag{9.48}$$

These equations show again that the molar flux of a component will generally be the sum of two contributions—one accounting for transport due to the combined motion of A and B (the first term on the right-hand side of each equation) and the other accounting for diffusive transport due to random walk (the second term). For a dilute mixture of A in an excess amount of B, with, say, $y_A < 0.01$, Eq. (9.47) will reduce to:

$$N_A \sim -\mathcal{D}_{A,B}\nabla C_A \tag{9.49}$$

which shows that the molar flux of A will become proportional to its molar concentration gradient. Note that since $n_A = M_{r,A} \times N_A$ and $\rho_A = M_{r,A} \times C_A$, $\mathcal{D}_{A,B}$ will be the same whether we use mass or molar fluxes. In either case, for dilute conditions, we can calculate the flux directly from this equation if we know the molar concentration gradient ∇C_A. This is not possible, however, in the general case (i.e., for nondilute conditions) since: (a) both fluxes appear in both Eqs. (9.47) and (9.48) and (b) combining the two equations will give an identity equation making a unique solution for the fluxes impossible without additional specifications. In order to handle these cases, we divide both sides of Eq. (9.47) by N_A, which gives:

$$1 = y_A\left(1 + \frac{N_B}{N_A}\right) - \frac{C\mathcal{D}_{A,B}}{N_A}\nabla y_A \tag{9.50}$$

With some rearrangement, this equation yields:

$$N_A = -\frac{C\mathcal{D}_{A,B}}{1 - y_A\left(1 + \frac{N_B}{N_A}\right)}\nabla y_A \tag{9.51}$$

We can now see that if we know the mole fraction and the mole fraction gradient at a certain point in the mixture, we can calculate N_A directly from this expression if we specify the ratio N_B/N_A. This ratio will depend on the particular physical situation. It should be noted that, regardless of whether a mass or a molar basis is used, the expression for the gradient will depend on the particular coordinate systems chosen to represent our physical situation. For example, in rectangular coordinates, we will have:

$$\nabla = \frac{\partial}{\partial x}\vec{i} + \frac{\partial}{\partial y}\vec{j} + \frac{\partial}{\partial x}\vec{k} \tag{9.52}$$

where $\vec{i}$, $\vec{j}$, and $\vec{k}$ are the unit vectors in the x, y, and z directions, respectively. The following examples show how the ratio N_B/N_A can be determined by looking at what happens at the system's boundaries for some simple one-dimensional diffusion problems.

Example 9.1. Equimolar counter-diffusion. In this example, we consider a case where a large tank containing initially pure hydrogen is connected to the atmosphere via a thin sampling tube as shown in Fig. 9.1. Temperature, T, and absolute pressure, P, are constant. If the tube is open, hydrogen will diffuse out while air will diffuse into the tank. Obviously, this could lead to a loss of hydrogen as well as to unsafe conditions. Thus, estimating the rate at which this will occur may be very important. Strictly speaking, this situation involves a three-component mixture (H_2, O_2, and N_2). However, we can still safely consider air as a single species without too much error since O_2 and N_2 behave so similarly. In this case, if A is H_2 and B is air, N_A will be positive (i.e., out of the tank) and N_B will be negative (i.e.,

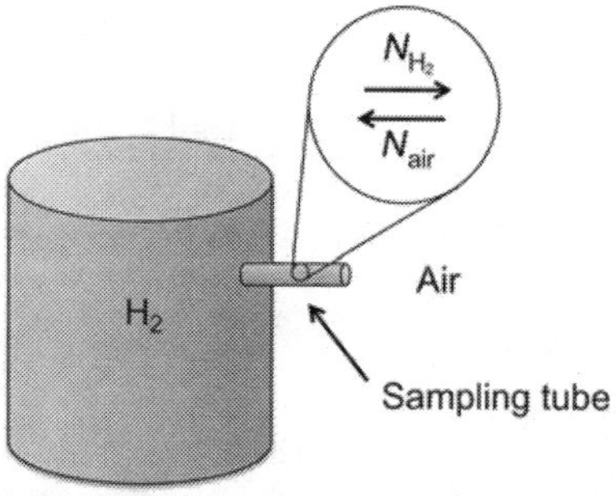

FIGURE 9.1 Equimolar counter-diffusion of hydrogen and air in a sampling tube.

into the tank). Since P and T are constant, the molar density $C = P/\mathcal{R}T$ will also be constant. Thus, for every mole of hydrogen that diffuses out, one mole of air will diffuse in. As a result, we will have:

$$N_A = -N_B \tag{9.53}$$

We call this situation equimolar counter-diffusion. In this case, since $N_B/N_A = -1$, Eq. (9.51) gives:

$$N_A = -C\mathcal{D}_{A,B}\nabla y_A = -\mathcal{D}_{A,B}\nabla C_A \tag{9.54}$$

This is the same result that we found, as an approximation, for a dilute mixture of A in B. However, for equimolar counter-diffusion this is an exact result valid for the entire range of mole fraction values, y_A, from 1 inside the tank to zero outside the sampling tube.

Example 9.2. Diffusion of *A* through stagnant *B*. This case is encountered in situations where mass transfer of component B cannot occur because a barrier at the system boundaries prevents it from moving. This can be a physical barrier, like a membrane or filter that is impervious to B, but it can also be because B is insoluble in an adjacent phase. Consider, for example, the evaporation of hexane from a liquid pool at the bottom of a tube open to the atmosphere as shown in Fig. 9.2. Air has almost no solubility in liquid hexane. Thus, in practice, it will be unable to cross the gas–liquid interface. Its mass transfer

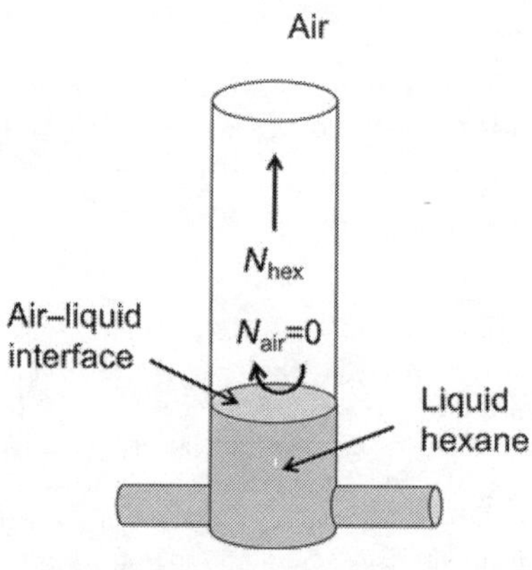

FIGURE 9.2 Diffusion of hexane through stagnant air.

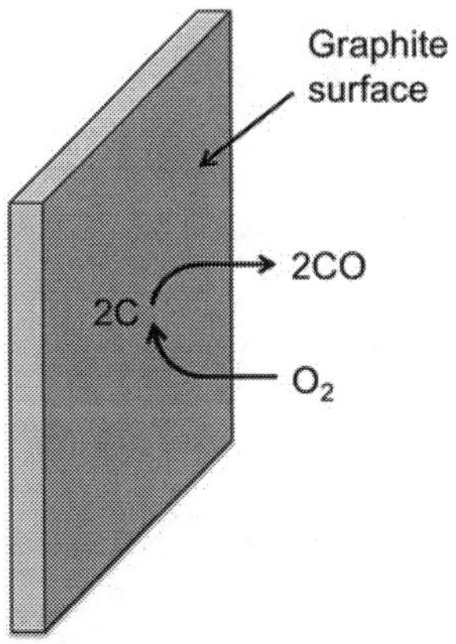

FIGURE 9.3 Fluxes constrained by reaction stoichiometry at gas–solid interface.

flux will thus be zero. On the other hand, hexane, which is fairly volatile, will go from liquid to vapor at the gas–liquid interface and will diffuse away toward the open end of the tube where it will be finally dispersed in the surrounding atmosphere. For this situation, if A is hexane and B is air, we will have:

$$N_B = 0 \tag{9.55}$$

Equation (9.51) will now give:

$$N_A = -\frac{C\mathcal{D}_{A,B}}{1-y_A}\nabla y_A \tag{9.56}$$

Example 9.3. Stoichiometric constraints. Let's consider the case of a reaction occurring at a gas–solid interface. An example could be the reaction of oxygen with carbon at a graphite surface as shown in Fig. 9.3. Under certain conditions, the reaction will form only CO according to the following stoichiometry:

$$2C_{(solid)} + O_{2(gas)} \rightarrow 2\,CO_{(gas)}$$

We can assume that the reaction occurs only at the graphite surface. In this case, for every mole of O_2 that diffuses toward the surface and reacts, 2 moles of CO will be formed and will diffuse away from the surface. Thus, if A is O_2 and B is CO, we will have:

$$N_B = -2N_A \tag{9.57}$$

Equation (9.51) now gives:

$$N_A = -\frac{C\mathcal{D}_{A,B}}{1-y_A(1-2)}\nabla y_A = -\frac{C\mathcal{D}_{A,B}}{1+y_A}\nabla y_A \tag{9.58}$$

9.4 Problems

P.9.1. Two glass bulbs are connected by a 0.2-m-long glass tube with a 0.001-m inner diameter. Initially, bulb 1 contains pure argon and bulb 2 contains pure xenon. Bulb 1 has a volume of 1 L. The absolute pressure and temperature are maintained at 1 atm and 110 °C, respectively. At a certain time, the contents of the two bulbs are analyzed and the argon mole fraction in bulb 1 is found to be 0.75, while the argon mole fraction in bulb 2 is found to be 0.20. (a) Determine the volume of bulb 2.

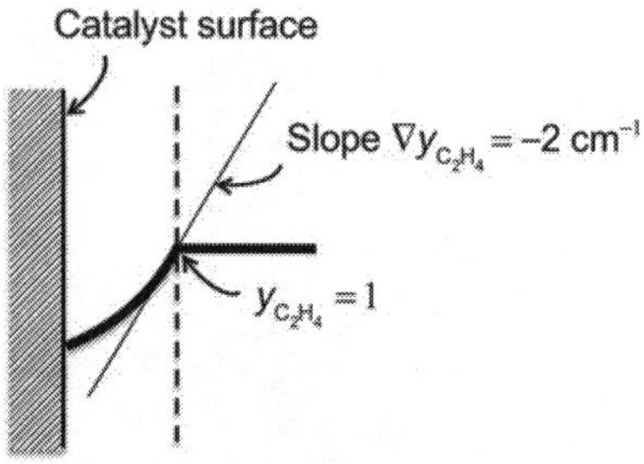

FIGURE P.9.3

(b) Assuming that only diffusion occurs in the connecting tube and that the mole fraction profile in the connecting tube is linear, for the given time determine the mass transfer fluxes of argon and xenon, the mass transfer rates, the component velocities v_i of argon and xenon in the tube at the bulb 1 end and at the bulb 2 end, and the molar average velocity of the mixture, v^*. The binary diffusivity of Ar/Xe mixtures at this temperature and pressure is $\mathcal{D}_{A,B} = 0.115\ cm^2/s$.

P.9.2. Under certain conditions, evaporation of a solvent A from freshly applied paint into air is controlled by diffusion. Assume that diffusion occurs through a thin film of stagnant air adjacent to the paint surface. The mole fraction of the solvent at a 2.5 mm distance from the paint surface is 0.08 and the mole fraction gradient is $-40\ m^{-1}$. Temperature and pressure are 20°C and 1 atm, respectively. The diffusivity of solvent in air is $0.091\ cm^2/s$. Determine the mass transfer flux of the solvent at this point.

P.9.3. Ethylene is dimerized to 1-butene on the surface of a catalyst. At high temperature the reaction kinetics becomes very fast and the rate of reaction becomes controlled by diffusional mass transfer. (a) Obtain an expression for the overall reaction rate of ethylene assuming that steady-state diffusion through a thin stagnant gas film adjacent to the catalytic surface controls the process. The reaction occurs only on the catalyst surface. (b) For the conditions shown in Fig. P.9.3, calculate the rate of reaction of ethylene per unit area of catalyst if the temperature is 100°C, the absolute pressure is 6.6 atm, and the binary diffusivity in ethylene –1-butene mixtures is estimated to be $\mathcal{D}_{A,B} = 0.0183\ cm^2/s$.

CHAPTER 10
Diffusivity

10.1 Introduction

In Chap. 9, we introduced the diffusivity as a physical property connected to a continuum description of mass transfer according to Fick's law. This continuum formulation is adequate for most practical problems by using $\mathcal{D}_{A,B}$ as a phenomenological descriptor of the rate at which atoms or molecules move in a given system. There is, however, a complementary description of diffusion that relies on a particle-based view of the process. This view can be helpful to understand the dependence of $\mathcal{D}_{A,B}$ on temperature and pressure and on the properties of the medium where transport takes place. It can also be helpful if we want to predict diffusivities from molecular simulations or to measure diffusivities directly by tracking individual molecules or particles; e.g., see Dickson et al., Science, 274 (1996) 966.

In a particle-based description of diffusion, we track the position of individual particles (representing atoms or molecules) as they move around a region of space. The movement of these particles is random and is affected by temperature and by collisions with other particles. We can imagine that, following each collision, each particle will change direction with equal probability of moving in any direction before the next collision. This type of movement is called Brownian motion. In the gas phase, for example, we can expect it to be faster at higher temperatures because the particles or molecules will have more energy. Increasing pressure will, however, increase the number of collisions, each with an ensuing change in direction. If we consider many particles undergoing many such steps, we will obtain a diffusion process. This type of motion can occur in one dimension, such as when a protein migrates along a DNA chain or moves along a microtubule, in two dimensions, such as when molecules migrate along a surface, or in three dimensions, such as when molecules diffuse in a bulk gas. Simulating particle movement is based on the same principle regardless of the number of dimensions, but one-dimensional simulations are, of course, simpler and are what we will consider next.

Let's first consider a single particle that can only move along a linear axis, x, and that is initially located at a position $x = 0$. We will assign equal probability that, at each step, the particle will move by one unit of length, ℓ, either in the positive direction or in the negative direction. After each step, we will reassign the same equal probability and repeat the process while keeping track of the cumulative displacement of the particle from its initial position at $x = 0$. Figure 10.1a shows, as an example, the position of 10 such particles over 100 such steps, with all particles starting at $x = 0$. Since we have assigned the same probability of moving in either direction at each step, each particle will be in a different randomly determined position. After just a few steps, most of the particles will still be very close to $x = 0$. However, as seen in Fig. 10.1a, as we increase the number of steps, the spread of the 10 particles around $x = 0$ increases. We can quantify this spreading by calculating the variance of the displacements from $x = 0$ according to the following equation:

$$\sigma^2 = \frac{\sum_{i=1}^{n_{\text{particles}}} (x_i - x_{\text{avg}})^2}{n_{\text{particles}} - 1} \tag{10.1}$$

This quantity is also plotted in Fig. 10.1a as a function of the number of steps. We can see that σ^2, also called the mean square displacement, increases linearly with the number of steps. With just 10 particles, the plot is scattered, but the scatter appears to be around a linear function. This function is shown by the regressed line with a slope of 1 in Fig. 10.1a. If we increase the number of particles to, say, 1,000, as shown in Fig. 10.1b, we see that the scatter around the straight line describing σ^2 decreases dramatically. The slope, however, remains the same, equal to 1. In order to relate this process to a physical system, we can assign a certain length of time, Δt, to each step and define the diffusion coefficient as follows:

$$\mathcal{D} = \frac{\ell^2}{2\Delta t} \tag{10.2}$$

Because we are considering only one type of particle, we call this a self-diffusivity, but we should realize that it could actually be measured experimentally if we were to look at a mixture of two molecules that are nearly identical, such as isotopes or

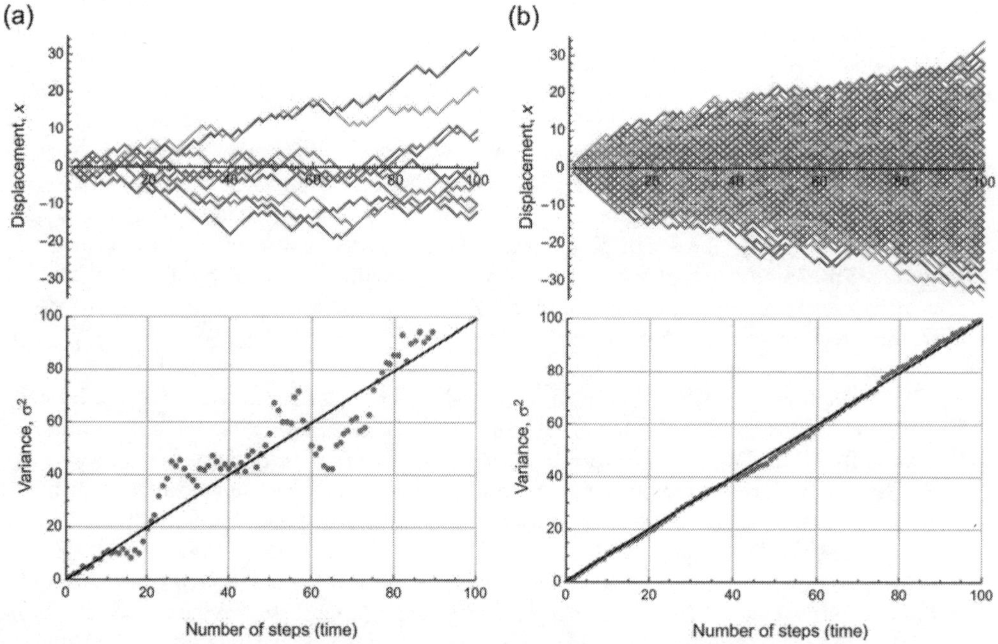

FIGURE 10.1 Simulation of random walk (Brownian motion) in one dimension with (a) 10 particles and (b) 1,000 particles. The lines in the bottom figures are regressed to the calculated mean square displacements. Mathematica notebook: (M.F.10.1).

enantiomers. If $\ell = 1$, as in the simulations in Fig. 10.1, and Δt is also set equal to 1, Eq. (10.2) gives $\mathcal{D} = 1/2$. Based on the slope of σ^2 vs. the number of steps and the value of $\mathcal{D}$, we will conclude that:

$$\sigma^2 = 2\mathcal{D}t \tag{10.3}$$

This means that the mean square displacement increases linearly with time with a proportionality constant equal to $2\mathcal{D}$. Note that if we had tracked particles moving in two dimensions, we would have found that the mean square displacement from the origin would have been described by the following equation:

$$\sigma^2 = 4\mathcal{D}t \tag{10.4}$$

whereas in three dimensions, it would have been:

$$\sigma^2 = 6\mathcal{D}t \tag{10.5}$$

These results differ from the 1-D case only by a numerical constant but show, otherwise, the same trends. It is important to recognize that while this particle-tracking approach is useful to understand the process, the continuum approach based on Fick's law is more useful from a practical viewpoint. Such an approach requires either experimentally determined or estimated diffusivities, which are discussed in the next sections.

10.2 Experimental Diffusivities

Diffusivities vary widely depending on the molecules involved, the medium where diffusion takes place, and the experimental conditions. Table 10.1 gives typical ranges. As seen from this table, there is an inverse correlation between the diffusivity and the density of the medium where transport takes place and between the diffusivity and the size of the diffusing molecule or particle. The diffusivities are larger in low-density phases compared to condensed phases and smaller for larger molecules compared to those observed for smaller molecules. Diffusivity data can be found in many sources but a good place to start is Section 2 of *Perry's Chemical Engineers' Handbook*, 9th edition.

Table 10.2 gives some representative values of the binary diffusivity $\mathcal{D}_{A,B}$ for a number of gas pairs at 1 atm absolute pressure. Note that these values depend on the properties of each component in the pair, but are otherwise independent of composition; i.e., they are generally the same for each system over the entire range of mole fraction from

System	Diffusivity, $\mathcal{D}_{A,B}$ in cm²/s
Gases and vapors at 1 atm	0.05–2
Small molecules ($M_r < 200$) in non-viscous liquids	1×10^{-5}–5×10^{-5}
Proteins and other polymers ($M_r > 5{,}000$) in water	1×10^{-7}–1×10^{-6}
Virus, virus-like particles, and plasmid DNA in water	1×10^{-9}–1×10^{-7}
Small organic molecules (solvents) in rubbery polymers	1×10^{-12}–1×10^{-8}
Small solutes in crystalline metals	1×10^{-30}–1×10^{-8}

TABLE 10.1 Ranges of Diffusivity Values at Room Temperature in cm²/s (divide by 10,000 to get m²/s)

A	B	T, K	$\mathcal{D}_{A,B}$, cm²/s
Air	Acetone	273	0.109
	Benzene	273	0.077
	Carbon dioxide	273	0.138
	Ethanol	273	0.102
	n-Hexane	328	0.093
	Hydrogen	273	0.611
	Methanol	273	0.132
	Phosgene	273	0.095
	Toluene	273	0.076
		303	0.088
	Water	273	0.220
Carbon dioxide	Methane	273	0.153
Hydrogen	Acetone	273	0.361
	Benzene	273	0.306
	Carbon dioxide	273	0.550
	Carbon monoxide	273	0.651
	Methane	288	0.694
	Oxygen	273	0.697
	Water	273	0.750

Table 10.2 Experimental Binary Diffusivities of Gas and Vapor Pairs at 1 atm Absolute Pressure

0 to 1. However, they vary with temperature and pressure, according to the following empirical relationship:

$$\mathcal{D}_{A,B} \propto \frac{T^{1.75}}{P} \tag{10.6}$$

where T is the absolute temperature and P is the absolute pressure. Thus, the values of $\mathcal{D}_{A,B}$ available at a particular temperature and pressure for a given gas pair can be used at different conditions for the same pair by scaling according to Eq. (10.6). We can see from Table 10.2 that the molecular size of both A and B affects the value of $\mathcal{D}_{A,B}$ with larger molecules resulting in smaller diffusivities. Mixtures with hydrogen have particularly high diffusivity.

Example 10.1. Effect of temperature and pressure on gas-phase diffusivity. Let's estimate the binary diffusivity of carbon dioxide–methane mixtures at 10 atm and 400 K based on the value $\mathcal{D}_{A,B} = 0.153$ cm²/s at 273 K and 1 atm in Table 10.2. Based on Eq. (10.6), at 400 K and 10 atm, we would predict:

$$\mathcal{D}_{A,B} = 0.153 \times \left(\frac{400}{273}\right)^{1.75} \frac{1}{10} = 0.0299 \text{ cm}^2/\text{s}$$

We can expect this value to be a good approximation. On the other hand, substantial errors can be expected when using this equation for conditions where the gas mixture is no longer treatable as an ideal gas, such as at higher pressures.

Table 10.3 provides some values of the diffusivity of a component A in a dilute solution in a solvent B. We refer to these values as diffusivities at infinite dilution, denoted by the symbol $\mathcal{D}^0_{A,B}$. Note that in the liquid phase, for a given pair, the diffusivity $\mathcal{D}_{A,B}$ can vary substantially with composition. Thus, the infinite dilution diffusivity of A in solvent B will often be quite different from the infinite dilution diffusivity of B in solvent A. In practice, in many cases, we can use the infinite dilution values at mass fractions up to 5% to 10%, but this is really dependent on the particular system. For example, for linear polymers like DNA in water, the solute concentration can affect the solution viscosity dramatically so that the infinite dilution values may be applicable only in extremely dilute mixtures. As seen from Table 10.3, $\mathcal{D}^0_{A,B}$ is influenced by the molecular mass of the solute and, strongly, by the viscosity of the solvent. Note, in particular, the data for the acetone–water pair. When acetone is the solute and water is the solvent, $\mathcal{D}^0_{A,B} = 1.3 \times 10^{-5}$ cm²/s. On the other hand, when water is the solute and acetone is the solvent, $\mathcal{D}^0_{A,B}$ becomes 5.2×10^{-5} cm²/s, which is four times larger. This occurs in large part because the viscosity of pure acetone is much lower than the viscosity of pure water.

Experimentally, infinite-dilution liquid-phase diffusivities for a given solute and solvent are correlated by the following equation, whose theoretical basis will be shown later:

$$\frac{\mathcal{D}^0_{A,B}\mu_B}{T} = \text{constant} \tag{10.7}$$

Solute (A)	Solute Molecular Mass, $M_{r,A}$ (g/mol)	Solvent (B)	Solvent Viscosity, μ_B (mPa s)	$\mathcal{D}^0_{A,B}$, cm²/s
Carbon dioxide	44	Water	0.89	2.0×10^{-5}
Ethanol	46	Water	0.89	1.3×10^{-5}
Acetone	58	Water	0.89	1.3×10^{-5}
Caffeine	194	Water	0.89	0.63×10^{-5}
Sucrose	342	Water	0.89	0.56×10^{-5}
Ribonuclease	14,000	Water	0.89	0.12×10^{-5}
Serum albumin	65,000	Water	0.89	0.060×10^{-5}
IgG	150,000	Water	0.89	0.040×10^{-5}
Adeno-associated virus (AAV)	3,900,000 (25 nm particle)	Water	0.89	0.015×10^{-5}
Water	18	Acetone	0.32	5.2×10^{-5}
Water	18	Ethanol	1.1	1.2×10^{-5}
Water	18	Ethylene glycol	20.0	0.18×10^{-5}
Water	18	Glycerol	120.0	0.013×10^{-5}

Table 10.3 Experimental Diffusivities of Different Solutes, A, in Different Solvents, B, at Infinite Dilution and 298 K

where μ_B is the solvent viscosity and T is the absolute temperature. Based on this equation, knowing $\mathcal{D}^0_{A,B}$ at a given temperature and in a given solvent, we can predict $\mathcal{D}^0_{A,B}$ at a different temperature just knowing how the solvent viscosity changes with temperature.

Example 10.2. Effect of temperature on liquid-phase diffusivity. Let's estimate $\mathcal{D}^0_{A,B}$ for acetone in water at 55°C based on the value $\mathcal{D}^0_{A,B} = 1.3 \times 10^{-5}$ cm²/s at 25°C from Table 10.3. The viscosity of water is 0.89 mPa s at 25°C and about 0.50 mPa s at 55°C (328 K). From Eq. (10.7), we have

$$\mathcal{D}^0_{A,B}\Big|_{55°\text{C}} = 1.3 \times 10^{-5} \times \frac{328}{298} \times \frac{0.89}{0.50} = 2.5 \times 10^{-5} \text{ cm}^2/\text{s}$$

The experimental value from Tyn and Calus [J. Chem. Eng. Data, 20 (1975) 310] at 55°C is 2.2×10^{-5} cm²/s; thus, our estimate is only about 14% off from this experimental value.

Diffusivities at infinite dilution in mixed solvents are important in a number of applications, such as extraction or chromatography, where solvent mixtures are commonplace. Figure 10.2 shows some illustrative data for the diffusion of benzaldehyde in acetonitrile/water mixtures and in methanol/water mixtures both at 298 K. For the acetonitrile/water system, the mixture viscosity decreases monotonically as the percentage of acetonitrile is increased. Correspondingly, $\mathcal{D}^0_{A,B}$ for benzaldehyde increases, also monotonically. On the other hand, for the methanol/water system, the mixture viscosity has a maximum at approximately 50% methanol and, correspondingly, $\mathcal{D}^0_{A,B}$ for benzaldehyde has a minimum at this percentage of methanol. Note that, as seen in this figure, in both cases, the product $\mathcal{D}^0_{A,B}\mu_B$ is relatively constant, which is

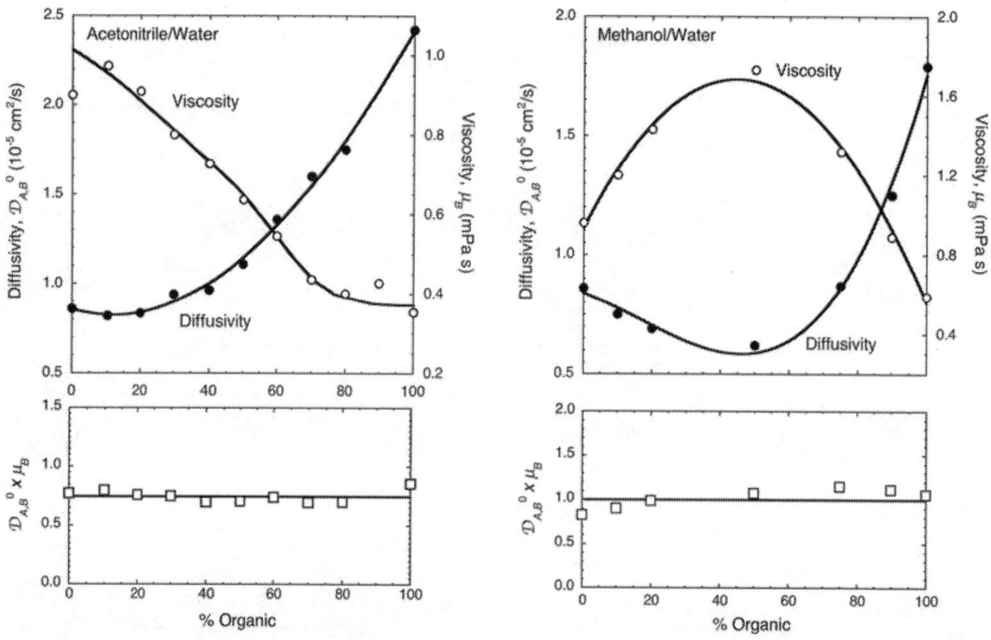

FIGURE 10.2 Diffusivity of benzaldehyde at infinite dilution at 25°C in acetonitrile/water mixtures (left) and in methanol/water mixtures (right). Data from Huss et al. [J. Chromatogr. 500 (1990) 241].

consistent with Eq. (10.7). From a processing viewpoint, 50% methanol would be a worst-case scenario. For instance, if we were to use this mixture as a chromatography solvent, we will have the highest viscosity and, thus, the highest operating pressure, combined with the lowest diffusivity and, thus, the slowest mass transfer. On the other hand, such a mixture could have other desirable properties with regards to other process performance, such as solubility and selectivity.

The diffusivity of macromolecules, such as proteins and DNA, in aqueous solution is especially important, since, as seen in Table 10.3, these molecules are large and have small diffusivities. As a result, diffusional mass transfer will often severely limit process performance when, for example, we try to purify them by chromatography, membrane processes, or other unit operations. The rates at which many naturally occurring processes occur will, of course, also be affected. Figure 10.3 shows some data for proteins and plasmid DNA, plotted as a function of their molecular mass. For the proteins shown, which are globular and, thus, have a compact, approximately spherical shape, the experimental $\mathcal{D}^0_{A,B}$ values decrease as a function of molecular mass with a $-1/3$ power dependence. On the other hand, for plasmid DNA molecules, which are not only larger in mass but which are also elongated, rod-shaped molecules, the $\mathcal{D}^0_{A,B}$ values are not only smaller than those for globular proteins but also decrease more rapidly with molecular mass with a $-2/3$ power dependence. Although not included in this graph, for linear polymers, such as dextran, the dependence on molecular mass is to a power approaching -1. Thus, it is empirically evident that not only molecular size but also molecular shape affects $\mathcal{D}^0_{A,B}$. As a rule, for the same molecular mass, more compact macromolecules will have a larger diffusivity compared to rod-shaped and linear molecules.

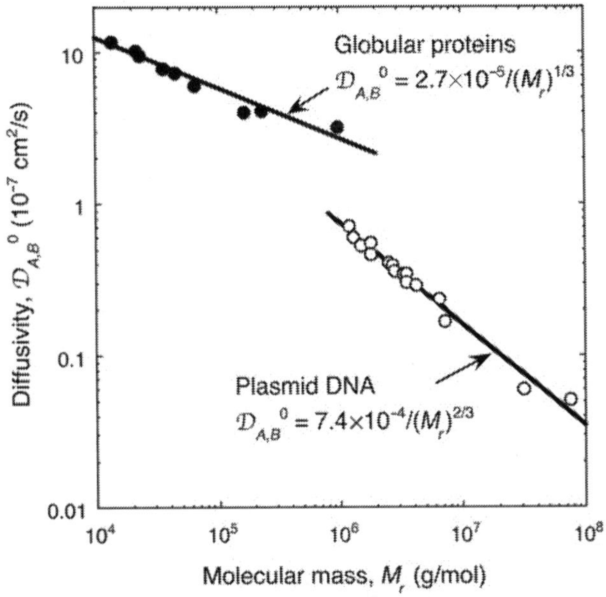

FIGURE 10.3 Diffusivity of globular proteins and plasmid DNA in aqueous solution at 20°C. Protein data are from Tyn and Gusek [Biotechnol. Bioeng. 35 (1990) 327]. Plasmid DNA data are from Prazeres [Biotechnol. Bioeng. 99 (2008) 1040]. The equations shown correlate the data with $\mathcal{D}^0_{A,B}$ in cm²/s and M_r in g/mol.

10.3 Prediction of Diffusivity

Because of the wide range over which diffusivities vary and the fact that $\mathcal{D}_{A,B}$ is a binary property, methods for estimating diffusivities are paramount and are used extensively. Moreover, measuring diffusivities accurately is not always easy and, more often than not, we rely on estimations. In this section, we provide an overview of commonly used estimation methods, but the reader is referred to the literature for a more comprehensive coverage. Section 5 of *Perry's Chemical Engineers' Handbook*, 9th edition, is a good place to start.

10.3.1 Gas-Phase Diffusivities

The foundation for the prediction of gas-phase diffusivities at low pressure is the kinetic theory of gases, which is known as the Chapman–Enskog theory of diffusion. Deriving the full theory is complex and beyond the scope of this book. However, enough information with regards to trends can be obtained even from the simplified model shown in Fig. 10.4, which illustrates a region of a gas phase where we have a mole fraction gradient of component A along the horizontal z axis. The molar density, C, is assumed to be constant throughout. The two vertical planes at positions a and c are located at distances equal to the mean free path, λ, from the center plane, located at b, where the mole fraction is y_A. The total number of molecules between a and b and those between b and c will be the same and will be proportional to the product of molar density and mean free path, $C\lambda$. If $\overline{v}$ is the mean molecule velocity, the mean time that elapses between collisions will be $\lambda/\overline{v}$. During this time, 1/3 of the molecules between a and b will move across the center plane from left to right while 1/3 of the molecules between b and c will move across the center plane from right to left. If we assume that the mole fraction profile is linear, which is reasonable over a short distance, the average mole fraction of A between plane a and plane b will be:

$$y_A - \frac{\lambda}{2}\frac{dy_A}{dz}$$

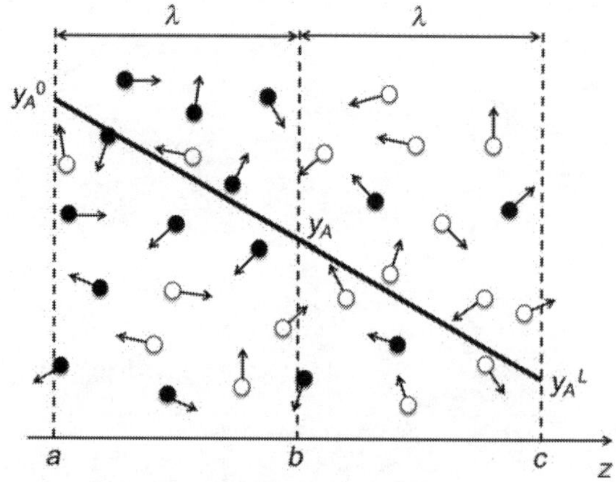

Figure 10.4 Schematic view of diffusion in an ideal gas.

while the average mole fraction of A between plane b and plane c will be:

$$y_A + \frac{\lambda}{2}\frac{dy_A}{dz}$$

As a result, there will be a net flux of A from left to right according to the following relationship:

$$N_A = \frac{1}{3}\frac{C\lambda\left[\left(y_A - \frac{\lambda}{2}\frac{dy_A}{dz}\right) - \left(y_A + \frac{\lambda}{2}\frac{dy_A}{dz}\right)\right]}{\lambda/\overline{v}} = -\frac{1}{3}\lambda\overline{v}C\frac{dy_A}{dz} \tag{10.8}$$

From Fick's law, we have:

$$N_A = -C\mathcal{D}\frac{dy_A}{dz} \tag{10.9}$$

As a result, combining Eqs. (10.8) and (10.9), we obtain the following expression for the diffusivity:

$$\mathcal{D} = \frac{1}{3}\lambda\overline{v} \tag{10.10}$$

From the kinetic theory of gases, we know that the mean free path increases with temperature and decreases with pressure according to the following equation:

$$\lambda \propto \frac{\mathcal{R}T}{P} \tag{10.11}$$

while the mean velocity varies with the square root of T and is independent of pressure according to:

$$\overline{v} \propto \sqrt{\frac{\mathcal{R}T}{M_r}} \tag{10.12}$$

As a result, combining Eqs. (10.10) to (10.12) yields:

$$\mathcal{D} \propto \frac{T^{3/2}}{P}\sqrt{\frac{1}{M_r}}$$

which is close to experimental observations.

The full Chapman–Enskog theory yields the following dimensional equation for the binary diffusion coefficient in a mixture of gases A and B as a function of the Lennard–Jones parameters σ and ε of the two components:

(M.E.10.13) $$\mathcal{D}_{A,B}\,(\text{cm}^2/\text{s}) = 1.86 \times 10^{-3} \times \frac{T^{3/2}\left(\frac{1}{M_{r,A}} + \frac{1}{M_{r,B}}\right)^{1/2}}{P\sigma_{AB}^2\Omega_D} \tag{10.13}$$

where

$$\sigma_{AB} = \frac{\sigma_A + \sigma_B}{2} \tag{10.14}$$

is in Å, T is in K, P is in atm, and M_r is in g/mol. Ω_D is the collision integral and is a function of the dimensionless temperature $T^* = k_B T/\varepsilon_{AB}$ where:

$$\frac{\varepsilon_{AB}}{k_B} = \left(\frac{\varepsilon_A}{k_B} \times \frac{\varepsilon_B}{k_B}\right)^{1/2} \tag{10.15}$$

The values of σ and ε/k_B are available in many standard handbooks, but can also be estimated from Eqs. (2.4) and (2.5) from the critical temperature and pressure of the two components. Note that Ω_D is different from the function $\Omega_\mu = \Omega_k$ used in Chap. 2 to predict gas viscosity and thermal conductivity and is approximated as a function of T^* by the following equation:

$$\begin{aligned}\Omega_D &= 1.06036(T^*)^{-0.15610} + 0.1930\exp(-0.47635T^*) \\ &\quad + 1.03587\exp(-1.52996T^*) + 1.76474\exp(-3.89411T^*)\end{aligned} \tag{10.16}$$

The following semi-empirical equation by Fuller et al. [J. Phys. Chem. 73 (1969) 3679] provides a practical and simpler alternative to the Chapman–Enskog theory for the prediction of binary gas diffusivities with comparable accuracy:

(M.E.10.17)
$$\mathcal{D}_{A,B}\ (\text{cm}^2/\text{s}) = 1.0 \times 10^{-3} \times \frac{T^{1.75}\left(\dfrac{1}{M_{r,A}} + \dfrac{1}{M_{r,B}}\right)^{1/2}}{P\left[\left(\Sigma v_A\right)^{1/3} + \left(\Sigma v_B\right)^{1/3}\right]^2} \tag{10.17}$$

with T in K and P in atm. The Σv_i values appearing in this equation are in cm^3/mol. They are available for a number of simple molecules or can be estimated for more complex molecules from atomic contributions based on the data in Table 10.4 or in the more extensive Table 5-13 of *Perry's Chemical Engineers' Handbook*, 9th edition.

Atomic volume contributions, v_i			
C	16.5	N	5.69
H	1.98	Cl	19.5
O	5.48	S	17.0
Aromatic ring	−20.2	Heterocyclic ring	−20.2
Molecular volumes, Σv_i			
H_2	7.07	CO	18.9
He	2.88	CO_2	26.9
N_2	17.9	N_2O	35.9
O_2	16.6	NH_3	14.9
Air	20.1	H_2O	12.7

TABLE 10.4 Atomic Volume Contributions and Molecular Volumes for Simple Gases for the Estimation of Diffusivities using the Fuller et al. equation, Eq. (10.17)

Example 10.3. Estimation of gas-phase diffusivity. Estimate the binary diffusion coefficient for the methane/hydrogen system at 288 K and 1 atm absolute pressure. Let's first use the Chapman–Enskog theory. The Lennard–Jones parameters for methane and hydrogen are as follows:

$$\sigma_{CH_4} = 3.76;\;\; \varepsilon_{CH_4}/k_B = 149$$

$$\sigma_{H_2} = 2.83;\;\; \varepsilon_{H_2}/k_B = 59.7$$

Thus,

$$\sigma_{AB} = \frac{3.76 + 2.83}{2} = 3.30$$

$$\frac{\varepsilon_{AB}}{k_B} = (149 \times 59.7)^{1/2} = 94.3$$

$$T^* = \frac{288}{94.3} = 3.05$$

Equation (10.16) gives $\Omega_D = 0.946$. Finally, we can calculate $\mathcal{D}_{A,B}$ from Eq. (10.13) as:

$$\mathcal{D}_{A,B} = 1.86 \times 10^{-3} \times \frac{288^{3/2}\left(\frac{1}{16} + \frac{1}{2}\right)^{1/2}}{1.0 \times 3.30^2 \times 0.946} = 0.664 \text{ cm}^2/\text{s}$$

For the Fuller et al. equation, Eq. (10.17), we need to estimate Σv_{CH_4} and Σv_{H_2}. We can get Σv_{H_2} = 7.07 cm^3/mol directly from Table 10.4. For methane, we need to estimate Σv_{CH_4} from the atomic and structural diffusion volume contributions of C and H. Table 10.4 or Table 5-13 in *Perry's Handbook* give the values $v_C = 16.5$ for carbon atoms and $v_H = 1.98$ for hydrogen atoms. Thus, for methane we will have $\Sigma v_{CH_4} = 16.5 + 4 \times 1.98 = 24.4$. Using these values, Eq. (10.17) yields the following result:

$$\mathcal{D}_{A,B} = 1.0 \times 10^{-3} \times \frac{288^{1.75}\left(\frac{1}{16} + \frac{1}{2}\right)^{1/2}}{1.0[(24.4)^{1/3} + (7.07)^{1/3}]^2} = 0.650 \text{ cm}^2/\text{s}$$

From Table 10.2, the experimental value at this temperature and pressure is 0.694 cm^2/s. We can see that, in this case, both equations provide values that are in close agreement with this experimental value (within 4% and 6%, respectively). Note that the temperature dependence in the Fuller et al. equation is completely explicit (1.75 power), while for the Chapman–Enskog theory we have both the explicit $T^{3/2}$ dependence as well as the temperature dependence of Ω_D, which decreases as a function of T. The 1.75-power dependence in the Fuller et al. equation captures this more complex dependence with reasonable accuracy.

As seen from either the Chapman–Enskog theory or the Fuller et al. equation, the product of diffusivity and absolute pressure, $\mathcal{D}_{A,B}P$, is predicted to be only a function of temperature for a given gas pair. In other words, $\mathcal{D}_{A,B}$ is predicted to be inversely proportional to the absolute pressure. This relationship is observed in practice at relatively low pressures and, more generally, for conditions that conform to the ideal gas law. Significant deviations can occur, however, at high pressures and moderate temperatures, leading to lower-than-predicted values of $\mathcal{D}_{A,B}$.

Takahashi [J. Chem. Eng. Japan 7 (1974) 417] has developed the generalized correlation shown in Fig. 10.5 to predict the effects of pressure on $\mathcal{D}_{A,B}$. In this chart,

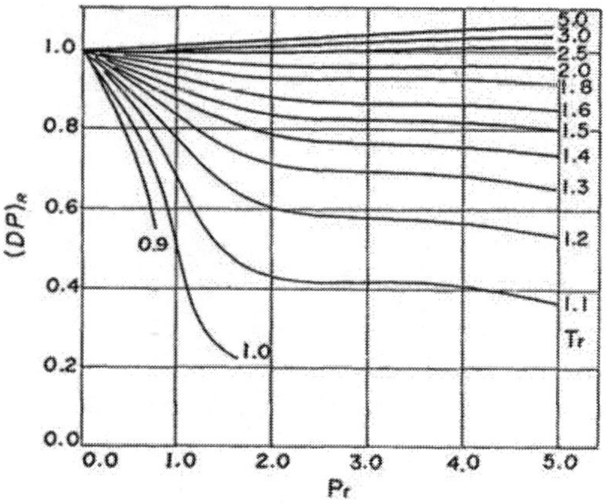

FIGURE 10.5 Takahashi correlation to predict the effect of pressure on the binary gas diffusivity. Reproduced from Takahashi [J. Chem. Eng. Japan 7 (1974) 417] with permission.

$P_r = P/P_c$ and $T_r = T/T_c$ are the reduced pressure and reduced temperature, respectively, and $(DP)_R$ is a correction factor defined as:

$$(DP)_R = \frac{\mathcal{D}_{A,B}P}{(\mathcal{D}_{A,B}P)_{1\text{atm}}} \tag{10.18}$$

where $(\mathcal{D}_{A,B}P)_{1\text{atm}}$ is the product of diffusivity and absolute pressure at 1 atm. Example 10.4 illustrates the use of this correlation.

Example 10.4. Effect of pressure on the binary-gas diffusivity. Let's consider the binary diffusivity for carbon dioxide/methane mixtures at 298 K and 40 atm. For mixtures containing 50 mol% CO_2 and 50 mol% CH_4 at 298 K and 40 atm, the experimental diffusivity value is $\mathcal{D}_{CO_2,CH_4} = 0.0022$ cm²/s [Jeffries and Drickamer, J. Chem. Phys. 54 (1971) 1881; *Perry's Chemical Engineers' Handbook*, 9th edition, Table 2-141]. If we use Eq. (10.13) with the values of $\sigma_{CH_4} = 3.76$, $\varepsilon_{CH_4}/k_B = 149$ for methane and $\sigma_{CO_2} = 3.929$, $\varepsilon_{CO_2}/k_B = 228$ for CO_2, we obtain $\mathcal{D}_{CO_2,CH_4} = 0.162$ cm²/s at 1 atm and 0.0041 cm²/s at 40 atm. The latter is almost double the experimental value, so a correction for the nonideal pressure effect is needed. In order to use the Takahashi correlation, we calculate the pseudocritical temperature and the pseudocritical pressure of the mixture according to Kay's mixing rule, which gives:

$$T_c' = y_{CO_2}T_{c,CO_2} + y_{CH_4}T_{c,CH_4} = 0.5 \times 304 + 0.5 \times 190 = 247 \text{ K}$$

$$P_c' = y_{CO_2}P_{c,CO_2} + y_{CH_4}P_{c,CH_4} = 0.5 \times 72.8 + 0.5 \times 45.8 = 59.3 \text{ atm}$$

Thus, $T_r = 298/247 = 1.2$ and $P_r = 40/59.3 = 0.67$. From Fig. 10.5, the correction factor is $(DP)_R \sim 0.85$. As a result, we have:

$$\mathcal{D}_{A,B} = (DP)_R \frac{(\mathcal{D}_{A,B}P)_{1\text{ atm}}}{P} = 0.85\frac{0.162}{40} = 0.0034 \text{ cm}^2/\text{s}$$

This is an improvement over the value predicted by the Chapman–Enskog theory with a correction in the expected direction, but the result is still a bit far from the experimental value of Jeffries and Drickamer. Takahashi suggests much better predictive accuracy based on different experimental data. It should be noted that it is difficult, in general, to measure gas-phase diffusivities at high pressure since even small, inadvertent pressure gradients can skew the data dramatically so that experimental values themselves can be affected by large errors.

10.3.2 Liquid-Phase Diffusivities

The theoretical basis for many of the methods used to estimate diffusivities in dilute solutions is the Stokes–Einstein theory. Although developed for colloidal particles and, thus, best suited to describe the diffusivity of macromolecules in low-molecular mass solvents, it also proves valuable as a foundation for small-molecule diffusion. The theory combines Einstein's expression for the force, F_C, acting on a spherical colloidal particle of radius r_A resulting from collisions from the solvent molecules B, with Stokes' expression for the drag force, F_D, experienced by a sphere moving in a viscous solvent. Einstein's equation predicts that

$$F_C = \frac{k_B T}{\mathcal{D}^0_{A,B}} \overline{v} \tag{10.19}$$

while Stokes' equation for creeping flow around a sphere predicts that

$$F_D = 6\pi r_A \mu_B \overline{v} \tag{10.20}$$

In both cases, $\overline{v}$ is the velocity of the colloidal particle. At steady state, the two forces will be equal. Thus, equating Eqs. (10.19) and (10.20) we obtain:

$$\mathcal{D}^0_{A,B} = \frac{k_B T}{6\pi r_A \mu_B} \tag{10.21}$$

or

$$\frac{\mathcal{D}^0_{A,B}\mu_B}{T} = \frac{k_B}{6\pi r_A} \tag{10.22}$$

Note that $k_B = \mathcal{R}/\mathcal{N}$ is the Boltzmann constant, where $\mathcal{R}$ is the ideal gas law constant and $\mathcal{N}$ is Avogadro's number. In SI units, $k_B = 1.38\times10^{-23}$ J/molecule K. This simple equation is remarkably accurate for macro-solutes that can be approximated as colloidal spheres. Most importantly, however, it predicts that the group $\mathcal{D}^0_{A,B}\mu_B/T$ will be constant for a given solute which is experimentally observed as noted in Section 10.2. An analogous expression has been derived by Perrin [J. Phys. Radium 7 (1936) 1] for a colloidal particle with the shape of a prolate ellipsoid with major and minor semi-axes r_a and r_b, respectively, and is as follows:

$$\frac{\mathcal{D}^0_{A,B}\mu_B}{T} = \frac{k_B}{6\pi r_a} \frac{\ln\left(\dfrac{1+[1-(r_b/r_a)^2]^{1/2}}{r_b/r_a}\right)}{[1-(r_b/r_a)^2]^{1/2}} \tag{10.23}$$

This expression coincides with Eq. (10.22) when $r_a = r_b = r_A$. Example 10.5 illustrates the use of these equations to predict the diffusivity of proteins in water.

Example 10.5. Liquid-phase diffusivity of a protein based on the Stokes-Einstein and Perrin equations. Let's use the equations presented in this section to estimate the diffusivity of the protein lysozyme in a dilute aqueous solution at 20°C. Lysozyme has a molecular mass of 14,500 g/mol and, as shown in Fig. 10.6, is an ellipsoid-shaped molecule with major and minor semi-axes of approximately 2.3 and 1.5 nm, respectively. The volume of the ellipsoid is $V = 4\pi r_a r_b^2/3 = 21.7$ nm^3. The radius of a

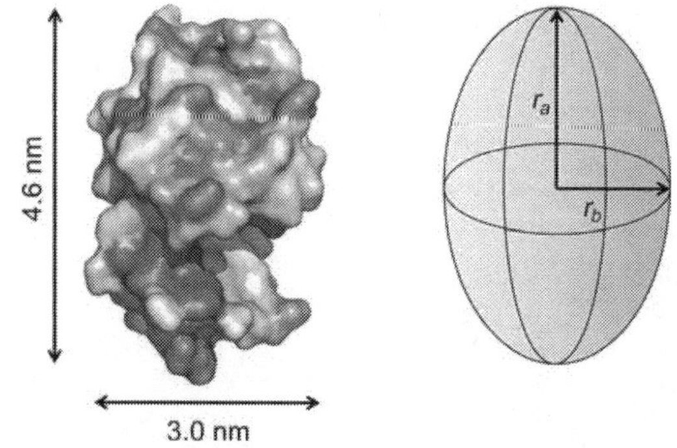

FIGURE 10.6 Molecular structure of lysozyme (left) and its prolate ellipsoid representation (right).

sphere with the same volume will then be $(3V/4\pi)^{1/3} = (3\times 21.7/4\pi)^{1/3} = 1.73$ nm. Using this value for r_A, Eq. (10.22) gives:

$$\mathcal{D}^0_{A,B} = \frac{(1.38\times 10^{-23})\times 293}{6\pi\times(1.73\times 10^{-9})\times 0.001} = 1.15\times 10^{-10}\ \text{m}^2/\text{s}$$

Using Eq. (10.23), instead, with $r_a = 2.3$ and $r_b = 1.5$ nm, gives $\mathcal{D}^0_{A,B} = 1.13\times 10^{-10}$ m^2/s, which is only slightly lower than the value predicted by the Stokes–Einstein equation. Based on the data in Fig. 10.3, the experimental value is $\mathcal{D}^0_{A,B} \sim 1.1\times 10^{-6}$ cm^2/s or 1.1×10^{-10} m^2/s. Thus, both equations are in remarkable agreement with actual data. We can see that, unless the molecule is highly elongated, using the volume-average radius with the Stokes–Einstein equation will generally be fairly accurate. Note that the molecular mass of the protein is proportional to its volume and the radius of a sphere is proportional to the cube root of the volume. Thus, Eq. (10.22) is also consistent with the dependence of $\mathcal{D}^0_{A,B}$ on $(M_r)^{-1/3}$ observed empirically for many proteins. Finally, note that if the diffusivity is measured experimentally, the molecular radius is often calculated according to Eq. (10.22) as the radius of a sphere with the same diffusivity as the measured value. In this case, the "experimental" molecular radius is called the hydrodynamic radius and is usually denoted by the symbol r_H. Diffusion data for proteins and other macromolecules are often reported as r_H.

In the case of small solutes, which can no longer be regarded as colloidal particles, the Stokes–Einstein theory is not valid for quantitative predictions since forces other than elastic collisions and viscous drag will affect the diffusivity. In this case, empirical and semi-empirical correlations are preferred. For a comprehensive coverage of the many available equations, the reader is referred to the literature. Section 5 of *Perry's Chemical Engineers' Handbook*, 9th edition, is, again, an excellent place to start. Here we limit ourselves to two popular expressions: the Wilke–Chang equation [AIChE J. 1 (1955) 264], which is valid for different solvents, and the Hyduk–Laudie equation [AIChE J. 20 (1974) 611], which is valid only for water as a solvent. Both equations are dimensional and require use of specified units as defined below. The Wilke–Chang (W-C) equation is given by:

(M.E.10.24)
$$\frac{\mathcal{D}^0_{A,B}\mu_B}{T} = 7.4\times 10^{-8}\frac{(\phi_B M_{r,B})^{1/2}}{(\hat{V}_A)^{0.6}} \qquad (10.24)$$

where $\mathcal{D}^0_{A,B}$ is in cm^2/s, μ_B is the solvent viscosity in mPa s (or cp), $M_{r,B}$ is the molecular mass of the solvent in g/mol, ϕ_B is a solvent association parameter, and $\hat{V}_A$ is the molar

Values of $\hat{V}_A$ for simple molecules in cm^3/mol				
$H_2 = 14.3$	$O_2 = 25.6$	$N_2 = 31.2$	$CO = 30.7$	$CO_2 = 34.0$
$SO_2 = 44.8$	$NO = 23.6$	$NH_3 = 25.8$	$H_2O = 18.9$	$H_2S = 32.9$

Additive atomic and structural contributions to $\hat{V}_A$ in cm^3/mol						
C = 14.8	H = 3.7	Br = 27.0	Cl = 24.6	F = 8.7	I = 37.0	S = 25.6

O	In methyl esters and ethers	9.1	N	Doubly bonded	15.6
	In ethyl esters and ethers	9.9		In primary amines	10.5
	In longer esters and ethers	11.0		In secondary amines	12.0
	In acids	12.0			
	Joined to S, P, N	8.3			
	Other cases	7.4			

Subtractions for ring structures: 3-member ring = –6.0, 4-member ring = –8.5, 5-member ring = –11.5, 6-member ring = –15.0, Naphtalene = –20.0, Anthracene = –47.5.

Table 10.5 Molar Volumes of Simple Molecules and LeBas Atomic and Structural Increments to Estimate $\hat{V}_A$ in cm^3/mol.

volume of the solute in its liquid state at the normal boiling point in cm^3/mol. ϕ_B has values of 2.6 for water, 1.9 for methanol, 1.5 for ethanol, 1.2 for propanol, and 1.0 for nonhydrogen bonded solvents. $\hat{V}_A$ can be estimated roughly from the liquid density of A or, more accurately, from the so-called LeBas volumes according to Table 10.5. In general, Eq. (10.24) may be in error by as much as ±20%. While this seems large, it is actually often acceptable for practical calculations.

The simpler Hyduk–Laudie (H-L) equation is valid only for diffusion in water and is given by:

$$\mathcal{D}^0_{A,B} = 1.33\times 10^{-4}(\mu_B)^{-1.14}(\hat{V}_A)^{-0.589} \tag{10.25}$$

where $\mathcal{D}^0_{A,B}$ is in cm^2/s, μ_B is the viscosity of water in mPa s (or cp), and $\hat{V}_A$ is in cm^3/mol.

Example 10.6. Estimation of diffusivity in water and in organic solvents. Estimate the diffusivity of benzaldehye in acetonitrile, in methanol, and in water at 298.K. Benzaldehyde has formula C_7H_6O and the following structure:

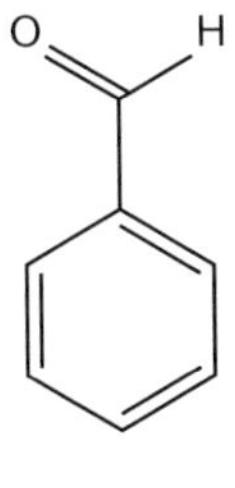

We can estimate $\hat{V}_A$ for benzaldehyde from Table 10.5 as $\hat{V}_A = 7\times 14.8 + 6\times 3.7 + 7.4 - 15.0 = 118\ cm^3/mol$. Subtracting the value of 15.0 accounts for the more compact structure of the aromatic ring compared to that of a linear hydrocarbon. Other parameters are summarized in the following table along with predictions based on the W-C equation for all three solvents as well as with predictions based on the H-L equation for water. Experimental values are also shown. As seen from this table, there is good agreement between experimental data and predictions from both the W-C equation for all three solvents and from the H-L equation for water. For this example, the W-C error is less than 6% for the two organic solvents. For water, the W-C error is 13% while the H-L error is about 6%. These errors are representative of what is often found in practice.

Solvent	ϕ_B	$M_{r,B}$	μ_B (mPa s)	$\mathcal{D}^0_{A,B}$ W-C (cm²/s)	$\mathcal{D}^0_{A,B}$ H-L (cm²/s)	$\mathcal{D}^0_{A,B}$ Exp.* (cm²/s)
Acetonitrile	1.0	41.1	0.35	2.3×10^{-5}	–	2.4×10^{-5}
Methanol	1.9	32.0	0.52	1.9×10^{-5}	–	1.8×10^{-5}
Water	2.6	18.0	0.89	0.97×10^{-5}	0.91×10^{-5}	0.86×10^{-5}

*Data from Huss et al., J. Chromatogr. 500 (1990) 241.

10.3.3 Diffusion in Porous Media

Many practical materials are naturally porous and mass transfer in such materials is often limiting. Porous materials are also used by design in many different applications including catalysis and adsorption, where the porosity provides a large surface area. The size of the pores is critical in these applications. If the pores are regarded as long cylinders, the relationship between the pore surface area per unit volume of material, a_s, and the pore diameter, d_{pore}, is given by the following equation:

$$a_s = \frac{4\varepsilon_p}{d_{pore}} \tag{10.26}$$

where ε_p is the porosity, which is defined as

$$\varepsilon_p = \frac{\text{Volume of the pores}}{\text{Total volume of the porous material}} \tag{10.27}$$

Since ε_p is dimensionless, a_s will have units of m² of surface area per m³ of material or m⁻¹. In general, as shown by Eq. (10.26), smaller pores and greater porosity will lead to greater surface areas, which will benefit catalytic reaction rates and adsorption capacity. On the other hand, greater porosity and small pore size will also have negative effects. For example, a greater porosity will usually lead to diminished mechanical strength, while a smaller pore diameter will negatively impact diffusion. Thus, in practice, a compromise between these quantities will often be needed.

In order to better understand the effect of pore size on diffusion, it is useful to refer to the following standard classification of pore sizes according to the International Union of Pure and Applied Chemistry (IUPAC):

	Pore Width or Diameter (nm)
Micropores	≤ 2
Mesopores	2–50
Macropores	> 50

Micropores are found, for example, in activated carbon and within crystalline materials like zeolites. Micropores are small enough that interactions between the diffusing molecule and the pore surface usually dominate transport. As a result, micropore diffusivities can be extremely small and are often unrelated to the diffusivity in a bulk fluid phase. Nevertheless, since partitioning in such pores from a surrounding fluid can be highly favorable when attractive interactions are present, micropore transport can still occur at a high rate despite the small diffusivity. At the other end of the spectrum, macropores are large enough that, in most cases, diffusion will occur in the fluid contained within the pores much like it occurs in the bulk fluid. Macropores are found in catalyst and adsorbent pellets that can be regarded as a conglomerate of microporous granules or crystals. Finally, in the intermediate case, diffusion in mesopores may result from a combination of ordinary diffusion and surface-mediated transport. Some materials, like activated carbon, may contain a broad, multimodal distribution of pore sizes including micropores, mesopores, and macropores. Other materials, like zeolite pellets, may contain a bimodal distribution of pore sizes including micropores and macropores, or even a monomodal distribution if they are based on a single crystal.

Predicting micropore diffusion is not straightforward as it is strongly influenced by the specific nature of molecule–surface interactions. Micropore diffusivities are also typically strongly dependent on the molecule concentration in the micropores and, thus, vary extensively with conditions. For details, the reader is referred to the literature on this topic which is quite vast. Section 16 of *Perry's Chemical Engineers' Handbook*, 9th edition, is a good place to start for adsorbent materials.

For diffusion in mesopores, different considerations will apply depending on whether we are dealing with diffusion in the gas or in the liquid phase. For gas-phase diffusion, the primary consideration will be whether the pores are larger or smaller than the molecule's mean free path, λ. For a bulk gas phase, λ is defined as the average distance travelled by a molecule before colliding with another molecule. There are two extreme cases. At one extreme, the pore size, d_{pore}, is much smaller than λ. In this case, molecules in the mesopore will have a much higher probability of colliding with the pore wall than with other molecules. As a result, molecule-wall collisions will determine diffusional transport. The corresponding transport mechanism is called Knudsen diffusion and is characterized by a Knudsen diffusivity given by the following dimensional equation:

$$\mathcal{D}_A^K(\text{cm}^2/\text{s}) = 4.85\times 10^{-4} d_{\text{pore}} \left(\frac{T}{M_{r,A}}\right)^{1/2} \tag{10.28}$$

where d_{pore} is in nm, T is in K, and $M_{r,A}$ is in g/mol. At the other extreme, the pore size is much larger than λ. In this case, transport will be dominated by ordinary diffusion, with a diffusivity $\mathcal{D}_{A,B}$ given, for example, by Eq. (10.13) or (10.17), just as in the case of ordinary diffusion in an unconfined bulk gas. In intermediate cases, Knudsen and ordinary diffusion will occur simultaneously both contributing to the overall transport process. In this case, transport will be described by an overall pore diffusivity, $\mathcal{D}_{A,\text{pore}}$, given by the following approximate relationship:

$$\frac{1}{\mathcal{D}_{A,\text{pore}}} = \frac{1}{\mathcal{D}_A^K} + \frac{1}{\mathcal{D}_{A,B}} \tag{10.29}$$

This equation suggests that, for a given system, the smallest of the two diffusivities $\mathcal{D}_A^K$ or $\mathcal{D}_{A,B}$ will dominate the overall process. Note that, based on Eq. (10.28), $\mathcal{D}_A^K$ is independent of pressure, proportional to the pores size, and only weakly dependent on temperature. Conversely, based on Eq. (10.13) or (10.17), $\mathcal{D}_{A,B}$ is independent of pore size, inversely proportional to pressure, and strongly dependent on temperature. As a result, Knudsen diffusion will tend to be dominant in smaller pores, at lower pressures, and at higher temperatures. Example 10.7 illustrates the use of these relationship to predict gas diffusion in mesopores.

Example 10.7. Gas diffusion in a mesopore. Consider diffusion in a carbon dioxide (A)–methane (B) mixture within a 25 nm pore at 273 K and 1 atm absolute pressure. From Table 10.1, the diffusivity in an unconfined gas mixture at this temperature and pressure is $\mathcal{D}_{A,B} = 0.153$ cm²/s. With $M_{r,A} = 44$ g/mol, Eq. (10.28) gives the following Knudsen diffusivity for CO_2:

$$\mathcal{D}_A^K = 4.85 \times 10^{-4} \times 25 \left(\frac{273}{44}\right)^{1/2} = 0.030 \text{ cm}^2/\text{s}$$

The overall diffusivity in the pore will thus be

$$\mathcal{D}_{A,\text{pore}} = \left(\frac{1}{\mathcal{D}_A^K} + \frac{1}{\mathcal{D}_{A,B}}\right)^{-1} = \left(\frac{1}{0.030} + \frac{1}{0.153}\right)^{-1} = (33 + 6.5)^{-1} = 0.025 \text{ cm}^2/\text{s}$$

Since this result is close to the Knudsen diffusivity, Knudsen diffusion will actually be dominant in this case. Note, however, that if we increased the pressure to 10 atm, $\mathcal{D}_{A,B}$ would become 0.015 cm²/s or less while $\mathcal{D}_A^K$ would remain the same. As a result, at 10 atm, $\mathcal{D}_{A,\text{pore}}$ would become 0.010 cm²/s, much closer to $\mathcal{D}_{A,B}$, indicating that ordinary diffusion would become dominant at this higher pressure.

For gas-phase diffusion in macropores and for liquid-phase diffusion of small molecules in mesopores and macropores, the pore diffusivity is typically close to the corresponding bulk diffusivity since the mean free path will be much smaller than the pore size. However, additional considerations will need to be made in the case of macromolecules and nanoparticles, including synthetic and natural polymers, proteins, DNA, and virus particles whose diameter may be in the range of 2 to 200 nm. In this case, both steric and hydrodynamic interactions will affect diffusion even in macropores. Figure 10.7 shows the case where a species A with radius r_A diffuses in a solvent B contained within a macropore. For simplicity, the pore is represented as a cylinder with radius r_{pore}. There are two important considerations. The first is the size-exclusion effect caused by the fact that the fraction of liquid in the annular space adjacent to the pore wall with thickness r_A is inaccessible by the diffusing species. This means that the volume available for diffusion

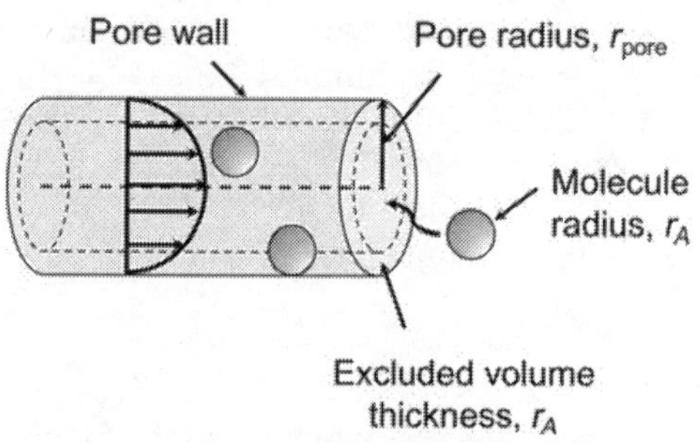

FIGURE 10.7 Hindered diffusion in a liquid-filled macropore.

will actually be smaller than the total pore volume. The second consideration is the drag force experienced by the diffusing species as the liquid leaves the pore to make room for it. The combination of these two effects is expressed by a hindrance factor, ψ_A, defined in such a way that:

$$\mathcal{D}_{A,\text{pore}} = \psi_A \times \mathcal{D}^0_{A,B} \tag{10.30}$$

where $\mathcal{D}_{A,\text{pore}}$ is the diffusivity of A in the pore and $\mathcal{D}^0_{A,B}$ is the diffusivity of species A in unconfined bulk liquid B. Since ψ_A is less than 1, the diffusivity in the pore will be smaller than in the bulk liquid as a result of these two effects. The accessible volume is a fraction of the pore volume defined by the following:

$$K_{D,A} = \frac{\text{Accessible pore volume}}{\text{Total pore volume}} = \frac{\pi\left(r_{\text{pore}} - r_A\right)^2 L}{\pi r^2_{\text{pore}} L} = \left(1 - \frac{r_A}{r_{\text{pore}}}\right)^2 = \left(1 - \varphi_A\right)^2 \tag{10.31}$$

where $\varphi_A = r_A / r_{\text{pore}}$. Obviously, if the size of the diffusing species is much smaller than the pore size, $K_{D,A}$ will be close to 1. On the other hand, as the size of the diffusing species approaches the pore size, $K_{D,A}$ will approach 0. A lower $K_{D,A}$ will reduce the diffusivity since less of the pore volume will be available for diffusion. Predicting the drag force effect is complex and a number of hydrodynamic theories have been developed to describe it. In all cases, the effect is dependent again on φ_A (e.g., see Deen, AIChE J. 33 (1987) 1410). The following relationships provide an approximate description of ψ_A based on the works of Brenner and Gaydos [Coll. Int. Sci., 58 (1977) 312] and Anderson and Quinn [Biophys. J. 14 (1974) 130], respectively:

$$\psi_A = \left(1 + \frac{9}{8}\varphi_A \ln \varphi_A - 1.539\varphi_A\right) \quad \text{for} \quad \varphi_A < 0.2 \tag{10.32}$$

$$\psi_A = 0.865\left(1 - \varphi_A\right)^2\left(1 - 2.104\varphi_A + 2.089\varphi_A^3 - 0.984\varphi_A^5\right) \quad \text{for} \quad \varphi_A > 0.2 \tag{10.33}$$

These functions are plotted in Fig. 10.8 over their respective ranges of validity. An important observation is that ψ_A decreases rapidly as φ_A increases, approaching 0 as φ_A approaches 0.5. This means that diffusion in a pore that is twice as large as the diffusing species will be highly hindered (i.e., $\mathcal{D}_{A,\text{pore}} << \mathcal{D}^0_{A,B}$). We see that, in practice, if we want to limit the effect of hindered diffusion, we will need to select materials whose pore size is many times larger than the size of the diffusing species. As a rule of thumb, values of $\varphi_A \sim 0.1$ or smaller will be needed. This means that a pore size 10 times larger than that of the diffusing species will be desirable. Note that if we are using a porous material as an adsorbent, adsorption at the pore wall will, effectively, restrict accessibility and increase drag, thereby further hindering diffusional transport. Thus, an even larger pore size may be needed to prevent excessive diffusional hindrance in these cases.

Example 10.8. Hindered diffusion in macropores. Let's use the equations introduced in this section to estimate the diffusivity of macrosolutes in a 150-nm-diameter pore ($r_{\text{pore}} = 75$ nm). Let's first consider diffusion of the protein serum albumin in aqueous solution at 20°C. Serum albumin is a globular

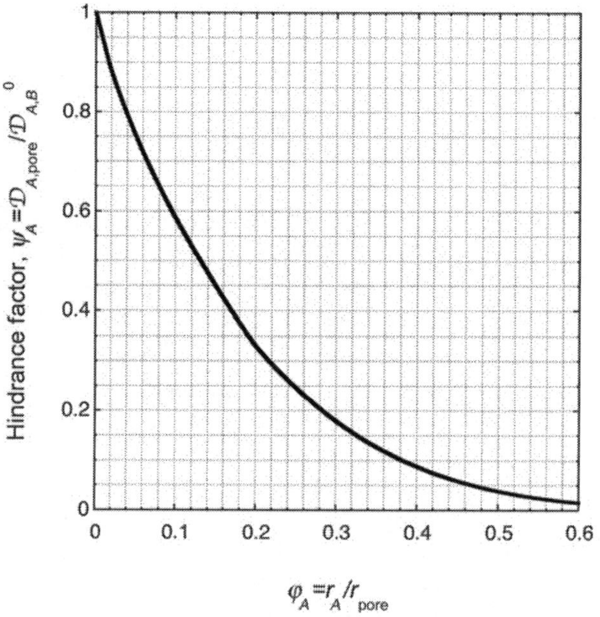

FIGURE 10.8 Hindrance factor for diffusion of a species A in a pore filled with a liquid solvent B.

protein with a 65,000 g/mol molecular mass. Based on Fig. 10.3, its solution diffusivity is estimated to be:

$$\mathcal{D}^0_{A,B} = \frac{2.7 \times 10^{-5}}{(M_{r,A})^{1/3}} = 6.7 \times 10^{-7} \text{ cm}^2/\text{s}$$

or 6.7×10^{-11} m^2/s. We can use this value to estimate the molecular radius of albumin assuming that it can be represented by the hydrodynamic radius, r_H. Based on the Stokes–Einstein equation, Eq. (10.22), we obtain:

$$r_A = \frac{k_B T}{6\pi\mu_B \mathcal{D}^0_{A,B}} = \frac{1.38 \times 10^{-23} \times 293}{6\pi \times 0.001 \times 6.7 \times 10^{-11}} = 3.2 \times 10^{-9} \text{ m}$$

or 3.2 nm. Thus, $\varphi_A = r_A/r_{\text{pore}} = 3.2/75 = 0.042$. From Eq. (10.32) or directly Fig. 10.8, we obtain $\psi_A = 0.77$ and, thus, $\mathcal{D}_{A,\text{pore}} = 0.77 \times 6.7 \times 10^{-11} = 5.2 \times 10^{-11}$ m^2/s. We see that, in this case, diffusional hindrance is moderate since φ_A is small. Let's consider next the diffusion of coronavirus in the same pore. Coronavirus is a spherical virus with a radius of about 50 nm (Note: coronavirus SARS-CoV-2 has been responsible for the COVID-19 pandemic). Since this virus is spherical, we can estimate $\mathcal{D}^0_{A,B}$ directly from Eq. (10.22), which gives $\mathcal{D}^0_{A,B} = 5.3 \times 10^{-12}$ m^2/s. In this case, we will have $\varphi_A = 50/75 = 0.66$, $\psi_A \sim 0$, and thus $\mathcal{D}_{A,\text{pore}} \sim 0$. If we wanted to design an adsorbent to remove this virus from a solution by adsorption in a porous particle, we would need larger pores. Using $\varphi_A \leq 0.1$ as a criterion, we would need a pore radius r_{pore} ~500 nm. In this case, we would estimate $\psi_A \sim 0.6$, and thus $\mathcal{D}_{A,\text{pore}} \sim 0.3 \times 10^{-11}$ m^2/s.

A final important consideration is how we can actually apply the relationships introduced to estimate the pore diffusivity $\mathcal{D}_{A,\text{pore}}$ to describe transport in practical materials. Most practical porous materials do not contain straight cylindrical pores such as that sketched in Fig. 10.7. Most will actually contain a distribution of pore sizes that

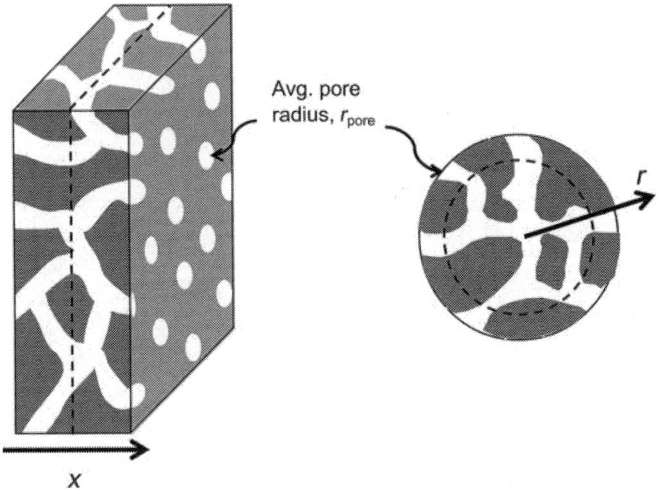

Figure 10.9 Sketches of idealized sections of a porous slab (left) and a porous sphere (right).

follow a tortuous path defining an interconnected porous network. In fact, practical materials will look more like those sketched in Fig. 10.9. In order to overcome the difficulties associated with a detailed description of these complicated and often poorly characterized or even unknown structures, problems involving diffusion in porous materials are typically formulated in terms of an effective pore diffusivity, $\mathcal{D}_{A,e}$, which is defined as follows:

$$\mathcal{D}_{A,e} = \frac{\varepsilon_p \mathcal{D}_{A,\text{pore}}}{\tau_p} \tag{10.34}$$

where ε_p is the porosity and τ_p is the tortuosity factor of the porous material. In order to understand the basis of this equation, let's imagine taking of slice of the material as shown in Fig. 10.9. The slice will have an open area where pore diffusion can take place and a solid area where diffusion cannot occur. If the pores are randomly distributed, we will find that the open area of the pores will also be about equal to a fraction ε_p of the total area of the slice. Furthermore, since the pores are not straight, the distance that a diffusing species will have to travel across the slice will be greater than the physical thickness of the slice itself. The tortuosity factor, τ_p, which is a number greater than 1, is used to describe this tortuous diffusion path.

According to this definition of $\mathcal{D}_{A,e}$, for dilute conditions, the flux across a porous slab will be given by:

$$N_A\big|_x = -\mathcal{D}_{A,e} \frac{dC_A}{dx} \tag{10.35}$$

where x is the axial coordinate. Analogously, the flux in a porous sphere will be given by:

$$N_A\big|_r = -\mathcal{D}_{A,e} \frac{dC_A}{dr} \tag{10.36}$$

where r is the radial coordinate. Note that, in both cases, the cross section onto which the flux is based is the total area, including both pores and solid. This area is defined by the vertical section of the slab or by a spherical shell for the sphere. In practice, for actual materials, τ_p is usually determined experimentally. For example, one can measure $\mathcal{D}_{A,e}$ using a test molecule for certain conditions and then use the same τ_p to predict diffusion of a different molecule in the same material. The advantage of this formalism is that, since ε_p and τ_p are, in theory, purely structural material properties, the same values can then be used to predict $\mathcal{D}_{A,e}$ for a different molecule or for different conditions. For practical materials, τ_p is typically in the range of 2 to 4 with larger values often found for materials with smaller values of ε_p. Some models are also available to predict τ_p (e.g., see Section 16 in *Perry's Chemical Engineers' Handbook*, 9th edition), but their accuracy is usually limited. If nothing is known about the porous structure, assuming a value of $\tau \sim 4$ is sometimes recommended for practical materials.

10.4 Problems

P.10.1. A component A spreads from a point source as a result of Brownian motion in one dimension. At time $t = 0$, all molecules are located at $x = 0$. During a 0.5 s time step, each molecule has the same probability of jumping a distance of 1.0 cm in either the positive or the negative x direction or to stay put. After 0.5 s, the process is repeated with each molecule having the same probability of traveling for another 0.5 s at the same speed in either positive or negative direction. Consider 1,000 molecules starting at $x = 0$ and plot their distribution as a function of distance from the source after 25, 50, 100, and 200 s, which correspond to 50, 100, 200, and 400 time steps, respectively. What is the diffusion coefficient? Can the simulated distribution be described by a Gaussian function?

P.10.2. Estimate the diffusion coefficient for the binary mixture N_2/acetone at 25°C and 1 atm absolute pressure using (a) the Chapman–Enskog theory and (b) the method of Fuller et al.

P.10.3. Estimate the binary diffusion coefficient for a ethane–nitrogen mixture containing 60 mol% ethane at 310 K amd 60 atm absolute pressure. First, estimate the diffusivity at 1 atm. Then, use the Takahashi correlation to determine the diffusivity at 60 atm.

P.10.4. Estimate the diffusion coefficient of acetic acid (A) at 25°C in dilute solutions of (a) benzene, (b) acetone, (c) ethyl acetate, and (d) water using the Wilke–Chang equation and, where appropriate, the Hyduk–Laudie equation and compare these estimates with the following experimental values. Note that while, in water, acetic acid is in monomer form, in organic solvents, it is associated and diffuses as a dimer.

Solvent (B)	T (°C)	$\mathcal{D}^0_{A,B}$ (10^{-5} cm²/s) Experimental	$\mathcal{D}^0_{A,B}$ (10^{-5} cm²/s) W-C	$\mathcal{D}^0_{A,B}$ (10^{-5} cm²/s) H-L
Acetone	25	2.92		–
Ethyl acetate	25	2.18		–
Benzene	25	2.09		–
Water	20	1.19		

P.10.5. Complete the following table by estimating the diffusion coefficient in dilute aqueous solution, $\mathcal{D}^0_{A,B}$, and the size (hydrodynamic radius, r_H). SARS-Co-V2 is spherical with a diameter of about 80 nm.

Molecule/Particle	M_r (g/mol)	T (°C)	Solution Viscosity (cP)	$\mathcal{D}^0_{A,B}$ (cm²/s)	Hydrodynamic Radius, r_H (nm)
Alanine	89	20			
Serum albumin monomer	65,000	20			
Serum albumin dimer	130,000	20			
Serum albumin dimer	130,000	4			
Plasmid DNA	1×10^7	4			
SARS-CoV-2	–	37			

P.10.6. Schneider and Smith [AIChE J. 14 (1968) 886] have measured the diffusivity of propane–helium mixtures in silica gel with a pore diameter of 2.2 nm. While helium is completely inert, propane is adsorbed at the silica surface. In this case, the overall transport of propane occurs as a result of two mechanisms in parallel: diffusion in the pores and diffusion along the pore surface. The overall diffusivity of propane (A) is given by:

$$\mathcal{D}_{A,o} = \mathcal{D}_{A,\text{pore}} + K_c \mathcal{D}_{A,s}$$

where $\mathcal{D}_{A,\text{pore}}$ is the diffusivity in the pores, $\mathcal{D}_{A,s}$ is the surface diffusivity, and K_c is an equilibrium constant describing the ad-sorption of propane. Based on the data in Schneider and Smith, at 50°C and 1 atm absolute pressure, $\mathcal{D}_{A,o} = 0.0106$ cm²/s and $K_c = 1{,}040$: (a) Estimate the pore diffusivity $\mathcal{D}_{A,\text{pore}}$ at this temperature and pressure. Note that the pores are small enough that Knudsen diffusion should be considered. (b) From your result in (a) estimate the surface diffusivity $\mathcal{D}_{A,s}$ and the fractional contribution of surface diffusion to the overall transport.

P.10.7. Estimate the effective pore diffusivity, $\mathcal{D}_{A,e}$, of a monoclonal antibody in a water-saturated porous particle with porosity $\varepsilon_p = 0.60$ at 20°C. The pore radius is $r_{\text{pore}} = 20$ nm and the tortuosity factor is $\tau_p = 2.0$. The diffusivity of the protein in water can be estimated from the Tyn & Gusek equation. The radius of the protein can be estimated from this diffusivity value using the Stokes–Einstein equation. The monoclonal antibody is a globular protein with molecular mass 150,000.

P.10.8. We want to compare the accuracy of various equations to predict the diffusivity of different solutes at infinite dilution in solvents with varying viscosity. The following values are given:

Solute (A)	Solvent (B)	T (°C)	$M_{r,B}$ (g/mol)	μ_B (cp)	$\hat{V}_A$ (cm³/mol)	$\hat{V}_B$ (cm³/mol)	$\mathcal{D}^0_{A,B}$ exp. (cm²/s)
Ethylene glycol	Water	20	18	1.0	67	19	1.04×10^{-5}
Water	Ethylene glycol	20	62	20	19	67	1.0×10^{-6}
Water	Glycerol	20	92	1410	19	110	1.33×10^{-7}
n-Hexanol	Glycerol	0	92	12100	153	110	1.37×10^{-8}

(a) Estimate the values of $\mathcal{D}^0_{A,B}$ for these conditions using the Stokes–Einstein and the Wilke–Chang equations and calculate the error relative to the experimental value given. (b) Note that these equations become less accurate for small solutes in very viscous solvents. In these cases, the Gainer–Metzner equation [Gainer and Metzner, AIChE Symp. Ser. 6 (1965) 74; Gainer, Ind. Eng. Chem. Res. 33 (1994) 2341] is recommended. This equation gives:

$$\frac{D^0_{A,B}\mu_B}{T} = \frac{k_B}{\xi}\left(\frac{\mathcal{N}}{\hat{V}_B}\right)^{1/3} \exp\left(\frac{E_{\mu,B} - E_{D,AB}}{RT}\right)$$

where k_B is the Boltzmann constant, $\mathcal{N}$ is Avogadro's number, ξ is the number of nearest neighbors (usually between 6 and 8), and $E_{\mu,B} - E_{D,AB}$ is the difference in activation energy for viscosity and diffusivity. The latter can be estimated from molecular properties using equations in Gainer and Metzner (1964) and Gainer (1994). The values of $E_{\mu,B} - E_{D,AB}$ are 1,940 cal/mol for the water–glycerol system and 1,310 cal/mol for the n-hexanol–glycerol system. Estimate the corresponding diffusivities and calculate the error relative to the experimental value using this equation.

CHAPTER 11

One-Dimensional Diffusion

11.1 Introduction

In this chapter, we consider one-dimensional diffusion problems where the composition is a function of a single spatial coordinate. We examine two different types of problems in this category:

A. Steady-state problems

Generally, we solve these problems where everything is independent of time by going through the following five steps:

1. Identify the appropriate coordinate systems (e.g., rectangular, cylindrical, or spherical).
2. Select a control-volume that is differential in the direction in which the composition varies.
3. Write a differential material balance for the control volume.
4. Establish the relationship between fluxes (e.g., equimolar counter-diffusion, diffusion of A through stagnant B, reaction stoichiometry, or other) by examining what happens at the system boundaries.
5. Combine the differential material balance with Fick's law and integrate over the system to determine the composition profiles and/or the mass transfer rate.

B. Pseudo-steady-state problems

These types of problems involve two different processes that occur simultaneously, but that have very different time constants: a "fast" process with a small time constant and a "slow" process with a large time constant. If the fast process were occurring by itself, it would reach a steady state in a very short time. However, when such a fast process is coupled to a slow process, the steady-state rate of the fast process will determine the rate at which the slow process actually occurs. We can solve these problems by first determining the steady-state solution of the fast process in the appropriate coordinate system following the steps 1 to 5 defined above. We then use the mass transfer rate determined from the steady-state solution in a transient material balance that describes the slow process. At this point, we can solve the transient balance to find the solution of the entire problem. Many practical situations are conveniently approximated as pseudo-steady-state problems. Their solution according to this approach is much simpler than obtaining an exact description of the entire transient problem and, in many cases, nearly equally accurate. Several examples are provided in the next sections.

11.2 Problems in Rectangular Coordinates

The general form of Fick's law for binary diffusion in one dimension, x, is given by Eq. (9.51) and is as follows:

$$N_{A,x} = -\frac{C\mathcal{D}_{A,B}}{1-y_A\left(1+\frac{N_{B,x}}{N_{A,x}}\right)}\frac{dy_A}{dx} \tag{11.1}$$

In order to use Eq. (11.1), we need to specify the relationship between the two fluxes $N_{A,x}$ and $N_{B,x}$. As seen in Sec. 9.3, this relationship will depend on the physical situation at hand. Several examples of application of Eq. (11.1) to one-dimensional diffusion problems are presented below.

Example 11.1. Gas-phase diffusion through a sampling tube. Let's consider again the diffusional transport problem previously illustrated in Example 9.1. In this problem, diffusion occurs through a thin sampling tube that connects a tank initially full of hydrogen at 1 atm to the atmosphere as sketched in Fig. 11.1. The tube length is $L = 10$ cm and its diameter is $D = 5$ mm. The tank volume is $V = 2$ m^3. The system is isothermal at 0°C and is isobaric. In this case, hydrogen will diffuse out of the tube and air will diffuse into the tube. The molar concentration $C = P/\mathcal{R}T$ will be constant throughout if T and P are constant. Thus, for every mole of hydrogen (A) that diffuses out, one mole of air (B) will diffuse in. As a result, for this problem, we will have equimolar counter-diffusion. Thus, $N_{A,x} = -N_{B,x}$ and

$$N_{A,x} = -C\mathcal{D}_{A,B}\frac{dy_A}{dx} \tag{11.2}$$

Let's first solve the underlying steady-state diffusion problem assuming that it will not take long to establish a steady mole fraction gradient over the length of the tube from the value in the tank, $y_{A,0}$, to the value at the end of the tube, $y_{A,L}$. Note that, in reality, $y_{A,0}$ will vary over time, but we will consider it constant for the time being.

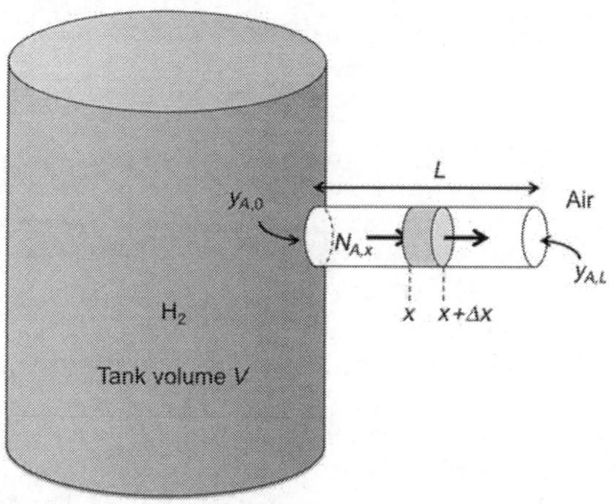

Figure 11.1 Gas diffusion through a thin sampling tube for Example 11.1.

The control volume over which we will write our material balance has a thickness Δx and a cross section $S = \pi D^2/4$ equal to that of the tube. The steady-state material balance over this volume is as follows:

$$SN_{A,x}\Big|_x = SN_{A,x}\Big|_{x+\Delta x}$$

Dividing the terms in this equation by Δx and taking the limit $\Delta x \to 0$, we obtain:

$$\lim_{\Delta x \to 0}\left(\frac{SN_{A,x}\big|_{x+\Delta x} - SN_{A,x}\big|_x}{\Delta x}\right) = 0$$

or:

$$\frac{dSN_{A,x}}{dx} = 0 \qquad (11.3)$$

Since the cross-section of the tube S is a constant independent of x, we can also write:

$$\frac{dN_{A,x}}{dx} = 0 \qquad (11.4)$$

This equation implies that the flux will be a constant along x; i.e., $N_{A,x}\big|_{x=0} = N_{A,x}\big|_{x=L}$. In this case, we can integrate Eq. (11.2) directly by separating the variables x and y_A, which yields:

$$N_{A,x}dx = -C\mathcal{D}_{A,B}dy_A \qquad (11.5)$$

or, integrating over the length of the vent tube:

$$N_{A,x}\int_0^L dx = -C\mathcal{D}_{A,B}\int_{y_{A,0}}^{y_{A,L}} dy_A \qquad (11.6)$$

Performing the integrals, this equation gives the following final result:

$$N_{A,x} = \frac{C\mathcal{D}_{A,B}}{L}(y_{A,0} - y_{A,L}) \qquad (11.7)$$

Equation (11.7) provides the steady-state flux for a given value of $y_{A,0} - y_{A,L}$. It has the same form that we found for steady-state heat conduction across a slab of thickness Δ, Eq. (3.4), with $N_{A,x}$ replacing the heat transfer flux, the quantity $C\mathcal{D}_{A,B}/L$ replacing k/Δ, and the mole fraction difference $y_{A,0} - y_{A,L}$ replacing the temperature difference $T_0 - T_1$. This analogy is not surprising once we realize that, for equimolar counter-diffusion, Fick's law has the same mathematical form as Fourier's law. For this problem, the molar density is $C = P/\mathcal{R}T = 101{,}300/(8.314 \times 273) = 44.6$ mol/m^3. From Table 10.2, $\mathcal{D}_{A,B} = 0.611 \times 10^{-4}$ m^2/s. Thus, the initial flux will be:

$$N_{A,x} = \frac{44.6 \times 0.611 \times 10^{-4}}{0.1}(1 - 0) = 0.0273 \text{ mol/m}^2\text{ s}$$

We obtain the initial rate of mass transfer by multiplying this flux by the sampling tube cross-sectional area, which gives:

$$SN_{A,x} = \frac{\pi D^2}{4} \times 0.0273 = \frac{\pi \times 0.005^2}{4} \times 0.0273 = 5.35 \times 10^{-7} \text{ mol/s}$$

Next, we consider the slow process involving the change in the number of moles of hydrogen in the tank. We assume that the hydrogen mole fraction in the tank is uniform (i.e., that the tank contents are well mixed) and write the following transient mole balance:

$$VC\frac{dy_{A,0}}{dt} = -SN_{A,x}\Big|_{x=0} \qquad (11.8)$$

This equation expresses that the change in the number of moles of A in the tank will be equal to the rate of mass transfer of A out of the tank. Combining Eqs. (11.8) and (11.7) yields the following result:

$$VC\frac{dy_{A,0}}{dt} = -S\frac{C\mathcal{D}_{A,B}}{L}(y_{A,0} - y_{A,L}) \tag{11.9}$$

The initial condition for this differential equation is $y_{A,0} = 1$ at $t = 0$. Assuming that the external mole fraction $y_{A,L}$ stays at 0, we can separate the variables $y_{A,0}$ and t in Eq. (11.9) and integrate as follows:

$$\int_1^{y_{A,0}} \frac{dy_{A,0}}{y_{A,0}} = -\frac{S}{V}\frac{\mathcal{D}_{A,B}}{L}\int_0^t dt \tag{11.10}$$

Performing the integrals, we obtain the mole fraction in the tank as a function of time as:

$$\ln(y_{A,0}) = -\frac{S}{V}\frac{\mathcal{D}_{A,B}}{L}t \tag{11.11}$$

or:

$$y_{A,0} = \exp\left(-\frac{S}{V}\frac{\mathcal{D}_{A,B}}{L}t\right) \tag{11.12}$$

We can now calculate the time required for the hydrogen mole fraction in the tank to drop to a certain value. For example, from Eq. (11.11), the time needed for the hydrogen mole fraction to drop from the initial mole fraction of 1 to a value $y_{A,0} = 0.9$ will be:

$$t = -\frac{V}{S}\frac{L}{\mathcal{D}_{A,B}}\ln(y_{A,0}) = -\frac{2.0}{\pi \times 0.005^2/4}\frac{0.1}{0.611\times 10^{-4}}\ln(0.9) = 1.76\times 10^7 \text{ s or } 4{,}890 \text{ h}$$

This is a long time. By comparison, as we will see later, the time needed to establish steady state for diffusion in the tube is on the order of $L^2/\mathcal{D}_{A,B} = 0.1^2/(0.611\times 10^{-4}) \sim 160$ s. This is the time constant of the fast process. The fact that this is much smaller than the time scale over which the slow process occurs justifies the assumption of pseudo-steady-state. Note that, in this problem, we have assumed that there is no pressure imbalance between the tank and the surrounding air. A pressure gradient would set up a bulk flow in the tube, which would likely result in transport rates that would be much faster than that obtained by diffusion alone.

Example 11.2. Gas permeation through a polymer film. Polyvinylidene chloride films, also known by the commercial name Saran Wrap, are used extensively to package foods and other perishable products in order to prevent rapid spoilage by keeping oxygen and/or moisture out. Let's first set up the equations that we can use to predict the rate of oxygen permeation through such a film. In this example problem, as sketched in Fig. 11.2, the film is used to extend the shelf life of a pharmaceutical product by keeping oxygen out. The film thickness is $\Delta = 0.05$ mm and the area of the film is $S = 2$ cm^2. The volume of the small tray containing the pharmaceutical is 3 cm^3 and is initially filled with pure N_2.

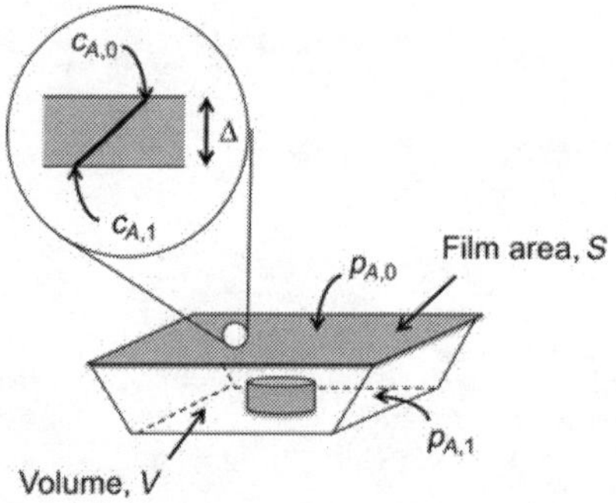

FIGURE 11.2 Diffusion through a thin polymer film used to seal a pharmaceutical product.

Oxygen is expected to have a low solubility in the polymer film. Thus, the O_2 concentration in the film is expected to be very low. We will assume that O_2 (A) partitions into the polymer at the gas–polymer interface according to the following equation:

$$c_A = H_s p_A \tag{11.13}$$

where c_A is the oxygen concentration in the polymer film, H_s is a solubility coefficient with dimensions of concentration/pressure, and p_A is the partial pressure of oxygen at the film surface. Note that this equation will apply at the interface on both sides of the film. Since the concentration of oxygen in the film is low, we can work this problem in either molar or mass units. In molar units, assuming steady-state transport across the film, the integrated form of Fick's law gives the following result:

$$N_{A,x} = \frac{\mathcal{D}^0_{A,B}}{\Delta}(c_{A,0} - c_{A,1}) \tag{11.14}$$

where $\mathcal{D}^0_{A,B}$ is the diffusivity of oxygen in the film. Inserting Eq. (11.13) into Eq. (11.14), we obtain:

$$N_{A,x} = \frac{\mathcal{D}^0_{A,B} H_s}{\Delta}(p_{A,0} - p_{A,1}) = \frac{\overline{P}}{\Delta}(p_{A,0} - p_{A,1}) \tag{11.15}$$

where the flux in the film is expressed as a function of the partial pressure difference between the two sides of the film. The quantity $\overline{P} = \mathcal{D}^0_{A,B} H_s$ in Eq. (11.15) is called the membrane permeability coefficient and has dimensions of (mole × distance)/(area × time × pressure). Since this expression combines $\mathcal{D}^0_{A,B}$ and H_s into the single parameter $\overline{P}$, only their product will matter for the steady-state solution. McKeen, *Permeability Properties of Plastics and Elastomers*, 4th Edition, Elsevier, 2017, p.198, gives a permeability coefficient for oxygen in polyvinylidene chloride of 1.8 in units of (cm^3 O_2 mm)/(m^2 day atm) at 20°C. Since, based on the ideal gas law, 1 cm^3 of O_2 at 20°C and 1 atm corresponds to 4.2×10^{-5} mol, we have $\overline{P} = 1.8 \times 4.2 \times 10^{-5} = 7.6 \times 10^{-5}$ (mol mm)/(m^2 day atm). We can now calculate the initial rate of permeation of oxygen by taking $p_{A,0} = 0.21$ atm and $p_{A,1} = 0$. We obtain the following result:

$$SN_{A,x} = \frac{S\overline{P}}{\Delta}(p_{A,0} - p_{A,1}) = \frac{(2.0 \times 10^{-4}) \times 7.6 \times 10^{-5}}{0.05}(0.21 - 0) = 6.4 \times 10^{-8}\ \text{mol/day}$$

Let's now find the time over which the partial pressure of oxygen will change inside the tray. A mole balance for the oxygen inside the tray yields:

$$VC\frac{dy_{A,1}}{dt} = SN_{A,x} \tag{11.16}$$

Since $C = P/\mathcal{R}T$ and $p_A = y_A P$, this equation can also be written as follows:

$$\frac{V}{\mathcal{R}T}\frac{dp_{A,1}}{dt} = \frac{S\overline{P}}{\Delta}(p_{A,0} - p_{A,1}) \tag{11.17}$$

Separating the variables $p_{A,1}$ and t in this equation and integrating gives:

$$\int_0^{p_{A,1}} \frac{dp_{A,1}}{p_{A,0} - p_{A,1}} = \frac{S\overline{P}\mathcal{R}T}{V\Delta}\int_0^t dt \tag{11.18}$$

Finally, performing the integrals we obtain:

$$\ln\left(\frac{p_{A,0} - p_{A,1}}{p_{A,0}}\right) = -\frac{S\overline{P}\mathcal{R}T}{V\Delta}t \tag{11.19}$$

Note that we have assumed that the gas in the tray is mixed. Let's calculate how long it will take for the oxygen concentration in the tray to reach 10% of the atmospheric value, i.e., $p_{A,1}/p_{A,0} = 0.1$. For this case, Eq. (11.19) gives:

$$t = -\frac{V\Delta}{S\overline{P}\mathcal{R}T}\ln\left(1 - \frac{p_{A,1}}{p_{A,0}}\right) = -\frac{(3.0 \times 10^{-6}) \times 0.05 \times \ln(1 - 0.1)}{(2.0 \times 10^{-4}) \times (7.6 \times 10^{-5}/101{,}300) \times 8.314 \times 293} = 44\ \text{days}$$

Example 11.3. Evaporation of a liquid in a Stefan tube. In this example, we consider the evaporation of a liquid, A, contained at the bottom of a tube that is open to the atmosphere as sketched in Fig. 11.3. This setup is often called a Stefan tube; Josef Stefan suggested its use to test gas diffusion theories and to experimentally determine gas phase diffusivities by measuring the change in the liquid level over time. As shown in Fig. 11.3, the total height of the tube is L_t, the height of liquid in the tube is ℓ, and the gas–liquid interface is at a distance L from the top of the tube. The process will take place as follows. Component A will change phase from liquid to vapor at the gas–liquid interface and will diffuse through the air in the tube to reach the top where it will be dispersed in the atmosphere. Over time, the liquid level will drop because of the evaporation and the distance L will increase thereby increasing the length over which diffusion of A occurs. Let's assume that, as an example, the liquid A is acetone, which is fairly volatile, and that temperature and absolute pressure are 293 K and 1 atm, respectively. The initial height of liquid in the tube is $\ell = 5$ cm and the initial distance from the surface of the liquid to the top of the tube is 10 cm. Since air (B) has little solubility in liquid acetone, this is a case of diffusion of A through stagnant B. Thus, Eq. (11.1) with $N_B = 0$ gives the following expression for the flux of acetone:

$$N_{A,x} = -\frac{C\mathcal{D}_{A,B}}{1-y_A}\frac{dy_A}{dx} \tag{11.20}$$

Intuitively, we can anticipate that the time needed for the liquid to drop will be long, compared to the time needed to attain a steady flux of A over the length L. Thus, let's treat this situation as a pseudo-steady-state problem. A steady-state mole balance for acetone in the air above the liquid along the tube length yields:

$$\frac{dN_{A,x}}{dx} = 0 \tag{11.21}$$

which implies that $N_{A,x}$ will be constant along the length x. As a result, separating the variables x and y_A in Eq. (11.20) and integrating we obtain:

$$N_{A,x}\int_0^L dx = -C\mathcal{D}_{A,B}\int_{y_{A,0}}^{y_{A,L}} \frac{dy_A}{1-y_A} \tag{11.22}$$

or:

$$N_{A,x} = \frac{C\mathcal{D}_{A,B}}{L}\ln\left(\frac{1-y_{A,L}}{1-y_{A,0}}\right) \tag{11.23}$$

Physically, the integration limits for this problem are defined as follows. At the top of the tube, where $x = L$, we can assume that $y_A = y_{A,L} = 0$, provided that the acetone that gets there is quickly dispersed in

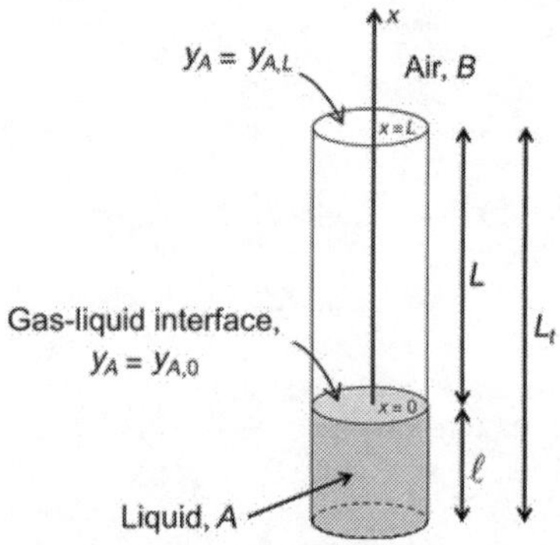

FIGURE 11.3 Evaporation of a liquid in a stagnant gas in a Stefan tube.

the atmosphere. At the gas–liquid interface, where $x = 0$, we can assume that the gas is in equilibrium with the liquid. In this case, Raoult's law gives:

$$y_{A,0} = \frac{p_A^*(T)}{P} \tag{11.24}$$

where $p_A^*(T)$ is the vapor pressure of acetone at temperature T. For the conditions of our problem, we have $p_A^* = 185$ mmHg. Thus, $y_{A,0} = 185/760 = 0.24$. The gas molar density is $C = P/\mathcal{R}T = 1/(82\times 293) = 4.2\times 10^{-5}$ mol/cm^3. We can estimate $\mathcal{D}_{A,B}$ from the experimental value of 0.109 cm^2/s at 273 K given in Table 10.2 by taking $\mathcal{D}_{A,B} = 0.109\times(293/273)^{1.75} = 0.123$ cm^2/s. Using these values, Eq. (11.23) yields the following value for the initial flux of acetone when $L = 10$ cm:

$$N_{A,x} = \frac{4.2\times 10^{-5}\times 0.123}{10}\ln\left(\frac{1-0}{1-0.24}\right) = 1.42\times 10^{-7}\ \text{mol/cm}^2\ \text{s}$$

The rate of evaporation will be $SN_{A,x}$ (mol/s) where S is the cross-sectional area of the tube. Note that, in this case, the mole fraction profile is a nonlinear function of distance. We can find it by integrating Eq. (11.20) from 0 to an arbitrary x and solving for y_A, which gives:

$$y_A = 1-(1-y_{A,0})\exp\left(\frac{N_{A,x}x}{C\mathcal{D}_{A,B}}\right) \tag{11.25}$$

Next, let's consider the variation of the liquid level resulting from evaporation, which is the slow process in this pseudo-steady-state formulation. The mass of liquid acetone present in the tube to a time t is given by $S\ell\rho_L$, where the product $S\ell$ is the volume of liquid and ρ_L is the mass density of liquid acetone. A transient mass balance for the liquid yields the following equation:

$$\frac{d(\text{Mass of liquid})}{dt} = S\rho_L\frac{d\ell}{dt} = -SN_{A,x}\times M_{r,A} \tag{11.26}$$

where $M_{r,A}$ is the molecular mass of acetone. The right-hand side of this equation is the mass rate of evaporation. Since $\ell = L_t - L$, we have $d\ell/dt = -dL/dt$. Thus, Eq. (11.26) becomes:

$$-S\rho_L\frac{dL}{dt} = -S\frac{C\mathcal{D}_{A,B}M_{r,A}}{L}\ln\left(\frac{1-y_{A,L}}{1-y_{A,0}}\right) \tag{11.27}$$

The final step is to separate the variables L and t and integrate Eq. (11.27), which gives:

$$\int_{L_0}^{L} L\,dL = \frac{C\mathcal{D}_{A,B}M_{r,A}}{\rho_L}\ln\left(\frac{1-y_{A,L}}{1-y_{A,0}}\right)\int_0^t dt \tag{11.28}$$

where L_0 is the initial value of L at $t = 0$. Performing the integrals in Eq. (11.28), we obtain the final result:

$$\frac{1}{2}(L^2 - L_0^2) = \frac{C\mathcal{D}_{A,B}M_{r,A}}{\rho_L}\ln\left(\frac{1-y_{A,L}}{1-y_{A,0}}\right)t \tag{11.29}$$

As an example, let's calculate the position of the gas–liquid interface after 10 h (36,000 s). For liquid acetone, we have $\rho_L = 0.784$ g/cm^3 while $M_{r,A} = 58$ g/mol. With these values, Eq. (11.29) gives:

$$\frac{1}{2}(L^2 - 10^2) = \frac{4.2\times 10^{-5}\times 0.123\times 58}{0.784}\ln\left(\frac{1-0}{1-0.24}\right)\times 36{,}000 = 3.78\ \text{cm}^2$$

from which we can find $L = 10.4$ cm. Note that the liquid level has changed very little (0.4 cm) over this long time. Conversely, as will be shown later, the time needed to reach a steady-state flux will be on the order of $L^2/\mathcal{D}_{A,B}$, which is only about $10^2/0.123 = 810$ s. Since this time is much shorter than the time scale over which the liquid level changes (say, 36,000 s), the pseudo-steady-state assumption is justified. Note that if $\mathcal{D}_{A,B}$ were unknown and we wished to determine it experimentally, we might measure how the level changes as a function of time. A plot of $L^2 - L_0^2$ vs. time would be expected to be a straight line, whose slope will be proportional to $\mathcal{D}_{A,B}$. The main experimental challenges, in this case, would be to ensure that (a) there is no convective current within the tube and that (b) the change in level is measured over times that are sufficiently long to render insignificant the effects of any initial transients.

11.3 Problems in Cylindrical and Spherical Coordinates

We now consider one-dimensional problems where diffusion occurs along the radial coordinate of either a cylindrical or spherical coordinate system. The choice of coordinate system will depend, of course, on the geometry of the physical situation. Fortunately, these problems are not fundamentally different from those that we have seen in rectangular coordinates. The main difference is that the steady-state flux will no longer be spatially uniform but will generally vary along the radial coordinate. In the following, we consider three different examples in such coordinate systems. For all three, we assume that conditions are dilute and that diffusion occurs only in the radial direction. In this case, Fick's law gives:

$$N_{A,r} = -\mathcal{D}^0_{A,B}\frac{dc_A}{dr} \tag{11.30}$$

For the dilute cases considered here, we can use either a molar or a mass basis interchangeably remembering that $n_{A,r} = M_{r,A} \times N_{A,r}$ where $n_{A,r}$ is the mass flux and $M_{r,A}$ is the molecular mass.

Example 11.4. Diffusion in a cylindrical shell. Sherwood et al. [*Mass Transfer*, McGraw-Hill, 1975, p. 53] have proposed a membrane-based process to separate N_2/He mixtures containing 30 mol% He by taking advantage of the higher permeability of He compared to that of N_2 in cellulose acetate–butyrate polymeric membranes. They provide values of the permeability coefficients $\overline{P}_{He} = 1.54 \times 10^{-12}$ and $\overline{P}_{N_2} = 0.171 \times 10^{-12}$ both in units of (mol cm)/(cm^2 s atm) at 21°C (see Example 11.2 for the definition of $\overline{P}$). Such a membrane device could be in the form of hollow fibers, which are thin cylindrical tubes made of this perm-selective polymer, as shown in Fig. 11.4a. In one particular design, the fibers have an internal diameter of 0.5 mm, an outside diameter of 1.0 mm, and are 100 cm long. The absolute pressure inside the fibers is 10 atm while the pressure outside the fibers is very close to 0. Let's calculate the expected rate of mass transfer of He and N_2 across the fiber wall by neglecting the composition change over the length of the hollow fiber and assuming steady state. Since the concentrations of both nitrogen and oxygen in the fiber wall will be functions of the radial coordinate, the control volume for our material balance will be a cylindrical shell with thickness Δr, as sketched in Fig. 11.4b. The surface area of this shell at r is $2\pi rL$. A mole balance for each component over this shell yields:

$$2\pi rLN_{A,r}\big|_r = 2\pi rLN_{A,r}\big|_{r+\Delta r}$$

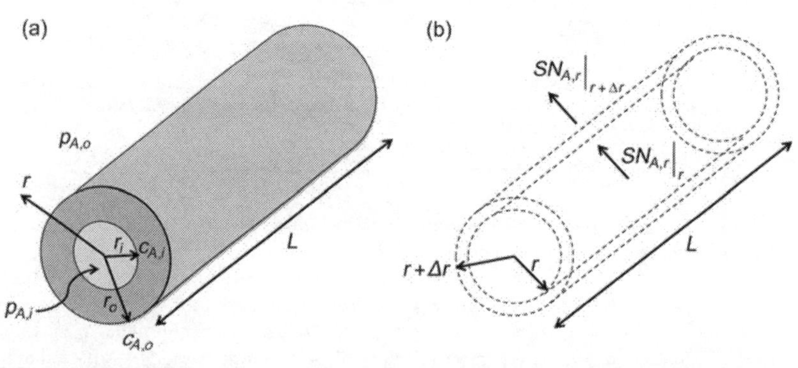

Figure 11.4 Diffusion of a component *A* through a cylindrical hollow fiber (a) and control volume (b) for Example 11.4. The control volume is a cylindrical shell with length *L* and faces located at *r* and *r* + Δ*r*. The surface area of the control volume at *r* is 2π*rL*.

Dividing this equation by Δr and taking the limit $\Delta r \to 0$ yields:

$$\lim_{\Delta r \to 0}\left(\frac{2\pi r L N_{A,r}\big|_{r+\Delta r} - 2\pi r L N_{A,r}\big|_{r}}{\Delta r}\right) = 0$$

or:

$$\frac{d(rN_{A,r})}{dr} = 0 \tag{11.31}$$

Integrating Eq. (11.31), we obtain:

$$N_{A,r} = \frac{a}{r} \tag{11.32}$$

where a is an integration constant. This equation shows that the flux will not be constant but will decrease along the r-coordinate from the inner surface of the fiber wall to its outer surface. Next, we introduce Fick's law in the radial coordinate system by assuming that He and N_2 do not affect each other because of their low solubility in the polymer. In this case, the flux is given by:

$$N_{A,r} = -\mathcal{D}^0_{A,B}\frac{dc_A}{dr} \tag{11.33}$$

where $\mathcal{D}^0_{A,B}$ and c_A are, respectively, the diffusivity and the concentration of component A dissolved in the polymer. Inserting Eq. (11.33) into Eq. (11.32), we obtain:

$$-\mathcal{D}^0_{A,B}\frac{dc_A}{dr} = \frac{a}{r} \tag{11.34}$$

Separating the c_A and r variables in Eq. (11.34) and integrating gives the following result:

$$-\mathcal{D}^0_{A,B}\int_{c_{A,i}}^{c_{A,o}} dc_A = a\int_{r_i}^{r_o}\frac{dr}{r} \tag{11.35}$$

Finally, performing the integrals we obtain:

$$\mathcal{D}^0_{A,B}(c_{A,i} - c_{A,o}) = a\ln\left(\frac{r_o}{r_i}\right) \tag{11.36}$$

or:

$$a = \frac{\mathcal{D}^0_{A,B}(c_{A,i} - c_{A,o})}{\ln\left(\frac{r_o}{r_i}\right)} \tag{11.37}$$

Next, we obtain the flux by inserting this expression for a into Eq. (11.32), which gives:

$$N_{A,r} = \frac{\mathcal{D}^0_{A,B}(c_{A,i} - c_{A,o})}{r\ln\left(\frac{r_o}{r_i}\right)} \tag{11.38}$$

The mass transfer rate (or permeation rate) is then found by multiplying the flux at a particular value of r times the corresponding area $S = 2\pi rL$. Thus,

$$SN_{A,r} = \frac{2\pi L\mathcal{D}^0_{A,B}}{\ln\left(\frac{r_o}{r_i}\right)}(c_{A,i} - c_{A,o}) \tag{11.39}$$

Not surprising, perhaps, this expression is the same as the expression that we found for the rate of conductive heat transfer across a cylindrical wall, Eq. (3.38), if we replace Q with $SN_{A,r}$, the thermal conductivity k with $\mathcal{D}^0_{A,B}$, and $T_i - T_o$ with $c_{A,i} - c_{A,o}$. This occurs because, for dilute conditions, Fourier's law and Fick's law have exactly the same mathematical form. Note that, since, at either surface of the hollow fiber membrane, we have $c_A = H_{s,A} p_A$, where $H_{s,A}$ is a solubility coefficient with dimensions of concentration/pressure, we can also express the permeation rate as follows:

$$SN_{A,r} = \frac{2\pi L\bar{P}_A}{\ln\left(\frac{r_o}{r_i}\right)}(p_{A,i} - p_{A,o})$$

where $\bar{P}_A = H_{s,A}\mathcal{D}^0_{A,B}$. We can now calculate the permeation rates for He and N_2, which are, rspectively:

$$SN_{\text{He},r} = \frac{2\pi L\bar{P}_{\text{He}}(p_{\text{He},i} - p_{\text{He},o})}{\ln\left(\frac{r_o}{r_i}\right)} = \frac{2\pi \times 100 \times 1.54 \times 10^{-12}(0.3 \times 10 - 0)}{\ln\left(\frac{1.0}{0.5}\right)} = 4.2 \times 10^{-9} \text{ mol/s}$$

$$SN_{N_2,r} = \frac{2\pi L\bar{P}_{N_2}\left(p_{N_2,i} - p_{N_2,o}\right)}{\ln\left(\frac{r_o}{r_i}\right)} = \frac{2\pi \times 100 \times 0.171 \times 10^{-12}(0.7 \times 10 - 0)}{\ln\left(\frac{1.0}{0.5}\right)} = 1.1 \times 10^{-9} \text{ mol/s}$$

The mole fraction in the permeate, obtained from a mole balance, will then be:

$$y_{\text{He}} = \frac{SN_{\text{He},r}}{SN_{\text{He},r} + SN_{N_2,r}} = \frac{4.2 \times 10^{-9}}{4.2 \times 10^{-9} + 1.1 \times 10^{-9}} = 0.79$$

This is much higher than the mole fraction inside the hollow fiber showing that a considerable enrichment in He is possible. Finally, as an example, let's calculate how many hollow fibers would be needed to obtain 1 g/s of the enriched permeate. As calculated above, one hollow fiber will produce a permeate flow comprising $4.2 \times 10^{-9} \times 4.0 = 1.7 \times 10^{-8}$ g/s of helium and $1.1 \times 10^{-9} \times 28 = 3.1 \times 10^{-8}$ g/s of nitrogen. Thus, in order to produce 1 g/s of the mixture, we would need $1/(1.7 \times 10^{-8} + 3.1 \times 10^{-8}) = 21 \times 10^6$ fibers. Obviously, this is a big number, but it could be achieved by densely packing the fibers in a proper device.

Example 11.5. Diffusion from a sphere in an infinite volume of a stagnant fluid. In this example, we consider the rate of sublimation of a 1-cm radius moth ball in still air at 25°C as sketched in Fig. 11.5. Moth balls are made of naphthalene (A), which has a sublimation pressure $p^*_A = 1.2 \times 10^{-4}$ atm at 25°C. The naphthalene mole fraction on the air side of the air–solid interface will thus be $y_{A,s} = p^*_A/P = 1.2 \times 10^{-4}$. The mole fraction will decrease from this value to essentially zero as we move away from the moth ball surface. Because the mole fraction of naphthalene in the air is so low, conditions are definitely dilute. Let's treat this situation as a pseudo-steady-state problem, assuming that the rate at which the moth ball shrinks will be small. The control volume for our mole balance will be a spherical shell with

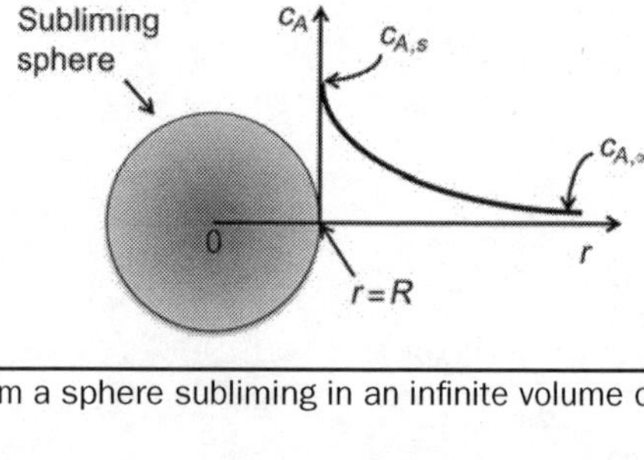

Figure 11.5 Diffusion from a sphere subliming in an infinite volume of still air.

faces located at r and $r + \Delta r$ in the air surrounding the moth ball. The surface area of this shell at r will be $4\pi r^2$. A naphthalene mole balance across this shell then yields:

$$4\pi r^2 N_{A,r}\Big|_r = 4\pi r^2 N_{A,r}\Big|_{r+\Delta r}$$

Dividing this equation by Δr and taking the limit for $\Delta r \to 0$ gives:

$$\lim_{\Delta r \to 0}\left(\frac{4\pi r^2 N_{A,r}\Big|_{r+\Delta r} - 4\pi r^2 N_{A,r}\Big|_r}{\Delta r}\right) = 0$$

or:

$$\frac{d(r^2 N_{A,r})}{dr} = 0 \tag{11.40}$$

Integrating Eq. (11.40), we obtain:

$$N_{A,r} = \frac{a}{r^2} \tag{11.41}$$

where a is an integration constant. As in the case of the cylindrical geometry, we see that the flux will decrease as we move along the r-coordinate also in this case. However, this time, the decrease will be in inverse proportion to the square of r. Next, we introduce Fick's law in the spherical coordinate system. For the dilute conditions of this problem, we have

$$N_{A,r} = -\mathcal{D}_{A,B}\frac{dC_A}{dr} \tag{11.42}$$

where $\mathcal{D}_{A,B}$ is the diffusivity in the naphthalene–air system. Combining Eqs. (11.42) and (11.41), we obtain:

$$-\mathcal{D}_{A,B}\frac{dC_A}{dr} = \frac{a}{r^2} \tag{11.43}$$

Equation (11.43) can be integrated as follows:

$$-\mathcal{D}_{A,B}\int_{C_{A,s}}^{C_{A,\infty}} dC_A = a\int_R^{\infty}\frac{dr}{r^2} \tag{11.44}$$

where $C_{A,s}$ and $C_{A,\infty}$ are the gas-phase concentrations of naphthalene at the moth ball surface and far away from it. Performing the integrals, we obtain the integration constant a as:

$$a = R\mathcal{D}_{A,B}(C_{A,s} - C_{A,\infty}) \tag{11.45}$$

We can now obtain the mass transfer flux by inserting this expression for a into Eq. (11.41), which gives the result:

$$N_{A,r} = \frac{R\mathcal{D}_{A,B}}{r^2}(C_{A,s} - C_{A,\infty}) \tag{11.46}$$

The flux at the moth ball surface is found by setting $r = R$, which gives:

$$N_{A,r}\Big|_{r=R} = \frac{R\mathcal{D}_{A,B}}{R^2}(C_{A,s} - C_{A,\infty}) = \frac{\mathcal{D}_{A,B}}{R}(C_{A,s} - C_{A,\infty}) \tag{11.47}$$

Finally, we get the sublimation rate by multiplying the flux at the surface by the surface area $S = 4\pi R^2$ as follows:

$$SN_{A,r}\Big|_{r=R} = 4\pi R^2\frac{\mathcal{D}_{A,B}}{R}(C_{A,s} - C_{A,\infty}) = 4\pi R\mathcal{D}_{A,B}(C_{A,s} - C_{A,\infty}) \tag{11.48}$$

Obviously, because of the pseudo-steady-state assumption, the rate of mass transfer will be the same regardless of position along r. In order to calculate this rate, we need to find $\mathcal{D}_{A,B}$ for the naphthalene–air system at 25°C. Using the methods in Chap. 10, we obtain $\mathcal{D}_{A,B} = 0.061$ cm²/s. At the moth ball surface, we have $y_{A,s} = 1.2\times10^{-4}$. Thus, $C_{A,s} = Cy_{A,s} = y_{A,s}P/\mathcal{R}T = 4.9\times10^{-9}$ mol/cm³. From Eq. (11.48), the initial rate of sublimation will thus be:

$$SN_{A,r}\big|_{r=R} = 4\pi\times1.0\times0.061(4.9\times10^{-9}-0) = 3.8\times10^{-9}\ \text{mol/s}$$

Let's now predict how the radius of the moth ball will change as a function of time. To do this, we write a transient material balance for the solid naphthalene. The mass of solid naphthalene at a time t will be $4\pi R^3\rho_s/3$, where ρ_S is the density of solid naphthalene and $4\pi R^3/3$ is the volume of the moth ball. A transient mass balance obtained by equating the rate of change of the amount of solid naphthalene to the rate of sublimation yields the following equation:

$$\frac{d(\text{Mass of solid naphthalene})}{dt} = \frac{d}{dt}\left(\frac{4}{3}\pi R^3\rho_S\right) = \frac{4}{3}\pi\rho_S\frac{dR^3}{dt} = -SN_{A,r}\big|_{r=R}\times M_{r,A} \tag{11.49}$$

where $M_{r,A}$ is the molecular mass of naphthalene. Since $dR^3 = 3R^2dR$, Eq. (11.49) yields:

$$\frac{4}{3}\pi\rho_S\left(3R^2\frac{dR}{dt}\right) = -4\pi R\mathcal{D}^0_{A,B}M_{r,A}(C_{A,s}-C_{A,\infty}) \tag{11.50}$$

Simplifying this equation by canceling common terms, we obtain:

$$R\frac{dR}{dt} = -\frac{\mathcal{D}^0_{A,B}M_{r,A}}{\rho_S}(C_{A,s}-C_{A,\infty}) \tag{11.51}$$

The final result is obtained by integrating with the condition that $R = R_0$ at $t = 0$, which gives:

$$\int_{R_0}^{R} R\,dR = -\frac{\mathcal{D}^0_{A,B}M_{r,A}}{\rho_S}(C_{A,s}-C_{A,\infty})\int_0^t dt \tag{11.52}$$

or:

$$\frac{1}{2}(R_0^2-R^2) = \frac{\mathcal{D}^0_{A,B}M_{r,A}}{\rho_S}(C_{A,s}-C_{A,\infty})t \tag{11.53}$$

Let's calculate how long it will take for the moth ball to shrink to one half of the initial size. For naphthalene, we have $\rho_S = 1.15$ g/cm³ and $M_{r,A} = 128$ g/mol. Using these values, from Eq. (11.53) we obtain:

$$t = \frac{1}{2}\frac{(R_0^2-R^2)\rho_S}{\mathcal{D}^0_{A,B}M_{r,A}(c_{A,s}-c_{A,\infty})} = \frac{1}{2}\frac{(1^2-0.5^2)1.15}{0.061\times128\times4.9\times10^{-9}} = 1.13\times10^7\ \text{s or about 130 days}$$

Let's now return to the expression for the flux at the particle surface derived in Example 11.5, which is given by Eq. (11.47). We can also write this expression for the flux as follows:

$$N_{A,r}\big|_{r=R} = \frac{\mathcal{D}_{A,B}}{R}(C_{A,s}-C_{A,\infty}) = k_c(C_{A,s}-C_{A,\infty}) \tag{11.54}$$

where k_c is a mass transfer coefficient with dimensions of length/time. For the problem at hand, since we have assumed that the air surrounding the sphere is completely stagnant, from Eq. (11.54) we have:

$$k_c = \frac{\mathcal{D}_{A,B}}{R} \tag{11.55}$$

In other cases, however, where the air moves relative to the sphere, we recognize that transport will be faster because of convection. Such a convective enhancement can then be described by a mass transfer coefficient k_c that becomes a function of air velocity. k_c is then analogous to the convective heat transfer coefficients that we learned about in Chap. 6. For mass transfer, in analogy to the Nusselt number, we define the following dimensionless group called the Sherwood number:

$$\mathrm{Sh} = \frac{k_c D}{\mathcal{D}_{A,B}} \tag{11.56}$$

where $D = 2R$ is the diameter of the sphere. If the air is still and only diffusion takes place, as in this example, since $k_c = \mathcal{D}_{A,B}/R$, Eq. (11.56) yields:

$$\mathrm{Sh} = \frac{k_c D}{\mathcal{D}_{A,B}} = \frac{\mathcal{D}_{A,B}}{R}\frac{D}{\mathcal{D}_{A,B}} = 2 \tag{11.57}$$

Movement of the fluid relative to the sphere will, of course, increase Sh above 2, making k_c larger. Thus, Sh = 2 will be the minimum value that can be observed for a single sphere in an infinite volume of a stagnant fluid.

Example 11.6. Design of a sustained drug delivery system. The sustained release of a drug is desirable for many applications since it prevents exposing us to a large initial bolus when the drug is first administered and avoids the need for repeated administrations. Without a sustained release design, a solid drug will usually dissolve at a rate controlled by diffusion in the surrounding fluid, which will typically be relatively fast if the drug is in direct contact with body fluids. On the other hand, we can slow down delivery of the drug and sustain it for a longer time, if the drug substance is incorporated into a porous matrix and allowed to diffuse out through the pores. Once at the surface of the porous matrix, the drug will be carried away by body fluids and circulated throughout the body. Figure 11.6 shows, schematically, a simple implementation of this concept using a spherical porous particle as the support matrix. Initially, the solid drug fills all of the pores in the particle. Over time, as the drug diffuses out, the solid core shrinks until the drug is eventually all delivered. The diffusion resistance in the pores slows down the process so that it may take a while to get complete dissolution. Let's develop a pseudo-steady-state model to describe this process. We will assume that the solubility of the drug (A) in the fluid, ρ_A^*, is small and that dissolution of the drug will occur only at the interface between the solid core and the fluid in the pores. We will also assume that the drug concentration in the external fluid will be practically zero. On a mass basis, a material balance for the drug in a spherical shell located within the particle but outside the solid core yields:

$$\frac{d(r^2 n_{A,r})}{dr} = 0 \tag{11.58}$$

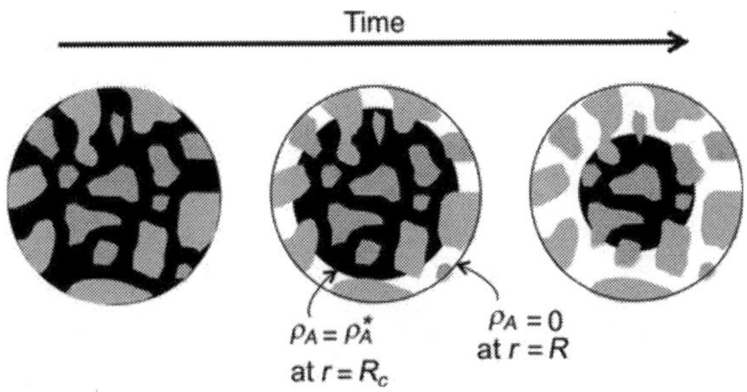

FIGURE 11.6 Dissolution of a solid contained in a spherical porous support matrix.

Integrating, this equation we obtain:

$$n_{A,r} = \frac{a}{r^2} \tag{11.59}$$

where a is an integration constant. Next, as in Example 11.5, we introduce Fick's law which, for this problem, takes on the form:

$$n_{A,r} = -\mathcal{D}_{A,e}\frac{d\rho_A}{dr} \tag{11.60}$$

where $\mathcal{D}_{A,e}$ is the effective pore diffusivity of the drug in the porous matrix. Combining Eqs. (11.59) and (11.60) and integrating gives the following result:

$$-\mathcal{D}_{A,e}\int_{\rho_A^*}^{0} d\rho_A = a\int_{R_c}^{R}\frac{dr}{r^2} \tag{11.61}$$

Note that the integration limits are different from those used in the Example 11.5. In this case, at the surface of the solid core ($r = R_c$), we assume that $\rho_A = \rho_A^*$, while at the external surface of the particle ($r = R$) we assume that $\rho_A = 0$. Integrating Eq. (11.61) we obtain:

$$\mathcal{D}_{A,e}\rho_A^* = a\left(\frac{1}{R_c} - \frac{1}{R}\right) \tag{11.62}$$

or, solving for a:

$$a = -\frac{\mathcal{D}_{A,e}\rho_A^*}{\dfrac{1}{R} - \dfrac{1}{R_c}} \tag{11.63}$$

As in the previous example, we obtain the flux by inserting this expression for a into the flux equation, Eq. (11.59), which gives:

$$n_{A,r} = -\frac{1}{r^2}\frac{\mathcal{D}_{A,e}\rho_A^*}{\dfrac{1}{R} - \dfrac{1}{R_c}} \tag{11.64}$$

The mass transfer rate is then obtained by multiplying the flux by the spherical area yielding:

$$Sn_{A,r}\Big|_{r=R} = -\frac{4\pi R^2}{R^2}\frac{\mathcal{D}_{A,e}\rho_A^*}{\dfrac{1}{R} - \dfrac{1}{R_c}} = -\frac{4\pi\mathcal{D}_{A,e}\rho_A^*}{\dfrac{1}{R} - \dfrac{1}{R_c}} \tag{11.65}$$

In order to find the rate at which the drug substance core shrinks, we need again a transient balance for the mass of drug in the particle. The mass of drug substance in the core at a certain time t will be $4\pi R_c^3\rho_{A,c}/3$, where $4\pi R_c^3/3$ is the core volume and $\rho_{A,c}$ is the density of the drug in the core (e.g., the g of drug per cm^3 of particle volume). The change in mass of drug in the core must be equal to the rate of mass transfer out of the particle. Hence:

$$\frac{d(\text{Mass of } A \text{ in core})}{dt} = -Sn_{A,r}\Big|_{r=R} \tag{11.66}$$

In combination with Eq. (11.65), Eq. (11.66) gives the following result:

$$\frac{4}{3}\pi\rho_{A,c}\frac{dR_c^3}{dt} = \frac{4\pi\mathcal{D}_{A,e}\rho_A^*}{\dfrac{1}{R} - \dfrac{1}{R_c}} \tag{11.67}$$

Note that in this case, the radius of the porous particle, R, is constant while the radius of the core, R_c, is a function of time. Since $dR_c^3 = 3R_c^2dR_c$, Eq. (11.67) becomes:

$$\left(\frac{1}{R} - \frac{1}{R_c}\right)R_c^2dR_c = \frac{\mathcal{D}_{A,e}\rho_A^*}{\rho_{A,c}}dt \tag{11.68}$$

We can now integrate this equation with the condition that $R_c = R$ at $t = 0$, which gives:

$$\int_R^{R_c}\left(\frac{1}{R}-\frac{1}{R_c}\right)R_c^2\,dR_c = \frac{\mathcal{D}_{A,e}\rho_A^*}{\rho_{A,c}}\int_0^t dt \tag{11.69}$$

Performing the integrals, we obtain:

$$\left(\frac{1}{3}\frac{R_c^3}{R}-\frac{1}{2}R_c^2\right)\Bigg|_R^{R_c} = \frac{\mathcal{D}_{A,e}\rho_A^*}{\rho_{A,c}}t \tag{11.70}$$

or:

$$\frac{1}{3}\frac{R_c^3}{R}-\frac{1}{2}R_c^2+\frac{1}{6}R^2 = \frac{\mathcal{D}_{A,e}\rho_A^*}{\rho_{A,c}}t \tag{11.71}$$

In a slightly more compact notation, this equation can be written as follows:

$$2\left(\frac{R_c}{R}\right)^3 - 3\left(\frac{R_c}{R}\right)^2 + 1 = \frac{t}{t_D} \tag{11.72}$$

where:

$$t_D = \frac{R^2\rho_{A,c}}{6\mathcal{D}_{A,e}\rho_A^*} \tag{11.73}$$

This is known as the "shrinking core model" and is applicable to many different diffusion-limited situations including gas–solid noncatalytic reactions, adsorption, and the leaching of material from porous matrices. Equation (11.72) is plotted in Fig. 11.7 in the form R_c/R vs. t/t_D, which gives the dimensionless position of the core–fluid interface as a function of time. As seen in Fig. 11.7, the core shrinks to zero ($R_c/R = 0$) when $t/t_D = 1$. Thus, t_D, defined by Eq. (11.73), is the time needed for the core to disappear and, thus, to deliver 100% of the drug.

Let's now use these equations to make some predictions if $R = 0.05$ cm, $\rho_{A,c} = 0.7\ \text{g/cm}^3$, and $\rho_A^* = 0.010\ \text{g/cm}^3$. Also, let's assume that the diffusivity of the drug in the fluid is $\mathcal{D}_{A,B}^0 = 0.8 \times 10^{-5}\ \text{cm}^2/\text{s}$

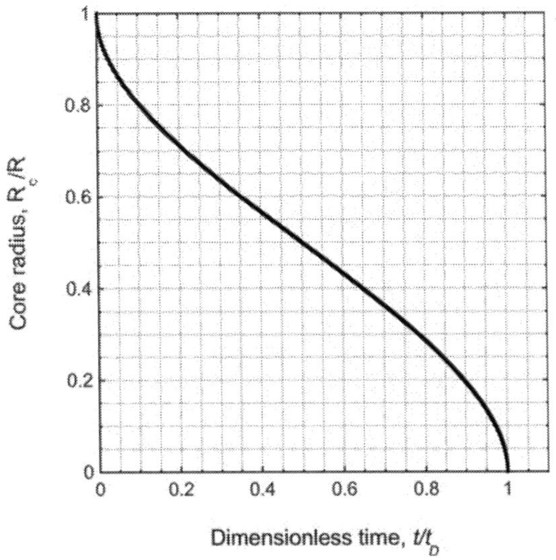

FIGURE 11.7 Shrinking core radius vs. dimensionless time for Example 11.6.

and that porosity ε_p and tortuosity factor τ_p of the porous matrix are 0.6 and 4, respectively. We can estimate the effective pore diffusivity of the drug in the pores from Eq. (10.34), which gives:

$$\mathcal{D}_{A,e} = \frac{\varepsilon_p \mathcal{D}^0_{A,B}}{\tau_p} = \frac{0.6 \times 0.8 \times 10^{-5}}{4} = 0.12 \times 10^{-5} \text{cm}^2/\text{s}$$

Therefore, from Eq. (11.73), we obtain:

$$t_D = \frac{0.05^2 \times 0.7}{6 \times 0.12 \times 10^{-5} \times 0.010} = 24{,}300 \text{ s or about 6.8 h}$$

which is the time to achieve 100% delivery. We can also calculate the cumulative amount of drug delivered at each time using Eq. (11.72) (or Fig. 11.7) from the following equation:

$$\text{Fraction delivered} = \frac{4\pi R^3 \rho_{A,c}/3 - 4\pi R_c^3 \rho_{A,c}/3}{4\pi R^3 \rho_{A,c}/3} = 1 - \left(\frac{R_c}{R}\right)^3 \qquad (11.74)$$

where $4\pi R^3 \rho_{A,c}/3$ is the initial amount of drug loaded in the particle. For example, after 3 h we will have $t/t_D = 0.44$ and $R_c/R \sim 0.54$. Thus, the fraction delivered will be $(1 - 0.54^3) = 0.84$ or 84%. Note that if we just had a solid sphere of the drug without a porous matrix, the time needed to completely dissolve the drug would be given by Eq. (11.53) with $R = 0$, or:

$$t = \frac{1}{2} \frac{0.05^2 \times 0.7}{0.8 \times 10^{-5} \times 0.01} = 10{,}900 \text{ s or about 3.0 h}$$

In reality, the actual time needed to dissolve the particle in this case would likely be much shorter than this value because convective mass transfer will almost certainly increase the dissolution rate. Note that there are many other ways to obtain sustained release. One alternative approach is to embed the drug into a matrix that is impermeable by the drug but that degrades at a certain rate. In this case, the drug will be released gradually as the matrix degrades with the rate being controlled by the degradation rate.

11.4 Problems

P.11.1. An open tank, 3.0 m in diameter, contains liquid benzene at 37°C exposed to air also at 37°C and 1 atm absolute pressure. Assume that the rate of evaporation is controlled by diffusion through a stagnant gaseous film, 0.50 cm thick, adjacent to the benzene surface. The mole fraction of benzene outside this film is zero. Determine the rate at which the benzene evaporates in kg per day and the rate at which the liquid level changes in m/day. Suggest some ways in which the loss of benzene from the tank could be mitigated.

P.11.2. A need arises in a laboratory to conduct experiments in an atmosphere of phosgene, which is highly toxic. The experiments will be conducted inside a plywood box designed to contain a mixture of phosgene (*A*) and air (*B*) having a constant 0.50 mole fraction of phosgene as shown in Fig. P.11.2. Because of the porosity of the plywood, phosgene will diffuse across the walls of the box. In order to reduce the leakage of phosgene, it is proposed to connect a small gas pump to the box to continuously pump out the air-phosgene–mixture while simultaneously feeding pure phosgene to the box to keep a constant mole fraction. The pump will create a slow flow of air into the box through the plywood, thereby reducing the leakage of phosgene. The pump will discharge into the laboratory hood exhaust system. (a) Determine the leakage of phosgene by diffusion through the plywood if no pumping occurs. Note that phosgene will have to be added to the box continuously to keep a constant mole fraction in the box. (b) Determine the inward flow rate of air and the pumping rate required to reduce the phosgene leakage to one-tenth of its value with no pumping. (c) Determine the rate at which pure phosgene must be supplied to the

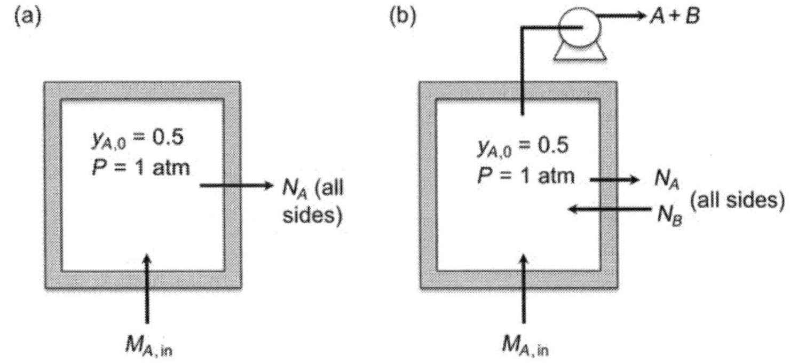

FIGURE P.11.2

box to continuously replace the phosgene that still diffuses out plus the phosgene that is pumped out along with air.

Data: The box is a cube with 60-cm-long sides and a wall thickness of 0.5 cm. The temperature is 25°C and the pressure both inside and outside the box is 1 atm. The phosgene mole fraction outside the box may be taken to be 0. The effective diffusivity of phosgene in the plywood is 0.011 cm^2/s.

P.11.3. It is March 13th and you just left the beach with your friends carrying along great memories of a fabulous Spring break. As your thoughts drift from sun-drenched beaches to mass transfer problems, you notice flashers behind you. You can't believe it—it is that same sheriff that you encountered on your way down—what now? You stop and reach for your license and registration, but the sheriff goes: "No need for that. I am sorry to have to stop you again, but you were so helpful a week ago. I am hoping you can help me again solve a missing person case." Eager to help (and, perhaps, avoid a fine for the broken taillight) you accompany the sheriff to the private office of a missing lawyer, one Billy Fick. It seems that he may have been involved in a money laundering operation and found himself in the wrong place at the wrong time. Determining when he was last in his office would help incriminate a suspect, one Stefan Schmidt. While looking around, you notice a half-empty coffee mug on the missing lawyer's desk. You measure the distance of the liquid surface from the top of the mug and find it to be 6 cm. A neat mark on the inner wall of the mug shows that the initial level of coffee was 1 cm below the rim and that no coffee had been drunk from it. Billy was the only person with access to his office and nothing has been touched since his disappearance. It is obvious that he left in a hurry after placing his mug on his desk and never returned. The temperature and relative humidity in the air-conditioned office are 20°C and 60%, respectively. When did Billy leave his office?

P.11.4. A volatile liquid A is placed in the tapered Stefan tube shown in Fig. P.11.4. The liquid evaporates and diffuses through the still air (B) in the tube until it reaches the top where it is dispersed in the atmosphere. Assuming pseudo-steady state and that the mole fraction of A in the space above the air–liquid interface is a function only of the vertical distance from the interface: (a) Develop an expression for the rate of evaporation of the liquid as a function of the height of liquid in the tube. (b) Use your expression in (a) to develop and expression that will predict how the liquid level will change as a function of time.

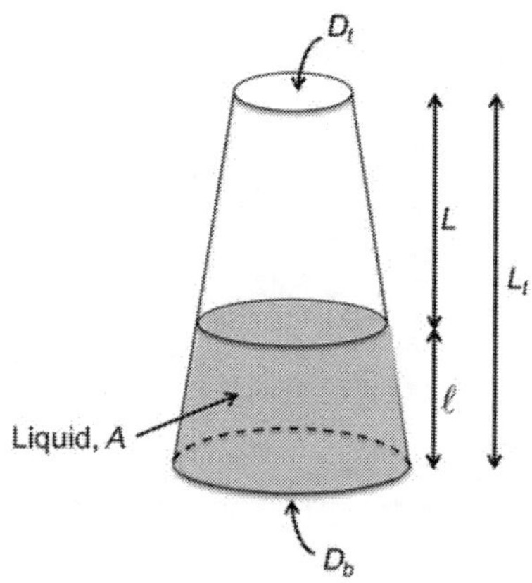

FIGURE P.11.4

P.11.5. In short-term artificial lungs (used, for example, during heart surgery) blood oxygenation is accomplished by bubbling pure oxygen directly through the blood. However, the high surface energy at the gas–liquid interface causes excessive damage to the blood and this method is not suitable for extended uses. To avoid this problem, the blood is separated from air by an oxygen-permeable membrane. In a particular design of such system, blood flows inside thin silicone rubber tubes and pure oxygen flows outside at 1 atm absolute pressure. The tubes have inside and outside diameters of 0.025 and 0.05 cm and the rate of oxygen transfer is 2.0×10^{-4} cm^3 (STP)/min per linear cm of tube. The diffusivity of O_2 in silicone rubber is 2.0×10^{-5} cm^2/s, and the solubility is 6.0×10^{-2} cm^3 (STP)/(atm cm^3 membrane). Determine the percentage by which the O_2 partial pressure will be reduced at the blood–membrane interface. Note: STP means at standard temperature and pressure.

P.11.6. A spherical naphthalene ball with an initial diameter of 1.2 cm slowly sublimes in a large volume of still air at 15°C and 1 atm. At this temperature, the sublimation pressure of naphthalene is 0.036 mmHg. Assuming pseudo-steady-state and neglecting convection, determine: (a) The initial, pseudo-steady-state rate of sublimation expressed as a percentage of the mass of the ball sublimed per day. (b) The time required to reduce the naphthalene ball diameter to one-tenth of the initial value. Note that the rate of sublimation will change over time. Thus, the mass balance equation for the naphthalene ball needs to be integrated to determine the correct time.

P.11.7. The solubility of hydrogen in a steel alloy at 80°F is given by the equation $c^*_{H_2} = b\sqrt{P_{H_2}}$ where $b = 0.0066$ lbmol/ft^3 psi$^{1/2}$ and P_{H_2} is the H_2 pressure in psi. Pure hydrogen is stored in a spherical vessel made of this steel alloy having an inside diameter of 8.0 in. and a wall thickness of 2.0 in. at an initial pressure of 200 psia. The diffusivity of H_2 in the steel alloy is 3.0×10^{-7} cm^2/s. Assuming pseudo-steady-state, determine: (a) The initial rate of mass transfer of hydrogen across the vessel wall in lbmol/s. (b) The initial rate of pressure change inside the sphere. (c) The time needed for the internal pressure to decrease to 50 psia. Assume that air does not diffuse in the steep alloy.

P.11.8. Macroporous adsorbent materials are frequently used in the form of spherical pellets in many different applications. Consider a case where the adsorption of a solute A is very favorable (i.e., A is strongly bound). In this case, when A is adsorbed from a dilute solution on a high-capacity adsorbent, an adsorption front develops within the particle separating a clean core from a saturated layer as shown in Fig. P.11.8. A diffuses through the saturated layer and binds at the adsorption front. Over time, the adsorption front moves toward the center of the particle until

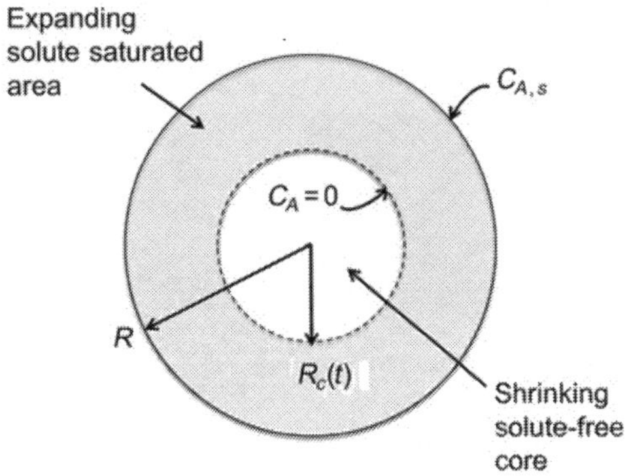

FIGURE P.11.8

the entire particle is fully saturated. Define q_m as the binding capacity in mg per cm^3 of particle volume, $C_{A,s}$ as the concentration of A in the fluid at the particle surface in mg per cm^3 of fluid, and $\mathcal{D}_e$ as the effective pore diffusivity of A in the saturated layer. All three are assumed to be constant. Assuming pseudo-steady-state and that $q_m >> C_{A,s}$: (a) Develop an expression to describe the rate mass transfer of A through the saturated layer when the adsorption front is at a radial distance R_c. Note that $C_A = 0$ at the adsorption front. (b) Write an expression for the amount of A adsorbed when the adsorption front is at a radial distance R_c. (c) Use your expressions in (a) and (b) to find an expression for the position of the adsorption front as a function of time by equating the rate of mass transfer to the rate of change of the amount of A adsorbed.

P.11.9. Perez-Almodovar, Tao, and Carta [Biotechnol. Progr. 27 (2011) 1264] have observed the progress of the adsorption on a monoclonal antibody (mAb) from an aqueous solution at 25°C into spherical adsorbent particles using confocal laser scanning microscopy (CLSM). CLSM allows us to image optical sections revealing the distribution of the bound protein at the particle equator as a function of time. Some of the experimental results are shown in Fig. P.11.9. In these experiments, the mAb concentration in solution was kept constant at 1.0 mg/cm^3 and the binding capacity, determined from independent measurements, was 110 mg/cm^3. The particles were around 54 μm

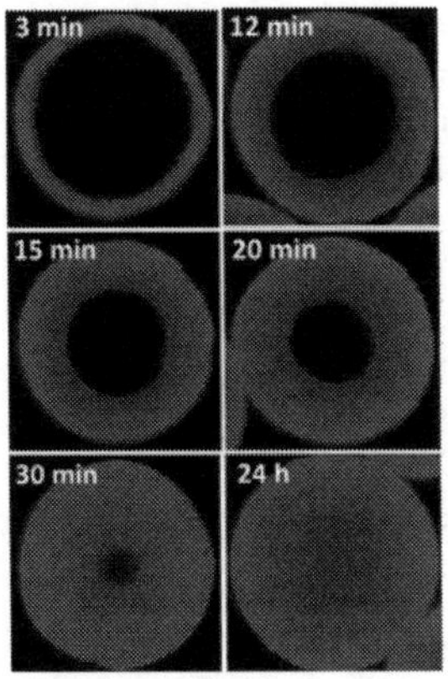

t/R^2 (10^7 s/cm^2)	R_c/R
2.78	0.864
5.64	0.781
8.54	0.722
8.93	0.675
14.6	0.597
16.9	0.506
20.2	0.457
22.1	0.404
23.0	0.385
32.0	0.250

FIGURE P.11.9

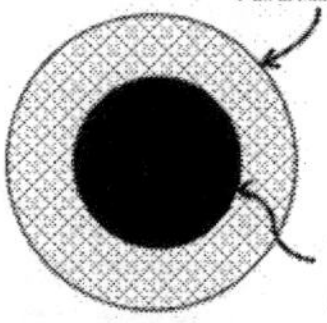

FIGURE P.11.12

in radius. Note that time is normalized by the square of the particle radius in order to account for some small differences between the different particles imaged. (a) Using these data, plot the dimensionless position of the adsorption front R_c/R as a function of t/R^2. (b) As suggested by the analysis conducted in Problem 11.8, replot the data by graphing the function $2(R_c/R)^3 - 3(R_c/R)^2 + 1$ vs. the quantity $6q_m t/C_{A,s}R^2$ and determine the effective pore diffusivity $\mathcal{D}_e$ of the mAb in these particles from the slope of this plot. (c) Assuming that the particles have a porosity $\varepsilon_p = 0.5$ and a pore radius of 50 nm, determine the tortuosity factor τ_p for these particles.

P.11.10. Extend the analysis conducted in Problem 11.8 by relaxing the assumption that the solute concentration $C_{A,s}$ remains constant. In a practical applications, $C_{A,s}$ will vary as a result of adsorption when, for example, a volume of adsorbent particles, V, is added to a finite volume, V_f, of solution with an initial concentration $C_{A,0}$ in a well-mixed tank. In order to find the solution of this problem, we need to combine the expression for the rate of adsorption obtained in Problem 11.8 with a transient material balance for the solute A in the fluid phase. Note that the rate of mass transfer will be equal to the rate at which the solute A disappears from the solution. Using this approach, develop an expression to predict the concentration of A in the fluid phase as a function of time.

P.11.11. Based on the data in Problem 11.9, we want to determine the course of adsorption of the mAb if 15 liters of the adsorbent particles are added to 2,000 L of a solution containing initially a 1.0 mg/cm^3 concentration of the mAb in a well-mixed tank. The particle radius is 54 μm, and the binding capacity is 110 mg/cm^3 of particle volume. Determine the time required to reduce the mAb concentration in solution to 25% of the initial value. What is the lowest possible concentration of the mAb that can be attained for these conditions? For this problem you can assume $\mathcal{D}_e = 5.0 \times 10^{-8}$ cm^2/s.

P.11.12. A "superficially porous" catalyst particle, sketched in Fig. P.11.12, consists of a solid spherical catalytic core surrounded by a spherical macroporous shell. The reaction $2\,C_2H_4 \rightarrow C_4H_8$ occurs only on the surface catalytic core. The particle is placed in a rapidly flowing stream of pure C_2H_4 gas, which diffuses in the porous shell and reacts instantly as soon as it reaches the catalytic core. The absolute pressure is 2 atm, and the temperature is 400 K. The shell porosity is 0.5, and the tortuosity factor is 4. The system is isothermal. Assuming that the reaction is irreversible and that the overall process is diffusion-controlled: (a) Write an equation to describe the steady-state flux of C_2H_4 in the shell for these conditions. (b) Write a steady-state differential material balance for C_2H_4 along with boundary conditions needed to describe diffusional mass transfer in the shell. (c) Integrate your material balance in part (b) to determine the rate of mass transfer of C_2H_4. (d) Calculate the rate of C_4H_8 production in mole/s.

CHAPTER 12

Multidimensional and Transient Diffusion

12.1 Introduction

In this chapter, we consider diffusional mass transfer processes where the composition within a given domain varies in more than one dimension and/or as a function of time in a manner analogous to what was presented in Chap. 5 for heat transfer. As we did for heat transfer, we also consider the existence of a velocity field within the domain where diffusion takes place to account for the possible simultaneous occurrence of convective processes. The basis for this analysis is a transient differential component material balance coupled with the continuity equation and with Fick's law. As in the case of heat transfer, boundary conditions are established at the boundaries of the spatial domain of interest either by specifying the composition at these boundaries or by imposing a condition on the diffusion flux. Although the approach is similar to that used for heat transfer, the way in which conditions are imposed at the domain's boundaries are different. For heat transfer, the temperature was assumed to be continuous across the interface between two phases. Conversely, for mass transfer, the composition is generally discontinuous at the interface with the compositions on the two sides of the interface being related to each other by the appropriate phase-equilibrium relationship. We will see that this difference can have very substantial impact on the results.

In the next sections, we go through the derivation of the differential component material balance in rectangular coordinates and provide the results for cylindrical and spherical coordinates. Although a completely general analysis is possible, we will limit ours to dilute conditions.

12.2 General Differential Component Material Balance

12.2.1 Rectangular Coordinates

The total differential mass balance, also called the continuity equation, was derived in Chap. 5 and is as follows:

$$\frac{\partial \rho}{\partial t}+\frac{\partial (v_x\rho)}{\partial x}+\frac{\partial (v_y\rho)}{\partial y}+\frac{\partial (v_z\rho)}{\partial z}=0 \tag{12.1}$$

where ρ is the mass density and v_x, v_y, and v_z are the components of the velocity vector. If the density is constant, this equation becomes:

$$\frac{\partial v_x}{\partial x} + \frac{\partial v_y}{\partial y} + \frac{\partial v_z}{\partial z} = 0 \tag{12.2}$$

The differential component mass balance can be derived starting with the control volume shown in Fig. 12.1, which is a cube with sides Δx, Δy, and Δz. The faces of the cube are located at x and $x + \Delta x$, y and $y + \Delta y$, and z and $z + \Delta z$. Within the domain of interest, the mass concentration of a component A is ρ_A, which is a scalar. Transport of a component A in and out of this cube will occur as a result of the combined effects of bulk flow and diffusion and is described by the mass flux, n_A, which is a vector. Thus, there will be $n_{A,x}$, $n_{A,y}$, and $n_{A,z}$ components of the flux. All of these quantities will be, in general, functions of time and of the three spatial coordinates.

The following component mass balance can be written for the control volume:

$$\text{Accumulation of } A = \text{Input of } A - \text{Output of } A + \text{Generation of } A$$

Accumulation is the time derivative of the mass of A in the control volume while input and outputs are found from the mass fluxes through each face of the control volume. Generation can be a result of chemical reactions and can be positive if A is formed or negative if A is consumed. The following expressions can be derived for each of the terms that appear in the component material balance:

Accumulation of A: $\frac{\partial}{\partial t}(\Delta x \Delta y \Delta z \times \rho_A)$

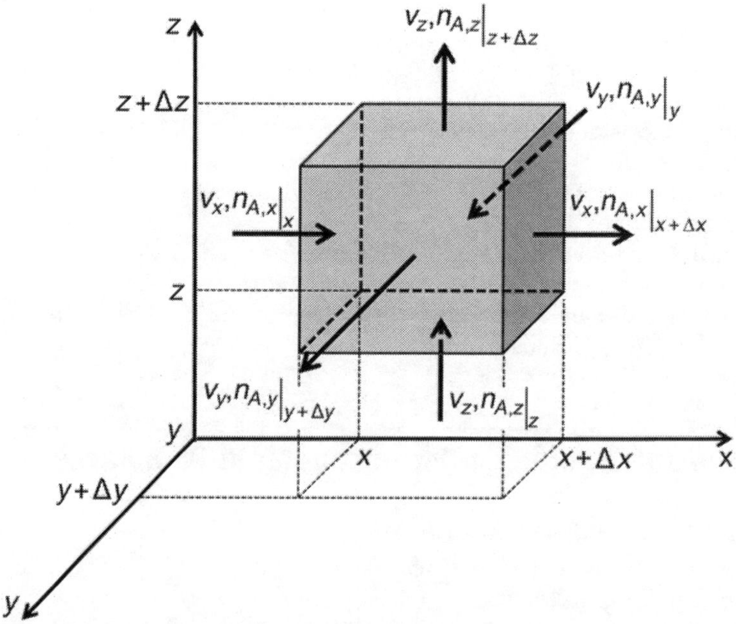

Figure 12.1 Control volume used to derive the general differential component material balance. v and n_A are the velocity and mass transfer flux vectors, respectively. Their x, y, and z components are shown on each face of the control volume.

Input of A: $\Delta y \Delta z \times n_{A,x}\big|_x + \Delta x \Delta z \times n_{A,y}\big|_y + \Delta x \Delta y \times n_{A,z}\big|_z$

Output of A: $\Delta y \Delta z \times n_{A,x}\big|_{x+\Delta x} + \Delta x \Delta z \times n_{A,y}\big|_{y+\Delta y} + \Delta x \Delta y \times n_{A,z}\big|_{z+\Delta z}$

Generation of A: $\Delta x \Delta y \Delta z \times r_A$

where r_A is the rate of generation of component A per unit volume. Balancing these terms yields the following equation:

$$\frac{\partial}{\partial t}(\Delta x \Delta y \Delta z \times \rho_A) = \left(\Delta y \Delta z \times n_{A,x}\big|_x + \Delta x \Delta z \times n_{A,y}\big|_y + \Delta x \Delta y \times n_{A,z}\big|_z\right)$$

$$-\left(\Delta y \Delta z \times n_{A,x}\big|_{x+\Delta x} + \Delta x \Delta z \times n_{A,y}\big|_{y+\Delta y} + \Delta x \Delta y \times n_{A,z}\big|_{z+\Delta z}\right)$$

$$+\Delta x \Delta y \Delta z \times r_A$$

Dividing all of the terms in this equation by $\Delta x \Delta y \Delta z$, we obtain the following result:

$$\frac{\partial \rho_A}{\partial t} + \frac{n_{A,x}\big|_{x+\Delta x} - n_{A,x}\big|_x}{\Delta x} + \frac{n_{A,y}\big|_{y+\Delta y} - n_{A,y}\big|_y}{\Delta y} + \frac{n_{A,z}\big|_{z+\Delta z} - n_{A,z}\big|_z}{\Delta z} = r_A$$

Finally, taking the limit of this equation for $\Delta x \to 0$, $\Delta y \to 0$, and $\Delta z \to 0$ we obtain:

$$\frac{\partial \rho_A}{\partial t} + \frac{\partial n_{A,x}}{\partial x} + \frac{\partial n_{A,y}}{\partial y} + \frac{\partial n_{A,z}}{\partial z} = r_A \tag{12.3}$$

Next, we introduce Fick's law. For dilute conditions with constant ρ, in vector form, we can express the total flux of A as follows:

$$n_A = \rho_A v - \mathcal{D}^0_{A,B} \nabla \rho_A \tag{12.4}$$

The components of the flux will then be given by the following expressions:

$$n_{A,x} = \rho_A v_x - \mathcal{D}^0_{A,B} \frac{\partial \rho_A}{\partial x} \tag{12.5}$$

$$n_{A,y} = \rho_A v_y - \mathcal{D}^0_{A,B} \frac{\partial \rho_A}{\partial y} \tag{12.6}$$

$$n_{A,z} = \rho_A v_z - \mathcal{D}^0_{A,B} \frac{\partial \rho_A}{\partial z} \tag{12.7}$$

Inserting these expressions in Eq. (12.3), assuming that $\mathcal{D}^0_{A,B}$ is a constant, gives the following result:

$$\frac{\partial \rho_A}{\partial t} + \frac{\partial v_x \rho_A}{\partial x} + \frac{\partial v_y \rho_A}{\partial y} + \frac{\partial v_z \rho_A}{\partial z} = \mathcal{D}^0_{A,B}\left(\frac{\partial^2 \rho_A}{\partial x^2} + \frac{\partial^2 \rho_A}{\partial y^2} + \frac{\partial^2 \rho_A}{\partial z^2}\right) + r_A \tag{12.8}$$

On the left-hand side of this equation, we can split the spatial derivatives as follows:

$$\frac{\partial v_x \rho_A}{\partial x}+\frac{\partial v_y \rho_A}{\partial y}+\frac{\partial v_z \rho_A}{\partial z}=\rho_A\left(\frac{\partial v_x}{\partial x}+\frac{\partial v_y}{\partial y}+\frac{\partial v_z}{\partial z}\right)+v_x\frac{\partial \rho_A}{\partial x}+v_y\frac{\partial \rho_A}{\partial y}+v_z\frac{\partial \rho_A}{\partial z} \quad (12.9)$$

Because of Eq. (12.2), the term in parentheses on the right-hand side of Eq. (12.9) is zero. Thus, Eq. (12.8) takes on the following simpler form:

$$\frac{\partial \rho_A}{\partial t}+v_x\frac{\partial \rho_A}{\partial x}+v_y\frac{\partial \rho_A}{\partial y}+v_z\frac{\partial \rho_A}{\partial z}=\mathcal{D}^0_{A,B}\left(\frac{\partial^2 \rho_A}{\partial x^2}+\frac{\partial^2 \rho_A}{\partial y^2}+\frac{\partial^2 \rho_A}{\partial z^2}\right)+r_A \quad (12.10)$$

Equation (12.10) can also be written in a more compact form in vector notation as follows:

$$\frac{D\rho_A}{Dt}=\mathcal{D}^0_{A,B}\nabla^2\rho_A+r_A \quad (12.11)$$

where:

$$\frac{D\rho_A}{Dt}=\frac{\partial \rho_A}{\partial t}+v\cdot\nabla\rho_A \quad (12.12)$$

is the "substantial" or "material" derivative of the mass concentration (note that $v\cdot\nabla\rho_A$ is the dot product of the velocity vector and the concentration gradient vector) and

$$\nabla^2=\frac{\partial^2}{\partial x^2}+\frac{\partial^2}{\partial y^2}+\frac{\partial^2}{\partial z^2} \quad (12.13)$$

is the Laplacian operator. We can see that Eq. (12.10) is term-by-term formally the same as the differential thermal energy balance, Eq. (5.11), with ρ_A replacing T, $\mathcal{D}^0_{A,B}$ replacing the thermal diffusivity α, and the rate of mass generation of A, r_A, replacing the rate of energy generation term, $G/\rho C_p$. Although $\mathcal{D}^0_{A,B}$ and α describe phenomena that are physically very different, they have the same dimensions (length squared over time) and, mathematically, play the same roles in the corresponding differential material and energy balances.

12.2.2 Cylindrical and Spherical Coordinates

The derivation of the general differential component material balance in cylindrical and spherical coordinates follows the same steps used for the rectangular coordinate system, but the control volume will no longer be a cube. In cylindrical coordinates, the control volume will be a cylindrical shell with sides of length Δr, Δz, and $\Delta r\theta$, where $\Delta\theta$ is the arc-length in radians, and r is the radial coordinate. Similarly, in spherical coordinates, the control volume will be a spherical shell with sides of length Δr, $\Delta r\theta$, and $r\sin\theta\Delta\phi$, where $\Delta\theta$ and $\Delta\phi$ are the arc-lengths in radians and r is the radial coordinate. The corresponding balances are given in Table 12.1 for the case of constant ρ and $\mathcal{D}^0_{A,B}$.

As we saw in the case of Table 5.1 for heat transfer, the equations in Table 12.1 are readily adapted to describe simpler situations. Simplified forms for five common cases are as follows:

1. Steady state processes: $\frac{\partial \rho_A}{\partial t}=0$
2. Systems with no fluid flow: $v=0$

A. Rectangular coordinates: $\rho_A = \rho_A(t, x, y, z)$
$\frac{\partial \rho_A}{\partial t} + v_x \frac{\partial \rho_A}{\partial x} + v_y \frac{\partial \rho_A}{\partial y} + v_z \frac{\partial \rho_A}{\partial z} = \mathcal{D}^0_{A,B}\left(\frac{\partial^2 \rho_A}{\partial x^2} + \frac{\partial^2 \rho_A}{\partial y^2} + \frac{\partial^2 \rho_A}{\partial z^2}\right) + r_A$
B. Cylindrical coordinates: $\rho_A = \rho_A(t, r, z, \theta)$
$\frac{\partial \rho_A}{\partial t} + v_r \frac{\partial \rho_A}{\partial r} + v_z \frac{\partial \rho_A}{\partial z} + \frac{v_\theta}{r}\frac{\partial \rho_A}{\partial \theta} = \mathcal{D}^0_{A,B}\left[\frac{1}{r}\frac{\partial}{\partial r}\left(r\frac{\partial \rho_A}{\partial r}\right) + \frac{\partial^2 \rho_A}{\partial z^2} + \frac{1}{r^2}\frac{\partial^2 \rho_A}{\partial \theta^2}\right] + r_A$
C. Spherical coordinates: $\rho_A = \rho_A(t, r, \theta, \phi)$
$\frac{\partial \rho_A}{\partial t} + v_r \frac{\partial \rho_A}{\partial r} + \frac{v_\theta}{r}\frac{\partial \rho_A}{\partial \theta} + \frac{v_\phi}{r \sin\theta}\frac{\partial \rho_A}{\partial \phi} = \mathcal{D}^0_{A,B}\left[\frac{1}{r^2}\frac{\partial}{\partial r}\left(r^2\frac{\partial \rho_A}{\partial r}\right) + \frac{1}{r^2 \sin\theta}\frac{\partial}{\partial \theta}\left(\sin\theta\frac{\partial \rho_A}{\partial \theta}\right) + \frac{1}{r^2 \sin^2\theta}\frac{\partial^2 \rho_A}{\partial \phi^2}\right] + r_A$

Table 12.1 General Differential Material Balance in Rectangular, Cylindrical, and Spherical Coordinate Systems with Constant ρ and $\mathcal{D}^0_{A,B}$

3. No y and z concentration gradients in rectangular coordinates:

$$\frac{\partial \rho_A}{\partial t} + v_x \frac{\partial \rho_A}{\partial x} = \mathcal{D}^0_{A,B}\frac{\partial^2 \rho_A}{\partial x^2} + r_A \tag{12.14}$$

4. No z and θ concentration gradients in cylindrical coordinates:

$$\frac{\partial \rho_A}{\partial t} + v_r \frac{\partial \rho_A}{\partial r} = \mathcal{D}^0_{A,B}\frac{1}{r}\frac{\partial}{\partial r}\left(r\frac{\partial \rho_A}{\partial r}\right) + r_A \tag{12.15}$$

5. No θ and ϕ concentration gradients in spherical coordinates:

$$\frac{\partial \rho_A}{\partial t} + v_r \frac{\partial \rho_A}{\partial r} = \mathcal{D}^0_{A,B}\frac{1}{r^2}\frac{\partial}{\partial r}\left(r^2\frac{\partial \rho_A}{\partial r}\right) + r_A \tag{12.16}$$

A general differential component balance can also be derived on a molar basis following the same approach that we used on a mass basis. However, for the dilute conditions that we are considering here, molar and mass fluxes and molar and mass concentrations are related to each through the molecular mass according to $N_A = n_A/M_{r,A}$ and $C_A = \rho_A/M_{r,A}$. As a result, the molar component balance will have the same form as the equations in Table 12.1 with the molar concentration C_A replacing the mass concentration ρ_A and the molar generation rate R_A replacing r_A. In vector notation, the molar component balance becomes:

$$\frac{DC_A}{Dt} = \mathcal{D}^0_{A,B}\nabla^2 C_A + R_A \tag{12.17}$$

where:

$$\frac{DC_A}{Dt} = \frac{\partial C_A}{\partial t} + v \cdot \nabla C_A \tag{12.18}$$

12.3 Transient Mass Transfer Processes

Just as in the case of heat transfer, multidimensional diffusional mass transfer problems are best handled by numerically solving the appropriate differential balances. Computational packages such as COMSOL and FLUENT are recommended for handling these simulations. Here we will focus on one-dimensional transient diffusion problems for dilute conditions, which can, in many cases, be handled with analytical mathematics. Despite these limitations, the corresponding analytical solutions can still approximate the actual behavior of many practical systems.

The transient problems that we consider in this section are fundamentally different from the pseudo-steady-state scenarios that we considered in Chap. 11. The assumption of pseudo-steady-state is appropriate when two processes that are characterized by time constants that differ substantially from each other are coupled. In these cases, the rate at which the "slow" process occurs (i.e., the process with a small time constant) is dependent on the steady-state rate of the "fast" process. As we have seen in Chap. 11, this assumption allows for great simplification of the mathematical treatment and is useful to describe many different practical situations. The transient situations considered here involve, instead, a single process where the concentration of a component varies in time and distance along an appropriate coordinate. Describing diffusional transport in such systems requires the solution of partial differential equations derived from the relevant mass or molar differential component balances. The initial and boundary conditions associated with these partial differential equations are critical and fundamentally different methods of solution are used depending on the nature of these conditions. Fortunately, since, as seen in Sec. 12.2, the differential component material balance and the differential energy balance are mathematically equivalent, many of the solutions obtained for transient heat conduction problems are directly applicable to analogous mass transfer problems provided that appropriate substitutions of variables and parameters are made.

In analogy to what we did for heat transfer, we consider cases where a solid object with an initial uniform concentration $c_{A,0}$ of a component A is suddenly exposed to a fluid with concentration C_A as shown in Fig. 12.2. Mass transfer of A will occur to or from the fluid in a direction that will depend on the driving force. Transport within the solid will occur by diffusion while transport in the surrounding fluid will normally occur as a result of convective mass transfer. Note that we are using different concentration variables: lower case c_A for the solid-phase concentration and capital C_A for the fluid-phase concentration. At the interface between the two phases, these two concentrations will be

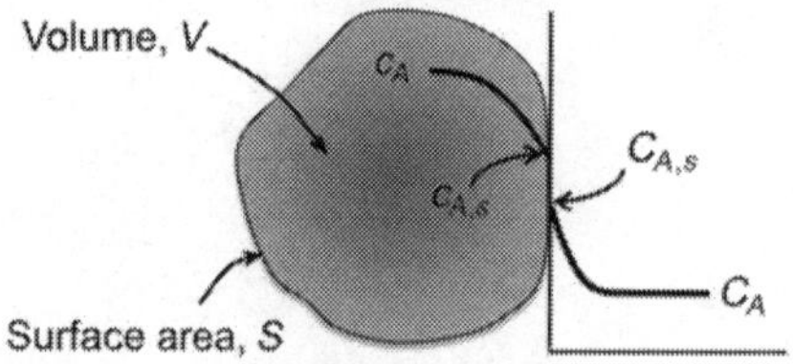

Figure 12.2 Concentration profiles near the interface between solid and fluid phases.

related to each other by the relevant phase equilibrium relationship. In general, at the interface we can write:

$$c_{A,s} = K_c C_{A,s} \tag{12.19}$$

where $c_{A,s}$ and $C_{A,s}$ are the interfacial concentrations and K_c is a distribution coefficient describing partitioning of component A between the two phases at equilibrium as a concentration ratio. The physical basis of K_c will depend on the nature of the phases involved in the actual process. For example, if the fluid is actually a gas and the "solid" object is a liquid, K_c could be described by Henry's law. If the "solid" and fluid are actually two immiscible liquids, then K_c will depend on liquid–liquid partitioning equilibria. Or if the solid is an adsorbent material, K_c could be described by an adsorption isotherm. In most cases, K_c will be a function of temperature and pressure. K_c can also be a function of concentration itself, in which case, the relationship between $c_{A,s}$ and $C_{A,s}$ would become nonlinear. This will often occur for more concentrated cases where Henry's law is no longer valid or when there are strong interactions between component A and the solid. For dilute conditions, assuming that K_c is a constant will often be a good assumption. Finally, it should be noted that instead of describing phase equilibrium in terms of a concentration ratio, a distribution coefficient can also be defined, for example, as a mole fraction ratio. In this case, equilibrium at the interface could be described by the following equation with a distribution coefficient K:

$$y_{A,s} = K x_{A,s} \tag{12.20}$$

Since $c_{A,s} = x_{A,s}c$ and $C_{A,s} = y_{A,s}C$, where c and C are the molar densities of the two phases, K_c and K will be related to each other by:

$$K = \frac{c}{C}\frac{1}{K_c} \tag{12.21}$$

As we saw for heat transfer, we can envision three distinct mass transfer scenarios depending on the relative importance of the internal diffusive resistance within the object and the external convective resistance in the surrounding fluid:

1. The mass transfer resistance within the object is very small—In this case, c_A will be uniform within the object and will vary with time only. At the interface, we will have $C_{A,s} = c_A/K_c$. Thus, at the interface, the fluid-side concentration will be in equilibrium with the concentration in the bulk solid.
2. The mass transfer resistance within the object is very large—In this case, c_A will vary both with time and with position within the object. At the interface, we will have $c_{A,s} = K_c C_A$, Thus, at the interface, the solid-side concentration will be in equilibrium with the concentration in the bulk fluid.
3. The internal and external mass transfer resistances are comparable in magnitude—In this case, c_A will vary both with time and with position within the object. Equilibrium will still prevail at the interface so that $c_{A,s} = K_c C_{A,s}$. In this case, $c_{A,s}$ will no longer in equilibrium with the bulk fluid but will be in equilibrium with the fluid-side interfacial concentration.

The relative importance of internal and external mass transfer resistances can be expressed again in terms of a Biot number, which now takes on the form:

$$\text{Bi} = \frac{\text{Internal diffusive resistance}}{\text{External convective resistance}}$$

In general, if Bi << 1, the external resistance will be dominant, and we will be in the first of the three scenarios outlined above. If Bi >> 1, the internal resistance will be dominant, and we will be in the second of the three scenarios. Finally, if Bi ~ 1, we will be in the third scenario and both resistances will be important. This is the same conclusion that we reached for heat transfer problems. An important difference exists, however, when this concept is applied to mass transfer problems regarding how we compare the magnitude of the two different resistances. For heat transfer, we could compare conductive and convective transport directly since, for both, the driving force is the temperature gradient. It is still true, of course, that diffusive and convective mass transfer are also both driven by a concentration gradient. However, near the interface, the solid phase concentration c_A and the fluid phase concentration C_A are, in general, different from each other according to the relevant phase equilibrium relationship. As a result, a comparison between the two resistances must take into account partitioning of component A at the interface, which is expressed, for example, by Eq. (12.19). In comparable terms, the order of magnitude of the two resistances can be estimated as follows:

$$\text{Internal diffusive resistance} \sim \frac{\ell}{\mathcal{D}^0_{A,B}}$$

$$\text{External convective resistance} \sim \frac{K_c}{k_c}$$

where ℓ is a characteristic length describing the size of the object, $\mathcal{D}^0_{A,B}$ is the diffusivity of A in the object, and k_c is the external convective mass transfer coefficient. Accordingly, we have the following expression for the Biot number:

$$\text{Bi} = \frac{k_c \ell}{\mathcal{D}^0_{A,B} K_c} \tag{12.22}$$

If k_c is small, for instance, because the surrounding fluid is a very viscous liquid and the flow velocity is small, Bi will be small and the external resistance will be large. Conversely, if the object is large in size and/or if the diffusivity of A in the object is small, Bi will be large and the internal diffusive resistance will be dominant. The distribution coefficient can play a big role, of course, regardless of the values of k_c, ℓ, and $\mathcal{D}^0_{A,B}$. For example, if K_c is very large, A will be preferentially partitioned into the solid making $C_{A,s}$ very small compared to $c_{A,s}$. This will make Bi very small and the external resistance dominant. Conversely, if K_c is very small, A will be preferentially partitioned into the fluid phase making $c_{A,s}$ very small compared to $C_{A,s}$. In turn, this will make Bi very large and the internal resistance dominant.

Note that, just as we saw for heat transfer, the actual value of Bi for mass transfer will depend on exactly how ℓ is defined. If the object is a semi-infinite slab with thickness $2B$, ℓ would normally be set equal to the half-thickness B. For a long cylinder or for a sphere with radius R, ℓ would normally be set equal to R. For irregular shapes, one can take again $\ell \sim V/S$ but only for estimation purposes. Examples 12.1 and 12.2 illustrate the impact of Bi on the controlling resistance for two practical cases.

Example 12.1. Relative importance of internal and external mass transfer in ion exchange. Ion exchange resins contain fixed charged functional groups attached to a fine-mesh cross-linked polymer network. Mobile ions (called the counter-ions) can diffuse in and out of the resin and exchange for each other but must satisfy a charge balance. Such resins are usually in the form of spherical beads and have many different practical uses including water de-ionization, water softening, and pharmaceutical purification. In cation exchangers, the fixed charged groups are negatively charged while the counter-ions are positively charged, such as, for example, H^+ and Na^+. Since the concentration of counter-ions in the resin must balance the concentration of the fixed functional groups, for each Na^+ ion that diffuses into the resin, one H^+ ion must diffuse out. Let's consider the exchange of H^+ and Na^+ from an alkaline solution with concentration C_{Na^+}. Since H^+ is practically zero in such solutions, at equilibrium, the Na^+ concentration in the resin will be equal to the concentration of the fixed functional groups, c_0. Thus, K_c for this case will be equal to c_0/C_{Na^+}. Let's now consider mass transfer effects. Roberts et al. [J. Chromatogr. A 1628 (2020) 461444] have reported a value of the external mass transfer coefficient $k_c = 0.0191$ cm/s for the mass transfer of Na^+ on 8% cross-linked sulfonated cation exchange resin beads having a diameter of 50 μm. For this resin, $c_0 \sim 2.8$ mol/L. The diffusivity of Na^+ in this resin has been reported to be 9.4×10^{-7} cm²/s. The corresponding Biot number is then as follows:

$$\mathrm{Bi} = \frac{k_c R C_{Na^+}}{\mathcal{D}^0_{A,B} c_0} = \frac{0.0191 \times 0.0025}{9.4 \times 10^{-7} \times 2.8} C_{Na^+} \sim 18 \times C_{Na^+}$$

with C_{Na^+} in mol/L. We can see that the relative importance of external and internal resistances will depend on the solution concentration. If the solution is very dilute, say $C_{Na^+} = 0.005$ mol/L (i.e., 5 mM), Bi will be less than 0.1 indicating that the external resistance will be dominant. On the other hand, if the solution is more concentrated, say $C_{Na^+} = 1$ mol/L, Bi will become greater than 10 indicating that internal diffusion will be the dominant resistance.

Example 12.2. Relative importance of internal and external mass transfer in protein adsorption. Monoclonal antibodies (mAbs) are important biopharmaceutical proteins with molecular mass around 150,000 g/mol. They are typically recovered and purified from cell culture fluids using selective adsorbents in packed-bed columns. Let's consider the adsorption of a mAb on macroporous spherical adsorbent particles from an aqueous solution at 20°C. The particles have a diameter of 90 μm, a porosity $\varepsilon_p = 0.6$, a tortuosity factor $\tau_p = 4$, and a pore radius of 30 nm (see Sec. 10.3.3 for the definition of these quantities). The mAb has a molecular radius $r_A = 5$ nm. The external mass transfer coefficient is estimated to be 0.0015 cm/s. Since the adsorbent particles are macroporous, protein transport within the particles will occur by diffusion in the liquid-filled pores. We can estimate the effective pore diffusivity according to the methods in Sec. 10.3.3. Following Example 10.5, the diffusivity of the mAb in solution is:

$$\mathcal{D}^0_{A,B} = \frac{2.7 \times 10^{-5}}{(M_{r,A})^{1/3}} = \frac{2.7 \times 10^{-5}}{(150{,}000)^{1/3}} = 5.1 \times 10^{-7}\ \mathrm{cm^2/s}$$

The hindrance factor can be estimated to be $\psi_A = 0.4$ from Fig. 10.8 with $\varphi_A = r_A/r_{pore} = 5/30 = 0.17$. Thus, the effective diffusivity is estimated to be:

$$\mathcal{D}_{A,e} = \frac{\varepsilon_p \mathcal{D}^0_{A,B}}{\tau_p} \psi_A = \frac{0.6 \times 5.1 \times 10^{-7}}{4} \times 0.4 = 3.1 \times 10^{-8}\ \mathrm{cm^2/s}$$

Note that, for this case, since diffusion occurs in liquid-filled macropores, we expect $K_c = c_A/C_A \sim 1$. As a result, for the Biot number we obtain:

$$\mathrm{Bi} = \frac{k_c R}{\mathcal{D}_{A,e} K_c} = \frac{0.0015 \times 45 \times 10^{-4}}{3.1 \times 10^{-8} \times 1} = 217$$

This large value of Bi indicates that the external mass transfer resistance will be negligible and that the process will be completely dominated by the internal diffusional resistance regardless of the protein concentration in solution. This will almost always be the case for transport in macroporous materials.

12.3.1 Processes Controlled by the External Convective Resistance

The analysis of processes where the external mass transfer resistance is controlling follows almost exactly the approach that we used to describe the analogous heat transfer processes. When the external resistance is dominant (Bi << 1), c_A will be uniform within the object and will vary only with time. Thus, the concentration on the fluid-side of the interface will be $C_{A,s} = c_A(t)/K_c$ in equilibrium with c_A. Assuming that there is no reaction and that the external fluid concentration far away from the surface is constant at $C_A = C_{A,b}$, a material balance for component A gives the following equation:

$$V\frac{dc_A}{dt} = -Sk_c(C_{A,s} - C_{A,b}) \tag{12.23}$$

or, replacing $C_{A,s}$ with c_A/K_c:

$$V\frac{dc_A}{dt} = -Sk_c\left(\frac{c_A}{K_c} - C_{A,b}\right) \tag{12.24}$$

Separating the variables c_A and t in this equation and integrating with the assumption that K_c is constant, we obtain:

$$\int_{c_{A,0}}^{c_A} \frac{dc_A}{c_A - K_c C_{A,b}} = -\frac{Sk_c}{VK_c}\int_0^t dt \tag{12.25}$$

The final result is obtained by carrying out the integrals and is as follows:

$$\frac{c_A - K_c C_{A,b}}{c_{A,0} - K_c C_{A,b}} = \exp\left(-\frac{Sk_c}{VK_c}t\right) \tag{12.26}$$

As in the heat transfer case (see Sec. 5.3), this result is independent of geometry and only the area-to-volume ratio, S/V, actually matters. Note, however, that K_c appears in this equation because of the assumption of phase equilibrium at the interface. Its value will obviously influence the result.

12.3.2 Processes Controlled by the Internal Diffusive Resistance

When the internal mass transfer resistance is controlling, the internal concentration c_A will vary with both time and position within the object while the concentration at the surface $c_{A,s}$ will be in equilibrium with the concentration in the bulk fluid. Thus, we will have $c_{A,s} = K_c C_{A,b}$. As in the heat transfer case, we consider four different situations: (a) diffusion in an infinite slab; (b) diffusion in a semi-infinite slab with thickness $2B$; (c) diffusion in a long cylinder with radius R; and (d) diffusion in a sphere with radius R. In all four cases, we will assume that there is no reaction, so that $R_A = 0$ in the differential component material balance. It is not difficult to see that the analytical solutions obtained in dimensionless form for heat conduction for these four cases shown in

Sec. 5.3.2 can also be used for the corresponding diffusion problems. For example, consider the case of one-dimensional diffusion of a component A in an infinite slab with initial concentration $c_{A,0}$ suddenly brought to a surface concentration $c_{A,s}$. The differential material balance for component A is easily obtained from item A in Table 12.1 by setting all velocity terms equal to 0 and retaining only the x-derivative, and is as follows:

$$\frac{\partial c_A}{\partial t} = \mathcal{D}^0_{A,B} \frac{\partial^2 c_A}{\partial x^2} \tag{12.27}$$

This equation is subject to the following initial and boundary conditions:

$$\begin{aligned} t = 0 &: \; c_A = c_{A,0} \quad \text{for any } x \\ x = 0 &: \; c_A = c_{A,s} \quad \text{for any } t > 0 \\ x \to \infty &: \; c_A = c_{A,0} \quad \text{for any } t \end{aligned} \tag{12.28}$$

Introducing the following dimensionless concentration:

$$Y = \frac{c_A - c_{A,s}}{c_{A,0} - c_{A,s}} \tag{12.29}$$

Eqs. (12.27) and (12.28) become:

$$\begin{aligned} &\frac{\partial Y}{\partial t} = \mathcal{D}^0_{A,B} \frac{\partial^2 Y}{\partial x^2} \\ t = 0 &: \quad Y = 1 \;\; \text{for any } x \\ x = 0 &: \quad Y = 0 \;\; \text{for any } t > 0 \\ x \to \infty &: \; Y = 1 \;\; \text{for any } t \end{aligned} \tag{12.30}$$

In this form, Eq. (12.30) and associated initial and boundary conditions are exactly the same as Eq. (5.35), which we derived to describe heat conduction into an infinite slab. The only difference is that, now, $\mathcal{D}^0_{A,B}$ replaces α. Since the dimensionless equation and boundary conditions are the same, in dimensionless form, the solution will, of course, also be the same and is given by:

$$Y = \text{erf}(\eta) \tag{12.31}$$

where $\eta = x/2\sqrt{\mathcal{D}^0_{A,B} t}$. This function is given in Fig. 5.4 along with the function $1 - Y = 1 - \text{erf}(\eta) = \text{erfc}(\eta)$. Similarly to the analogous heat conduction problem, $1 - Y$, now equal to $(c_A - c_{A,0})/(c_{A,s} - c_{A,0})$, represents the fractional change in concentration that occurs at specified values of x and t relative to the maximum possible change, $c_{A,s} - c_{A,0}$, that would be seen after a very long time or, mathematically, at $t \to \infty$.

Based on this result, we can now predict the order of magnitude of the time needed for a component A to diffuse over a certain distance δ. From Fig. 5.4, we see that $Y \sim 0.995$ (i.e., $1 - Y \sim 0.005$) when $\eta \sim 2$. The corresponding time will thus be:

$$t_{0.5\%} = \frac{1}{16} \frac{\delta^2}{\mathcal{D}^0_{A,B}} \tag{12.32}$$

We might consider this time to be how long we would have to wait to see any significant change in concentration at a distance δ from a surface that has been exposed to a different concentration. Analogously, for $Y = 0.5$, which corresponds to one-half of the maximum concentration change possible, we find $\eta \sim 0.5$ and $t_{50\%} \sim \delta^2/\mathcal{D}^0_{A,B}$.

As we did for heat transfer, we can also find the mass transfer flux at the surface from Fick's law, which is given by the following equation:

$$N_{A,x}\Big|_{x=0} = -\mathcal{D}^0_{A,B}\frac{\partial c_A}{\partial x}\Big|_{x=0} \tag{12.33}$$

In terms of Y and η, Eq. (12.33) can be written as follows:

$$N_{A,x}\Big|_{x=0} = -\mathcal{D}^0_{A,B}(c_{A,0} - c_{A,s})\frac{\partial \eta}{\partial x}\frac{dY}{d\eta}\Big|_{\eta=0} = -\mathcal{D}^0_{A,B}(c_{A,0} - c_{A,s})\frac{1}{2\sqrt{\mathcal{D}^0_{A,B}t}}\frac{dY}{d\eta}\Big|_{\eta=0} \tag{12.34}$$

Since, as seen in Fig. 5.4, $(dY/d\eta)_{\eta=0} = 2/\sqrt{\pi}$, we have:

$$N_{A,x}\Big|_{x=0} = \sqrt{\frac{\mathcal{D}^0_{A,B}}{\pi t}}(c_{A,s} - c_{A,0}) \tag{12.35}$$

Finally, by integrating Eq. (12.35) over time, we can find the cumulative amount of component A that will have diffused into the slab over a period of time t per unit area of the slab S. The following result is obtained:

$$\frac{M_A}{S} = \int_0^t N_{A,x}\Big|_{x=0} dt = \int_0^t \sqrt{\frac{\mathcal{D}^0_{A,B}}{\pi t}}(c_{A,s} - c_{A,0})\,dt = \sqrt{\frac{4\mathcal{D}^0_{A,B}t}{\pi}}(c_{A,s} - c_{A,0}) \tag{12.36}$$

Dividing M_A/S by t will give us the flux at the surface averaged over time as:

$$N_{A,\text{avg}} = \frac{1}{t}\int_0^t N_{A,x}\Big|_{x=0} dt = \sqrt{\frac{4\mathcal{D}^0_{A,B}}{\pi t}}(c_{A,s} - c_{A,0}) \tag{12.37}$$

Example 12.2 demonstrates the use of these relationships.

Example 12.3. Diffusion into a large slab. A large pool of stagnant liquid water is suddenly exposed to nitrous oxide (N_2O) gas at a partial pressure of 0.05 atm (5,070 Pa) and 303 K. What will be the concentration of N_2O in the water at the gas–liquid interface? How far below the surface will N_2O have diffused over a 1-h period? How far below the surface will we find an N_2O concentration equal to one-half of its surface value? How much N_2O will have diffused into the water in 1 h? In order to answer these questions, we need to know the Henry's law constant and the diffusivity of N_2O in water. Li and Lee [J. Chem. Eng. Data 41 (1996) 551] report a solubility coefficient H_s = 0.00022 mol/m³ Pa and $\mathcal{D}^0_{A,B} = 2.1 \times 10^{-9}$ m²/s for N_2O in water at 303 K. At a partial pressure of N_2O of 5,070 Pa, the surface concentration of dissolved N_2O (A) will thus be $c_{A,s} = 0.00022 \times 5{,}070 =$ 1.1 mol/m³. From Eq. (12.32), the distance over which N_2O will diffuse in 1 h will be on the order of:

$$\delta \sim 4\sqrt{\mathcal{D}^0_{A,B}t} = 4\sqrt{2.1 \times 10^{-9} \times 3{,}600} = 0.011 \text{ m}$$

From Eq. (12.31) or Fig. 5.4, we can find that η will have a value of 0.46 when $Y = (c_A - c_{A,s})/(c_{A,0} - c_{A,s})$ is equal to 0.5. Thus, the distance where c_A will reach one-half of the surface concentration will be:

$$x = 2\eta\sqrt{\mathcal{D}^0_{A,B}t} = 2\times 0.46\times\sqrt{2.1\times 10^{-9}\times 3{,}600} = 0.0025 \text{ m}$$

Finally, from Eq. (12.36), the cumulative amount of N_2O that will have diffused into the water, normalized by the surface area, in 1 h will be:

$$\frac{M_A}{S} = \sqrt{\frac{4\mathcal{D}^0_{A,B}t}{\pi}}(c_{A,s} - c_{A,0}) = \sqrt{\frac{4\times 2.1\times 10^{-9}\times 3{,}600}{\pi}}(1.1-0) = 0.0034 \text{ mol/m}^2$$

Let's now consider the case of diffusion of a component A in a semi-infinite slab with thickness $2B$ when both sides of the slab are suddenly brought to the same concentration $c_{A,s}$ starting with a uniform initial concentration $c_{A,0}$. The process is described by the following equation and initial and boundary conditions:

$$\begin{aligned} &\frac{\partial c_A}{\partial t} = \mathcal{D}^0_{A,B}\frac{\partial^2 c_A}{\partial x^2} \\ t=0:\quad & c_A = c_{A,0} \quad \text{for any } x \\ x=0:\quad & \frac{\partial c_A}{\partial x} = 0 \quad \text{for any } t \\ x=B:\quad & c_A = c_{A,s} \quad \text{for any } t>0 \end{aligned} \tag{12.38}$$

where $x = 0$ is the slab centerline. In a manner analogous to the equivalent heat transfer case, Eq. (12.38) also describes diffusion in a semi-infinite slab with thickness B where one side is suddenly brought to $c_{A,s}$ while the other side is impervious to A. In this case, setting $x = 0$ at the impervious surface where the flux of A is 0, we will have $(\partial c_A/\partial x)_{x=0} = 0$, which is the same as the boundary condition at the centerline of a slab of thickness $2B$ with both sides suddenly brought to $c_{A,s}$. By introducing the following dimensionless variables:

$$Y = \frac{c_A - c_{A,s}}{c_{A,0} - c_{A,s}} \tag{12.39}$$

$$\tau = \frac{\mathcal{D}^0_{A,B}t}{B^2} \tag{12.40}$$

$$\xi = \frac{x}{B} \tag{12.41}$$

Eq. (12.38) and its boundary conditions become:

$$\begin{aligned} &\frac{\partial Y}{\partial \tau} = \frac{\partial^2 Y}{\partial \xi^2} \\ \tau=0:\quad & Y=1 \quad \text{for any } \xi \\ \xi=0:\quad & \frac{\partial Y}{\partial \xi} = 0 \quad \text{for any } \tau \\ \xi=1:\quad & Y=0 \quad \text{for any } \tau>0 \end{aligned} \tag{12.42}$$

Equation (12.42) and associated conditions are exactly the same as those obtained for the corresponding heat conduction problem, Eq. (5.52). The solution is thus given by item *A* in Table 5.2. Note that, just like in the case of heat transfer, where we calculated the average temperature in the slab, it is also useful to calculate the average concentration in the slab according to the following equation:

$$c_{A,\text{avg}}(t) = \frac{1}{B}\int_0^B c_A(t,x)dx \tag{12.43}$$

In dimensionless form, Eq. (12.43) can be written as follows:

$$Y_{\text{avg}}(\tau) = \frac{c_{A,\text{avg}} - c_{A,s}}{c_{A,0} - c_{A,s}} = \int_0^1 Y(\tau,\xi)d\xi \tag{12.44}$$

Equation (12.44) is, of course, mathematically the same as expression that we derived for the average temperature and is also given in Table 5.2. The analogous results for Y and Y_{avg} for radial heat conduction into a long cylinder and into a sphere, both of radius R, included in Table 5.2, can also be used to describe radial diffusion in these geometries. In these cases, the dimensionless time and the dimensionless distance will be expressed by:

$$\tau = \frac{\mathcal{D}^0_{A,B}t}{R^2} \tag{12.45}$$

and

$$\xi = \frac{r}{R} \tag{12.46}$$

Figure 5.8 shows Y_{avg} for slabs, cylinders, and spheres in graphical form as a function of τ. Note that for cylinders and for spheres the average dimensionless concentrations are calculated as cylindrical and spherical averages, respectively, according to Eqs. (5.72) and (5.75). Thus, knowing Y_{avg}, we can calculate the total amount of component A in a cylinder or in a sphere at a certain time simply by calculating $c_{A,\text{avg}}$ from Y_{avg} and multiplying its value by the volume of the object.

Newman's rule, described in Sec. 5.3.2.d for the analogous heat conduction problem, can also be used as an approximation for diffusion in other geometries. For example, for a three-dimensional object shaped like a brick with sides $2A$, $2B$, and $2C$, the dimensionless concentration at a point with coordinates x, y, and z can be found from the following approximate equation:

$$Y = Y_A(\tau_A,\xi_A) \times Y_B(\tau_B,\xi_B) \times Y_C(\tau_C,\xi_C) \tag{12.47}$$

where the dimensionless times and distances along the three dimensions are calculated as follows:

$$\begin{gathered}\tau_A = \frac{\mathcal{D}^0_{A,B}t}{A^2},\quad \tau_B = \frac{\mathcal{D}^0_{A,B}t}{B^2},\quad \tau_C = \frac{\mathcal{D}^0_{A,B}t}{C^2}\\ \xi_A = \frac{x}{A},\quad \xi_B = \frac{y}{B},\quad \xi_C = \frac{z}{C}\end{gathered} \tag{12.48}$$

Examples 12.3 and 12.4 illustrate the application of the differential material balance to describe diffusional transport in a few additional cases.

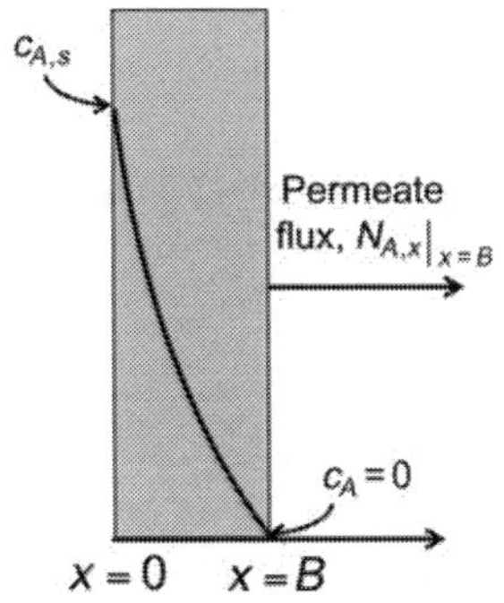

FIGURE 12.3 Concentration profile during transient diffusion of a component A across a slab of thickness B for Example 12.3. The slab is initially at $c_A = 0$. The surface at $x = 0$ is suddenly brought to $c_{A,s}$ while keeping the surface at $x = B$ at $c_A = 0$.

Example 12.4. Diffusion in a slab of thickness *B* with its two faces held at different concentrations. In this example, we consider diffusion of a component A in a slab of thickness B initially free of A when the surface at $x = 0$ is suddenly brought to a concentration $c_{A,s}$ while keeping the surface at $x = B$ at $c_A = 0$. This situation is shown schematically in Fig. 12.3. We obtain a different result from the solution of Eq. (12.42) because, in this case, the system will approach a steady state with a finite flux across the slab rather than zero flux and equilibrium. When a steady state is eventually reached, the concentration profile across the slab will be linear going from $c_A = c_{A,s}$ at $x = 0$ to $c_A = 0$ at $x = B$. Such a situation is encountered, for example, during the start-up of a membrane-based separation process. In this case, after exposing one side of the membrane to a component A, it will take some time before we will see A emerge on the other side and even longer before a steady-state flux is attained. Systems of this type are also used to measure, simultaneously, both the solubility and the diffusivity of a solute in homogeneous matrices, such as membranes. The equations and boundary conditions describing this problem are as follows:

$$\frac{\partial c_A}{\partial t} = \mathcal{D}^0_{A,B}\frac{\partial^2 c_A}{\partial x^2}$$

$$t = 0: \quad c_A = 0 \quad \text{for any } x \qquad (12.49)$$

$$x = 0: \quad c_A = c_{A,s} \quad \text{for any } t > 0$$

$$x = B: \; c_A = 0 \quad \text{for any } t$$

where $\mathcal{D}^0_{A,B}$ is the diffusivity of component A in the slab. In dimensionless form, Eq. (12.49) and associated initial and boundary conditions become:

$$\frac{\partial Y}{\partial \tau} = \frac{\partial^2 Y}{\partial \xi^2}$$

$$\tau = 0: \; Y = 1 \quad \text{for any } \xi \qquad (12.50)$$

$$\xi = 0: \; Y = 0 \;\text{ for any } \tau$$

$$\xi = 1: \quad Y = 1 \;\text{ for any } \tau$$

where $Y = (c_A - c_{A,s})/(0 - c_{A,s})$ is the dimensionless concentration in the slab, $\tau = \mathcal{D}^0_{A,B}t/B^2$ is the dimensionless time, and $\xi = x/B$ is the dimensionless distance. The solution can be obtained as a convergent series by separation of variables and is given by the following expression, analogous to Eq. (5.60) for the equivalent heat conduction problem:

(M.E.12.51)
$$Y = \frac{c_{A,s} - c_A}{c_{A,s}} = \xi + \frac{2}{\pi}\sum_{n=1}^{\infty}\frac{\sin(n\pi\xi)}{n}\exp(-n^2\pi^2\tau) \qquad (12.51)$$

which gives the concentration profile in the slab. Since $c_A = c_{A,s}(1-Y)$, we can find the flux of component A at $x = B$ based on Fick's law, which gives:

$$N_{A,x}\Big|_{x=B} = -\mathcal{D}^0_{A,B}\left.\frac{\partial c_A}{\partial x}\right|_{x=B} = \frac{\mathcal{D}^0_{A,B}c_{A,s}}{B}\left.\frac{\partial Y}{\partial \xi}\right|_{\xi=1} \tag{12.52}$$

The derivative $(\partial Y/\partial \xi)_{\xi=1}$ in Eq. (12.52) is obtained from Eq. (12.51) and is as follows:

$$\left.\frac{\partial Y}{\partial \xi}\right|_{\xi=1} = \frac{\partial}{\partial \xi}\left[\xi + \frac{2}{\pi}\sum_{n=1}^{\infty}\frac{\sin(n\pi\xi)}{n}\exp(-n^2\pi^2\tau)\right]\Bigg|_{\xi=1} = 1 + 2\sum_{n=1}^{\infty}\cos(n\pi)\exp(-n^2\pi^2\tau) \tag{12.53}$$

The final expression for the flux is thus:

$$N_{A,x}\Big|_{x=B} = \frac{\mathcal{D}^0_{A,B}c_{A,s}}{B}\left[1 + 2\sum_{n=1}^{\infty}(-1)^n\exp(-n^2\pi^2\tau)\right] \tag{12.54}$$

Note that $N_{A,x}\big|_{x=B}$ represents the number of moles of component A that emerge or permeate through the slab per unit time and unit cross-sectional area at $x = B$. Using this expression for the flux, we can now calculate the cumulative number of moles of component A that permeate through a slab with area S during a period of time t. Integrating the flux from $t = 0$ to time t we obtain the following:

$$M_A = S\int_0^t N_{A,x}\Big|_{x=B}dt = \frac{S\mathcal{D}^0_{A,B}c_{A,s}}{B}\left[\int_0^t dt + 2\sum_{n=1}^{\infty}(-1)^n\frac{B^2}{\mathcal{D}^0_{A,B}}\int_0^{\tau}\exp(-n^2\pi^2\tau)d\tau\right] \tag{12.55}$$

The final result is obtained by carrying out the integrals and is as follows:

$$\begin{aligned} M_A &= \frac{S\mathcal{D}^0_{A,B}c_{A,s}}{B}\left\{t + \frac{2B^2}{\mathcal{D}^0_{A,B}}\sum_{n=1}^{\infty}\frac{(-1)^n}{\pi^2n^2}[1-\exp(-n^2\pi^2\tau)]\right\} \\ &= \frac{S\mathcal{D}^0_{A,B}c_{A,s}}{B}t + 2SBc_{A,s}\sum_{n=1}^{\infty}\frac{(-1)^n}{\pi^2n^2}[1-\exp(-n^2\pi^2\tau)] \end{aligned} \tag{12.56}$$

We can also write this result in dimensionless form as follows:

(M.E.12.51)

$$\frac{M_A}{SBc_{A,s}} = \tau + 2\sum_{n=1}^{\infty}\frac{(-1)^n}{\pi^2n^2}\left[1-\exp\left(-n^2\pi^2\tau\right)\right] \tag{12.57}$$

Figure 12.4 shows the dimensionless concentration profiles in the slab, $1-Y$, vs. distance, ξ, at different values of the dimensionless time, τ, calculated from Eq. (12.51), along with the dimensionless cumulative number of moles that permeate across the slab, $M_A/SBc_{A,s}$, vs. τ, calculated from Eq. (12.57). We can see that for values of τ larger than 0.5, the solution reaches steady-state conditions with a linear concentration profile. For these conditions, the instantaneous flux becomes simply:

$$N_{A,x}\Big|_{x=B} = \frac{\mathcal{D}^0_{A,B}c_{A,s}}{B} \tag{12.58}$$

which is exactly the result that we expect for steady-state diffusion. The cumulative number of moles permeated also becomes linear as a function of τ for $\tau > 0.5$, approaching an asymptote with slope equal to 1 and an intercept on the x-axis equal to $1/6$. In dimensional form, this asymptote is given by the following equation:

$$M_A = \frac{S\mathcal{D}^0_{A,B}c_{A,s}}{B}t - \frac{SBc_{A,s}}{6} \tag{12.59}$$

Let's do a sample calculation by considering the following permeation data presented by Eusebio et al. [Membranes 10 (2020) 8] for CO_2 diffusing in a 0.0107-cm-thick polyurethane membrane

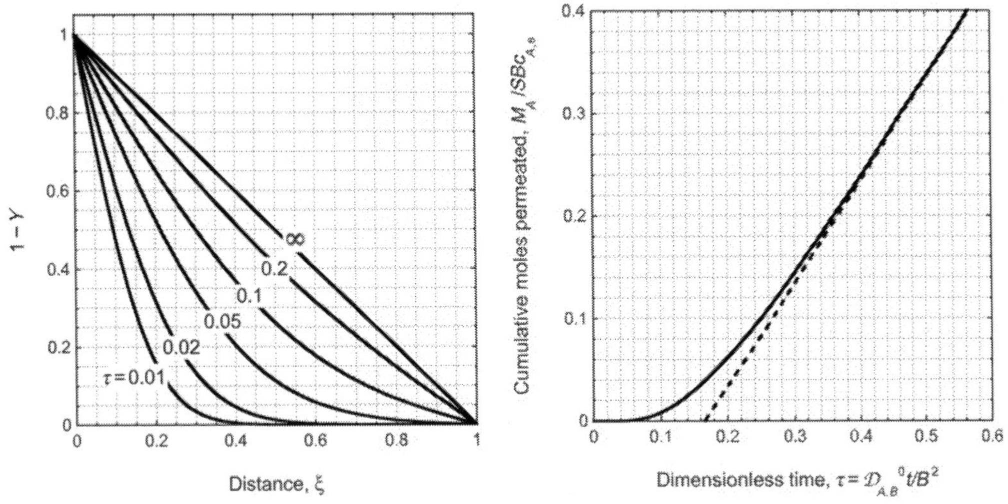

FIGURE 12.4 Concentration profiles at different times (left) and normalized cumulative permeation (right) for a slab with thickness B and cross section S for Example 12.3. $c_{A,s}$ is the concentration of the permeant A at the slab surface. $\mathcal{D}^0_{A,B}$ is the diffusion coefficient of A in the slab. Mathematica notebook: (M.F.12.4)

(PU-10) in a test cell at 310 K shown schematically in Fig. 12.5. The test cell used consisted of a high-pressure compartment and a low-pressure compartment that were separated by the membrane initially free of any solute. The membrane area was $S = 9.62\ \text{cm}^2$, and the volume of the low-pressure compartment was $V = 12.6\ \text{cm}^3$. In their experiment, the high-pressure compartment was rapidly filled with CO_2 gas at an absolute pressure of 269 cmHg at time zero and kept at this pressure. As the CO_2 began to permeate through the membrane, the pressure in the low-pressure compartment rose gradually. After an initial lag, the pressure in this compartment was found experimentally to be described by the following equation:

$$P(t) = 0.033 \times t - 0.446$$

where $P(t)$ is the low-pressure compartment pressure in cmHg, and t is the time elapsed from the beginning of the experiment in s. Note that this pressure remained very low compared to that in the

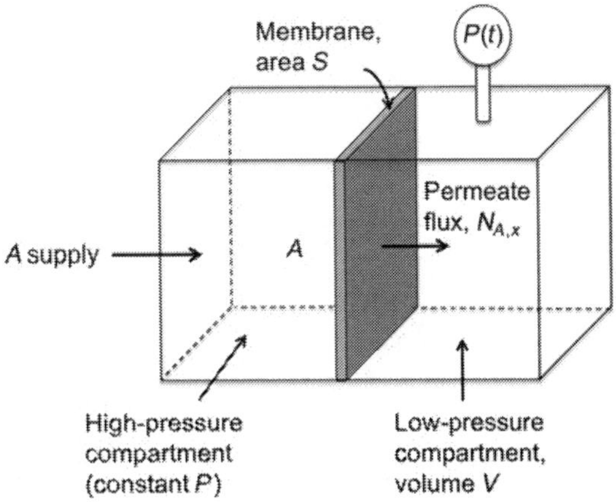

FIGURE 12.5 Schematic of a diffusion cell for gas-permeation studies.

high-pressure side of the test cell for the duration of the experiment, never exceeding 1.4 cmHg. We can determine both the solubility and the diffusivity of CO_2 in the membrane from these data by comparing this experimental result with Eq. (12.59). To do this, we first convert the recorded pressure to moles of gas permeated using the ideal gas law, which gives:

$$M_A = \frac{PV}{\mathcal{R}T} = \frac{[P(t)/76]\times 12.6}{82\times 310} = 6.52\times 10^{-6}P(t)$$

From the intercept of the experimental curve on the time axis, we have:

$$\frac{SBc_{A,s}}{6} = 0.446\times 6.52\times 10^{-6} = 2.91\times 10^{-6}$$

Thus, $c_{A,s} = 6\times 2.91\times 10^{-6}/0.0107\times 9.62 = 1.70\times 10^{-4}\ \text{mol/cm}^3$, which is the solubility of CO_2 in the membrane at 269 cmHg absolute pressure. From the slope of the experimental pressure curve, we obtain:

$$\frac{S\mathcal{D}^0_{A,B}c_{A,s}}{B} = 0.033\times 6.52\times 10^{-6} = 2.15\times 10^{-7}$$

from which we calculate $\mathcal{D}^0_{A,B} = 2.15\times 10^{-7}\times 0.0107/9.62\times 1.70\times 10^{-4} = 1.41\times 10^{-6}\ \text{cm}^2/\text{s}$. Gas solubilities are often reported in terms of cm^3 at STP (i.e., at standard temperature, $T = 0°C$, and standard pressure, $P = 1$ atm,) rather than in mol. In this case, the solubility at 269 cmHg absolute pressure becomes equal to $4.8\times 10^{-5}\times \mathcal{R}T/P = 1.07\ \text{cm}^3\ CO_2\ \text{STP/cm}^3$. We should recognize that if only a steady-state measurement of the flux had been made, only the product of diffusivity and solubility could have been determined according to Eq. (12.58). It is only because of the transient measurement of the time lag that it has been possible to determine the individual values of solubility and diffusivity. This occurs because, while the product $\mathcal{D}^0_{A,B}c_{A,s}$ will determine the steady-state result, separate values of the capacity of the membrane to hold the solute, which is measured by $c_{A,s}$, along with the diffusivity, $\mathcal{D}^0_{A,B}$, will determine the transient behavior.

Example 12.5. Diffusion in spheres suspended in a finite volume of fluid. In Sec. 5.3.2.e, we considered the case of heat transfer to or from spheres immersed in a finite volume of a fluid at a given initial temperature different from the initial temperature of the spheres. We now consider the case of spheres that are immersed in a finite volume of a well-mixed fluid containing an initial concentration of a component A as illustrated in Fig. 12.6. This component will diffuse into the spheres and, at each time, the amount

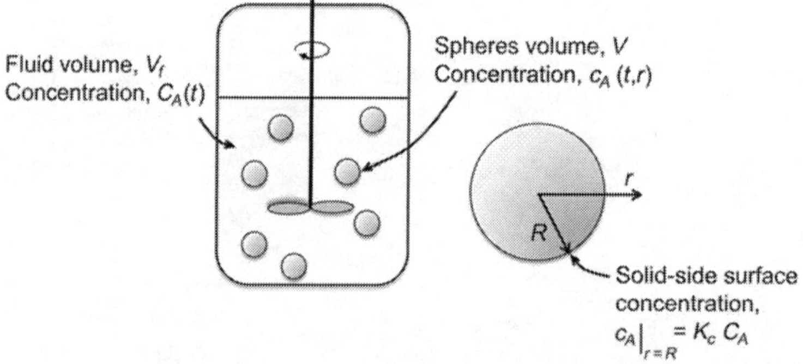

FIGURE 12.6 Diffusion of a component A in spheres at an initial concentration $c_{A,0}$ immersed in a finite volume of fluid containing an initial concentration of A equal to $C_{A,0}$.

of A accumulated in the spheres will be equal to the amount of A removed from the fluid. The reverse process can also occur, where A initially in the spheres diffuses out and accumulates in the fluid. In order to describe these processes, we will need to track two concentrations—the concentration in the spheres, c_A, which will be a function of time and radial position within the spheres, and the concentration in the external fluid, C_A, which will be a function of time. The approach to solve this problem is the same as that used for the analogous heat transfer problem. Let's denote the volume of spheres as V, the initial concentration of A in the spheres as $c_{A,0}$, the volume of fluid as V_f, and the initial concentration of A in the fluid as $C_{A,0}$. Two different material balances are needed in order to find the two concentrations. For the spheres, we write a transient differential material balance while for the fluid we write a transient macroscopic balance. The corresponding equations and boundary conditions are as follows:

For the spheres:

$$\frac{\partial c_A}{\partial t} = \frac{\mathcal{D}^0_{A,B}}{r^2}\frac{\partial}{\partial r}\left(r^2\frac{\partial c_A}{\partial r}\right)$$

$$t = 0: \quad c_A = c_{A,0} \quad \text{for any } r$$

$$r = 0: \quad \frac{\partial c_A}{\partial r} = 0 \ \text{ for any } t \tag{12.60}$$

$$r = R: \ c_A = K_c C_A(t) \ \text{ for any } t > 0$$

For the fluid:

$$V_f \frac{dC_A}{dt} = 4\pi R^2 \times N_{\text{spheres}} \times \left(-\mathcal{D}^0_{A,B}\frac{\partial c_A}{\partial r}\bigg|_{r=R}\right) \tag{12.61}$$

$$t = 0, \quad C_A = C_{A,0}$$

where $\mathcal{D}^0_{A,B}$ is the diffusivity of component A in the spheres and N_{spheres} is the number of spheres. We can calculate the latter from the total volume of spheres, V, since $N_{\text{spheres}} = V/(4\pi R^3/3)$. Equation (12.60) is easily derived from Table 12.1 and is analogous to Eq. (5.80) for the corresponding heat transfer problem. Equation (12.61) is analogous to Eq. (5.81) and is derived from a transient macroscopic material balance that relates the rate of change of the concentration of A in the fluid to the rate of diffusion of A to or from the spheres. The quantity $-\mathcal{D}^0_{A,B}\,\partial c_A/\partial r\big|_{r=R}$ on the right-hand-side of Eq. (12.61) is the mass transfer flux at the sphere surface. When multiplied by the surface area of the spheres, this quantity provides the rate of mass transfer. Equations (12.60) and (12.61) are coupled to each other by the boundary condition at the sphere surface which relates the concentration on the solid-side of the interface to the fluid-phase concentration. Note that the only substantial difference between the mathematical description of this problem and that of the analogous heat transfer problem lies in what we assume at the sphere–fluid interface. For the heat transfer problem, in the absence of an external resistance, we assumed that the temperature at the sphere surface was the same as the fluid temperature. For the present mass transfer problem, we also assume equilibrium. However, in this case, equilibrium will be represented by a phase-equilibrium relationship rather than by an equality of concentrations. This relationship is given in terms of the distribution coefficient K_c and is $c_{A,s} = K_c C_A$. An analytical solution of this system of two equations with associated initial and boundary conditions (12.60) and (12.61) can be found if K_c is a constant. In dimensionless form, the solution is the same as Eq. (5.82) and is:

(M.5.82)

$$Y_f = \frac{\Lambda}{1+\Lambda} + 6\Lambda\sum_{n=1}^{\infty}\frac{\exp(-b_n^2\tau)}{\Lambda^2 b_n^2 + 9(1+\Lambda)} \tag{12.62}$$

where the dimensionless variables now take on the following form:

$$Y_f = \frac{K_c C_A - c_{A,0}}{K_c C_{A,0} - c_{A,0}}$$

$$\Lambda = \frac{V_f}{VK_c} \tag{12.63}$$

$$\tau = \frac{\mathcal{D}^0_{A,B} t}{R^2}$$

As in the case of Eq. (5.82), the b_n terms in Eq. (12.62) are obtained as the roots of the following equation:

$$\tan(b_n) = \frac{3(b_n)}{3 + \Lambda(b_n)^2} \tag{12.64}$$

Figure 5.11 provides a graph based on Eq. (5.82) for the analogous heat conduction problem using a range of values of the parameter Λ. The same graph can be used for the mass transfer problem if Y_f, Λ, and τ are defined according to Eq. (12.63). Note that, in this case, Y_f will give the dimensionless concentration in the fluid. The average concentration in the spheres at a particular time can be found from this value based on following overall material balance:

$$V(c_{A,\text{avg}} - c_{A,0}) = V_f(C_{A,0} - C_A) \tag{12.65}$$

or

$$c_{A,\text{avg}} = c_{A,0} + \frac{V_f}{V}(C_{A,0} - C_A) \tag{12.66}$$

Large values of the parameter Λ will correspond to cases where the fluid volume is large compared to the volume of spheres or, equivalently, to cases where K_c is small. In these cases, the fluid concentration, C_A, would remain very close to its initial value, $C_{A,0}$. As a result, the concentration in the spheres could also be obtained directly from Table 5.2 with $c_{A,s} = K_c C_{A,0}$.

12.3.3 Processes Where Both the Internal and External Resistances Are Important

Mass transfer processes where both internal and external resistances are important are characterized by intermediate values of the Biot number. In general, as in the case of heat transfer, the external mass transfer resistance will be dominant when Bi < 0.1, while the internal resistance will be dominant when Bi > 10. At values of Bi between 0.1 and 10, both resistances will be important. In these cases, while the differential material balances used to describe diffusive mass transfer are the same as those used in Sec. 12.3.2, the boundary conditions at the surface of the object will be different.

Before providing a quantitative treatment of this case, let's first focus on understanding qualitatively what happens at the fluid–solid interface when both resistances are important. At the fluid–solid interface, we will need to consider two different concentrations: $c_{A,s}$, on the solid side of the interface, and $C_{A,s}$, on the fluid side. These two concentrations will be related to each other by the relevant phase-equilibrium relationship and will vary in time even if the concentration in the bulk fluid, $C_{A,b}$, is kept constant. Imagine, for example, dropping an adsorbent particle into a fluid containing a concentration $C_{A,b}$ of an adsorbable component A. The fluid-phase concentration right at the particle surface will suddenly drop to zero as A is adsorbed and diffuses into the particles. Convective mass transfer will replace the A that diffuses into the particles. However, since the diffusion rate will initially be very high, the fluid-phase concentration at the particle surface will be very low early on in this process providing the fluid-phase concentration driving force needed to sustain convective transport at a rate equal to the rate of diffusion inside the particle. As time goes on, more and more A will have diffused into the particles, thereby reducing the intraparticle concentration gradient and, thus, reducing the rate of diffusion within the particles. During this process, $C_{A,s}$ will gradually increase since convective transport will require a smaller and smaller driving force to balance the lower rate of diffusion within the solid. Eventually, the surface concentration $C_{A,s}$ will come to equilibrium with the bulk fluid concentration $C_{A,b}$. Figure 12.7 illustrates, schematically, the concentration profiles expected during such a process if the particle is represented as a slab of thickness $2B$. The initial concentration of component A in the slab is $c_{A,0}$ while the bulk fluid concentration is $C_{A,b}$. Generally, these concentration profiles are analogous to the temperature profiles sketched in Fig. 5.12 for the corresponding heat transfer process. The difference lies again in the interfacial condition, which is equality of temperatures for the heat transfer case and phase equilibrium for the mass transfer case. The differential material balance describing diffusion in the slab is derived again from Table 12.1 and is as follows along with the associated initial and boundary conditions:

$$\frac{\partial c_A}{\partial t} = \mathcal{D}^0_{A,B} \frac{\partial^2 c_A}{\partial x^2}$$

$$t=0: \quad c_A = c_{A,0} \quad \text{for any } x$$

$$x=0: \quad \frac{\partial c_A}{\partial x} = 0 \quad \text{for any } t \tag{12.67}$$

$$x=B: \quad -\mathcal{D}^0_{A,B} \left.\frac{\partial c_A}{\partial x}\right|_{x=B} = k_c(C_{A,s} - C_{A,b}) \quad \text{for any } t>0$$

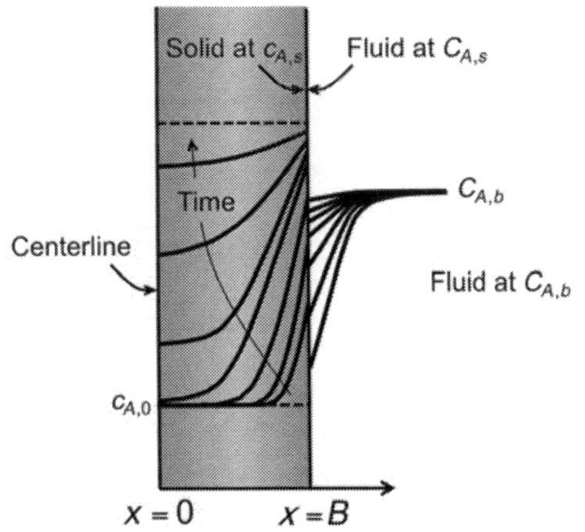

FIGURE 12.7 Evolution of concentration profiles for mass transfer of a component A in a semi-infinite slab with thickness $2B$ with initial concentration $c_{A,0}$ in contact on both sides with a fluid at a constant concentration $C_{A,b}$ when both internal and external mass transfer resistances are important. Only one-half of the slab is shown. $x = 0$ represents the centerline of the slab. Note that the interfacial concentrations $c_{A,s}$ on the solid-side and $C_{A,s}$ on the fluid-side are, in general, different from each other.

where $\mathcal{D}^0_{A,B}$ is the diffusivity of component A in the solid and k_c is the convective mass transfer coefficient. The differential balance and the first two of these conditions are the same as those that we wrote in the absence of an external mass transfer resistance, Eq. (12.38). However, the last of these conditions is different. This condition now imposes that, at the interface, the rate of diffusion into the solid is equal to the rate of convection from the bulk fluid to the solid surface. Since the fluid-side concentration at the interface, $C_{A,s}$, will be in equilibrium with the solid-side concentration $c_{A,s} = c_A\big|_{x=B}$, this condition can also be written as follows:

$$-\mathcal{D}^0_{A,B}\left.\frac{\partial c_A}{\partial x}\right|_{x=B} = k_c\left(\frac{c_A\big|_{x=B}}{K_c} - C_{A,b}\right) = \frac{k_c}{K_c}(c_A\,|_{x=B} - K_c C_{A,b}) \tag{12.68}$$

Let's now introduce the following dimensionless variables:

$$Y = \frac{c_A - K_c C_{A,b}}{c_{A,0} - K_c C_{A,b}} \tag{12.69}$$

$$\tau = \frac{\mathcal{D}^0_{A,B} t}{B^2} \tag{12.70}$$

$$\xi = \frac{x}{B} \tag{12.71}$$

Note that while the definitions of τ and ξ are the same as those used for the case without external resistance, as seen in Eqs. (12.40) and (12.41), the definition of the dimensionless concentration Y is now based on $K_c C_{A,b}$ rather than $c_{A,s}$. Since $c_A = (c_{A,0} - K_c C_{A,b})Y + K_c C_{A,b}$, substituting these variables in Eqs. (12.67) and (12.68) with the assumption that K_c and $C_{A,b}$ are constant, we obtain:

$$\begin{aligned}
&(c_{A,0} - K_c C_{A,b})\frac{\mathcal{D}^0_{A,B}}{B^2}\frac{\partial Y}{\partial \tau} = \mathcal{D}^0_{A,B}(c_{A,0} - K_c C_{A,b})\frac{1}{B^2}\frac{\partial^2 Y}{\partial \xi^2} \\
&\tau = 0: \quad Y = 1 \quad \text{for any } \xi \\
&\xi = 0: \quad (c_{A,0} - K_c C_{A,b})\frac{1}{B}\frac{\partial Y}{\partial \xi} = 0 \quad \text{for any } \tau \\
&\xi = 1: \quad -\mathcal{D}^0_{A,B}(c_{A,0} - K_c C_{A,b})\frac{1}{B}\left.\frac{\partial Y}{\partial \xi}\right|_{\xi=1} \\
&\qquad = \frac{k_c}{K_c}[(c_{A,0} - K_c C_{A,b})Y\,|_{\xi=1} + K_c C_{A,b} - K_c C_{A,b}] \quad \text{for any } \tau > 0
\end{aligned} \tag{12.72}$$

or:

$$\frac{\partial Y}{\partial \tau}=\frac{\partial^2 Y}{\partial \xi^2}$$

$$\begin{aligned} &\tau=0: \quad Y=1 \quad \text{for any } \xi \\ &\xi=0: \quad \frac{\partial Y}{\partial \xi}=0 \quad \text{for any } \tau \\ &\xi=1: \quad \left.\frac{\partial Y}{\partial \xi}\right|_{\xi=1}=-\text{Bi}\times \left.Y\right|_{\xi=1} \quad \text{for any } \tau>0 \end{aligned} \tag{12.73}$$

where:

$$\text{Bi}=\frac{k_c B}{\mathcal{D}^0_{A,B} K_c} \tag{12.74}$$

is the Biot number. In this dimensionless form, Eq. (12.73) and associated initial and boundary conditions are now exactly the same as those describing heat conduction in a semi-infinite slab with external resistance given by Eqs. (5.95) to (5.98). As a result, the solution of Eq. (12.73) is also the same as the solution of the analogous heat transfer problem. This is also the case for mass transfer in a long cylinder of radius R or in a sphere of radius R provided that we define τ, ξ, and Bi as follows:

$$\tau=\frac{\mathcal{D}^0_{A,B} t}{R^2} \tag{12.75}$$

$$\xi=\frac{r}{R} \tag{12.76}$$

$$\text{Bi}=\frac{k_c R}{\mathcal{D}^0_{A,B} K_c} \tag{12.77}$$

For all three geometries, in this dimensionless form, $Y=(c_A-K_c C_{A,b})/(c_{A,0}-K_c C_{A,b})$ is given by same functions of τ, ξ, and Bi that we found for the analogous heat transfer problems. These functions are given in Table 5.3 in terms of series expressions, and in Figs. 5.13 and 5.14, graphically, in the Gurney-Lurie format. When using these expressions for mass transfer, we must remember that they are valid only for cases where the distribution coefficient K_c is a constant. Cases where K_c is not constant and, thus, where the equilibrium relationship between $c_{A,s}$ and $C_{A,s}$ is nonlinear, will generally require a numerical solution of the relevant differential balances at intermediate values of Bi. Example 12.5 illustrates the application of these relationships for cases where the relative importance of internal and external resistances varies.

Example 12.6. Transient mass transfer for the extraction of caffeine from coffee beans. Let's use the equations developed in this chapter to predict the time required to extract 80% of the caffeine (A) from coffee beans using a solvent for which $K_c=c_{A,s}/C_{A,s}=0.2$. Note that K_c is defined as the equilibrium concentration of caffeine in the beans divided by the concentration of caffeine in the solvent. For simplicity, let's assume that the coffee beans are spheres with a radius $R=0.3$ cm. Let's also assume that the diffusivity of caffeine in the coffee beans is $\mathcal{D}^0_{A,B} = 1.8 \times 10^{-6}$ cm²/s, that the external mass transfer coefficient is $k_c = 2.4 \times 10^{-5}$ cm/s, and that the initial concentration in the solvent, $C_{A,b}$, is zero. We will solve this problem for three different scenarios assuming: (a) no internal resistance with an infinite volume of solvent; (b) no external resistance with an infinite volume of solvent; and (c) no external resistance but using 2 L of solvent per liter of coffee beans.

(a) Solution for no internal resistance with an infinite solvent volume. No internal resistance means that the concentration inside the beans will be spatially uniform and that external mass transfer will be controlling. The relevant equations were developed in Sec. 12.2.1 and give the result:

$$t = -\frac{VK_c}{Sk_c}\ln\left(\frac{c_A - K_c C_{A,b}}{c_{A,0} - K_c C_{A,b}}\right) \tag{12.78}$$

For a sphere, $V = 4\pi R^3/3$ and $S = 4\pi R^2$. Thus, $V/S = R/3$ and

$$t = -\frac{RK_c}{3k_c}\ln\left(\frac{c_A - K_c C_{A,b}}{c_{A,0} - K_c C_{A,b}}\right) = -\frac{0.3 \times 0.2}{3 \times 2.4 \times 10^{-5}}\ln\left(\frac{0.2-0}{1-0}\right) = 1{,}340 \text{ s}$$

(b) Solution for no external resistance with an infinite solvent volume. No external resistance means that the bean surface will be in equilibrium with the caffeine concentration in the bulk solvent. In this case, since $C_{A,b} = 0$, we will also have $c_{A,s} = 0$. The solution can be found in Table 5.2 (item C) or, graphically, in Fig. 5.8 or in Fig. 5.14C for Bi = ∞. For this problem, we want to find the time at which $c_{A,\text{avg}} = 0.2c_{A,0}$. In dimensionless form, this means finding the value of τ that gives $Y_{\text{avg}} = (c_{A,\text{avg}} - c_{A,s})/(c_{A,0} - c_{A,s}) = c_{A,\text{avg}}/c_{A,0} = 0.2$. From the series solution or, more quickly, from Fig. 5.8, we find $\tau = 0.11$. Thus, the time needed will be:

$$t = \frac{R^2}{\mathcal{D}^0_{A,B}}\tau = \frac{0.3^2}{1.8 \times 10^{-6}}0.11 = 5{,}500 \text{ s}$$

Let's now verify which of these two scenarios is likely correct by calculating the Biot number. For this problem we have:

$$\text{Bi} = \frac{k_c R}{\mathcal{D}^0_{A,B}K_c} = \frac{2.4 \times 10^{-5} \times 0.3}{1.8 \times 10^{-6} \times 0.2} = 20$$

This value is sufficiently large that the external resistance can indeed be considered negligible. The reason why the time calculated in (a) is much smaller than the time calculated in (b) is that, in reality, the internal resistance is much greater than the external resistance. An interesting question is what would happen if, hypothetically, we switched to a different solvent for which $K_c = 2.0$ (this would be a much poorer solvent for caffeine). In this case, the Biot number would become:

$$\text{Bi} = \frac{k_c R}{\mathcal{D}^0_{A,B}K_c} = \frac{2.4 \times 10^{-5} \times 0.3}{1.8 \times 10^{-6} \times 2.0} = 2$$

which is now pretty small suggesting that both internal and external resistances will be important. In this case, from the series solution in Table 5.3 or from Fig. 5.14C, for the same $Y_{\text{avg}} = 0.2$ and Bi = 2, we would get $\tau = 0.35$, which would give $t = 17{,}500$ s. This much longer time is due to the increased contribution of the external resistance that would come about because of the larger K_c. The larger K_c would make partitioning into the fluid less favorable. In turn, this would lower $C_{A,s}$ thereby reducing the driving force for convective transport from the surface of the coffee beans to the bulk solvent.

(c) Solution for no external resistance with a finite volume of solvent. If the volume of solvent used is finite, the caffeine extracted from the coffee beans will cause $C_{A,b}$ to increase significantly above zero. If the external resistance is negligible (i.e., if Bi is large, as we have already ascertained), we can use Eq. (12.62) to predict the time needed to extract 80% of the caffeine. To do this, we first calculate:

$$\Lambda = \frac{V_f}{VK_c} = \frac{2}{1 \times 0.2} = 10$$

which is not so large that we can neglect the change in $C_{A,b}$ (see Fig. 5.11). Next, from the overall material balance, Eq. (12.65), we calculate what the value of $C_{A,b}$ will be when we have extracted 80% of the caffeine. Rearranging Eq. (12.65) we obtain:

$$C_{A,b} = C_{A,0} + \frac{V}{V_f}(c_{A,0} - c_{A,\text{avg}}) = \frac{Vc_{A,0}}{V_f}\left(1 - \frac{c_{A,\text{avg}}}{c_{A,0}}\right) = \frac{1}{2}(1 - 0.2)c_{A,0} = 0.4c_{A,0}$$

Thus, the value of Y_f will be as follows:

$$Y_f = \frac{0.2 \times 0.4c_{A,0} - c_{A,0}}{0 - c_{A,0}} = 0.92$$

Finally, from Fig. 5.11, we want to find the value of τ at which

$$(1 + \Lambda)(1 - Y_f) = (1 + 10)(1 - 0.92) = 0.88$$

which yields $\tau = 0.18$ and $t = R^2\tau/\mathcal{D}^0_{A,B} = 0.3^2 \times 0.18/1.8 \times 10^{-6} = 9{,}000$ s. This time is longer than that calculated for case (b) since the finite accumulation of caffeine in the solvent will increase $C_{A,b}$ and, thus, $c_{A,s}$, thereby reducing the driving force for diffusion out of the coffee beans.

12.4 Problems

P.12.1. The device shown in Fig. P.12.1 is used to measure the diffusion coefficient of oxygen (O_2) in water and various solutions at 10°C. The O_2 sensor is used to measure the O_2 mole fraction at the bottom of the pool as a function of time. In one experiment, water, initially pure, rests in a shallow pool with a depth of 0.3 cm. The device is initially purged with pure nitrogen. At time zero, the pure nitrogen purge stream is replaced by nitrogen containing 15.0 mole% O_2. The absolute pressure in the device is 1.0 atm. At time $t = 30$ min, the O_2 sensor at the bottom of the water pool indicates an O_2 mole fraction of 1.9×10^{-6}. (a) Write down the equation and boundary conditions needed to describe mass transfer of O_2 to the bottom of the water pool if the process is controlled by diffusion in the liquid. (b) Using the data above, determine the diffusion coefficient of O_2 in water for these conditions. (c) Determine the average mole fraction of O_2 in the water at $t = 30$ min. (d) Compare the value determined in part (b) with the diffusion coefficient of O_2 estimated from the Wilke-Chang equation.

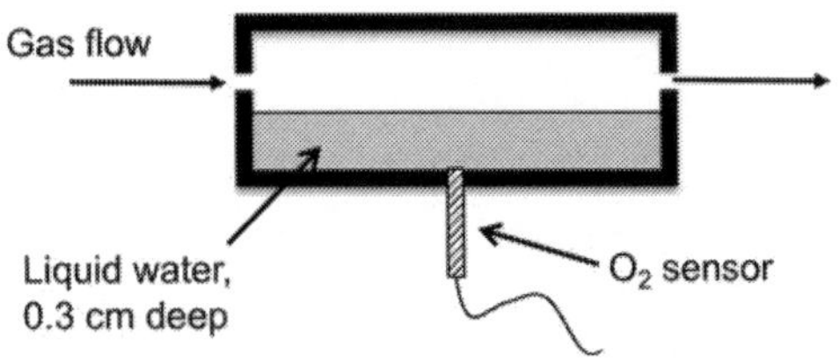

FIGURE P.12.1

P.12.2. Lasagna pasta (wide flat noodles 0.3 cm in thickness) is produced industrially by continuously extruding well-mixed dough to form a 10 cm wide ribbon. The ribbon emerges from the extruder with a water content of 0.05 g/cm^3 and passes through a tunnel drier at a velocity of 0.5 cm/s. Dry air at 1 atm absolute pressure and 300 K flows in the drier at a high flow rate to reduce the average moisture content of the pasta to 20% of the initial value. After leaving the drier, the lasagna ribbon is cut to length for packaging. The distribution coefficient of water over the pasta at equilibrium is measured and found to be $K_c = 4{,}000$ defined as the concentration of water in the pasta divided by the concentration of water vapor in air. The diffusivity of water in the pasta is estimated to be 2.2×10^{-5} cm^2/s. The average mass transfer coefficient at the past-air interface k_c is 1.2 cm/s. (a) Sketch the lasagna as it flows into the drier and write the differential material balance equation and initial and boundary conditions needed to determine the required length of the drier assuming that both internal and external mass transfer resistances are important and if only the external resistance is considered. (b) Calculate the length of the tunnel drier required if external mass transfer is assumed to be controlling. (c) Calculate the length of the tunnel drier required if internal mass transfer is assumed to be controlling. (d) Calculate the length of the tunnel drier required if both internal and external resistances are considered. (e) Calculate the concentration of water in the centerline of the lasagna and at its surface when it emerges from the drier, considering both internal and external mass transfer resistances.

P.12.3. A large amount of a heavy oil has been accidentally spilled on a large, elevated parking lot and settled, coating all surfaces, including both flat areas and sloping access ramps. Following the spill, a brief, but intense, storm, lasting 20 min hit the area and we are concerned with contamination of the rain water. The temperature is constant at 10°C and the solubility of the oil in water at this temperature is 1.5×10^{-5} mol/cm^3. The molar volume of the oil at the normal boiling point is estimated to be 200 cm^3/mol. (a) Estimate the diffusion coefficient of the oil in water for these conditions. (b) In a flat area of the parking lot, the rain water quickly pooled on top of the oil forming a 1.0 cm thick stagnant layer 5 m in diameter. Determine the concentration of the oil (in mol/cm^3) at the top of the water layer and the total amount of oil (in mol) contaminating the pooled water in this area after 48 h. (c) In a covered sloping access ramp, 3 m wide and 20 m long, the rain water did not pool over the oil but ran over it as a 0.5-cm thick layer at a velocity of 2 m/s. The water flow is turbulent for these conditions and the mass transfer coefficient at the oil–water interface is 5.0×10^{-3} cm/s. In this case, there is no significant diffusive resistance in the flowing water. Estimate the average concentration of the oil in the water running off the ramp, and the total amount of oil contaminating the water run-off from this ramp.

P.12.4. Long, cylindrical homogeneous nylon fibers with a radius of 0.05 cm contain a 1.2×10^{-5} mol/cm^3 concentration of a solvent A. It is proposed to remove this solvent from the fibers by placing them in a rapidly flowing air stream at 343 K and at an absolute pressure $P = 1$ atm. The diffusion coefficient of the solvent A in the fibers at 343 K is 1.9×10^{-6} cm^2/s. The equilibrium of the solvent A between air and the fibers is described by the equation $y_A P = H'_A$ where y_A is the gas phase mole fraction of A, c_A is the concentration of A dissolved in the polymer, and $H' = 9.0$ atm cm^3/mol. The mass transfer coefficient for transport from the fibers to the air is $k_c = 0.6$ cm/s. Determine: (a) The time required to reduce the centerline concentration of A in the fibers to 10% of the initial value if the external mass transfer resistance is neglected. (b) The time required to reduce the centerline concentration of A in the fibers to 10% of the initial value if the external mass transfer resistance is considered. (c) The time required to reduce the average concentration of A in the fibers to 50% of the initial value if the external mass transfer resistance is considered.

P.12.5. An ophthalmic drug is dissolved in a polymer and cast in the form of tiny spheres for delivery to the eye with a dropper. Once in the eye, the drug diffuses out providing longer-term delivery. Under physiological conditions (37°C), once out of the spheres and into the lacrimal fluid, the drug is quickly absorbed into the eye. The initial drug content of the spheres is 1.0 wt%. The diffusion coefficient of the drug in water at 37°C is 6.0×10^{-6} cm^2/s and the diffusivity in the polymer is 1.1×10^{-12} cm^2/s. The distribution coefficient describing partitioning between the polymer and the lacrimal fluid is $K_c = 5.0 \times 10^6$. (a) Determine the radius of the spheres needed if it is desired to deliver 50% of the drug over a 7-day period (168 h) and the external mass transfer resistance is neglected. (b) Determine the radius of the spheres needed if it is desired to deliver 50% of the drug over a 7-day period (168 h) and the external mass transfer resistance is considered assuming that the Sherwood number is equal to 2 as suggested by Eq. (11.56).

P.12.6. Macroporous glass spheres with 60% porosity and a 0.3 cm diameter are impregnated with a 2.0 wt% solution of methanol in water. 100 L of these particles are dropped into 250 L of water and kept in suspension by agitation at 20°C. The water is well mixed and external mass transfer can be neglected. Assuming that the tortuosity factor for the particles is 2, determine (a) The amount of methanol remaining in the particles, expressed as a percentage of the initial value and the methanol concentration in the surrounding solution after 15 min, and (b) The final concentration of methanol in the solution when $t = \infty$. Note that the methanol–water mixture is not adsorbed but just fills the pores.

P.12.7. Various chemicals are often added to polymers, such as Polyethylene terephthalate (PET), to improve their properties, including flexibility and UV resistance. However, leakage of these additives into foods placed in containers made from plastics containing such additives can be a concern because of potential toxicity. Li et al. [Food Packaging and Shelf Life, 21 (2019) 100332] have measured the leakage of various additives from PET into various food simulants including 50% ethanol. For the UV stabilizer 2-(2-Hydroxy-5-methylphenyl) benzotriazole (UV-P) at 70°C they report a diffusivity in PET of 1.1×10^{-10} cm^2/s and an equilibrium distribution coefficient $K_c = 1{,}158$. The latter describes the equilibrium ratio of the UV-P concentration in PET and that in 50% ethanol. Consider a cylindrical PET container with a 10 cm diameter and a wall thickness of 0.2 mm filled to a 26 cm height with 50% ethanol. The initial UV-P content of the PET is 0.05 g/cm^3. (a) What is the maximum concentration of UV-P that we can expect in the liquid after a long time? (b) Develop a model to describe the leakage of UV-P into the liquid assuming that the PET can be described as a flat wall and that diffusion in the PET is controlling. Plot the expected concentration of UV-P in the liquid as a function of time assuming that leaching is controlled by diffusion in PET. What will be the UV-P concentration in the liquid and the percentage of the UV-P will have leaked after 20,000 s? (c) What will be the average concentration of UV-P in the liquid after 20,000 s if we assume that leaching is completely controlled by diffusion in the liquid and that the liquid is perfectly still?

Hint: For Part (b), Crank, *The Mathematics of Diffusion*, Clarendon Press, 1975, p. 57, Eq. 4.37, provides the solution of the problem of diffusion in a slab of volume V and half-thickness B placed in a volume V_f of a well-mixed fluid.

P.12.8. Diffusion of a component A from a line source in a linear dimension x in an infinite medium as shown in Fig. P.12.8 is described by the following equation and boundary conditions:

$$\frac{\partial C_A}{\partial t} = \mathcal{D}^0_{A,B} \frac{\partial^2 C_A}{\partial x^2}$$

$$C_A(0, x) = Q_A \delta(x)$$

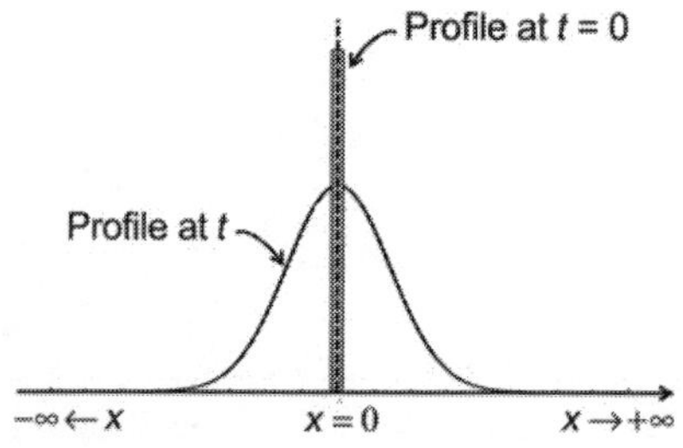

FIGURE P.12.8

$$C_A(t, -\infty) = 0$$

$$C_A(t, +\infty) = 0$$

where Q_A is the amount of A initially at $x = 0$ in mol per cm^2 of surface perpendicular to x and $\delta(x)$ is the Dirac delta function. (a) Obtain an analytical solution of this equation and boundary conditions. Use of the Laplace transform is recommended. (b) Compare quantitatively your solution with the results of the simulation of random walk obtained in Problem P.10.1.

CHAPTER 13
Convective Mass Transfer

13.1 Introduction

As introduced in Chap. 8, we use the term convective mass transfer to describe mass transfer processes that occur at the interface between two phases in relative motion, such as the interface between a fluid and a solid or between two immiscible fluids. This process is analogous to convective heat transfer without a change of phase and is not to be confused with the bulk transport of mass by a flowing fluid. Just like convective heat transfer is driven by the temperature gradient at the interface, convective mass transfer is driven by a concentration gradient in the fluid at the fluid-solid interface. When a fluid flows over a solid surface, a velocity gradient exists adjacent to the surface with the velocity varying from zero at the solid surface to a bulk or free-stream value away from the surface. The combination of diffusive transport and fluid motion near the interface is what we call convective mass transfer. As already noted, the convective mass transfer flux of a component A at a fluid-solid interface can be written as follows:

$$N_A = k_c(C_{A,s} - C_{A,b}) \tag{13.1}$$

where k_c is the convective mass transfer coefficient, typically measured in units of m/s or ft/h, $C_{A,s}$ is the concentration of A on the fluid-side of the interface, and $C_{A,b}$ is the bulk- or free-stream concentration of A away from the interface. We can also write an equivalent expression for the mass flux on a mass basis as follows:

$$n_A = k_\rho(\rho_{A,s} - \rho_{A,b}) \tag{13.2}$$

where $\rho_{A,s}$ is the mass concentrations of A on the fluid-side of the interface and $\rho_{A,b}$ is the mass concentration of A in the bulk away from the interface. Since $n_A = M_{r,A} \times N_A$ and $\rho_A = M_{r,A} \times C_A$, for dilute conditions, the two coefficients k_c and k_ρ must be identical. Note that convective mass transfer is not limited to the interface between a solid and a fluid and can also occur at the interface between two fluid phases such as a gas-liquid interface or the interface between two immiscible liquids.

Some examples of convective mass transfer are the dissolution of a solid into a flowing fluid, the absorption of a gas into a flowing liquid, the evaporation of component in a liquid mixture into a gas, the transport of a solute from a pipe wall to or from a flowing fluid, and the transport of an adsorbable species at the interface between a fluid and an adsorbent particle in relative motion. All of these are examples of mass transfer by forced convection, and they are analogous to heat transfer by forced convection. In general, however, there is not such a close analogy between heat transfer by natural convection or heat transfer by change of phase and mass transfer. It should be noted that Eq. (13.1) is strictly valid only for transport in dilute mixtures where the mass or mole fraction of component A does not exceed, say, 5%. In this case, k_c will depend only on the diffusivity of A in the mixture, on the geometry on the surface, and on the fluid flow, but will be otherwise independent of the concentration of A. Convective mass transfer coefficients in more concentrated mixtures will also depend on composition, and more complicated expressions will relate the flux to the driving force $C_{A,s} - C_{A,b}$ as discussed, for example, in Sec. 14.3.1 for the case of transport at gas-liquid interfaces.

In addition to the flux expressions given by Eqs. (13.1) and (13.2), there are alternative expressions based on different driving forces. Each will require a different mass transfer coefficient. Two commonly used expressions for gas-phase transport are:

$$N_A = k_y(y_{A,s} - y_{A,b}) \tag{13.3}$$

which is based on a gas-phase mole fraction driving force, and

$$N_A = k_p(p_{A,s} - p_{A,b}) \tag{13.4}$$

which is based on a partial pressure driving force. For liquid-phase transport, the following expression in terms of liquid-phase mole fractions is also used:

$$N_A = k_x(x_{A,s} - x_{A,b}) \tag{13.5}$$

In all three cases, subscripts s and b refer, respectively, to conditions on the fluid-side of the interface and in the bulk fluid far away from the interface. The mass transfer coefficients k_y, k_x, and k_p have physical meaning analogous to k_c, but their dimensions and units are different. If the flux, N_A, is in units of mol/m^2 s, the units of k_y or k_x will also be mol/m^2 s while if the partial pressure is in Pa, the typical units of k_p will be mol/m^2 s Pa. Any of these definitions must, of course, give us the same flux for the same situation. Thus, for gas-phase transport, we have the following identities:

$$k_c(C_{A,s} - C_{A,b}) = k_y(y_{A,s} - y_{A,b}) = k_p(p_{A,s} - p_{A,b}) \tag{13.6}$$

As a result, since $y_A = C_A/C$ and $p_A = y_A P$, the coefficients k_y and k_p are directly related to k_c according to the following relationship:

$$k_c = \frac{k_y}{C} = \frac{k_p P}{C} \tag{13.7}$$

where C is the molar density of the gas where transport occurs and P is the absolute pressure. Similarly, for liquid-phase transport we must have:

$$k_c(c_{A,s} - c_{A,b}) = k_x(x_{A,s} - x_{A,b}) \tag{13.8}$$

In this case, since $c_A = x_A c$, k_x will be related to k_c by the equation:

$$k_c = \frac{k_x}{c} \tag{13.9}$$

where c is the molar density of the liquid.

As we saw for convective heat transfer, the film model provides again a simple description useful to conceptualize convective mass transfer. In this case, we represent the convective mass transfer resistance as the resistance offered at steady state by a stagnant film of fluid with thickness Δ that we hypothesize exists adjacent to the solid surface. If we assume that only diffusion takes place through this hypothetical film, based on Fick's law, we have:

$$N_A = \frac{\mathcal{D}^0_{A,B}}{\Delta}(C_{A,s} - C_{A,b}) \tag{13.10}$$

where $\mathcal{D}^0_{A,B}$ is the diffusivity of species A in the fluid. Comparing Eqs. (13.1) and (13.10), we obtain:

$$k_c = \frac{\mathcal{D}^0_{A,B}}{\Delta} \tag{13.11}$$

As in the case of the equivalent expression for the heat transfer coefficient, Eq. (6.3), Eq. (13.11) is useful, but only to some extent. It does show that k_c will be larger for a solute with a larger $\mathcal{D}^0_{A,B}$. We can also say, on the basis of this equation, that faster flow over a surface will result in a higher k_c because the thickness of the hypothetical film Δ will be smaller. However, since, in fact, in most practical system there really isn't a stagnant film, we cannot predict what Δ is. As a result, Eq. (13.11) by itself will have no real predictive value. Nevertheless, we will see that this model will still help us to conceptually understand convective transport and make certain practically useful extrapolations.

As seen from some of the examples in the previous chapters, convective mass transfer coefficients vary widely in magnitude depending on whether the fluid is a gas or a liquid, on the specific physical properties of the fluid, on the flow velocity, on the geometry of the surface, and on the diffusivity of component A in the fluid. Thus, correlation and prediction of mass transfer coefficients are important for practical applications. There are two distinct approaches to obtain such predictions for dilute cases depending on the system at hand:

1. For laminar flow of Newtonian fluids, k_c can be predicted either analytically, for simple geometries, or using computational methods (e.g., computational fluid dynamics or CFD) for complex systems.
2. For complex geometries and in turbulent flow, k_c can be predicted using generalized semi-empirical correlations of mass transfer or heat transfer data based on analogies between momentum transfer, heat transfer, and mass transfer at fluid-solid interfaces.

As in the case of heat transfer, purely empirical approaches are also useful, but they are limited to specific fluids or to specific conditions and are difficult to generalize.

In this chapter, we consider first analytical solutions for some laminar flow systems, which have both conceptual and practical value. We then explore methods for the prediction of k_c in different systems based on analogies. Again, rather than providing a comprehensive treatment of all available correlations, we focus on conceptual aspects and illustrative examples. Section 5 in *Perry's Chemical Engineers' Handbook*, 9th edition, is a useful reference with extensive coverage of convective mass transfer coefficients for many practical cases.

13.2 Forced Convection in Laminar Flow

13.2.1 Laminar Flow Over a Flat Plate

In Sec. 6.2.1, we considered the problem of steady-state convective heat transfer for the case of a fluid flowing tangentially past the edge of a wide flat plate. We now use an analogous approach to analyze mass transfer between a flowing fluid and a wide flat plate but limiting ourselves to dilute conditions. With reference to Fig. 13.1, the fluid approaches the leading edge of the plate with velocity v_∞ and with a concentration $C_{A,\infty}$ of a component A while the fluid at the surface is kept at a constant, uniform concentration $C_{A,s}$. A practical example might be the dissolution of a slab of solid A into a flowing fluid. In this case, $C_{A,s}$ would be the solubility of A in the fluid. We will be able to treat this problem as two-dimensional if the slab is wide, and transport can be described in terms of the x and y coordinates only as shown in Fig. 13.1.

Since, at the surface, the fluid velocity is zero and the concentration is fixed, velocity and concentration profiles will develop starting at the leading edge of the plate with the velocity going from 0 at the surface to v_∞ and the concentration going from $C_{A,s}$ at the surface to $C_{A,\infty}$ far away from it. As we did in Chap. 6, we refer to the distance δ from the surface where the fluid velocity approaches v_∞ as the thickness of the momentum transport boundary layer and to the distance δ_D from the surface where the concentration of component A approaches $C_{A,\infty}$ as the thickness of the diffusive boundary layer. In general, δ and δ_D will be different from each other, but they will both vary with distance downstream from the edge of the plate; i.e., $\delta = \delta(x)$ and $\delta_D = \delta_D(x)$. We found the same situation for heat transfer with a thermal boundary layer thickness $\delta_T = \delta_T(x)$ also different, in general, from δ.

The equations needed to describe transport in this system are the differential mass balance (or continuity equation), the differential momentum balance, and the differential component material balance, which are as follows:

$$\frac{\partial v_x}{\partial x} + \frac{\partial v_y}{\partial y} = 0 \tag{13.12}$$

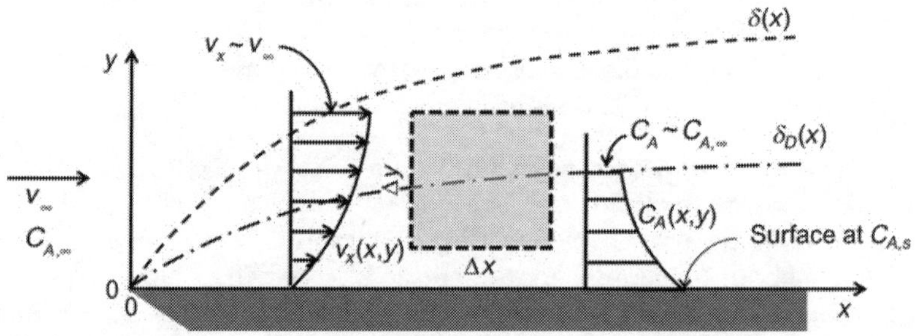

Figure 13.1 Convective mass transfer of a component A for flow past the edge of a flat plate. The control volume used to derive differential momentum and material balances has dimensions Δx and Δy. $\delta(x)$ is the thickness of the momentum transport boundary layer while $\delta_D(x)$ is the thickness of the diffusive boundary layer.

$$v_x \frac{\partial v_x}{\partial x} + v_y \frac{\partial v_x}{\partial y} = \nu \left(\frac{\partial^2 v_x}{\partial x^2} + \frac{\partial^2 v_x}{\partial y^2} \right) \tag{13.13}$$

$$v_x \frac{\partial C_A}{\partial x} + v_y \frac{\partial C_A}{\partial y} = \mathcal{D}^0_{A,B} \left(\frac{\partial^2 C_A}{\partial x^2} + \frac{\partial^2 C_A}{\partial y^2} \right) \tag{13.14}$$

where $\nu = \mu/\rho$ is the kinematic viscosity of the fluid. Equation (13.14) can be obtained directly from Table 12.1 by setting the time-derivative and the derivatives with respect to z all equal to zero. Since we are limiting our analysis to dilute conditions, it does not matter whether we use mass or molar concentration since $\rho_A = C_A M_{r,A}$, where $M_{r,A}$ is the molecular mass of component A.

Following our previous analysis of the analogous heat transfer problem in Sec. 6.2.1, here we also consider only the x-component of the momentum balance. It is important to note again that, while the main component of the velocity vector is v_x, because of Eq. (13.12), the velocity will also have a finite y-component, v_y, everywhere except right at the surface where both v_x and v_y are zero. With minimal loss of generality, we can neglect the first term in parenthesis on the right-hand side of both Eq. (13.13) and Eq. (13.14). In Eq. (13.13), the term $\partial^2 v_x / \partial x^2$ arises from the contribution of viscous transport to total momentum transport in the x-direction. This contribution is typically negligible, since momentum transport in this direction will be dominated by bulk flow and, thus, by v_x. Similarly, in Eq. (13.14), the term $\partial^2 C_A / \partial x^2$ arises from the contribution of diffusion to total mass transport of A in the x-direction. Again, because flow is mainly along the x axis, transport of A in this direction will be dominated by bulk flow and, thus, by v_x. This will not be the case, of course, for either momentum or mass transfer in the y-direction along which viscous momentum transport and diffusion will both be important. The simplified forms of these equations are as follows:

$$v_x \frac{\partial v_x}{\partial x} + v_y \frac{\partial v_x}{\partial y} = \nu \frac{\partial^2 v_x}{\partial y^2} \tag{13.15}$$

$$v_x \frac{\partial C_A}{\partial x} + v_y \frac{\partial C_A}{\partial y} = \mathcal{D}^0_{A,B} \frac{\partial^2 C_A}{\partial y^2} \tag{13.16}$$

These equations are subject to the following boundary conditions:

$$\begin{aligned} &x = 0 : v_x = v_\infty, C_A = C_{A,\infty} \\ &y = 0 : v_x = 0, C_A = C_{A,s} \\ &y \to \infty : v_x = v_\infty, C_A = C_{A,\infty} \end{aligned} \tag{13.17}$$

As we did in our analysis of the analogous heat transfer problem, we introduce the following dimensionless variables:

$$u_x = \frac{v_x}{v_\infty} \tag{13.18}$$

$$u_y = \frac{v_y}{v_\infty} \tag{13.19}$$

$$Y = \frac{C_A - C_{A,s}}{C_{A,\infty} - C_{A,s}} \tag{13.20}$$

Inserting these variables in Eqs. (13.15) to (13.17) gives the following set of transformed equations and boundary conditions:

$$u_x \frac{\partial u_x}{\partial x} + u_y \frac{\partial u_x}{\partial y} = \frac{\nu}{v_\infty} \frac{\partial^2 u_x}{\partial y^2} \tag{13.21}$$

$$u_x \frac{\partial Y}{\partial x} + u_y \frac{\partial Y}{\partial y} = \frac{\mathcal{D}^0_{A,B}}{v_\infty} \frac{\partial^2 Y}{\partial y^2} \tag{13.22}$$

$$\begin{aligned} &x = 0 : u_x = Y = 1 \\ &y = 0 : u_x = Y = 0 \\ &y \to \infty : u_x = Y = 1 \end{aligned} \tag{13.23}$$

Note that, in this form, the boundary conditions are exactly the same for both u_x and Y and, in fact, they are also exactly the same as the boundary conditions for the corresponding heat transfer problem formulated in terms of the dimensionless temperature $\theta = (T - T_s)/(T_\infty - T_s)$ as seen from Eqs. (6.12) to (6.14). Moreover, Eqs. (13.21) and (13.22) are also exactly identical to each other in the case of fluids for which $\nu = \mathcal{D}^0_{A,B}$. In these cases, since the two equations are the same and they have the same boundary conditions, the momentum and the mass transfer problems will also have the same solution. In other words, the solution of Eq. (13.21) for the dimensionless velocity, $u_x(x, y)$, will also be the solution of Eq. (13.22) for the dimensionless concentration $Y(x, y)$. We reached the same conclusion for the analogous heat transfer problem for fluids for which $\nu = \alpha$. For heat transfer, this equivalence was represented in terms of the Prandtl number, defined as $\mathrm{Pr} = \nu/\alpha$. We do the same for mass transfer by defining the Schmidt number as follows:

$$\mathrm{Sc} = \frac{\nu}{\mathcal{D}^0_{A,B}} = \frac{\mu}{\rho \mathcal{D}^0_{A,B}} \tag{13.24}$$

The Schmidt number, Sc, represents the ratio of momentum transport associated with viscous forces, determined by the magnitude of the kinematic viscosity ν, and of diffusional transport, determined by the magnitude of the diffusivity $\mathcal{D}^0_{A,B}$. When $\mathrm{Sc} = 1$, the momentum transport and the component material balance equations will have the same solution. We will see that for gases, for which $\mathrm{Pr} \sim 1$, Sc is also close to 1. In this case, the momentum transport equation, the differential energy balance equation, and the differential component material balance equation will all have the same solution; i.e., $u_x = \theta = Y$. This solution is shown in Fig. 6.2 for u_x as a function of the dimensionless variable:

$$\eta = \frac{y}{2}\left(\frac{v_\infty}{\nu x}\right)^{1/2} \tag{13.25}$$

and leads to the following expression for the thickness of the momentum transport boundary layer:

$$\delta = 5\left(\frac{\nu x}{v_\infty}\right)^{1/2} \tag{13.26}$$

As already noted, if Pr = 1 and Sc = 1, the same solution will apply to heat transfer and to mass transfer. As a result, we will have $\delta = \delta_T = \delta_D$; i.e., the momentum, thermal, and diffusive boundary layers will all have the same thickness.

In Sec. 6.2.1, we learned how we could find the heat transfer flux and the heat transfer coefficient from the dimensionless temperature profile, θ. An analogous approach can be used to find the mass transfer flux and the mass transfer coefficient. In this case, by definition, we have:

$$N_A = k_{c,x}\left(C_{A,s} - C_{A,\infty}\right) \tag{13.27}$$

where $k_{c,x}$ is the local mass transfer coefficient at a distance x from the leading edge of the plate. Since the fluid velocity is zero at the surface, we can also write the mass transfer flux based on Fick's law, which gives:

$$N_A = -\mathcal{D}^0_{A,B}\left.\frac{\partial C_A}{\partial y}\right|_{y=0} \tag{13.28}$$

Equations (13.27) and (13.28) must give the same result. Thus, comparing these two expressions for N_A, we obtain:

$$k_{c,x} = -\frac{\mathcal{D}^0_{A,B}}{C_{A,s} - C_{A,\infty}}\left(\frac{\partial C_A}{\partial y}\right)_{y=0} \tag{13.29}$$

or, in terms of dimensionless concentration,

$$k_{c,x} = \mathcal{D}^0_{A,B}\left(\frac{\partial Y}{\partial y}\right)_{y=0} \tag{13.30}$$

For Sc = 1, $Y(\eta)$ will be equal to $u_x(\eta)$. As a result, we can write:

$$k_{c,x} = \mathcal{D}^0_{A,B}\frac{\partial \eta}{\partial y}\left(\frac{dY}{d\eta}\right)_{y=0} = \frac{\mathcal{D}^0_{A,B}}{2}\left(\frac{v_\infty}{\nu x}\right)^{1/2}\left(\frac{dY}{d\eta}\right)_{\eta=0} = \frac{\mathcal{D}^0_{A,B}}{2}\left(\frac{v_\infty}{\nu x}\right)^{1/2}\left(\frac{du_x}{d\eta}\right)_{\eta=0} \tag{13.31}$$

Finally, as seen in Fig. 6.2, the slope $(du_x/d\eta)_{\eta=0}$ is equal to 0.644. Thus, inserting this value in Eq. (13.31) we obtain:

$$k_{c,x} = 0.332\mathcal{D}^0_{A,B}\left(\frac{v_\infty}{\nu x}\right)^{1/2} \tag{13.32}$$

The mass transfer coefficient averaged over a distance $x = L$ from the leading edge of the plate, k_c, is of interest and is found by integrating this equation along x as follows:

$$k_c = \frac{1}{L}\int_0^L k_{c,x}(x)dx = \frac{1}{L}\int_0^L 0.332\mathcal{D}^0_{A,B}\left(\frac{v_\infty}{\nu x}\right)^{1/2} dx = 0.664\mathcal{D}^0_{A,B}\left(\frac{v_\infty}{\nu L}\right)^{1/2} \tag{13.33}$$

Equation (13.33) is useful in practice since we can use this result to find the mass transfer rate to or from the plate surface simply multiplying this average k_c by the relevant concentration difference driving force and by the surface area of the plate.

In analogy to heat transfer, it is useful to also express these quantities in terms of dimensionless variables. Using a Reynolds number defined as follows:

$$\mathrm{Re}_x = \frac{v_\infty x}{\nu} \tag{13.34}$$

we obtain:

$$\delta = \delta_D = 5x(\mathrm{Re}_x)^{-1/2} \tag{13.35}$$

and

$$\mathrm{Sh}_x = \frac{k_{c,x} x}{\mathcal{D}^0_{A,B}} = 0.332(\mathrm{Re}_x)^{1/2} \tag{13.36}$$

The quantity $\mathrm{Sh}_x = k_{c,x} x/\mathcal{D}^0_{A,B}$ is the local Sherwood number. When Sc = Pr = 1, the local Sherwood number, Sh_x, is predicted to be equal to the local Nusselt number, Nu_x, for the analogous heat transfer problem. Similarly, averaging over the distance L, with the Reynolds number defined as:

$$\mathrm{Re} = \frac{v_\infty L}{\nu} \tag{13.37}$$

we obtain:

$$\mathrm{Sh} = \frac{k_c L}{\mathcal{D}^0_{A,B}} = 0.664(\mathrm{Re})^{1/2} \tag{13.38}$$

where $\mathrm{Sh} = k_c L/\mathcal{D}^0_{A,B}$ is the average Sherwood number. Again, for cases where Sc = Pr = 1, Sh and Nu are predicted to be identical at the same value of Re.

In Sec. 6.2.1, we saw that for conditions where Pr is not equal to one, the ratio of thermal and momentum transport boundary layer thicknesses can be described approximately by the following empirical relationship:

$$\frac{\delta_T}{\delta} = (\mathrm{Pr})^{-1/3} \tag{13.39}$$

Analogously, for mass transfer, we can use the equivalent expression:

$$\frac{\delta_D}{\delta} = (\mathrm{Sc})^{-1/3} \tag{13.40}$$

In turn, this equation leads to the following expressions for the local and average Sherwood numbers:

$$\mathrm{Sh}_x = 0.332(\mathrm{Re}_x)^{1/2}(\mathrm{Sc})^{1/3} \tag{13.41}$$

$$\mathrm{Sh} = 0.664(\mathrm{Re})^{1/2}(\mathrm{Sc})^{1/3} \tag{13.42}$$

These expressions are analogous to the corresponding relationships for the Nusselt number and are found to be consistent with experimental results. Note that, while for gases Pr~Sc~1, indicating that momentum and thermal boundary layers have the essentially same thickness, for liquids, Pr and Sc can be vastly different. For example, while Pr is about 10 for liquid water, Sc is on the order of 1,000 for small molecules in water and much larger (e.g., >10,000) for macromolecules as a result of their much smaller diffusivity. As a consequence, while for gases δ_T and δ_D are comparable in magnitude, for liquids the diffusive boundary layer is usually thinner than the thermal boundary layer and much thinner than the momentum transfer boundary layer at the same Reynolds number. Example 13.1 illustrates the magnitude of these boundary layers for some practical cases.

Example. 13.1. Mass transfer coefficient and boundary layers for laminar flow over a flat plate. Let's use the equations presented in this section to predict the thickness of the boundary layers and the mass transfer coefficient for the sublimation of a wide slab of solid benzoic acid (*A*) into air (*B*) flowing at 1.0 m/s over the slab at 1 atm and 25°C. The length of the slab in the direction of flow is 0.5 m. Based on Example 6.1, for air at this temperature and pressure we have $\nu = 1.56\times10^{-5}$ m²/s, $\alpha = 2.2\times10^{-5}$ m²/s, Pr = 0.71, and Re = 3.2×10^4, which corresponds to laminar flow. Thus, we obtain:

$$\delta = 5\times0.5(3.2\times10^4)^{-1/2} = 0.014 \text{ m}$$

$$\delta_T = 0.014(0.71)^{-1/3} = 0.016 \text{ m}$$

$$\text{Nu} = 0.664(3.2\times10^4)^{1/2}(0.71)^{1/3} = 106$$

For benzoic acid in air at 1 atm and 25°C, the Fuller et al. equation, Eq. (10.17), predicts $\mathcal{D}_{A,B} = 0.070\times10^{-4}$ m²/s. Thus, $\text{Sc} = \nu/\mathcal{D}_{A,B} = 1.56\times10^{-5}/0.070\times10^{-4} = 2.2$. Accordingly:

$$\delta_D = 0.014(2.2)^{-1/3} = 0.011 \text{ m}$$

$$\text{Sh} = 0.664(3.2\times10^4)^{1/2}(2.2)^{1/3} = 154$$

$$k_c = \text{Sh}\frac{\mathcal{D}_{A,B}}{L} = 154\times\frac{0.070\times10^{-4}}{0.5} = 0.0022 \text{ m/s}$$

As expected for convective mass transfer to a gas, we see that δ, δ_T, and δ_D are similar in magnitude since both Pr and Sc are on the order of 1. As a consequence, Nusselt and Sherwood numbers are also similar in magnitude to each other.

Let's now compare these results with the values that we would obtain if instead of air we had liquid water flowing over the plate at 25°C and at the same velocity. For water at 25°C, we have $\nu = 0.89\times10^{-6}$ m²/s, $\alpha = 1.4\times10^{-7}$ m²/s, Pr = 6.4, and Re = 5.6×10^5, which give:

$$\delta = 5\times0.5(5.6\times10^5)^{-1/2} = 0.0033 \text{ m}$$

$$\delta_T = 0.0033(6.4)^{-1/3} = 0.0018 \text{ m}$$

$$\text{Nu} = 0.664(5.6\times10^5)^{1/2}(6.4)^{1/3} = 920$$

For benzoic acid in water 25°C, Noulty and Leaist [J. Chem. Eng. Data 32 (1987) 418] give $\mathcal{D}^0_{A,B} \sim 0.9\times10^{-9}$ m²/s. Thus, $\text{Sc} = \nu/\mathcal{D}^0_{A,B} = 0.89\times10^{-6}/0.9\times10^{-9} = 990$. Therefore, we have:

$$\delta_D = 0.0033(990)^{-1/3} = 0.00033 \text{ m}$$

$$\text{Sh} = 0.664(5.6\times10^5)^{1/2}(990)^{1/3} = 4,950$$

$$k_c = \text{Sh}\frac{\mathcal{D}^0_{A,B}}{L} = 4,950\frac{0.9\times10^{-9}}{0.5} = 8.9\times10^{-6} \text{ m/s}$$

We can see that, as expected, compared to the case where air flows over the plate, the boundary layers are much thinner for flow of water. In this case, since Sc is large, the diffusive boundary layer is also much thinner than the thermal boundary layer and the Sherwood number is much larger than the corresponding Nusselt number. For heat transfer, these results come about because of the combination of higher thermal conductivity and higher viscosity of liquid water compared to air. For mass transfer, this occurs through the combination of the much smaller diffusivity of benzoic acid in water compared to that in air and of the higher viscosity of water. Note that in all of these calculations, we have used the physical properties of the fluid. The heat transfer coefficient is independent of what material the plate is made of provided that the boundary conditions are satisfied at its surface. Also note that the mass transfer coefficients calculated in this example are, in both cases, averaged over the length of the plate. The average mass transfer flux can then be calculated simply multiplying k_c by $C_{A,s} - C_{A,\infty}$, where $C_{A,s}$ and $C_{A,\infty}$ are, respectively, the fluid-side benzoic acid concentrations at the slab surface and far away from it. For mass transfer in air, the surface concentration is calculated from the sublimation pressure of benzoic acid, which is $p_A^* = 7 \times 10^{-4}$ mmHg at 25°C. Thus, $C_{A,s} = p_A^*/RT = (7 \times 10^{-4}/760)/(82 \times 10^{-6} \times 298) = 3.8 \times 10^{-5}$ mol/m³. For mass transfer in water, the surface concentration will be the solubility of benzoic acid in water, which is $C_{A,s} = 29$ mol/m³ at 25°C. With $C_{A,\infty} = 0$, the corresponding fluxes will be $0.0022 \times 3.8 \times 10^{-5} = 8.4 \times 10^{-8}$ mol/m² s for sublimation in air and $8.9 \times 10^{-6} \times 29 = 2.6 \times 10^{-4}$ mol/m² s for dissolution in water. We can see that the flux in water will be greater than the flux in air despite the lower k_c for the former because the solubility and, thus, the driving force for mass transfer is greater in water.

13.2.2 Laminar Flow in Different Geometries

In general, as demonstrated above for flow past the edge of a flat plate, solutions obtained for heat transfer in laminar flow can also be used for mass transfer in dilute solutions since the dimensionless governing equations are the same in the two cases once α is replaced with $\mathcal{D}^0_{A,B}$. Thus, the expressions obtained for the Nusselt number for convective heat transfer can be used to predict the Sherwood number simply by replacing Pr with Sc. One example is the Graetz solution for heat transfer from a pipe wall with a uniform temperature to a flowing fluid with fully developed laminar flow presented in Sec. 6.2.2. This solution can also describe mass transfer of a component A to or from a pipe wall with a uniform surface concentration $C_{A,s}$. In this case, the local Sherwood number at a distance z from the entrance of a pipe with diameter $D = 2R$ is described by the function of ζ given by Eq. (6.56) and shown in Fig. 6.4 with Sh_z, used in lieu of Nu_z, and ζ now defined as follows:

$$\text{Sh}_z = \frac{k_{c,x} D}{\mathcal{D}^0_{A,B}} \tag{13.43}$$

and

$$\zeta = \frac{z}{R} \frac{1}{\text{Re}\,\text{Sc}} \tag{13.44}$$

Similarly, the Sherwood number averaged over the length of the pipe L and defined as:

$$\text{Sh} = \frac{k_c D}{\mathcal{D}^0_{A,B}} = \frac{1}{\text{Z}} \int_0^1 \text{Sh}_z(\zeta)\, d\zeta \tag{13.45}$$

is predicted to be the same as the average Nusselt number from Eq. (6.58) or from Fig. 6.5 provided that the dimensionless pipe length Z is now defined as:

$$\text{Z} = \frac{L}{R} \frac{1}{\text{Re}\,\text{Sc}} \tag{13.46}$$

Note that the associated asymptotic expressions for the Nusselt number discussed in Sec. 6.2.2 will also be applicable to mass transfer over the appropriate ranges of values of ζ and Z with Sh replacing Nu and Sc replacing Pr. Other situations can be described in a similar manner as shown, for instance, in Example 13.2.

Example 13.2. Mass transfer to/from a falling liquid film. In this example, we consider mass transfer of a component A from a gas phase into a liquid film that falls over a vertical surface as shown in Fig. 13.2. The surface can be a flat wall, the outer surface of a solid cylinder, or the inner surface of a hollow cylinder. Liquid fed from the top flows down under gravity clinging to the wall. Viscous drag at the solid wall surface will balance the force of gravity, which, in turn, will result in a liquid film with constant thickness. If the liquid film is thin, the analysis will be the same whether we have a flat wall or a cylindrical surface since, in this case, the curvature can be neglected. Under laminar flow conditions, the velocity in the film will have a parabolic profile, being 0 at the wall and maximum at the gas-liquid interface where shear is minimal. This "wetted wall" or falling film scenario is a model

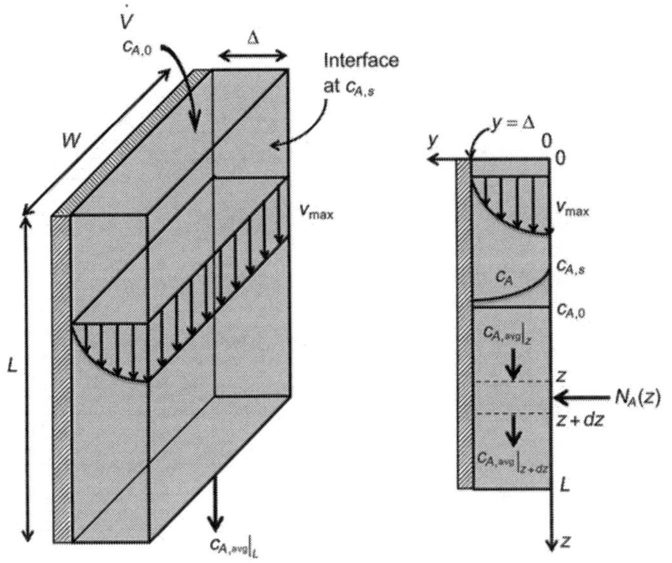

Figure 13.2 Mass transfer of a component *A* into a falling laminar liquid film. The figure on the right shows a vertical cross section of the film.

for many practical systems including gas absorption or stripping columns and other devices that are designed to facilitate mass exchange between a flowing liquid and a gas.

Let's consider the case of absorption of component *A* from the gas as sketched in Fig. 13.2. The liquid has a volumetric flow rate $\dot{V}$ and is uniformly distributed over the width of the wall *W*. If the film thickness is Δ, the average liquid velocity in the film will be:

$$v_{\text{avg}} = \frac{\dot{V}}{W\Delta} \tag{13.47}$$

Furthermore, if the flow is laminar and the liquid is Newtonian, the velocity profile will be parabolic with a maximum at the gas-liquid interface. Thus, v_z will be a function of *y* given by:

$$v_z(y) = v_{\max}\left[1 - \left(\frac{y}{\Delta}\right)^2\right] \tag{13.48}$$

Neglecting the shear stress at the gas-liquid interface because of the very low viscosity of the gas compared to that of the liquid, we can write the following force balance for the liquid film:

$$WL\Delta \times \rho g = WL \times \tau_W \tag{13.49}$$

where ρ is the liquid density. The left-hand side of this equation is the weight of the liquid film (mass of liquid times gravity constant) while the right-hand side is the opposing shear force at the wall. For a Newtonian fluid, the shear stress at the wall is given by:

$$\tau_w = -\mu \left.\frac{dv_z}{dy}\right|_{y=\Delta} \tag{13.50}$$

where μ is the viscosity of the liquid. Inserting the derivative $(dv_z/dy)_{y=\Delta}$ calculated from Eq. (13.48) into Eq. (13.50), we obtain:

$$\tau_w = \frac{2\mu v_{\max}}{\Delta} \tag{13.51}$$

Finally, substituting this result into the force balance, Eq. (13.49), we obtain the following expression for the film thickness:

$$\Delta = \left(\frac{2\mu v_{\max}}{\rho g}\right)^{\frac{1}{2}} \tag{13.52}$$

We can also express Δ as a function of the average velocity or of the volumetric flow rate by recognizing that

$$v_{\text{avg}} = \frac{1}{\Delta}\int_0^{\Delta} v_z(y)\,dy = \frac{1}{\Delta}\int_0^{\Delta} v_{\max}\left[1-\left(\frac{y}{\Delta}\right)^2\right]dy \tag{13.53}$$

Integrating this equation gives:

$$v_{\text{avg}} = \frac{2}{3}v_{\max} \tag{13.54}$$

Therefore:

$$\Delta = \left(\frac{3\mu \dot{V}}{\rho g W \Delta}\right)^{\frac{1}{2}} \tag{13.55}$$

Solving this equation for Δ, we obtain:

$$\Delta = \left(\frac{3\mu \dot{V}}{\rho g W}\right)^{\frac{1}{3}} \tag{13.56}$$

The Reynolds number for the film is defined as follows:

$$\text{Re} = \frac{\rho v_{\max}\Delta}{\mu} = \frac{3}{2}\frac{\rho v_{\text{avg}}\Delta}{\mu} \tag{13.57}$$

Note that if the wall is not vertical but is inclined at an angle β over vertical, the same relationships apply by replacing g with $g \cos \beta$.

We now turn our attention to mass transfer and develop first an exact solution of the problem and then an approximate solution. In order to obtain an exact solution of this problem, we need to consider that the steady-state concentration of component A in the liquid, c_A, will be a function of y and z. The differential component material balance for this case is easily obtained from item A in Table 12.1 and is as follows:

$$v_z\frac{\partial c_A}{\partial z} = \mathcal{D}^0_{A,B}\left(\frac{\partial^2 c_A}{\partial y^2} + \frac{\partial^2 c_A}{\partial z^2}\right) \tag{13.58}$$

where $\mathcal{D}^0_{A,B}$ is the diffusivity of A in the liquid. Provided that the vertical length of the film is not too short, we can neglect the second term in parenthesis on the right-hand side of Eq. (13.58). This term represents diffusive transport in the vertical direction, which is likely small compared to transport associated with the bulk flow. Thus, we obtain the following simplified equation:

$$v_z\frac{\partial c_A}{\partial z} = \mathcal{D}^0_{A,B}\frac{\partial^2 c_A}{\partial y^2} \tag{13.59}$$

The boundary conditions for this equation are

$$\begin{aligned} &z = 0, \quad c_A = c_{A,0} \\ &y = 0, \quad c_A = c_{A,s} \\ &y = \Delta, \quad \frac{\partial c_A}{\partial y} = 0 \end{aligned} \tag{13.60}$$

and are derived as follows. At $z=0$, where the liquid is fed to the wall, we assume that the concentration is uniform and equal to $c_{A,0}$. At $y = 0$, which defines the location of the gas-liquid interface, if there is no resistance to mass transfer on the gas side, we assume that there will be equilibrium with the surrounding gas. In this case, if Henry's law is applicable, we will have:

$$y_A P = Hx_{A,s} \tag{13.61}$$

where y_A is the mole fraction of A in the gas, P is the absolute pressure, and H is the Henry's constant in units of pressure. Since $c_A = x_A c$, where c is the molar density of the liquid, we will have:

$$c_{A,s} = \frac{c}{H} p_A \tag{13.62}$$

where $p_A = y_A P$ is the partial pressure of A in the gas. Finally, at $y = \Delta$, we assume that the wall is impermeable to A so that the mass transfer flux and, hence, $\partial c_A / \partial y$ is zero. Note that v_z in Eq. (13.59) is a function of y. Thus, the solution of this equation is not so simple. However, it has been obtained by Olbrich and Wild [Chem. Eng. Sci. 24 (1969) 25] in the form of the following infinite series:

$$Y_{avg} = \frac{c_{A,avg} - c_{A,s}}{c_{A,0} - c_{A,s}} = \sum_{n=1}^{\infty} A_n \exp(-\lambda_n \zeta) \tag{13.63}$$

where:

$$\zeta = \frac{z\mathcal{D}^0_{A,B}}{v_{max}\Delta^2} = \frac{z}{\Delta}\frac{1}{\text{Re}\,\text{Sc}} \tag{13.64}$$

is a dimensionless distance down the wall. The first 10 values of A_n and λ_n, obtained numerically, are as follows:

n	1	2	3	4	5	6	7	8	9	10
A_n	0.7897	0.097260	0.036093	0.018686	0.011400	0.007670	0.005518	0.004157	0.003244	0.002601
λ_n	5.122	39.661	106.245	204.855	335.460	498.090	632.715	919.350	1,178.00	1,468.63

Note that the expression of Olbrich and Wild gives the mixing-cup average concentration of component A in the liquid, $c_{A,avg}$. This is the concentration that we would get if we collected the outflow from the film at a distance z from the top into a cup and mixed it. The value of $c_{A,avg}$ can be related to the actual concentration profile $c_A(y,z)$ by the following material balance:

$$\dot{V} c_{A,avg}(z) = W\int_0^{\Delta} v_z(y) c_A(y,z)\,dy \tag{13.65}$$

which gives:

$$c_{A,avg}(z) = \frac{1}{v_{avg}\Delta}\int_0^{\Delta} v_z(y) c_A(y,z)\,dy \tag{13.66}$$

We can now obtain the mass transfer coefficient k_c by considering a differential balance for the film based on the change in the average concentration as we move down the wall. With reference to the right-hand side of Fig. 13.2, we can write the following material balance on component A for a differential control volume of thickness dz as follows:

$$v_{avg} W\Delta \times c_{A,avg}\big|_z + Wdz \times N_A(z) = v_{avg} W\Delta \times c_{A,avg}\big|_{z+dz}$$

where $N_A(z)$ is the local flux of A into the film at the gas-liquid interface. This balance equates the change in the amount of component A in the liquid between z and $z + dz$ to the rate of mass transfer at the gas-liquid interface, which is given by $Wdz \times N_A(z)$. Dividing all terms in this equation by dz we obtain the following result:

$$v_{avg}\Delta \frac{dc_{A,avg}}{dz} = N_A(z) \tag{13.67}$$

Next, we consider the following expression for the mass transfer flux at the gas-liquid interface in terms of a local mass transfer coefficient $k_{c,z}$:

$$N_A = k_{c,z}(c_{A,s} - c_{A,\text{avg}}) \tag{13.68}$$

Since, at a given value of z, the flux N_A that appears in Eq. (13.67) must be the same as that given by Eq. (13.68), we can obtain the local mass transfer coefficient by equating the two, which gives:

$$k_{c,z} = \frac{v_{\text{avg}}\Delta}{c_{A,s} - c_{A,\text{avg}}} \frac{dc_{A,\text{avg}}}{dz} \tag{13.69}$$

where $dc_{A,\text{avg}}/dz$ is obtained from Eq. (13.63). In practice, we are more interested in the mass transfer coefficient averaged over the length of the wall, which is given from the following expression:

$$k_c = \frac{1}{L}\int_0^L k_{c,z}(z)dz \tag{13.70}$$

Inserting Eq. (13.69) into Eq. (13.70) and integrating, we obtain the following result:

$$k_c = \frac{v_{\text{avg}}\Delta}{L}\int_{c_{A,0}}^{c_{A,\text{avg}}} \frac{dc_{A,\text{avg}}}{c_{A,s} - c_{A,\text{avg}}} = -\frac{v_{\text{avg}}\Delta}{L}\ln\left(\frac{c_{A,\text{avg}} - c_{A,s}}{c_{A,0} - c_{A,s}}\right) \tag{13.71}$$

In terms of the Olbrich and Wild solution, Eq. (13.63), Eq. (13.71) can be written as follows:

$$k_c = -\frac{v_{\text{avg}}\Delta}{L}\ln(Y_{\text{avg}}) = -\frac{v_{\text{avg}}\Delta}{L}\ln\left[\sum_{n=1}^{\infty} A_n \exp(-\lambda_n \mathrm{Z})\right] \tag{13.72}$$

where

$$\mathrm{Z} = \frac{L\mathcal{D}^0_{A,B}}{v_{\max}\Delta^2} = \frac{L}{\Delta}\frac{1}{\mathrm{Re}\,\mathrm{Sc}} \tag{13.73}$$

We can also conveniently express this result in terms of a Sherwood number as follows:

$$\mathrm{Sh} = -\frac{2}{3\mathrm{Z}}\ln\left[\sum_{n=1}^{\infty} A_n \exp(-\lambda_n \mathrm{Z})\right] \tag{13.74}$$

where

$$\mathrm{Sh} = \frac{k_c\Delta}{\mathcal{D}^0_{A,B}} \tag{13.75}$$

This function is plotted in Fig. 13.3 and can be used to estimate the mass transfer coefficient for a falling film in a general case. We can see that the Sherwood number decreases as the wall length increases (i.e., as Z increases). This occurs because as A penetrates deeper into the film, the concentration gradient of A in the liquid at the gas-liquid interface becomes shallower, thereby reducing the flux. At high values of Z, the Sherwood number approaches the horizontal asymptote Sh ~ 3.4. The mass transfer rate becomes small for these conditions since $c_{A,\text{avg}}$ will approach $c_{A,s}$ and Y_{avg} will approach 0.

Let's now consider a useful approximate solution of this mass transfer problem by assuming that the depth of penetration of component A in the liquid film is small. We expect that this will be the case when L is small or the fluid velocity $v_{\max}$ is high, which will leave little time for A to diffuse into the film before the liquid reaches the bottom of the wall and exits the system. As shown in Fig. 13.4, if the depth of penetration of component A is confined to a thin layer near the gas-liquid interface, we can assume that the fluid velocity will be approximately constant and equal to $v_{\max}$. In this case, Eq. (13.59) and the associated boundary conditions will simplify to the following:

$$v_{\max}\frac{\partial c_A}{\partial z} = \mathcal{D}^0_{A,B}\frac{\partial^2 c_A}{\partial y^2} \tag{13.76}$$

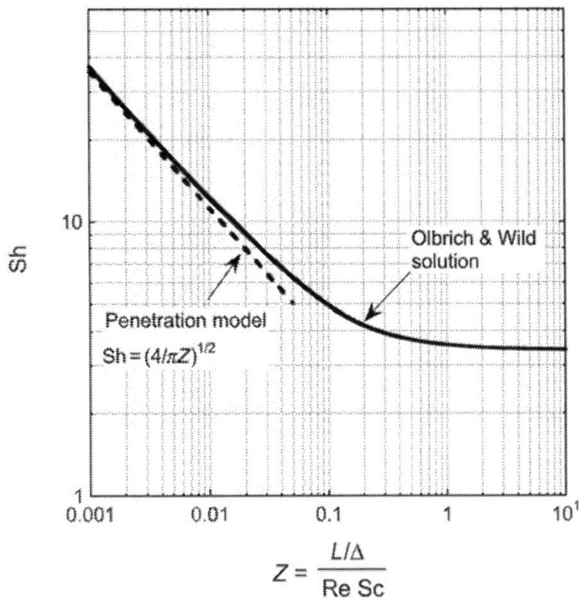

Figure 13.3 Sherwood number ($Sh = k_c\Delta/\mathcal{D}^0_{A,B}$) for mass transfer of a gaseous component A into a laminar liquid film falling on a vertical wall. Mathematica notebook: (M.F.13.3).

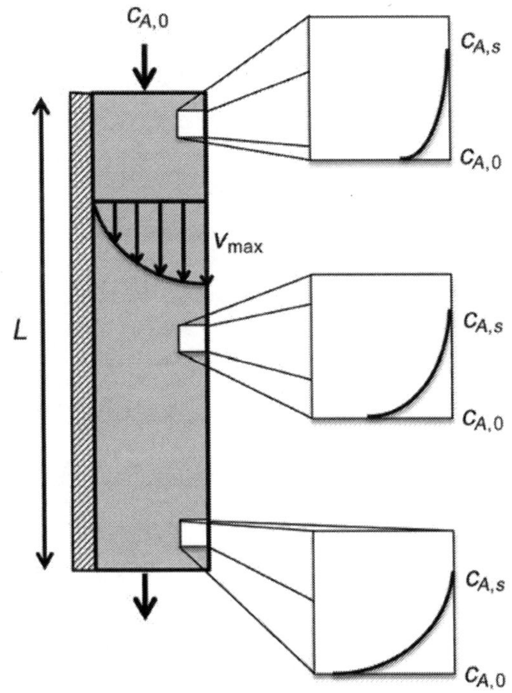

Figure 13.4 Penetration model for mass transfer of a gaseous component A into a laminar falling liquid film with initial concentration $c_{A,0}$ and interfacial concentration $c_{A,s}$. The graphs on the right-hand side illustrate the concentration profiles in the liquid at a short distance from the gas-liquid interface.

$$z = 0, \quad c_A = c_{A,0}$$
$$y = 0, \quad c_A = c_{A,s} \tag{13.77}$$
$$y \to \infty, \quad c_A = c_{A,0}$$

Note that the difference between the original equation, Eq. (13.59), and its boundary conditions, Eq. (13.60), and this approximate formulation is that, on the one hand, we have replaced $v_z(y)$ with $v_{\max}$ while, on the other hand, we have replaced the boundary condition at $y = \Delta$ with a new boundary condition set at $y \to \infty$. This is appropriate if the depth of penetration of the concentration profile is small and diffusion of component A is not significantly affected by the wall itself. We can now convert Eqs. (13.76) and (13.77) into a time-dependent problem by defining time as:

$$t = \frac{z}{v_{\max}} \tag{13.78}$$

t is now the time during which an element of liquid at the surface is exposed to the gas before reaching a distance z from the top of the wall. In terms of this time, Eq. (13.76) and its initial and boundary conditions, Eq. (13.77), become:

$$\frac{\partial c_A}{\partial t} = \mathcal{D}^0_{A,B} \frac{\partial^2 c_A}{\partial y^2} \tag{13.79}$$

$$t = 0, \quad c_A = c_{A,0}$$
$$y = 0, \quad c_A = c_{A,s} \tag{13.80}$$
$$y \to \infty, \quad c_A = c_{A,0}$$

This is exactly the same partial differential equation and boundary conditions that we have written for diffusion of a component A in an infinite slab. The solution, given in Sec. 12.3.2, is thus as follows:

$$Y = \frac{c_A - c_{A,s}}{c_{A,0} - c_{A,s}} = \operatorname{erf}(\eta) \tag{13.81}$$

where

$$\eta = \frac{y}{\sqrt{4\mathcal{D}^0_{A,B} t}} \tag{13.82}$$

Since $t = z/v_{\max}$, for the falling film, η becomes:

$$\eta = \frac{y}{\sqrt{4\mathcal{D}^0_{A,B} z / v_{\max}}} \tag{13.83}$$

As shown in Sec. 12.3.2, we can use Eq. (13.81) to find the instantaneous flux of A as well as to find the flux of A averaged over time. The latter is given by Eq. (12.37) as follows:

$$N_{A,\text{avg}} = \sqrt{\frac{4\mathcal{D}^0_{A,B}}{\pi t_E}} (c_{A,s} - c_{A,0}) \tag{13.84}$$

where, now,

$$t_E = \frac{L}{v_{\max}} \tag{13.85}$$

is the total time of exposure of a liquid element to the gas; i.e., the time taken by an element of fluid on the film surface to travel the entire length of the wall. Since, in this case, we can write:

$$k_c = \frac{N_{A,\text{avg}}}{c_{A,s} - c_{A,0}} \tag{13.86}$$

(note that $c_{A,avg} \sim c_{A,0}$ when the depth of penetration is small), we end up with the following result for the mass transfer coefficient:

$$k_c = \sqrt{\frac{4\mathcal{D}^0_{A,B}}{\pi t_E}} = \sqrt{\frac{4\mathcal{D}^0_{A,B} v_{max}}{\pi L}} \tag{13.87}$$

We can also express this result in terms of the dimensionless variables Sh and Z as follows:

$$\text{Sh} = \frac{k_c \Delta}{\mathcal{D}^0_{A,B}} = \sqrt{\frac{4}{\pi Z}} \tag{13.88}$$

which can be compared directly with the exact solution as shown in Fig. 13.3. As seen from this figure, this approximate solution is in good agreement with the exact result if Z is less than about 0.01. This approximate solution is widely known as the "penetration model" or "penetration theory" and is applicable to many practical cases. The critical parameter in this model is the time of exposure of a surface element to the gas at the gas-liquid interface, t_E.

Let's now apply these results to a practical case where we have $v_{max} = 0.1$ m/s, $\mu = 1.0$ mPa s, $\rho = 1{,}000$ kg/m^3, and $\mathcal{D}^0_{A,B} = 1.0 \times 10^{-9}$ m^2/s. From Eq. (13.55), we can calculate the thickness of the falling film to be:

$$\Delta = \left(\frac{2 \times 0.001 \times 0.1}{1{,}000 \times 9.8}\right)^{1/2} = 0.00014 \text{ m}$$

If the length of the wall is $L = 0.1$ m, from Eq. (13.73) we have:

$$\text{Z} = \frac{0.1 \times 1.0 \times 10^{-9}}{0.1 \times (0.00014)^2} = 0.050$$

Finally, from Eq. (13.74) or, directly, from Fig. 13.3, the Olbrich and Wild solution gives Sh ~ 6. For these conditions, the penetration model, Eq. (13.88), gives $\text{Sh} = \sqrt{4/\pi Z} = \sqrt{4/(\pi \times 0.050)} = 5$. This result is close to the exact solution since, in this case, the time of exposure $t_E = L/v_{max} = 0.1/0.1 = 1$ s is relatively short. Note that, for the case at hand, the time of exposure can be predicted from basic fluid mechanics. In other instances with more complex geometries, the time of exposure can be determined from empirical measurements and used to predict k_c based on Eq. (13.87). Example 13.3 provides and illustration of this approach.

We are now in the position of ascertaining if the assumption made early on in the analysis of this problem that diffusive transport in the z-direction can be neglected relative to bulk flow in the falling film is in fact correct. This assumption, which led us from Eq. (13.58) to (13.59) by neglecting the term $\partial^2 c_A / \partial z^2$ on the right-hand side of Eq. (13.58), was justified, intuitively, by assuming that the wetted wall is not too short. We can now obtain a quantitative criterion by comparing the time scales of these two transport mechanisms. Based on Eq. (12.32), the time scale for diffusional transport of component A along the length L of the falling film is on the order of $L^2/\mathcal{D}^0_{A,B}$. Conversely, the time scale for bulk transport of A along the same length is on the order of L/v_{max}. The ratio of these two time scales is $\text{Pe}_L = (L^2/\mathcal{D}^0_{A,B})/(L/v_{max}) = Lv_{max}/\mathcal{D}^0_{A,B}$ and is called the Peclet number. A large value of Pe_L would indicate that bulk transport of component A along L is much faster than diffusion. For the conditions of our problem, we have $\text{Pe}_L = 0.1 \times 0.1/1.0 \times 10^{-9} = 1.0 \times 10^7$. This very large value confirms that our assumption is definitely correct.

Example 13.3. Absorption into a liquid film falling over a sphere. In this example, we consider absorption of a component A from a gas into a laminar liquid film falling over a vertical string of spheres each with radius R as sketched in Fig. 13.5. The liquid flows first on the top sphere, spreads over its surface and flows down as a thin film onto the next sphere. This type of apparatus is called a "string of spheres absorber" and has been used as a model for packed absorption columns. Let's consider first the top sphere where the liquid enters with an initial concentration $c_{A,0}$ of dissolved A. The situation is similar to that of a liquid film flowing over a flat wall. The surface of the liquid is in equilibrium with the surrounding gas and has a uniform concentration $c_{A,s}$ determined by the solubility of A in the liquid. The velocity profile in the film is also parabolic with a maximum at the gas-liquid interface. However,

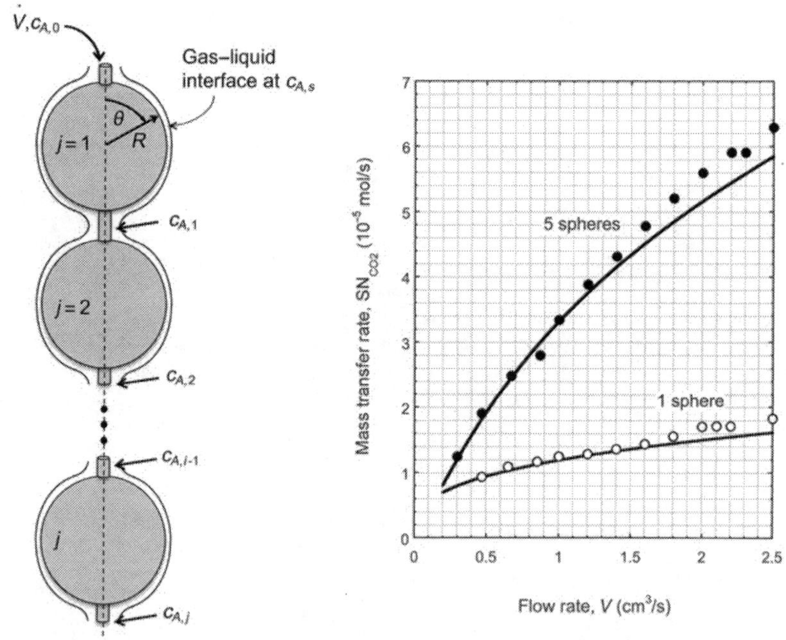

Figure 13.5 Mass transfer of a gaseous component *A* into a liquid film falling on a string of spheres. The graph on the right shows experimental data for CO_2 absorption in water at 20°C and 1 atm over 1.88 cm radius spheres [Carta and Pigford, Ind. Eng. Chem. Fundam. 22 (1983) 329]. The lines in the graph show model predictions based on Eq. (13.96).

in this case, the film thickness Δ and the maximum velocity at the gas-liquid interface, v_s, will vary with angular position along the sphere according to the following equations from Sherwood et al. [*Mass Transfer*, McGraw-Hill, 1975, pp. 86–89]:

$$\Delta = \left(\frac{3\nu\dot{V}}{2\pi g R}\right)^{1/3} \sin^{-2/3}(\theta) \tag{13.89}$$

$$v_s = \frac{g}{2\nu}\left(\frac{3\nu\dot{V}}{2\pi g R}\right)^{2/3} \sin^{-1/3}(\theta) \tag{13.90}$$

where $\nu = \mu/\rho$ is the kinematic viscosity of the liquid and $\dot{V}$ is its volumetric flow rate. From these equations, we can see that the thickness of the falling film will be minimum and, correspondingly, the surface fluid velocity will be maximum at the equator of the sphere where $\theta = \pi/2$. Based on the penetration model, which assumes that diffusion of component *A* is limited to a thin layer near the liquid surface, Davidson and Cullen [Trans. Inst. Chem. Eng. (London) 35 (1957) 51] derived the following expression for the mass transfer coefficient for one sphere:

$$k_c = \frac{5.09}{4\pi} g^{1/6}\nu^{-1/6}R^{-5/6}(\mathcal{D}^0_{A,B})^{1/2}\dot{V}^{1/3} \tag{13.91}$$

Accordingly, the mass transfer rate for the top sphere with area $S = 4\pi R^2$ will be:

$$SN_{A,1} = Sk_c(c_{A,s} - c_{A,0}) = 4\pi R^2 k_c(c_{A,s} - c_{A,0}) = 5.09 g^{1/6}\nu^{-1/6}R^{7/6}(\mathcal{D}^0_{A,B})^{1/2}\dot{V}^{1/3}(c_{A,s} - c_{A,0}) \tag{13.92}$$

The next sphere will behave exactly like the top sphere, except that incoming liquid will have a concentration higher than $c_{A,0}$ because of the amount of component A already absorbed on the top sphere. If we assume that the liquid is mixed between adjacent spheres, by material balance we obtain:

$$c_{A,1} = c_{A,0} + \frac{SN_{A,1}}{\dot{V}} = c_{A,0} + \frac{Sk_c}{\dot{V}}(c_{A,s} - c_{A,0}) \tag{13.93}$$

We can generalize this result for the jth sphere in a string of spheres by writing:

$$c_{A,j} = c_{A,j-1} + \frac{Sk_c}{\dot{V}}\left(c_{A,s} - c_{A,j-1}\right) = \left(1 - \frac{Sk_c}{\dot{V}}\right)c_{A,j-1} + \frac{Sk_c}{\dot{V}}c_{A,0} \tag{13.94}$$

This is a linear difference equation (not a differential equation) in $c_{A,j}$, which has the following solution valid for any j:

$$c_{A,j} = c_{A,s} - \left(c_{A,s} - c_{A,0}\right)\left(1 - \frac{Sk_c}{\dot{V}}\right)^j \tag{13.95}$$

Finally, we can find the total rate of absorption over all the spheres, $M_{A,n}$, again from an overall material balance, which gives the following result:

$$M_{A,n} = \dot{V}\left[c_{A,s} - \left(c_{A,s} - c_{A,0}\right)\left(1 - \frac{Sk_c}{\dot{V}}\right)^n\right] \tag{13.96}$$

where n is the total number of spheres.

Let's now apply these equations to analyze the experimental data for the absorption of CO_2 in water at 20°C and 1 atm obtained by Carta and Pigford [Ind. Eng. Chem. Fundam. 22 (1983) 329] that are shown on the right-hand side of Fig. 13.5. For the conditions of these experiments, $\nu = 0.01$ cm^2/s and $R = 1.88$ cm (the spheres in this apparatus were actually standard ping-pong balls drilled through and skewered with a thin metal rod). The diffusivity of CO_2 in water at this temperature is $\mathcal{D}^0_{A,B} = 1.7 \times 10^{-5}$ cm^2/s and its solubility is $c_{A,s} = 4.0 \times 10^{-5}$ mol/cm^3 at 1 atm. Let's first find the thickness of the liquid film at the sphere equator ($\theta = \pi/2$). At $\dot{V} = 1$ cm^3/s, in the middle of the experimental range of Fig. 13.5, Eq. (13.89) gives:

$$\Delta = \left(\frac{3 \times 0.01 \times 1}{2\pi \times 980 \times 1.88}\right)^{1/3} \sin^{-2/3}\left(\frac{\pi}{2}\right) = 0.0137 \text{ cm}$$

At this flow rate, Eq. (13.91) gives $k_c = 0.0067$ cm/s. Thus, from Eq. (13.87) the time of exposure will be:

$$t_E = \frac{4\mathcal{D}^0_{A,B}}{\pi k_c^2} = \frac{4 \times 1.7 \times 10^5}{\pi(0.0067)^2} = 0.48 \text{ s}$$

which is relatively short. The corresponding depth of penetration of A (CO_2) into the liquid at the sphere equator will thus be:

$$\delta \sim 4\sqrt{\mathcal{D}^0_{A,B} t_E} = 4\sqrt{1.7 \times 10^{-5} \times 0.24} = 0.0081 \text{ cm}$$

This is small enough compared to Δ that we can consider the penetration model valid. Finally, let's compare experimental and predicted mass transfer rates. For one sphere at 1 cm^3/s, from Eq. (13.92), we have:

$$SN_{A,1} = 4\pi R^2 k_c(c_{A,s} - c_{A,0}) = 4\pi(1.88)^2 0.0067(4.0 \times 10^{-5} - 0) = 1.20 \times 10^{-5} \text{ mol/s}$$

while for 5 spheres, from Eq. (13.96), we find:

$$M_A = \dot{V}\left[c_{A,s} - (c_{A,s} - c_{A,0})\left(1 - \frac{Sk_c}{\dot{V}}\right)^5\right]$$

$$= 1\left[4\times10^5 - (4\times10^5 - 0)\left(1 - \frac{4\pi(1.88)^2\,0.0067}{1}\right)^5\right]$$

$$= 3.3\times10^{-5}\ \text{mol/s}$$

As seen in Fig. 13.5, both values are in good agreement with the experimental data. Note that if we just assumed that the total rate is five times the rate for the top sphere, we would get $M_A = 6.0\times10^{-5}$ mol/s. This value would not be possible, however, since if it were true, the outlet concentration would exceed the solubility value. The reason why the actual rate is smaller is that with each mixing between spheres the inlet concentration to the next sphere increases, which reduces the driving force $c_{A,s} - c_{A,j-1}$, thereby reducing the rate of absorption on each successive sphere. Even so, the model does appear to deviate somewhat from the experimental data at flow rates higher than about 1.5 cm^3/s whether we have one sphere or five. These deviations likely occur as a result of the onset of ripples in the film that create turbulence and increase mass transfer.

13.3 Forced Convection in Turbulent Flow

13.3.1 Analogies Between Momentum, Heat, and Mass Transfer

As we saw for heat transfer, turbulent flow conditions are also likely to prevail in many practical mass transfer processes. In this section, we extend the development of analogies between momentum and heat transfer to the case of mass transfer leading to generalized correlations that connect the three phenomena together despite the vastly different underlying physical basis. As noted for heat transfer, mass transfer in turbulent flow can also be thought of as a result of the movement of turbulent eddies.

With reference to Fig. 13.6, we can describe the shear stress, the heat transfer flux, and the mass transfer flux of a component A at the solid-fluid interface based on the following equations:

$$\tau_w = \frac{1}{2}\rho v_\infty^2 f \tag{13.97}$$

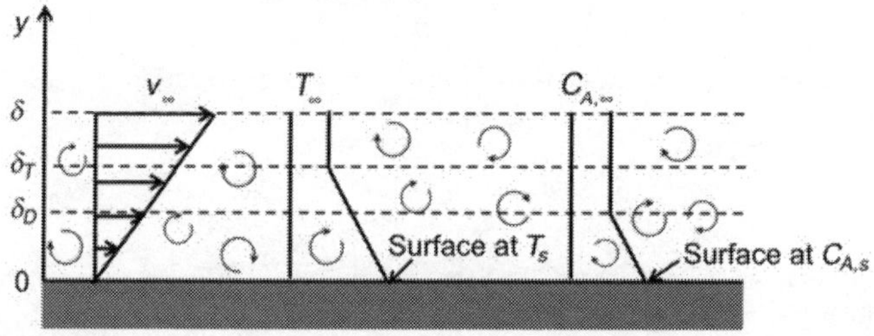

FIGURE 13.6 Momentum, heat, and mass transfer between a surface and a fluid in turbulent flow. δ, δ_T, and δ_D are, respectively, the thicknesses of the momentum transfer boundary layer, the thermal boundary layer, and the diffusive boundary layer.

$$q = h(T_s - T_\infty) \tag{13.98}$$

$$N_A = k_c(C_{A,s} - C_{A,\infty}) \tag{13.99}$$

where f is the friction factor, h is the heat transfer coefficient, and k_c is the mass transfer coefficient. Let's now assume that turbulent momentum, heat, and mass transfer can be described in terms of eddy transport coefficients according to the following relationships:

$$\tau_w = \rho E_M \frac{dv_x}{dy} \tag{13.100}$$

$$q = -\rho C_p E_H \frac{dT}{dy} \tag{13.101}$$

$$N_A = -E_D \frac{dC_A}{dy} \tag{13.102}$$

These equations have the same forms as Newton's law for viscous momentum transport, Fourier's law for conductive heat transfer, and Fick's law for diffusive mass transfer in a dilute solution, respectively. However, instead of viscosity, thermal conductivity, and diffusivity, they are based on, respectively, the eddy viscosity E_M, the eddy thermal diffusivity E_H, and the eddy mass diffusivity E_D. All of these quantities have dimensions of length squared over time, just like the kinematic viscosity ν, the thermal diffusivity α, and the diffusivity, $\mathcal{D}^0_{A,B}$. However, while ν, α, and $\mathcal{D}^0_{A,B}$ are molecular properties, E_M, E_H, and E_D represent transport associated with turbulent eddies and are, thus, dependent on the local level of turbulence. Combining Eqs. (13.97) and (13.100), Eqs. (13.98) and (13.101), and Eqs. (13.99) and (13.102), respectively, we obtain:

$$\frac{1}{2}\rho v_\infty^2 f = \rho E_M \frac{dv_x}{dy} \tag{13.103}$$

$$h(T_s - T_\infty) = -\rho C_p E_H \frac{dT}{dy} \tag{13.104}$$

$$k_c(C_{A,s} - C_{A,\infty}) = -E_D \frac{dC_A}{dy} \tag{13.105}$$

or

$$dy = \frac{E_M}{\frac{1}{2} v_\infty^2 f} dv_x \tag{13.106}$$

$$dy = -\frac{\rho C_p E_H}{h(T_s - T_\infty)} dT \tag{13.107}$$

$$dy = -\frac{E_D}{k_c(C_{A,s} - C_{A,\infty})} dC_A \tag{13.108}$$

Equating these three expressions for dy to each other gives the following result:

$$\frac{E_M}{\frac{1}{2}v_\infty^2 f}dv_x = -\frac{\rho C_p E_H}{h(T_s - T_\infty)}dT = -\frac{E_D}{k_c(C_{A,s} - C_{A,\infty})}dC_A \tag{13.109}$$

Now, having eliminated dy, we can integrate this result from the interface, where $v_x = 0$, $T = T_s$, and $C_A = C_{A,s}$, to a distance from the surface where $v_x = v_\infty$, $T = T_\infty$, and $C_A = C_{A,\infty}$ as follows:

$$\frac{E_M}{\frac{1}{2}v_\infty^2 f}\int_0^{v_\infty} dv_x = -\frac{\rho C_p E_H}{h(T_s - T_\infty)}\int_{T_s}^{T_\infty} dT = -\frac{E_D}{k_c(C_{A,s} - C_{A,\infty})}\int_{C_{A,s}}^{C_{A,\infty}} dC_A \tag{13.110}$$

Carrying out the integrations and canceling common terms, we obtain the following result:

$$\frac{E_M}{\frac{1}{2}v_\infty f} = \frac{\rho C_p E_H}{h} = \frac{E_D}{k_c} \tag{13.111}$$

or, after some rearrangement:

$$\frac{f}{2} = \frac{h}{\rho C_p v_\infty}\frac{E_M}{E_H} = \frac{k_c}{v_\infty}\frac{E_M}{E_D} \tag{13.112}$$

We can also rewrite these two equalities as follows:

$$\frac{f}{2} = \mathrm{St_H}\frac{E_M}{E_H} = \mathrm{St_D}\frac{E_M}{E_D} \tag{13.113}$$

where

$$\mathrm{St_H} = \frac{h}{\rho C_p v_\infty} \tag{13.114}$$

$$\mathrm{St_D} = \frac{k_c}{v_\infty} \tag{13.115}$$

are the Stanton numbers for heat transfer and for mass transfer, respectively. Equation (13.112) establishes a quantitative relationship between the friction factor, f, which is a purely fluid mechanical concept and the Stanton numbers, which describe convective heat transfer and convective mass transfer. In cases where form drag is unimportant, if the friction factor, f, is measured or if it is estimated as a function of the Reynolds number, we can use this relationship to predict heat and mass transfer rates for the same conditions. As was shown in Sec. 6.3.1 for the case of heat transfer alone, the result for heat and mass transfer will depend on what assumptions are made for the relationship between E_M, E_H, and E_D. The simplest assumption that we can make is to take $E_M = E_H = E_D$, which leads to the following Reynolds analogy:

$$\frac{f}{2} = \mathrm{St_H} = \mathrm{St_D} \tag{13.116}$$

Another assumption is to take $E_M/E_H = \mathrm{Pr}^{2/3}$ and $E_M/E_D = \mathrm{Sc}^{2/3}$, which leads to the following Chilton–Colburn analogy:

$$\frac{f}{2} = \mathrm{St_H Pr}^{2/3} = \mathrm{St_D\,Sc}^{2/3} \tag{13.117}$$

Many practical systems where f depends solely on surface drag are found experimentally to conform to this relationship. We can also express this relation in terms of a Reynolds number defined according to the appropriate length scale. For example, for turbulent flow in a pipe, where the length-scale for Re is the pipe diameter D and v_∞ is equal to the average velocity $v_{\mathrm{avg}} = \dot{V}/(\pi D^2/4)$, we will have:

$$\mathrm{St_H} = \frac{h}{\rho C_p v_{\mathrm{avg}}} = \frac{hD}{k}\frac{\nu}{v_{\mathrm{avg}}D}\frac{\alpha}{\nu} = \frac{\mathrm{Nu}}{\mathrm{RePr}} \tag{13.118}$$

$$\mathrm{St_D} = \frac{k_c}{v_{\mathrm{avg}}} = \frac{k_c D}{\mathcal{D}^0_{A,B}}\frac{\nu}{v_{\mathrm{avg}}D}\frac{\mathcal{D}^0_{A,B}}{\nu} = \frac{\mathrm{Sh}}{\mathrm{Re\,Sc}} \tag{13.119}$$

Accordingly, Eq. (13.117) can be written as follows:

$$\frac{f}{2} = \mathrm{J_H} = \mathrm{J_D} \tag{13.120}$$

where:

$$\mathrm{J_H} = \frac{\mathrm{Nu}}{\mathrm{Re\,Pr}^{1/3}} \tag{13.121}$$

and:

$$\mathrm{J_D} = \frac{\mathrm{Sh}}{\mathrm{Re\,Sc}^{1/3}} \tag{13.122}$$

$\mathrm{J_H}$ and $\mathrm{J_D}$ are known as the Chilton–Colburn factors for heat transfer and for mass transfer, respectively. Obviously, for fluids for which Pr ~ Sc ~ 1, such as gases at low pressure, the Reynolds and Chilton–Colburn results will coincide.

13.3.2 Sherwood Number Correlations

Based on the analogies developed in the previous section, in most cases, correlations developed for convective heat transfer can also be used for an approximate prediction of convective mass transfer by replacing Nu with Sh and Pr with Sc. Thus, the equations presented in Table 6.1 for forced convection heat transfer will take the form:

$$\mathrm{Sh} = \mathrm{Sh(Re, Sc)} \tag{13.123}$$

for mass transfer. An extensive list of correlations specific for mass transfer in the form of Eq. (13.123) can be found in Tables 5-17 to 5-23 of *Perry's Chemical Engineers Handbook*, 9th edition. These tables should be consulted for more precise calculations or to handle more complex cases. In general, as we saw in Chap. 6 for heat transfer, if the Sherwood number is known for a certain fluid at a given Re, estimates for a different fluid at the same Re can be made by scaling Sh with $\mathrm{Sc}^{1/3}$. Because of the 1/3 power, the effect of Sc on Sh is generally modest. However, while Pr varies over a relatively modest range for most fluids, very large ranges are found for Sc. For example, while for low-pressure gases Sc ~ 1, for liquids Sc will vary from about 1,000 for small molecules in water to tens or even hundreds of thousands depending on the size of the diffusing solute and the viscosity of the liquid.

Dimensionless Number	Definition	Meaning	Range*
Reynolds	$\mathrm{Re} = \frac{\rho v \ell}{\mu} = \frac{\rho v^2}{\mu v/\ell}$	$\frac{\text{Inertial momentum transport}}{\text{Viscous momentum transport}}$	
Prandtl	$\mathrm{Pr} = \frac{\nu}{\alpha} = \frac{\mu}{k/C_p}$	$\frac{\text{Viscous momentum transport}}{\text{Conductive heat transport}}$	~1 for gases ~10–100 for liquids
Nusselt	$\mathrm{Nu} = \frac{h\ell}{k} = \frac{h}{k/\ell}$	$\frac{\text{Convective heat transport}}{\text{Conductive heat transport}}$	
Schmidt	$\mathrm{Sc} = \frac{\nu}{\mathcal{D}^0_{A,B}} = \frac{\mu}{\rho\mathcal{D}^0_{A,B}}$	$\frac{\text{Viscous momentum transport}}{\text{Diffusive mass transport}}$	~1 for gases ~1,000–100,000 for liquids
Sherwood	$\mathrm{Sh} = \frac{k_c\ell}{\mathcal{D}^0_{A,B}} = \frac{k_c}{\mathcal{D}^0_{A,B}/\ell}$	$\frac{\text{Convective mass transport}}{\text{Diffusive mass transport}}$	

*Order of magnitude only.

TABLE 13.1 Summary of Dimensionless Groups Used for Convective Heat and Mass Transfer Problems. ℓ Is a Characteristic Length Used Consistently in the Definitions of Re, Nu, and Sh and Defining the Scale Over Which Velocity, Temperature, and Concentration Changes Drive the Corresponding Transport Process

Before proceeding with some examples, it is useful to take a look at Table 13.1, which summarizes the various dimensionless groups that appear in these equations, their physical meaning, and, where appropriate, the range of values observed in typical cases. The magnitude of these dimensionless ratios will tell us which mechanism is dominant. For instance, large values of Nu and Sh will tell us that convective transport is dominant compared to, respectively, conductive and diffusive mechanisms. Examples 13.4, 13.5, and 13.6 illustrate use of these equations for a few different practical cases.

Example 13.4. Mass transfer to a sphere falling through a stagnant fluid. Consider a process where benzoic acid is extracted from water droplets that fall in a column of stagnant liquid benzene at 25°C. The droplets have a diameter $D = 3$ mm and initially contain a small concentration of benzoic acid ($c_{A,0} \sim 3$ g/L). Benzoic acid is partitioned between the two phases and is, thus, extracted into the benzene. Determine the convective mass transfer coefficient at the interface between the two liquids. Since the droplets are small, we can treat them as solid spheres. In this case, based on the Rantz and Marshall equation, item D in Table 6.1, convective mass transfer to spheres will be described by the following equation:

$$\mathrm{Sh} = 2.0 + 0.60\,\mathrm{Re}^{1/2}\,\mathrm{Sc}^{1/3} \tag{13.124}$$

where $\mathrm{Sh} = k_c D/\mathcal{D}^0_{A,B}$, $\mathrm{Re} = v_\infty D/\nu$, and $\mathrm{Sc} = \nu/\mathcal{D}^0_{A,B}$. Note that this equation assumes that the spheres can be considered as isolated in a large volume of fluid. In order to use this equation, we need to estimate: (a) the velocity, v_∞, at which the spheres fall and (b) the diffusivity of benzoic acid (A) in benzene (B), $\mathcal{D}^0_{A,B}$. We can estimate v_∞ by assuming that it is equal to the terminal velocity v_t. In turn, v_t is obtained by balancing the forces due to gravity, F_G, and buoyancy, F_B, with the drag force, F_D, acting on the sphere. At steady state, these forces are given, respectively, by the following equations:

$$F_G = \frac{\pi D^3}{6}\rho_s g \tag{13.125}$$

$$F_B = \frac{\pi D^3}{6}\rho g \tag{13.126}$$

$$F_D = \frac{1}{2}\rho v_t^2 C_D A_p \tag{13.127}$$

where ρ_s is the density of the spheres (water in this case), ρ is the density of the fluid (benzene in this case), C_D is the drag coefficient, and $A_p = \pi D^2/4$ is the sphere area projected in the direction of the velocity (see Sec. 6.2 in *Perry's Chemical Engineers' Handbook*, 9th edition). $C_D = C_D(\text{Re})$ is a function of the Reynolds number, $\text{Re} = v_t D/\nu$, given by Fig. 13.7. Balancing the three forces above gives:

$$F_D = F_G - F_B \tag{13.128}$$

Substituting Eqs. (13.125) to (13.127) into Eq. (13.128), we obtain:

$$\frac{\pi D^2}{8}\rho C_D v_t^2 = \frac{\pi D^3}{6}(\rho_s - \rho)g \tag{13.129}$$

From this Eq. (13.129), we obtain:

$$C_D = \frac{4gD(\rho_s - \rho)}{3\rho v_t^2} = \frac{4gD^3(\rho_s - \rho)\rho}{3\mu^2}\frac{1}{\text{Re}^2} \tag{13.130}$$

and, with some rearrangement,

$$v_t = \sqrt{\frac{4gD(\rho_s - \rho)}{3\rho C_D}} \tag{13.131}$$

The terminal velocity can be found at the intercept of Eq. (13.130) with the function $C_D = C_D(\text{Re})$ given by Fig. 13.7. For this problem, the density of the sphere will be the density of water, or $\rho_s = 1{,}000\ \text{kg/m}^3$. The density and viscosity of the fluid will be the values for liquid benzene at 25°C, which are $\rho = 876\ \text{kg/m}^3$ and $\mu = 0.6 \times 10^{-3}$ Pa s, respectively. As shown in Fig. 13.7, the intercept between the two lines is at Re = 428, where $C_D = 0.58$. Using this value, Eq. (13.131) gives $v_t = 0.098$ m/s. Next, we estimate $\mathcal{D}^0_{A,B}$ for benzoic acid in benzene from the Wilke–Chang equation, Eq. (10.24), with $\mu_B = 0.60$ cp, $\phi_B = 1$, and $M_{r,B} = 78.1$. For benzoic acid (C_6H_5COOH with an aromatic ring), we estimate $\hat{V}_A$ from the LeBas volumes in Table 10.5 to be $\hat{V}_A = 7 \times 14.8 + 6 \times 3.7 + 2 \times 12.0 - 15.0 = 135\ \text{cm}^3/\text{mol}$. Using these values, Eq. (10.24) predicts:

$$\mathcal{D}^0_{A,B} = 7.4 \times 10^{-8}\frac{298}{0.60}\frac{(1.0 \times 78.1)^{1/2}}{(135)^{0.6}} = 1.7 \times 10^{-5}\ \text{cm}^2/\text{s} = 1.7 \times 10^{-9}\ \text{m}^2/\text{s}$$

The Schmidt number is thus $\text{Sc} = 0.6 \times 10^{-3}/(876 \times 1.7 \times 10^{-9}) = 403$. Finally, from Eq. (13.124) we have:

$$\text{Sh} = 2.0 + 0.60(428)^{1/2}(403)^{1/3} = 94$$

and

$$k_c = \text{Sh}\frac{\mathcal{D}^0_{A,B}}{D} = 94 \times \frac{1.7 \times 10^{-9}}{3.0 \times 10^{-3}} = 5.3 \times 10^{-5}\ \text{m/s}$$

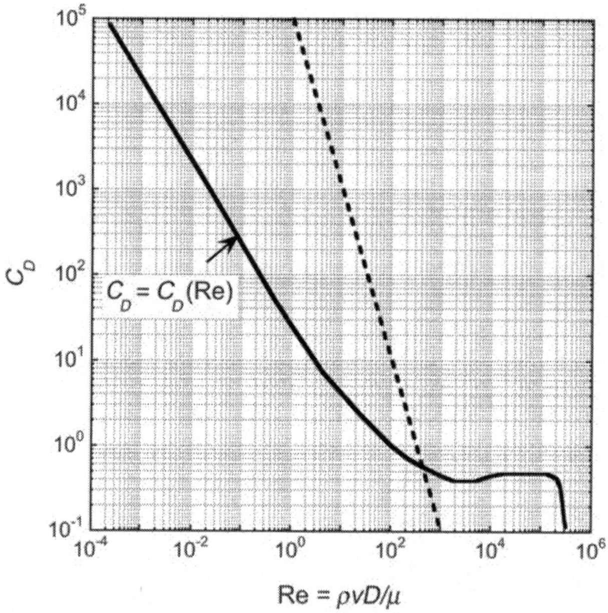

FIGURE 13.7 Drag coefficient, C_D, vs. Reynolds number for spheres. The dashed line represents Eq. (13.130) for the conditions of this problem. The intercept of the two lines gives Re and C_D at the terminal velocity.

Example 13.5. Mass transfer coefficient in a packed adsorption bed. Packed adsorption beds are used extensively, for example, to remove organic vapors from air and other gaseous streams. Let's estimate the mass transfer coefficient for the adsorption of hexane from a dilute hexane–air mixture at 300 K and 1 atm in a bed of granular activated carbon. The granules are approximately spherical and 5 mm in diameter. The diameter of the adsorption bed is 0.5 m. The feed air flows at the rate of 0.1 m³/s. The Nusselt number correlation in Table 6.1 (item F) can be used to estimate the Sherwood number by replacing Pr with Sc. For packed beds, the Reynolds number is defined based on the fluid superficial velocity in the bed, v_{avg}, and the particle diameter, D, according to the following equation:

$$\text{Re} = \frac{v_{avg} D}{\nu} \tag{13.132}$$

v_{avg} is obtained by diving the volumetric flow rate by the bed cross-section, which, in this case, gives $v_{avg} = 0.1/(\pi D^2/4) = 0.1/(\pi \times 0.5^2/4) = 0.51\ \text{m/s}$. For air at 300 K and 1 atm, we have $\nu = 1.6 \times 10^{-5}\ \text{m}^2/\text{s}$. Thus, $\text{Re} = 0.51 \times 0.005/1.6 \times 10^{-5} = 159$. The following equation from item F in Table 6.1, replacing Nu with Re and Pr with Sc, is applicable at this value of Re:

$$\text{Sh} = \frac{0.25}{\varepsilon} \text{Re}^{0.69}\, \text{Sc}^{1/3} \tag{13.133}$$

where ε is the bed void fraction (or bed porosity). Absent specific information, a reasonable estimate is $\varepsilon = 0.35$ (packed-bed void fractions are usually in the range 0.3–0.4, regardless of particle size). Next, we need to estimate $\mathcal{D}_{A,B}$ for the hexane-air system. Table 2-141 in *Perry's Chemical Engineer's Handbook*, 9th edition, gives $\mathcal{D}_{A,B} = 0.075 \times 10^{-4}\ \text{m}^2/\text{s}$ for this system at 15°C. The value at 25°C can be estimated as $0.075 \times 10^{-4}(298/288)^{1.75} = 8.0 \times 10^{-6}\ \text{m}^2/\text{s}$ based on Eq. (10.6). Thus, $\text{Sc} = 1.6 \times 10^{-5}/8.0 \times 10^{-6} = 2.0$. Finally, from Eq. (13.133) we obtain:

$$\text{Sh} = \frac{0.25}{0.35} \times (159)^{0.69} (2.0)^{1/3} = 29.7$$

and

$$k_c = \text{Sh}\frac{\mathcal{D}_{A,B}}{D} = 29.7 \times \frac{8.0\times10^{-6}}{0.005} = 0.048 \text{ m/s}$$

Ruthven, *Principles of Adsorption and Adsorption Processes*, Wiley, 1984, recommends the following equation by Wakao and Funazkri [Chem. Eng. Sci. 33 (1978) 1375] for adsorption beds that gives Sh without an explicit dependence on ε:

$$\text{Sh} = 2 + 1.1\text{Re}^{0.6}\,\text{Sc}^{1/3} \quad \text{for } 3 < \text{Re} < 10{,}000 \tag{13.134}$$

For our case, this equation gives Sh = 31, in pretty good agreement (~ 4% difference) with the estimate obtained from Eq. (13.133) with $\varepsilon = 0.35$.

Example 13.6. Mass transfer to particles suspended in an agitated tank. In Example 6.5 we estimated the heat transfer coefficient for particles of diameter D suspended in a liquid in an agitated tank. Let's do the analogous calculation for mass transfer. Consider the use of 50 μm adsorbent beads to recover a valuable protein with 150,000 molecular mass from a dilute aqueous solution at 20°C having a viscosity of 5 mPa s. Let's assume that the tank has the same geometry and agitation rate as in Example 6.5. For those conditions, we estimated a power input per unit mass $\bar{\varepsilon} = 0.12$ W/kg. Mass transfer is predicted by the following correlation presented as Eq. (6.89) for the analogous heat transfer process:

$$\text{Sh} = \frac{k_c D}{\mathcal{D}^0_{A,B}} = 2 + a\left(\frac{\bar{\varepsilon}^{1/3}D^{4/3}}{\nu}\right)^b \text{Sc}^{1/3} \tag{13.135}$$

where D is the particle diameter and the constants a and b are a = 0.5 and b = 0.62 according to Levins and Glastonbury [Trans. Inst. Chem. Eng., 50 (1972) 132] and a = 0.52 and b = 0.52 according to Armenante and Kirwan [Chem. Eng. Sci., 44 (1989) 2781] (*cf.* Fig. 6.10). For our case, we have:

$$\frac{\bar{\varepsilon}^{1/3}D^{4/3}}{\nu} = \frac{0.12^{1/3}\times\left(50\times10^{-6}\right)^{4/3}}{5.0\times10^{-6}} = 0.18$$

From Fig. 6.10, we obtain $(\text{Sh}-2)/\text{Sc}^{1/3} \sim 0.3$. The diffusivity of a 150,000 molecular mass protein in water at 20°C can be estimated from Fig. 10.3 (assuming that the protein is globular), which gives $\mathcal{D}^0_{A,B} = 5.1\times10^{-7}$ cm²/s or 5.1×10^{-11} m²/s. Correcting for the higher viscosity, this value becomes 1.0×10^{-11} m²/s. Accordingly, $\text{Sc} = \nu/\mathcal{D}^0_{A,B} = 5.0\times10^{-6}/1.0\times10^{-11} = 500{,}000$. Finally, we obtain $\text{Sh} = 2 + 0.3\times(500{,}000)^{1/3} = 26$, which, in turn, gives $k_c = 26\mathcal{D}^0_{A,B}/D = 26\times1.0\times10^{-11}/50\times10^{-6} = 5.2\times10^{-6}$ m/s.

13.4 Problems

P.13.1. A wetted-wall column consists of a vertical solid cylinder, 2.5 cm in diameter and 10.0 cm tall. Pure water at 25°C flows down the outer surface of the cylinder at the rate of 10.0 cm³/s. The column is placed in an atmosphere of pure CO_2 at 25°C and 1 atm absolute pressure. Assuming steady state and that the flow is laminar, determine: (a) The average mass transfer coefficient for CO_2 absorption in the wetted-wall column. (b) The mixing-cup average concentration of CO_2 in the liquid leaving the wetted-wall column (see Perry's Table 2-21 for gas solubility data). (c) The rate of mass transfer of CO_2 into the film.

P.13.2. Following a textile operation, residual liquid n-hexane initially present in 1.0-mm-thick cloth is to be removed by pulling the cloth continuously through a drying oven where the n-hexane is vaporized and removed. The dry cloth is continuously rolled up at the outlet of the drying oven. The cloth is 0.5 m wide and the length of the oven is 2.5 m. In the oven, air at 20°C and 1 atm flows over both sides across the width of the cloth at a velocity of 1.5 m/s. Assuming that the process is controlled by the external mass transfer resistance, determine the velocity at which the cloth can be pulled through the oven if it is desired to remove 95% of the n-hexane and the initial concentration of n-hexane in the cloth is 10 kg/m^3. Neglect the effects of the cloth velocity on the mass transfer coefficient.

P.13.3. Water at 25°C flows at 5 ft/s inside a straight cylindrical tube made of benzoic acid, with a 1-in. inside diameter. If the tube is 20 ft long, estimate the mixing-cup average concentration of benzoic acid in the water leaving the tube. The Schmidt number for these conditions is Sc = 970 and the solubility of benzoic acid in water is 2.5 kg/m^3.

P.13.4. Acetic acid is to be removed from an aqueous solution by allowing droplets of the acetic acid-water solution to fall through a column of benzene. The droplets initially contain 1.0 weight % of acetic acid and this has to be reduced to 0.2%. The droplets have a diameter of 1 mm and the temperature is 25°C. At this temperature, the diffusivity of acetic acid in water is 0.88×10^{-9} m^2/s while the diffusivity of acetic acid in benzene is 1.92×10^{-9} m^2/s, both in dilute conditions. The distribution coefficient of acetic acid between water and benzene is 45.5, defined as the weight fraction of acetic acid in water divided by the weight fraction of acetic acid in benzene. (a) Determine the distance the drops must fall if they are treated as solid spheres and both internal and external mass transfer resistances are considered. (b) Repeat part (a) assuming that the droplets fall at the same velocity as in part (a) but that now they are completely mixed by internal recirculation so that the external mass transfer resistance is controlling. Assume that the volume of benzene is very large so that the weight fraction of acetic acid in the bulk benzene phase remains zero in both part (a) and part (b).

P.13.5. An impurity is to be removed from an aqueous solution by adsorption on a selective adsorbent in a batch system. The solution, at 25°C and containing 0.1 kg/m^3 of the impurity, will be loaded into a 1 m diameter baffled cylindrical tank to a 1 m height agitated by a 0.33 m diameter six-blade turbine at 30 rpm. All other geometric ratios correspond to curve 1 in Figure 18-1 of *Perry's Chemical Engineers' Handbook*, 9th edition. The adsorbent particles are 0.05 cm in diameter, have a density of 1,100 kg/m^3, and are completely suspended in the liquid. The diffusivity of the impurity in the solution at 25°C is 6.6×10^{-10} m^2/s. (a) Estimate the mass transfer coefficient for transport of the impurity to the surface of the adsorbent particles. (b) Predict the time needed to remove 90% of the impurity if 1 kg of adsorbent particles is added to the tank and the adsorption process is controlled by the external mass transfer resistance. Assume that the impurity concentration at the particle surface is kept at zero because the adsorption is very favorable.

P.13.6. You have developed a reputation as an expert in transport phenomena and this time Vincent Gambini (who made quite a career for himself after the eponymous movie "My Cousin Vinny" and is now working for the DA office in that small Alabama town) is asking for your help on a case he is working on. Some months ago, people fishing on a river 2 miles below the Reynolds & Schmidt chemical plant noted small particles bobbing around and being carried downstream by the river flow. Suspicious that there could be a problem, they collected some of these particles and handed them to the local DEQ office, which determined that they are made of a chemical that is very toxic to aquatic life. The case is now being tried in court with Reynolds & Schmidt accused of illegally dumping these toxic particles in the river. Vinny, as the lead prosecutor, has argued that Reynolds & Schmidt produces this chemical in the form of solid spherical pellets, 2 mm in diameter, and a witness has testified that he saw some Reynolds & Schmidt employees unloading drums containing what looked like the same particles along the river bank at the Reynolds & Schmidt location exactly 2 h before the fisherman observed these particles. "Surely, Reynolds & Schmidt is responsible!" says Vinny. The defense does not dispute these facts, but has argued that whatever was in those drums could not have been the toxic particles in question since, surely, they would have been completely dissolved after 2 h in the river and the fisherman could not have seen them. "Those particles must have come from somewhere else and Reynolds & Schmidt is not responsible. The prosecution case does not hold water!" says the lead defense lawyer. There is no time for measurements, but we know that the chemical structure of the chemical in question is:

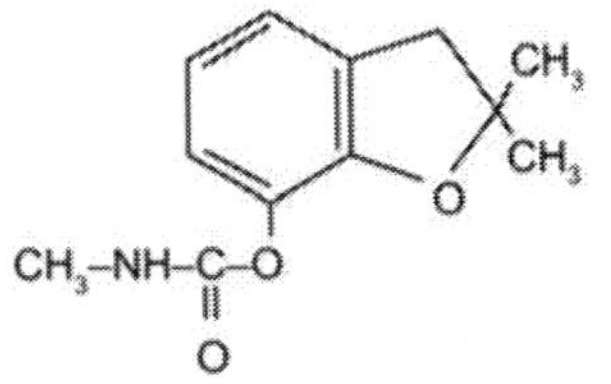

The solid particles are essentially neutrally buoyant and the solubility of the chemical in water is 0.4 kg/m^3. That faithful day the river temperature was 20°C and the average stream velocity was 1 m/s. Vinny wants you to testify as an "expert witness" to rebut the defense's argument. What would your expert testimony be?

P.13.7. The controlled release of pheromones can be used to attract flying crop-damaging insects to a trap as a safer alternative to pesticides. A simple pheromone release device is shown in the figure below (not to scale). The pheromone is a waxy solid cylinder, which sublimes and is transported away into the flowing air. The sublimation pressure of the pheromone at 27°C is 0.50 mmHg, its molecular mass is 200, and its diffusion volume is $\Sigma v = 80$ cm^3/mol. The air temperature is 27°C. The device will be placed in a location where the wind speed is a constant 2.0 m/s. The density of the solid pheromone is 1,200 kg/m^3, the initial diameter of the pheromone cylinder is 0.02 m, and its length is 0.4 m. (a) Estimate as accurately as possible the diffusion coefficient of the pheromone in air at 27°C and 1 atm. (b) Calculate the initial value of the mass transfer coefficient for pheromone transport into the air. (c) Calculate the initial rate of pheromone release. (d) Derive an explicit equation to predict how the radius of the cylinder will change as a function of time. Note that the mass transfer coefficient will vary with the cylinder radius. (e) Integrate your equation in part (d) and determine the time required for the cylinder radius to decrease to one half of the initial value.

P.13.8. A bubble of pure carbon monoxide (CO) gas at 1 atm, initially 0.2 cm in diameter, is held stationary in liquid water 25°C flowing at a constant velocity of 10 cm/s. (a) Estimate the diffusion coefficient of CO in liquid water for these conditions. (b) Estimate the convective mass

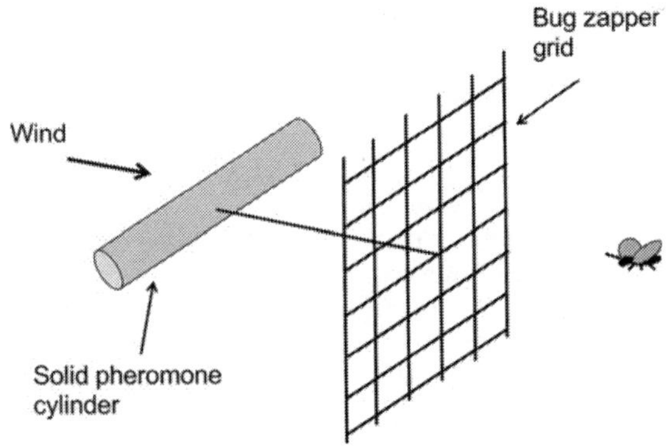

FIGURE P.13.7.

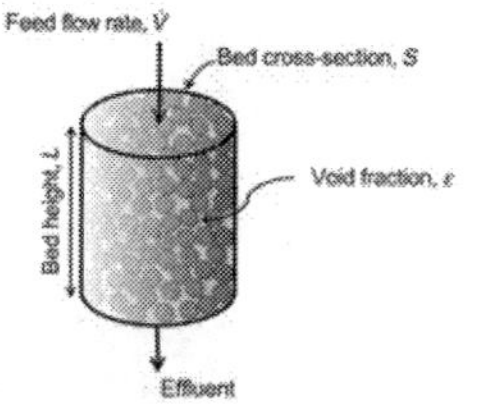

FIGURE P.13.10.

transfer coefficient for CO at the gas-liquid interface assuming that the CO bubble can be treated as a solid sphere. (c) Determine the initial rate of mass transfer of CO from the bubble to the liquid in mol/s. (d) Write an equation to predict the rate of change of the diameter of the gas bubble. (e) Calculate the initial rate of change of the bubble diameter in cm/s. (f) Integrate your equation in (d) to obtain an expression for the bubble diameter as a function of time. (g) Use your equation in (f) to predict the time needed for the bubble diameter to become one half of the initial value. See Perry's Table 12-21 for gas solubility data.

P.13.9. The free-fall terminal velocity u (cm/s) of spherical water droplets in quiet air at 298 K and 1 atm is given approximately by:

$$u = 3{,}930 \times D$$

where D is the droplet diameter in cm. The mass transfer coefficient at the droplet-air interface can be predicted from the following approximate correlation:

$$\mathrm{Sh} = \frac{k_c D}{D^0_{A,B}} \sim 0.6\,\mathrm{Re}^{1/2}\mathrm{Sc}^{1/3}$$

$$\mathrm{Re} = \frac{\rho u D}{\mu}$$

Assume that a droplet initially 0.3 cm in diameter falls in quiet, air at 298 K, 1 atm, and 40% relative humidity and that it can be treated as isotherm. (a) Write down an explicit expression to predict the rate of evaporation from the droplet (mol/s) as a function of the droplet diameter D. (b) Calculate the mass transfer coefficient when the droplet diameter is 0.3 cm. (c) Calculate the initial rate of evaporation of water (mol/s) when the droplet diameter is 0.3 cm. (d) Derive an ordinary differential equation relating the droplet diameter D to time. (e) Integrate your equation in (d) and determine the time required for the falling droplet to attain a diameter of 0.1 cm.

P.13.10. A bed of granular activated carbon is to be used to remove chlorobenzene from contaminated water at 20°C. The carbon consists of roughly spherical particles with a 3-mm average diameter which are packed in a column with a void fraction $\varepsilon = 0.35$ operated at a superficial velocity $u = \dot{V}/S = 0.0014$ m/s as shown in Fig. P.13.10. (a) Estimate the diffusivity of chlorobenzene in water for these conditions. (b) Estimate the mass transfer coefficient. (c) Assuming pseudo steady state, determine the minimum bed height to achieve 99.9% removal of the chlorobenzene if mass transfer to the surface of the particles is assumed to be controlling. Note that the adsorption capacity is very high and adsorption is very favorable which keep the chlorobenzene concentration at the particle surface at essentially zero throughout the adsorption bed. (d) Repeat your calculations in (a) to (c) if the same activated carbon is used to remove chlorobenzene from air at 20°C and 1 atm if, in this case, the adsorption bed is designed to operate at a superficial velocity of 0.4 m/s. Approaches for the complete design of adsorption beds can be found in specialized textbooks as well as in Sec. 16 of *Perry's Chemical Engineer's Handbook*, 9th edition.

CHAPTER 14

Design of Packed Gas Absorption and Stripping Columns

14.1 Introduction

In this chapter, we consider the design of gas absorption and stripping columns as an example of application of mass transfer principles to the design of industrial units. Both absorption and stripping columns are used extensively in a variety of applications. Absorption columns (or just absorbers) are used to remove or recover one or more components from a gas stream by absorbing them into a liquid. Stripping columns (or strippers) are used to remove one or more components from a liquid by volatilizing them in a gas stream. In both cases, mass transfer at the gas-liquid interface is critical. Thus, the equipment used to perform either absorption or stripping is designed to: (a) generate a large gas-liquid interfacial area and (b) promote interphase transport by reducing the mass transfer resistance. Figure 14.1 shows, schematically, a typical packed column used for absorption or stripping operations. The column itself is usually a cylindrical vessel. The gas enters from the bottom and exits from the top. The liquid enters from the top and flows down by gravity leaving from the bottom of the column. The column itself is filled to a certain height with a packing material whose purpose is to create a large interfacial area between the gas phase and the liquid phase. The packing is supported by a grid or mesh at the bottom while the liquid is distributed at the top evenly across the column diameter and flows down over the packing forming films and drops. The gas phase flows upwards through the packing in close contact with the liquid. Convective transport occurs at these interfaces from the gas to the liquid in an absorber and from the liquid to the gas in a stripper.

There is an enormous variety of practical column packings used in gas absorption and stripping columns varying in size, shape, structure, and materials of construction. Section 14.4 in *Perry's Chemical Engineers' Handbook*, 9th edition, gives many examples. There are two basic types: random packings and structured packings. **Random packings** consist of individual pieces, typically 1 to 9 cm in size, that fill the column in a random arrangement. Figure 14.1 shows an example of this type of packing called a Raschig ring. Since each piece of packing supports the weight of the pieces above, there are limits as to the maximum packing height that can be attained. If a greater height is

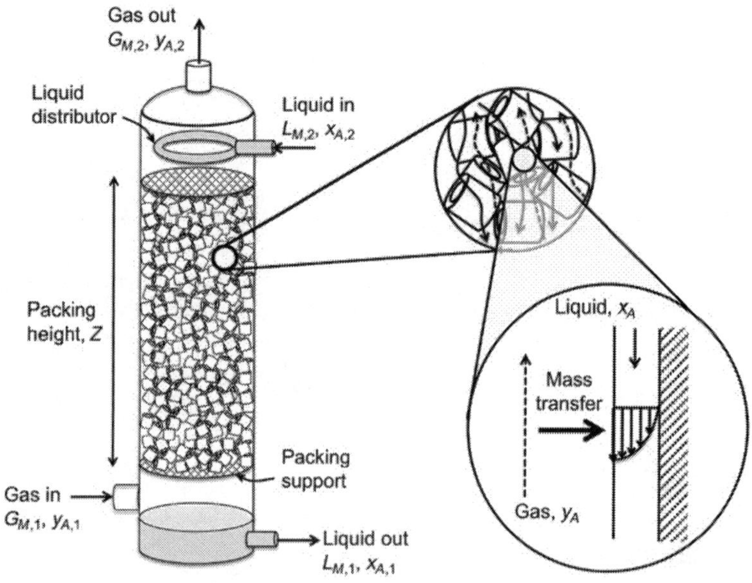

FIGURE 14.1 Sketch of a counter-current packed column used to absorb a component *A* from the entering gas or to strip a component *A* from the entering liquid. The column shown contains random packing. Columns with structured packing are similar except that no packing support grid is needed. G_M and y_A are the molar flow rate of the gas and the gas-phase mole fraction of *A*, respectively. L_M and x_A are the molar flow rate of the liquid and the liquid-phase mole fraction of *A*, respectively.

needed than that which can be self-supporting, the packing can be divided into multiple sections, each with its own packing support grid and liquid distributor. **Structured packings** consist of self-supporting structures that are manufactured into sections, which are neatly arranged within the column. In both cases, whether random or structured, the design of the packing must provide a high surface area along with open areas for the gas to flow. The gas flow is pressure-driven. Thus, the pressure is higher at the bottom of the column providing the driving force for the gas to flow. If the resistance to flow through the packing is high, then there will be a substantial pressure drop, which may be undesirable. Such a pressure drop is particularly challenging if the column is operated near or below atmospheric pressure. In these cases, only very small pressure drops can be tolerated.

The packing design, whether random or structured, also affects the liquid flow, which is driven by gravity. In general, the smaller the packing size, the greater the gas-liquid interfacial area will be, which is favorable for mass transfer. Conversely, smaller packing pieces will lead to higher pressure drops and to a reduced ability to handle the flow of liquid. Packing designed to provide a high void fraction will also result in a lower pressure drop and better liquid-handling capability. However, this will also often reduce the mechanical strength and ability to pack tall columns. Thus, a compromise of these trends will often be needed. Typically, structured packings will provide a better compromise of gas-liquid interfacial area, pressure drop, and liquid-handling capability than random packing. Thus, structured packings will often be the type chosen for low pressure operation.

Structure packings are, however, also typically more expensive compared to random packings. Thus, the choice between the two types of packing will ultimately depend on economic factors. A final important consideration pertains to the materials of construction. Early on, random packings made of ceramics were commonplace. Such packings are not very efficient since they do not allow the complex geometries needed to promote larger interfacial areas and are, of course, heavy. However, they have the advantage of being resistant to corrosive conditions and are low-cost. Metals or plastics, on the other hand, whether in random or structured form, provide much greater flexibility for the design of open structures with high surface area, high void fraction, and high liquid-handling capability.

The design of packed absorption or stripping columns entails three major tasks. The first task is the determination of the flow rate of absorbent liquid required for an absorber or of the flow rate of gas required for a stripper. The second task is the determination of the height of packing needed. The third and final major task is the determination of the diameter of the column. These three tasks are somewhat interdependent. In general, however, we can say that the feed liquid flow rate in an absorber or the feed gas flow rate in a stripper will be determined, largely, by material balances and phase equilibrium relationships. The height of packing will be determined by transport considerations that account for the effects of phase equilibrium on the driving force for mass transfer. Finally, the column diameter will be determined by the liquid-handling capability of the packing and by the allowable pressure drop. The liquid-handling capability will be determined by flooding and pressure drop characteristics of the packing used. Flooding occurs at elevated gas velocities when the liquid is no longer able to flow down under gravity because of the resistance brought about by the gas flow. Reliable performance will require that we operate below flooding conditions. Although flooding and pressure drop are specific to the particular type of packing used, its size, and the material of constructions, generalized correlations that can be used to predict stable operating conditions exist and are discussed in Sec. 14.4. In the following sections, we address the determination of the required feed flow rate of absorbent liquid or stripping gas and of the packing height. We do this initially for the case of dilute conditions and then for concentrated cases. First, however, we discuss how to describe interphase mass transfer at the gas-liquid interface.

14.2 Column Design for Dilute Conditions

We consider first dilute conditions where, for example, the gas and liquid mole fractions do not exceed about 0.05. This is not a hard-and-fast rule, however, and should be viewed only as a guideline. The analysis becomes increasingly less accurate if the mole fractions rise above this value, but can still be used as an approximation for many cases.

14.2.1 Mass Transfer at the Gas-Liquid Interface

Figure 14.2 shows, schematically, the mole fraction profiles that we can expect in the vicinity of the gas-liquid interface at a particular point within the column. y_A and x_A are the mole fractions of a component A in the bulk gas and in the bulk liquid, respectively. $y_{A,s}$ and $x_{A,s}$ are the corresponding values at the interface itself. We already know that,

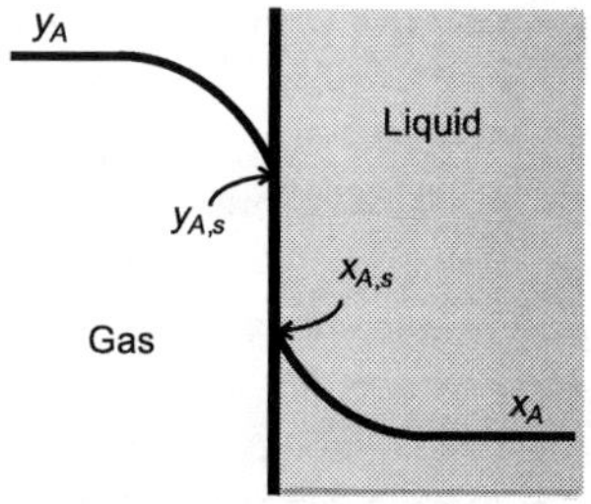

FIGURE 14.2 Mole fraction profiles of a component A near the gas-liquid interface in an absorption column.

in general, $y_{A,s}$ and $x_{A,s}$ are not equal to each other. Instead, they are related to each other by the phase equilibrium relationship which we can write as follows:

$$y_{A,s} = Kx_{A,s} \tag{14.1}$$

where K is a distribution coefficient based on mole fractions. In general, K will be a function of temperature and pressure. For example, if Henry' law is applicable, Eq. (14.1) can be written as follows:

$$y_{A,s} = \frac{H(T)}{P} x_{A,s} \tag{14.2}$$

where $H(T)$ is the Henry's law constant (units of pressure) and P is the absolute pressure. Thus, in this case, $K = H(T)/P$. According to this relationship, at constant T and P, the relationship between $y_{A,s}$ and $x_{A,s}$ is linear. We refer to this as a straight equilibrium line. More generally, however, the relationship between $y_{A,s}$ and $x_{A,s}$ at constant T and P will be non-linear. For a given system, this will often occur at higher values of the mole fractions where deviations from Henry's law become significant. We refer to this as a curved equilibrium line. We will see that somewhat different considerations will apply depending on whether the equilibrium line is straight or curved.

Regardless of the shape of the equilibrium line, the mass transfer flux of component A within each phase can be written in terms of the corresponding mass transfer coefficient and driving force. For dilute conditions, on the gas-side of the interface we can write:

$$N_A = k_y(y_A - y_{A,s}) \tag{14.3}$$

while on the liquid side of the interface, we can write:

$$N_A = k_x(x_{A,s} - x_A) \tag{14.4}$$

In these equations, k_y and k_x are the convective mass transfer coefficients of component A in the gas and liquid phase, respectively. In general, these coefficients will depend on the gas and liquid flow rates, on the packing size and structure, and on the physical properties of the two phases and of the component A being transferred across the interface. Typical units of k_y and k_x will be mol/m^2 s or lbmol/ft^2 h. Note that, as discussed in Sec. 13.1, alternate driving forces could be used (e.g., based on concentration or partial pressure), but, for simplicity, here we will conduct our analyses just in terms of mole fractions. We can always interconvert between different driving forces based on Eqs. (13.7) and (13.9).

Since the mass transfer flux is the same on the two sides of the interface, we can find the interfacial composition by setting Eqs. (14.3) and (14.4) equal to each other, which gives:

$$k_y(y_A - y_{A,s}) = k_x(x_{A,s} - x_A) \tag{14.5}$$

With some rearrangement, Eq. (14.5) yields the following result:

$$y_{A,s} = -\frac{k_x}{k_y}x_{A,s} + \left(y_A + \frac{k_x}{k_y}x_A\right) \tag{14.6}$$

For given values of the bulk mole fractions y_A and x_A and knowing the ratio of the mass transfer coefficients, k_x/k_y, we can use Eq. (14.6) in conjunction with the equilibrium line to find $y_{A,s}$ and $x_{A,s}$. As is illustrated below, there are two different scenarios for this determination depending on whether the equilibrium line is straight or curved.

A. Straight Equilibrium Line

If the equilibrium line is straight, we will have $x_{A,s} = y_{A,s}/K$, where K is a constant at given T and P. In this case, Eq. (14.6) becomes:

$$y_{A,s} = -\frac{k_x}{k_y}\frac{y_{A,s}}{K} + \left(y_A + \frac{k_x}{k_y}x_A\right) \tag{14.7}$$

or, solving for $y_{A,s}$,

$$y_{A,s} = \frac{y_A + \dfrac{k_x}{k_y}x_A}{1 + \dfrac{k_x}{Kk_y}} \tag{14.8}$$

This equation shows that if k_x/k_y is very large, we will have $y_{A,s} \sim Kx_A$. In this case, mass transfer on the liquid side will be sufficiently fast that the interfacial gas-side mole fraction will be essentially in equilibrium with the bulk liquid. On the other hand, if k_x/k_y is very small, we will have $y_{A,s} \sim y_A$. In this case, mass transfer on the gas side will be sufficiently fast that the interfacial liquid-side mole fraction will be essentially in equilibrium with the bulk gas. As we have also done for heat transfer resistances in series, we can also rewrite Eqs. (14.3) and (14.4) as follows:

$$\frac{N_A}{k_y} = y_A - y_{A,s} \tag{14.9}$$

$$\frac{N_A}{k_x} = x_{A,s} - x_A \tag{14.10}$$

Multiplying both sides of Eq. (14.10) by K, we obtain:

$$\frac{N_A K}{k_x} = Kx_{A,s} - Kx_A = y_{A,s} - y_A^* \tag{14.11}$$

where $y_A^* = Kx_A$ is the mole fraction that the gas would have if it were in equilibrium with the bulk liquid. Adding Eqs. (14.9) and (14.11) together gives:

$$N_A\left(\frac{1}{k_y}+\frac{K}{k_x}\right)=y_A-y_A^* \tag{14.12}$$

We can re-write this equation as follows:

$$N_A = K_y(y_A - y_A^*) \tag{14.13}$$

where

$$K_y=\left(\frac{1}{k_y}+\frac{K}{k_x}\right)^{-1} \tag{14.14}$$

is an overall gas-side mass transfer coefficient. This result is similar to what we obtained for the combination of heat transfer resistances in series, as shown, for example, by Eq. (3.28). The difference is that, for mass transfer, the combination of resistances will depend not only on the individual coefficients k_y and k_x, but also on the distribution coefficient K. From Eq. (14.14), we can see that if $k_x/Kk_y >> 1$, we will have $K_y \sim k_y$, which will imply a dominance of the gas-side resistance. This will occur if k_y is very small, if k_x is very large, or if K is very small. Since a small K corresponds to a high solubility of component A in the liquid, we see that, often, absorption of highly soluble gases will be dominated by the gas-side resistance. We will obtain the opposite result if $k_x/Kk_y << 1$. In this case, we will have $K_y \sim k_x/K$, which will imply a dominance of the liquid-side resistance. This will occur if k_y is very large, if k_x is very small, or if K is very large. Since a large K corresponds to a low solubility of component A in the liquid, we see that, often, absorption of sparingly soluble gases will be dominated by the liquid-side resistance.

Following the same procedure that we just used to derive an expression for K_y, we can also derive the following alternative expression for the flux in terms of an overall liquid-side mass transfer coefficient:

$$N_A = K_x(x_A^* - x_A) \tag{14.15}$$

where $x_A^* = y_A/K$ is the mole fraction that the liquid would have if it were in equilibrium with the bulk gas and

$$K_x=\left(\frac{1}{Kk_y}+\frac{1}{k_x}\right)^{-1} \tag{14.16}$$

is the overall liquid-side mass transfer coefficient. Consistently with our previous finding, if $k_x/Kk_y >> 1$, we will have $K_x \sim Kk_y$ and a dominance of the gas-side resistance. On the other hand, if $k_x/Kk_y << 1$, we will have $K_x \sim k_x$ and a dominance of the liquid-side resistance.

B. Curved Equilibrium Line

If the equilibrium line is curved, i.e., if K is not a constant independent of composition, it is still possible to solve Eq. (14.6) numerically to find the interfacial composition. It is also instructive, in this case, to use a graphical representation of this solution as shown in the y-x plot of Fig. 14.3. This plot shows a hypothetical curved equilibrium line and a

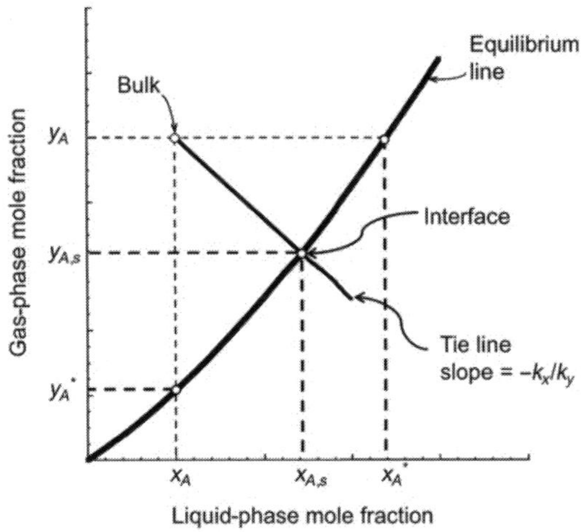

FIGURE 14.3 Determination of the composition at the gas-liquid interface from the bulk mole fractions x_A and y_A. The tie line is given by Eq. (14.6) and connects the points representing the bulk and interfacial compositions.

point with coordinates x_A, y_A representing the composition of the two bulk phases. This could represent a point at a certain height within a packed absorption column. On this plot, treating $y_{A,s}$ and $x_{A,s}$ as variables, Eq. (14.6) describes a straight line with slope equal to $-k_x/k_y$ that passes through the point representing the bulk composition. The intercept of this line with the equilibrium line will correspond to the interfacial composition $x_{A,s}$, $y_{A,s}$. Thus, this line, called the "tie line," will connect the point representing the bulk composition with the point representing the interfacial composition. If k_x/k_y is very large, the tie line will become nearly vertical and, in turn, $y_{A,s}$ will become equal to y_A^*, which is the gas-phase mole fraction in equilibrium with the bulk liquid. Conversely, if k_x/k_y is very small, the tie line will become nearly horizontal and, in turn, $x_{A,s}$ will become equal to x_A^*, which is the liquid-phase mole fraction in equilibrium with the bulk gas. These two extreme cases will correspond, respectively, to a dominance of the gas-side mass transfer resistance if the tie line is vertical and to a dominance of the liquid-side mass transfer resistance if the tie line is horizontal. We can also find the overall mass transfer coefficients by considering the following identities:

$$\frac{1}{K_y}=\frac{y_A-y_A^*}{N_A}=\frac{y_A-y_{A,s}}{N_A}+\frac{y_{A,s}-y_A^*}{N_A}=\frac{1}{k_y}+\frac{y_{A,s}-y_A^*}{x_{A,s}-x_A}\frac{1}{k_x} \tag{14.17}$$

Thus, the overall gas-side mass transfer coefficient can be expressed as:

$$K_y=\left(\frac{1}{k_y}+\frac{y_{A,s}-y_A^*}{x_{A,s}-x_A}\frac{1}{k_x}\right)^{-1} \tag{14.18}$$

In a similar way, we can find the following expression for the overall liquid-side mass transfer coefficient:

$$K_x=\left(\frac{1}{k_x}+\frac{x_A^*-x_{A,s}}{y_A-y_{A,s}}\frac{1}{k_y}\right)^{-1} \tag{14.19}$$

Equations (14.18) and (14.19) will reduce to the corresponding Eqs. (14.14) and (14.16) when the equilibrium line is straight. In this case, K_y and K_x will be constants independent of the bulk composition y_A and x_A. On the other hand, if the equilibrium line is curved as in Fig. 14.3, Eqs. (14.18) and (14.19) will predict values of K_y and K_x that vary with the bulk composition values of y_A and x_A. Example 14.1 illustrates the use of these relationships and their impact on the interfacial composition.

Example 14.1. Determination of the interfacial composition [Mathematica notebook: (M.EX.14.1)]. The equilibrium line for the absorption of ammonia in water at 333 K and 10 atm with $x_{NH_3} \leq 0.3$ is represented approximately by the following empirical equation based on data in Table 2-123 of *Perry's Chemical Engineers' Handbook*, 7th edition:

$$y_{NH_3} = 0.355x_{NH_3} + 1.602x_{NH_3}^2 - 1.357x_{NH_3}^3$$

Let's assume that at a certain point in an absorption column designed to absorb ammonia in water, the gas phase contains 12.0 mol % ammonia while the liquid phase contains 8.0 mol % ammonia. For this example, the gas-side and liquid-side mass transfer coefficients are $k_x = 6.65$ mol/m² s and $k_y = 0.986$ mol/m² s. Let's use these data to determine the mole fractions at the gas-liquid interface. Since, for this problem, $k_x/k_y = 6.65/0.986 = 6.74$, the tie line, Eq. (14.6) is given by:

$$y_{NH_3,s} = -6.74x_{NH_3,s} + (0.12 + 6.74 \times 0.08) = -6.74x_{NH_3,s} + 0.659$$

We can solve this equation in conjunction with the equilibrium line to find, numerically, the interfacial values $x_{NH_3,s} = 0.0912$ and $y_{NH_3,s} = 0.0446$. The corresponding graphical construction is shown in Fig. 14.4. Note that, in this case, since k_x/k_y is relatively large, the tie line is steep, which makes $x_{NH_3,s}$ end up being close to the bulk value of 0.08.

14.2.2 Column Material Balances and Operating Line

The operating line is based on a material balance for the absorbable component, A, across the height of the packing and relates the bulk-gas and bulk-liquid mole fractions

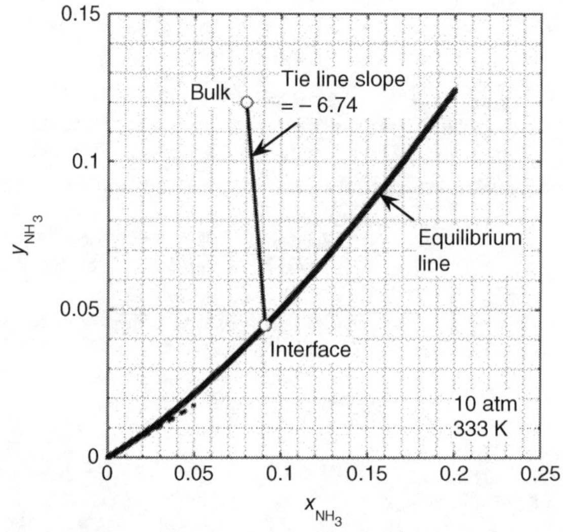

FIGURE 14.4 Determination of the composition at the gas-liquid interface for Example 14.1. The dashed line shows the linear limit of the equilibrium line.

at a certain point in the column to their respective values at one end of the column or the other. Let's consider first an absorption column. In this case, it is convenient to write this material balance for a control volume defined between the top of the column, where we generally know $y_{A,2}$ and $x_{A,2}$, and an arbitrary point in the column where the bulk-gas and bulk-liquid mole fractions are y_A and x_A, respectively. With reference to Fig. 14.5, we obtain the following balance:

$$G_M y_A + L_{M,2} x_{A,2} = G_{M,2} y_{A,2} + L_M x_A \tag{14.20}$$

For dilute conditions, we can neglect changes in the molar flow rates. Thus, we can take $G_M = G_{M,2}$ and $L_M = L_{M,2}$. In this case, Eq. (14.20) simplifies to:

$$y_A = \frac{L_M}{G_M} x_A + \left(y_{A,2} - \frac{L_M}{G_M} x_{A,2} \right) \tag{14.21}$$

It is convenient to work in terms of gas and liquid molar velocities, which are defined as follows:

$$G = \frac{G_M}{S} \tag{14.22}$$

$$L = \frac{L_M}{S} \tag{14.23}$$

where S is the cross-sectional area of the column. Typical units of G and L will be mol/m^2 s or lbmol/ft^2 h. In terms of G and L, Eq. (14.21) becomes:

$$y_A = \frac{L}{G} x_A + \left(y_{A,2} - \frac{L}{G} x_{A,2} \right) \tag{14.24}$$

Note that Eq. (14.24) is satisfied if we set $y_A = y_{A,2}$ and $x_A = x_{A,2}$. Thus, on the y-x plot, the operating line representing this equation will pivot at the point $x_{A,2}, y_{A,2}$ (representing the composition at the top of the column) with a slope equal to L/G. As seen in Fig. 14.5, there will be a minimum value of this slope, $(L/G)_{\min}$, at which the operating line will cross

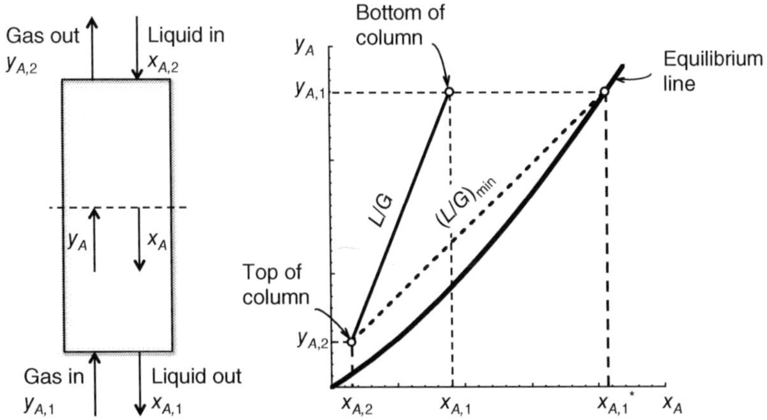

FIGURE 14.5 Operating lines for an absorption column.

the equilibrium line at $y_A = y_{A,1}$. If we were to operate the column at this value of L/G, the gas entering the column and the liquid leaving the column would be in equilibrium with each other. This could only be approximated but not actually attained in practice since, as the gas and liquid mole fractions get closer and closer to equilibrium with each other, the mass transfer rate decreases. As a result, attaining equilibrium would require an infinitely tall column, which, of course, would not be possible. Thus, we will need to operate with a value of L/G greater than the minimum. Of course, there will also be drawbacks of using too large a value of L/G. For a given gas flow rate, increasing L/G will require greater and greater liquid flow rates, which will be expensive in itself since the absorbent liquid is not free. Moreover, increasing L/G will increase the diameter that the column will need to have to handle the higher flow. We can see that, in practice, there will be an optimum L/G: too small a value of L/G will result in a very tall column while too large a value of L/G will result in a very large absorbent flow and a very large column diameter. The exact value of the optimum L/G will depend on economic factors. However, values of L/G that are 1.2 to 1.5 times greater than $(L/G)_{min}$ are often found in practice.

Let's now consider a stripping column as shown in Fig. 14.6. In this case, we usually know the inlet-gas mole fraction and we know or can easily calculate the outlet-liquid mole fraction. Thus, it is convenient to write the material balance for component A around a control volume defined between the bottom of the column, where we know $y_{A,1}$ and $x_{A,1}$, and an arbitrary point in the column, where the bulk-gas and bulk-liquid mole fractions are, respectively, y_A and x_A. Such a balance leads to the following expression:

$$y_A = \frac{L}{G} x_A + \left(y_{A,1} - \frac{L}{G} x_{A,1} \right) \tag{14.25}$$

On the y-x plot, Eq. (14.25) is again represented by straight line with slope equal to L/G. In this case, however, this line will pivot around the point $x_{A,1}$, $y_{A,1}$ as we vary L/G. As seen in Fig. 14.6, there will be a maximum $(L/G)_{max}$ (or a minimum G for a given L) at which the operating lines will come in contact with the equilibrium line. This can occur at $x_A = x_{A,2}$ or, depending on the curvature of the equilibrium line, at a value of x_A where the operating line becomes tangent to the equilibrium line. In either case, we will

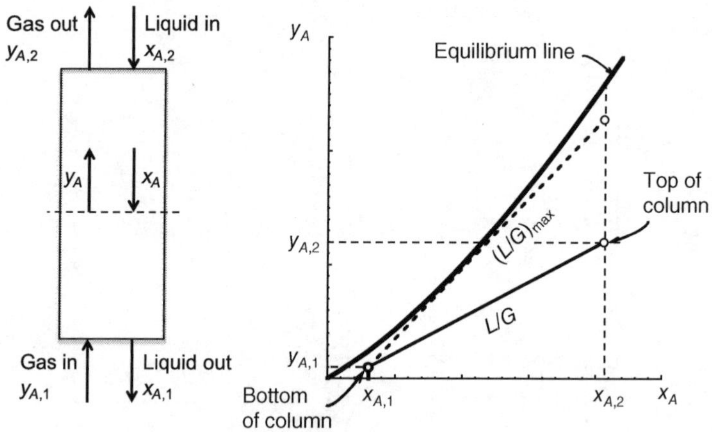

FIGURE 14.6 Operating lines for a stripping column.

need to operate the stripper with an L/G smaller than the maximum (or an actual G greater than the minimum) in order to avoid having an infinitely tall column. As we saw for an absorber, there will also be an optimum L/G for a stripper: too large a value of L/G will result in a very tall column, while too small a value of L/G will result in a very large flow of the stripping gas which, again, will result in a very large column diameter.

If the equilibrium line is straight, we can find $(L/G)_{min}$ analytically for an absorber or $(L/G)_{max}$ for a stripper, without the need for a graphical construction. For an absorber, when $L/G = (L/G)_{min}$, equilibrium is reached at the bottom of the column where x_A becomes equal to $y_{A,1}/K$. In this case, Eq. (14.24) yields:

$$\left(\frac{L}{G}\right)_{min} = \frac{y_{A,1} - y_{A,2}}{\dfrac{y_{A,1}}{K} - x_{A,2}} \tag{14.26}$$

Conversely, for a stripper, if $L/G = (L/G)_{max}$, equilibrium is reached at the top of the column where y_A becomes equal to $Kx_{A,2}$. For this case, we obtain:

$$\left(\frac{L}{G}\right)_{max} = \frac{Kx_{A,2} - y_{A,1}}{x_{A,2} - x_{A,1}} \tag{14.27}$$

If the equilibrium line is curved, either a graphical construction or the numerical solution of a system of two equations consisting of the operating line coupled with an analytical expression of the equilibrium line will be needed to find $(L/G)_{min}$ for an absorber or $(L/G)_{max}$ for a stripper.

14.2.3 Absorption and Stripping Columns as Mass Exchangers

An alternative analysis of absorption and stripping columns can be done by regarding them as mass exchangers in a manner analogous to how we analyzed heat exchangers. For an absorber, for example, we can plot mole fractions vs. the cumulative number of moles absorbed, just like for a heat exchanger we plotted temperature vs. the cumulative amount of heat exchanged. Material balances for the gas and the liquid give the following expressions for the total rate of mass transfer of a component A in the column:

$$M_{t,A} = G_M(y_{A,1} - y_{A,2}) \tag{14.28}$$

$$M_{t,A} = L_M(x_{A,1} - x_{A,2}) \tag{14.29}$$

Similarly, material balances from the bottom of the column to an arbitrary height where the bulk-gas mole fraction is y_A and the bulk-liquid mole fraction is x_A give:

$$M_A = G_M(y_{A,1} - y_A) \tag{14.30}$$

$$M_A = L_M(x_{A,1} - x_A) \tag{14.31}$$

where M_A is the cumulative rate of absorption of component A from the bottom of the column up to the point in the column where the bulk mole fractions are y_A and x_A. With a slight rearrangement, Eq. (14.30) gives:

$$y_A = y_{A,1} - \frac{1}{G_M} M_A \tag{14.32}$$

which is a straight line with slope $-1/G_M$ when plotted vs. M_A. We could also plot x_A in the same manner. However, to make the plot useful, it is preferable to plot $y_A^* = Kx_A$ instead of x_A. This way, we will be able to compare directly y_A and y_A^*. From Eq. (14.31) we can write:

$$x_A = x_{A,1} - \frac{1}{L_M} M_A \tag{14.33}$$

or

$$y_A^* = Kx_{A,1} - \frac{K}{L_M} M_A \tag{14.34}$$

When K is constant, i.e., when the equilibrium line is straight, Eq. (14.34) is also represented by a straight line with slope $-K/L_M$ when plotted vs. M_A. Figure 14.7 shows an example plot of y_A and y_A^* vs. the cumulative absorption rate. The vertical distance between the two lines $y_A - y_A^*$ is the overall driving force for mass transfer. When multiplied by the overall gas-side mass transfer coefficient, K_y, this difference will give the local mass transfer flux. This is analogous to the case of a heat exchanger where the difference $T_H - T_C$ between the hot and the cold streams multiplied by the overall heat transfer coefficient, U_o, gave us the local heat transfer flux. For the mass transfer case, we see from Fig. 14.7 that if G_M, $y_{A,2}$, and $x_{A,2}$ are fixed, the minimum L_M will correspond to the dashed line obtained when $y_{A,1} = y_{A,1}^*$. This graphical representation is also possible if the equilibrium line is curved. In this case, however, the y_A^* vs. M_A line will also become curved making the graphical analysis less useful.

14.2.4 Packing Height

The height of packing required for an absorption or stripping column can be determined starting from a differential material balance on the absorbable component A over

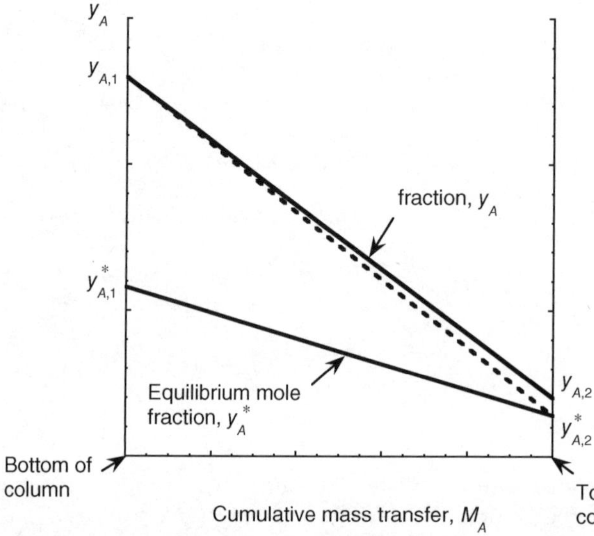

FIGURE 14.7 Mole fraction vs. cumulative mass transfer plot for an absorption column. The vertical distance between the two solid lines is the overall driving force for mass transfer $y_A - y_A^*$. The dashed line shows the y_A^*-line at $L/G = (L/G)_{min}$.

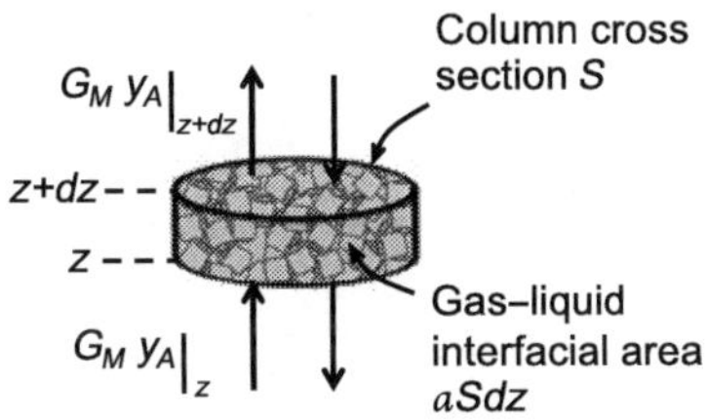

FIGURE 14.8 Differential slice of a packed absorption or stripping column used to derive the mass transfer rate equation.

a slice of packing of thickness dz as illustrated in Fig. 14.8. We define the specific gas-liquid interfacial area as follows:

$$a = \frac{\text{Gas-liquid interfacial area}}{\text{Volume of packing}} \tag{14.35}$$

The value of a will depend on the type of packing, on its size, on its specific design, and on the material of construction, as well as on the actual gas and liquid velocities. Note that the units of a will typically be m^2 per m^3 of column and are thus m^{-1}. In general, a will either have to be determined empirically or will be estimated from suitable correlations. In either case, the differential gas-liquid interfacial area contained within the slice of column shown in Fig. 14.8 will be:

$$dA = Sadz$$

where S is the column cross-sectional area. The differential rate of absorption over this section of the column is $dM_A = -G_M dy_A$. A component material balance on the gas phase can then be written as follows:

$$G_M y_A\big|_z = G_M y_A\big|_{z+dz} + SaN_A dz \tag{14.36}$$

where

$$N_A = K_y(y_A - y_A^*) \tag{14.37}$$

Dividing Eq. (14.36) by dz gives:

$$G_M \frac{dy_A}{dz} = -SaK_y(y_A - y_A^*) \tag{14.38}$$

Finally, since $G_M/S = G$, we can write this result as follows:

$$dz = -\frac{G}{aK_y}\frac{dy_A}{y_A - y_A^*} \tag{14.39}$$

Equation (14.39) can be integrated directly if we make the assumption that G/aK_y is a constant. This is correct if the equilibrium line is straight, but is also often found to be approximately true for curved equilibrium lines. The integration limits for Eq. (14.39) are $y_A = y_{A,1}$ at $z = 0$ and $y_A = y_{A,2}$ at $z = Z$. Thus, the following result is obtained:

$$Z = \frac{G}{aK_y}\int_{y_{A,2}}^{y_{A,1}} \frac{dy_A}{y_A - y_A^*} \tag{14.40}$$

This result is often written in the following easy-to-remember form:

$$Z = H_{o,G} N_{o,G} \tag{14.41}$$

(the mnemonic is "hog-nog") where:

$$H_{o,G} = \frac{G}{aK_y} \tag{14.42}$$

is the height of an overall gas-side mass transfer unit and

$$N_{o,G} = \int_{y_{A,2}}^{y_{A,1}} \frac{dy_A}{y_A - y_A^*} \tag{14.43}$$

is the number of overall gas-side mass transfer units.

In addition to providing a convenient mnemonic, expressing the packing height Z as the product of these two quantities is also physically meaningful. $N_{o,G}$ depends only on the inlet and outlet mole fractions, the operating line, and the equilibrium line. Thus, it is unaffected by the type of packing or by its mass transfer efficiency. Since $y_A - y_A^*$ appears in the denominator of the expression for $N_{o,G}$, column designs where the operating line closely approaches the equilibrium line will result in a larger $N_{o,G}$, reflecting a more challenging absorption task. On the other hand, $H_{o,G}$ is, to a first approximation, independent of the difficulty of the absorption task. In this case, less-efficient packings that give smaller a and slower mass transfer will result in larger values of $H_{o,G}$ and, ultimately, in a greater packing height regardless of the intrinsic difficulty of the absorption task. In practice, $H_{o,G}$ is typically estimated from engineering correlations. $N_{o,G}$, on the other hand, is calculated from the integral in Eq. (14.43) as illustrated in the following sections. How we proceed with the calculation of $H_{o,G}$ and $N_{o,G}$ will depend on whether the equilibrium line is straight or not.

The equations presented in this section are also valid for stripping columns. The only difference is the meaning of the mole fractions at the column entrance and exit. For a stripper, the bottom of the column is the dilute end, where the stripping gas is fed and mole fractions are small, while the top of the column is where the liquid to be stripped is fed and where the mole fractions are larger.

A. Calculation of $N_{o,G}$ and $H_{o,G}$ with a Straight Equilibrium Line

When the equilibrium line is straight, we can obtain $N_{o,G}$ in a manner completely analogous to how we derived the expression for the heat transfer area of a counter-current heat exchanger in Sec. 7.2.1. From Eqs. (14.30) and (14.34), the differential change in mole fraction, dy_A, over a differential height of packing dz is related to the differential rate of absorption, dM_A. Thus, we have:

$$dy_A = -\frac{1}{G_M} dM_A \tag{14.44}$$

and

$$dy_A^* = -\frac{K}{L_M} dM_A \tag{14.45}$$

which are analogous to Eq. (7.33) and (7.34) for a counter-current heat exchanger. Subtracting Eq. (14.45) from Eq. (14.44), we obtain:

$$\frac{d\Delta y_A}{dM_A} = -\frac{1}{G_M} + \frac{K}{L_M} \tag{14.46}$$

where $\Delta y_A = y_A - y_A^*$. Next, let's return to the material balances, Eqs. (14.28) and (14.29), which can be written as follows:

$$\frac{1}{G_M} = \frac{y_{A,1} - y_{A,2}}{M_{t,A}} \tag{14.47}$$

and

$$\frac{K}{L_M} = \frac{y_{A,1}^* - y_{A,2}^*}{M_{t,A}} \tag{14.48}$$

Subtracting Eq. (14.48) from Eq. (14.47) gives the following result:

$$\frac{1}{G_M} - \frac{K}{L_M} = \frac{(\Delta y_A)_1 - (\Delta y_A)_2}{M_{t,A}} \tag{14.49}$$

where $(\Delta y_A)_1 = y_{A,1} - y_{A,1}^* = y_{A,1} - Kx_{A,1}$ and $(\Delta y_A)_2 = y_{A,2} - y_{A,2}^* = y_{A,2} - Kx_{A,2}$. Finally, combining Eq. (14.49) with Eq. (14.46) gives:

$$\frac{d\Delta y_A}{dM_A} = -\frac{(\Delta y_A)_1 - (\Delta y_A)_2}{M_{t,A}} \tag{14.50}$$

Let's now return to Eq. (14.38), which yields:

$$dM_A = -G_M dy_A = SaK_y(y - y_A^*)dz = SaK_y \Delta y_A dz \tag{14.51}$$

Combining this Eq. (14.51) with Eq. (14.50) and integrating the result over the height of packing, we obtain:

$$\int_1^2 \frac{d\Delta y_A}{\Delta y_A} = -\frac{(\Delta y_A)_1 - (\Delta y_A)_2}{M_{t,A}} SaK_y \int_0^Z dz \tag{14.52}$$

After integration, we obtain the final result, which is given by:

$$M_{t,A} = ZSaK_y(\Delta y_A)_{lm} \tag{14.53}$$

where:

$$(\Delta y_A)_{lm} = \frac{(\Delta y_A)_1 - (\Delta y_A)_2}{\ln \dfrac{(\Delta y_A)_1}{(\Delta y_A)_2}} \tag{14.54}$$

Not surprisingly, this result is analogous to Eqs. (7.37) and (7.38) for a countercurrent heat exchanger with the total mass transfer rate $M_{t,A}$ replacing the total heat transfer rate Q_t, the overall mass transfer coefficient K_y replacing the overall heat transfer coefficient U_o, the product ZSa replacing the heat transfer area A_o, and the Δy_A logarithmic mean replacing the ΔT logarithmic mean.

In terms of $N_{o,G}$, this result can also be expressed by either of the two following equivalent equations:

$$N_{o,G} = \frac{y_{A,1} - y_{A,2}}{(\Delta y_A)_{lm}} \tag{14.55}$$

or:

(M.E.14.56)
$$N_{o,G} = \frac{1}{1-\gamma} \ln\left[(1-\gamma)\frac{y_{A,1} - Kx_{A,2}}{y_{A,2} - Kx_{A,2}} + \gamma\right] \tag{14.56}$$

where

$$\gamma = \frac{KG}{L} = \frac{K}{L/G} \tag{14.57}$$

is called the stripping factor. This quantity is the slope of the equilibrium line divided by the slope of the operating line. Equations (14.55) and (14.56) are equivalent and give the same result for $N_{o,G}$. However, Eq. (14.56), which is known as the Colburn equation, has the advantage that it does not explicitly involve $x_{A,1}$. Thus, it allows for a quicker calculation of $N_{o,G}$ given the gas inlet and outlet mole fractions and the value of γ just by knowing the mole fraction in the inlet liquid $x_{A,2}$. The latter will often be zero for an absorber. The Colburn equation is plotted in Fig. 14.9 for different values of γ. We can see that for a given ratio $y_{A,1}/y_{A,2}$, increasing γ results in a sharp increase of $N_{o,G}$ when γ rises above 1. This happens because, since $\gamma = K/(L/G)$, a larger value of γ corresponds to a smaller slope of the operating line. In turn, with a smaller slope, the operating line will be closer to the equilibrium line thereby reducing the vertical distance between the two lines and, thus, reducing the driving force for mass transfer.

For a stripping column, the following modified form of the Colburn equation is more convenient:

$$N_{o,G} = \frac{1}{\gamma - 1} \ln\left[\left(1 - \frac{1}{\gamma}\right)\frac{x_{A,2} - y_{A,1}/K}{x_{A,1} - y_{A,1}/K} + \frac{1}{\gamma}\right] \tag{14.58}$$

This expression contains only the liquid inlet and outlet mole fractions and the inlet-gas mole fractions, all of which are generally known.

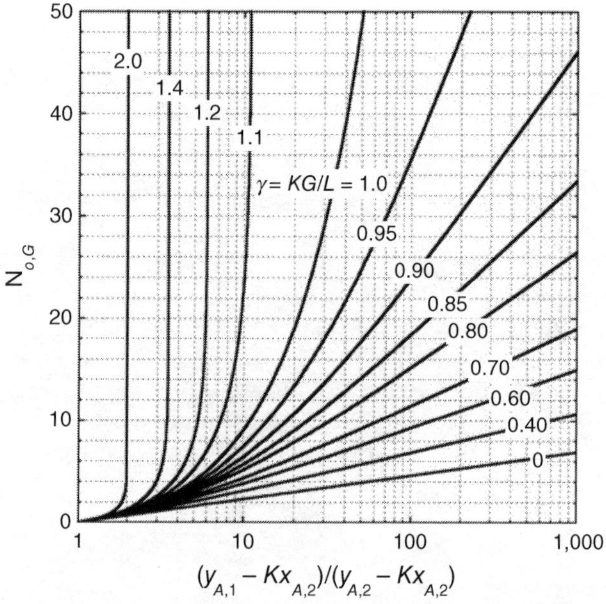

FIGURE 14.9 Plot of the Colburn equation, Eq.(14.56), for different values of γ. Mathematica notebook: (M.F.14.9).

The last item to consider for the case of a straight equilibrium line is how we calculate $H_{o,G}$. Considering Eq. (14.14), we can write:

$$H_{o,G} = \frac{G}{aK_y} = \frac{G}{a}\left(\frac{1}{k_y} + \frac{K}{k_x}\right) = \frac{G}{ak_y} + \frac{KG}{L}\frac{L}{ak_x} = H_G + \gamma H_L \tag{14.59}$$

where the quantities

$$H_G = \frac{G}{ak_y} \tag{14.60}$$

$$H_L = \frac{L}{ak_x} \tag{14.61}$$

are the height of a gas-phase mass transfer unit and the height of a liquid-phase mass transfer unit, respectively. Both H_G and H_L have units of m or ft. Since they are both independent of the equilibrium line, they can be estimated from generalized correlations. Note, however, that the stripping factor does affect the overall result since it appears in Eq. (14.59). For instance if H_G and H_L are comparable in magnitude, a small value of γ (corresponding to a small K and, thus, to a high solubility of the absorbable component A in the liquid) will give $H_{o,G} \sim H_G$, which means that the gas-side mass transfer resistance will be dominant. On the other hand, a large value of γ (corresponding to a large K and, thus, to a low solubility of component A in the liquid) will give $H_{o,G} \sim \gamma H_L$, which means that the gas-side mass transfer resistance will be dominant instead. We can see that, although they describe completely different phenomena, H_G and H_L are analogous to the individual heat transfer resistances in a heat exchanger while $H_{o,G}$ is analogous to the overall heat transfer resistance, $1/U_o$.

Various correlations are available to estimate H_G and H_L. For random packings, Sec. 14 in *Perry's Chemical Engineers' Handbook*, 9th edition, recommends the Onda method described by Bravo and Fair, Ind. Eng. Chem. Process Des. Dev. 21 (1982) 162. This method provides estimates of the gas-liquid interfacial area and of the gas-side and liquid-side mass transfer coefficients as functions of gas and liquid velocities, gas and liquid physical properties, and packing properties according to the following equations:

(M.E.14.62) $$a = a_p\left\{1 - \exp\left[-1.45\left(\frac{\sigma_c}{\sigma_L}\right)^{0.75}\left(\frac{L'}{a_p\mu_L}\right)^{0.1}\left(\frac{L'^2 a_p}{\rho_L^2 g}\right)^{-0.05}\left(\frac{L'^2}{\rho_L\sigma_L a_p}\right)^{0.2}\right]\right\} \tag{14.62}$$

(M.E.14.63) $$k_y = 5.23(a_p C_G \mathcal{D}_G)\left(\frac{G'}{a_p\mu_G}\right)^{0.7}(\mathrm{Sc}_G)^{1/3}(a_p d_p)^{-2} \tag{14.63}$$

(M.E.14.64) $$k_x = 0.0051\left[\frac{\mu_L g (C_L)^3}{\rho_L}\right]^{1/3}\left(\frac{L'}{a\mu_L}\right)^{2/3}(\mathrm{Sc}_L)^{-1/2}(a_p d_p)^{0.4} \tag{14.64}$$

where σ_c = 0.061 N/m for ceramic packings, 0.075 N/m for metal packings, and 0.033 N/m for plastic packings. In these equations, σ_L is the surface tension of the liquid (N/m), μ is the viscosity (Pa s), $\mathcal{D}$ is the diffusivity (m^2/s), Sc is the Schmidt number, and C is the molar density (mol/m^3). Subscripts L and G indicate the gas and liquid phase, respectively. The quantities L' and G' are the gas and liquid mass velocities (kg/m² s), which are defined by the following equations:

$$L' = M_{r,L}L \tag{14.65}$$

$$G' = M_{r,G}G \tag{14.66}$$

where $M_{r,L}$ and $M_{r,G}$ are the molecular masses of the liquid phase and gas phase, respectively. Finally, a_p and d_p are packing-specific parameters. a_p is the surface area of the dry packing in m²/m³ while d_p is the packing size in m. Table 14-13 in *Perry's Chemical Engineers' Handbook*, 9th edition, provides the values of a_p and d_p for many different packing designs and materials of construction. Example 14.2 illustrates the use of these equations.

Example 14.2. Packing height for a SO_2 absorber. Let's use what we have learned so far to calculate the height of packing needed to remove 95% of the SO_2 (component A) in an air stream containing 2.0 mol % SO_2 and flowing at the rate of 200 kg/h using pure water as the absorbent with a value of L/G equal to 1.3 times greater than the minimum. The column is 0.3 m in diameter, is packed with 38-mm ceramic Berl Saddles, and is operated at 298 K and 1 atm absolute pressure. According to Sander [Atms. Chem. Phys. 15 (2015) 4399], the solubility coefficient for SO_2 in water at 298 K is 1.3×10^{-2} in units of mol SO_2/m³ Pa-. In order to obtain the Henry constant H as defined by Eq. (14.2), we need to multiply the reciprocal of this value by the molar density of water c = 55,600 mol/m³, which gives $H = 55{,}600/1.3 \times 10^{-2} = 4.3 \times 10^6$ Pa. Finally, we obtain $K = H/P = 4.3 \times 10^6/101{,}300 = 42$ at 1 atm absolute pressure. In this example, the gas is dilute ($y_{A,1} = 0.02$) and the solubility of the absorbable component, SO_2, in the liquid is low. Thus, we can assume that the gas and liquid flow rates will be essentially constant along the column. The outlet mole fraction of SO_2 is then:

$$y_{A,2} = (1 - 0.95) \times 0.02 = 0.001$$

Next, we find the minimum liquid flow rate. From Eq. (14.26), we obtain:

$$\left(\frac{L}{G}\right)_{\min} = \frac{0.02 - 0.001}{\dfrac{0.02}{42} - 0} = 40.0$$

So, the actual L/G and the stripping factor will be, respectively,

$$\frac{L}{G} = 1.3 \times 40.0 = 52$$

and:

$$\gamma = \frac{K}{L/G} = \frac{42}{52} = 0.81$$

From Fig. 14.9 or, more precisely, directly from Eq. (14.56) we obtain:

$$N_{o,G} = \frac{1}{1 - 0.81}\ln\left[(1 - 0.81)\frac{0.02 - 0}{0.001 - 0} + 0.81\right] = 8.04$$

In order to estimate H_{oG}, we need to first calculate the mass velocities G' and L' in kg/m² s. The gas molar flow rate is:

$$G_M = \frac{200}{29/1{,}000}\frac{1}{3{,}600} = 1.92 \text{ mol/s}$$

The liquid molar flow rate is found from this value from the L/G ratio giving:

$$L_M = 52.0 \times 1.92 = 99.6 \text{ mol/s}$$

The column cross-section is:

$$S = \frac{\pi D_c^2}{4} = 0.0707 \text{ m}^2$$

Thus, the gas and liquid molar velocities are as follows:

$$G = \frac{G_M}{S} = \frac{1.92}{0.0707} = 27.1 \text{ mol/m}^2 \text{ s}$$

$$L = \frac{L_M}{S} = \frac{99.6}{0.0707} = 1{,}410 \text{ mol/m}^2 \text{ s}$$

Finally, the gas and liquid mass velocities are:

$$G' = G \times M_{r,\text{air}} = 27.1 \times \frac{29}{1{,}000} = 0.786 \text{ kg/m}^2 \text{ s}$$

$$L' = L \times M_{r,\text{water}} = 1{,}410 \times \frac{18}{1{,}000} = 25.4 \text{ kg/m}^2 \text{ s}$$

The Onda method used to predict a, k_y, and k_x requires several physical properties of the gas (air) and the liquid (water), which are estimated to be as follows at 298 K and 1 atm absolute pressure (see Appendix A for the bulk properties and the methods in Chapter 10 for the diffusivities):

Gas Phase	Liquid Phase
$\mu_G = 1.8 \times 10^{-5}$ Pa s $\rho_G = 1.2$ kg/m³ $\mathcal{D}_G = \mathcal{D}_{SO_2,\text{air}} = 0.12 \times 10^{-4}$ m²/s	$\mu_L = 0.91 \times 10^{-3}$ Pa s $\rho_L = 990$ kg/m³ $\mathcal{D}_L = \mathcal{D}_{SO_2,\text{water}} = 2.1 \times 10^{-9}$ m²/s $\sigma_L = 0.070$ N/m

For 38-mm ceramic Berl Saddles ($d_p = 0.038$ m), Table 14-13 in *Perry's Chemical Engineers' Handbook*, 9th edition, gives $a_p = 150$ m²/m³. With these values, Eqs. (14.62) to (14.64) yield:

$$a = 118 \text{ m}^2/\text{m}^3$$

$$k_y = 0.68 \text{ mol/m}^2 \text{ s}$$

$$k_x = 21.3 \text{ mol/m}^2 \text{ s}$$

H_G and H_L are then obtained from Eqs. (14.60) and (14.61), which give:

$$H_G = \frac{G}{ak_y} = \frac{27.1}{118 \times 0.68} = 0.34 \text{ m}$$

$$H_L = \frac{L}{ak_x} = \frac{1{,}410}{118 \times 21.3} = 0.56 \text{ m}$$

Finally, $H_{o,G}$ and the packing height are calculated as follows:

$$H_{o,G} = H_G + \gamma H_L = 0.34 + 0.81 \times 0.56 = 0.79 \text{ m}$$

$$Z = H_{oG} N_{oG} = 0.79 \times 8.04 = 6.4 \text{ m}$$

Note that, in this case, gas-side and liquid-side resistances are comparable since $H_G = 0.34$ m and the product $\gamma H_L = 0.45$ m are similar in magnitude.

B. Calculation of $N_{o,G}$ and $H_{o,G}$ with a Curved Equilibrium Line

In general, the calculation of $N_{o,G}$ with a curved equilibrium line requires a numerical evaluation of the integral in Eq. (14.43). This can be done, for example, by dividing the y_A-axis between $y_{A,2}$ and $y_{A,1}$ into a number of discrete increments. Starting at $y_{A,2}$, we can read $y^*_{A,2}$ from the graph by drawing a vertical line to the equilibrium line and then calculate the value of $1/(y_{A,2}-y^*_{A,2})$. We can repeat this determination at each y_A-increment and then plot the values of $1/(y_A-y^*_A)$ vs. y_A. The area under the curve defined by $1/(y_A-y^*_A)$ vs. y_A between $y_{A,2}$ and $y_{A,1}$ will be $N_{o,G}$. The difficulty with this procedure is that, sometimes, the operating line gets so close to the equilibrium line that it becomes challenging to determine $y_A-y^*_A$ with sufficient precision. For an absorber, this will often occur in the low-mole fraction region near the top of the packing. In this case, it is convenient to break down the calculation into two parts. In the dilute region, up to an intermediate value of $y_A=y'_A$, where the equilibrium line is essentially linear, we calculate $N_{o,G}$ analytically from Eq. (14.55) or (14.56). Between the intermediate value of y'_A and $y_{A,1}$, where the equilibrium line is significantly curved, we calculate $N_{o,G}$ numerically from Eq. (14.43) with $y_{A,2}=y'_A$. We then add the two values of $N_{o,G}$ to find the total $N_{o,G}$ for the entire packing. Alternatively, as shown in Example 14.3, if an analytical expression is available to accurately describe the equilibrium line, the calculation of $N_{o,G}$ can be done numerically without having to resort to a graphical approach.

The calculation of H_{oG} when the equilibrium line is curved will still follow the approach that led us to Eq. (14.59). In this case, considering Eq. (14.18), we have:

$$H_{o,G}=\frac{G}{aK_y}=\frac{G}{a}\left(\frac{1}{k_y}+\frac{y_{A,s}-y^*_A}{x_{A,s}-x_A}\frac{1}{k_x}\right)=\frac{G}{ak_y}+\frac{y_{A,s}-y^*_A}{x_{A,s}-x_A}\frac{G}{L}\frac{L}{ak_x}=H_G+\frac{y_{A,s}-y^*_A}{x_{A,s}-x_A}\frac{G}{L}H_L \tag{14.67}$$

From this result we can see that, in general, H_{oG} will vary along the packing height. We can handle this variation in different ways. One way, which is more accurate but also more complicated, is to carry out the entire calculation numerically based on the equation:

$$Z=\int_{y_{A,2}}^{y_{A,1}}\frac{H_{oG}dy_A}{y_A-y^*_A} \tag{14.68}$$

This would require determining the interfacial composition at each step as shown in Example 14.1 by noting that the slope of the tie line can be expressed as follows in terms of H_G and H_L:

$$\frac{k_x}{k_y}=\frac{L}{G}\frac{H_G}{H_L} \tag{14.69}$$

Another way, which is simpler, is to divide again the calculation into two different parts—one for the dilute end of the packing, where the equilibrium line is essentially straight and H_{oG} is constant, and the other for the more concentrated region, where the equilibrium line is significantly curved, using a different value of H_{oG} averaged over this portion. Given the uncertainties in estimating H_G and H_L (errors can be ±20% or even larger) such approximations are often sufficient for preliminary design. Example 14.3 illustrates an application of these relationships.

Example 14.3. Calculation of packing height for ammonia absorption in water [Mathematica notebook: (M.EX.14.3)]. In this example, we consider an absorber to be designed to remove 95% of the ammonia (component A) from an air stream flowing at the rate of 1.0 kg/s and containing 5.0 mole %

ammonia using pure water as the absorbent in a column operated at 10 atm and 333 K with L/G equal to 1.2 times the minimum value. The column is packed with 38-mm ceramic Berl Saddles and has a diameter of 0.5 m. The equilibrium line is given by the following approximate empirical equation plotted in Fig. 14.10:

$$y_{NH_3} = 0.355x_{NH_3} + 1.602x_{NH_3}^2 - 1.357x_{NH_3}^3$$

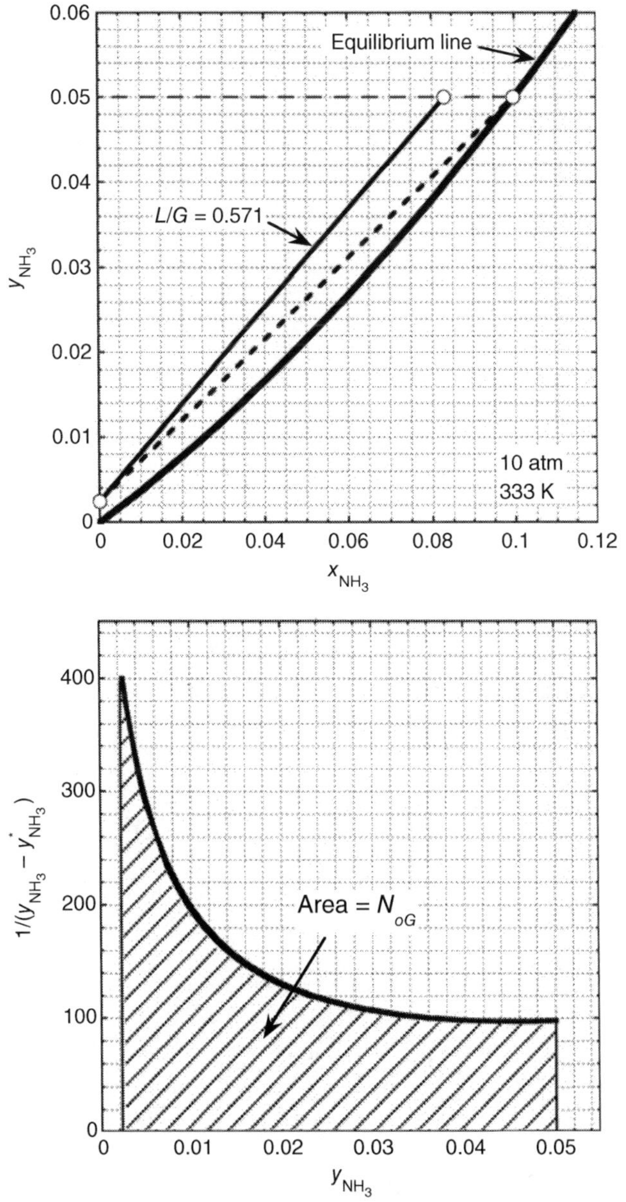

FIGURE 14.10 y-x plot (top) and $N_{o,G}$ plot (bottom) for the ammonia absorber in Example 14.3.

We first find the outlet-gas mole fraction. For 95% removal, neglecting the change in the gas total flow rate, we have:

$$y_{NH_3,2} = (1-0.95)y_{NH_3,1} = 0.05 \times 0.05 = 0.0025$$

Next, we find $(L/G)_{min}$ by determining the value of x_{NH_3} for which the equilibrium line gives $y_{NH_3} = 0.05$. This occurs when $x_{NH_3} = 0.0997$, as is shown by the dashed line in Fig. 14.10 (top). Thus:

$$\left(\frac{L}{G}\right)_{min} = \frac{0.05-0.0025}{0.0997-0} = 0.476$$

Since the actual value is to be 20% larger, we will have $L/G = (1+0.2)\times 0.476 = 0.571$. The operating line, Eq. (14.24), is, thus:

$$y_{NH_3} = 0.571x_{NH_3} + (y_{NH_3,2} - 0.571x_{NH_3,2}) = 0.571x_{NH_3} + 0.0025$$

or

$$x_{NH_3} = \frac{y_{NH_3}}{0.571} - 0.00438$$

We can now calculate $y^*_{NH_3}$ at each point in the column by substituting this value of x_{NH_3} into the expression for the equilibrium line. Finally, we can calculate the function $1/(y_{NH_3} - y^*_{NH_3})$ and integrate to find the area from $y_{NH_3,2}$ to $y_{NH_3,1}$ as shown in Fig. 14.10 (bottom). The result is $N_{oG} = 6.78$. Note that if we neglect the curvature of the equilibrium line and use $K = 0.355$, which is its initial slope, from the Colburn equation we would get $\gamma = 0.355/0.571 = 0.622$ and

$$N_{o,G} = \frac{1}{1-0.622}\ln\left[(1-0.622)\frac{0.05-0}{0.0025-0} + 0.622\right] = 5.56$$

This is about 18% lower than the more accurate value of 6.78, indicating that the curvature of the equilibrium line plays a significant role in this case. Let's now estimate $H_{o,G}$. For our conditions, the gas mass velocity is $G' = 1.0/(\pi D_c^2/4) = 5.1$ kg/m² s. Thus, the molar gas velocity is $G = 5.1/(29/1{,}000) = 176$ mol/m² s. From the L/G value, we obtain $L = 0.571 \times G = 100$ mol/m² s and $L' = 100 \times 18/1{,}000 = 1.8$ kg/m² s. Using these values along with the relevant physical properties, the Onda method predicts $a = 69.6$ m², $k_y = 0.88$ mol/m² s, and $k_x = 12.0$ mol/m² s. From these values, we obtain $H_G = 2.87$ m and $H_L = 0.12$ m. Now, at the top of the column (which is the bottom of the y-x plot) where the equilibrium line is essentially straight, we have $H_{o,G} = H_G + \gamma H_L = 2.87 + 0.622 \times 0.12 = 2.94$ m. At the bottom of the column, where the equilibrium line is curved, we have $x_A = x_{A,1} = 0.083$ and $y_A = y_{A,1} = 0.05$. For these conditions, following Example 14.1, we find $x_{A,s} = 0.0837$ and $y_{A,s} = 0.040$. Finally, with $y^*_A = 0.0397$, Eq. (14.67) gives:

$$H_{o,G} = 2.87 + \frac{0.0400-0.0397}{0.0837-0.0830} \times \frac{0.120}{0.571} = 2.96 \text{ m}$$

Obviously, the variation in $H_{o,G}$ is hardly significant in this case, so that we can safely estimate the packing height to be $Z = 2.93 \times 6.78 = 20$ m.

14.3 Column Design for Non-Dilute Conditions

When the mole fraction of the absorbable component exceeds 0.1 in either the gas or the liquid, the error incurred when using the procedures described in Sec. 14.2 can become unacceptably high. In this case, we need to account for two important effects. The first effect is that the mass transfer coefficients will become functions of composition, which, in turn, will affect the rate of mass transfer. The second effect is that the gas and liquid flow rates will vary along the length of the column, which, in turn, will affect the material balances and, thus, the determination of L/G.

14.3.1 Mass Transfer

Let's consider again the expected mole fraction profiles near the gas-liquid interface during absorption of a component A from an inert gas B as shown in Fig. 14.2, but this time in the context of the film model. In this model, we assume that the mass transfer resistance in each phase is represented by a film with thickness Δ where only diffusion occurs. There will be a gas-side film thickness, Δ_G, and a liquid-side film thickness, Δ_L, each representing the corresponding mass transfer resistance. If the mole fraction of A is small on the gas-side, Fick's law can be written as follows:

$$N_A = -C\mathcal{D}_{A,B}\frac{dy_A}{dx} \tag{14.70}$$

where x is the distance through the film. Integrating this equation across the thickness of the gas film Δ_G gives:

$$N_A = \frac{C\mathcal{D}_{A,B}}{\Delta_G}(y_A - y_{A,s}) \tag{14.71}$$

In the framework of the film model, the quantity $C\mathcal{D}_{A,B}/\Delta_G$ corresponds to the gas-side mass transfer coefficient, k_y. If the mole fraction of component A is not small and we assume that component B is stagnant because it is not soluble in the liquid, Fick's law is given by the following expression according to Eq. (9.56) or (11.20):

$$N_A = -\frac{C\mathcal{D}_{A,B}}{1-y_A}\frac{dy_A}{dx} \tag{14.72}$$

This time, separating the variables y_A and x in Eq. (14.72) and integrating the result across the gas film, we obtain:

$$N_A = \frac{C\mathcal{D}_{A,B}}{\Delta_G}\ln\left(\frac{1-y_{A,s}}{1-y_A}\right) = k_y \ln\left(\frac{1-y_{A,s}}{1-y_A}\right) \tag{14.73}$$

Equation (14.73) shows that the flux is no longer just proportional to $y_A - y_{A,s}$ as it was in the dilute case. We can rearrange Eq. (14.73) to the following form:

$$N_A = \frac{k_y}{(1-y_A)_{lm}}(y_A - y_{A,s}) \tag{14.74}$$

where:

$$(1-y_A)_{lm} = \frac{(1-y_A)-(1-y_{A,s})}{\ln\left(\dfrac{1-y_A}{1-y_{A,s}}\right)} \tag{14.75}$$

Note that $(1-y_A)_{lm}$ is the logarithmic mean of the values of $1-y_A$ in the bulk gas and at the gas-liquid interface and can be thought of as a correction factor that accounts for diffusion of A through a stagnant gas. Accordingly, we can define a corrected mass transfer coefficient k_y' as follows:

$$k_y' = \frac{k_y}{(1-y_A)_{lm}} \tag{14.76}$$

In terms of this corrected mass transfer coefficient, the mass transfer flux can be written simply as:

$$N_A = k_y'(y_A - y_{A,s}) \tag{14.77}$$

The following table shows the effect of y_A on the ratio k_y'/k_y if, as an example, we take $y_{A,s}$ to be 50% of y_A. We can see that the $1/(1-y_A)_{lm}$ correction amounts to less than

4% if $y_A \le 0.05$ but that it exceeds 10% if y_A is larger than about 0.1. Obviously, in many instances, neglecting this correction will have relatively small consequences on the accuracy of the results.

y_A	$\frac{k'_y}{k_y} = \frac{1}{(1-y_A)_{lm}}$
0	1
0.05	1.04
0.10	1.08
0.15	1.13
0.20	1.18

The same considerations also apply to the liquid phase for which we can write:

$$N_A = \frac{k_x}{(1-x_A)_{lm}}(x_{A,s} - x_A) = k'_x(x_{A,s} - x_A) \tag{14.78}$$

with:

$$k'_x = \frac{k_x}{(1-x_A)_{lm}} \tag{14.79}$$

Finally, we can combine Eqs. (14.77) and (14.78) to find the interfacial composition just as we did for the case of dilute conditions. In this case, we obtain the following expression:

$$\frac{k_y}{(1-y_A)_{lm}}(y_A - y_{A,s}) = \frac{k_x}{(1-x_A)_{lm}}(x_{A,s} - x_A) \tag{14.80}$$

Which, in turn, gives:

$$y_{A,s} = -\frac{k_x}{k_y}\frac{(1-y_A)_{lm}}{(1-x_A)_{lm}} x_{A,s} + y_A + \frac{k_x}{k_y}\frac{(1-y_A)_{lm}}{(1-x_A)_{lm}} x_A \tag{14.81}$$

This is equivalent to the tie line that we found for dilute conditions connecting the points representing the bulk and interfacial compositions on the y-x plot, Eq. (14.6). However, because of the dependence of the mass transfer coefficients on composition, the tie line is no longer straight in this case. As a result, an iterative procedure will be needed to find the intersection of this line with the equilibrium line where we will find the interfacial composition.

14.3.2 Material Balance and Operating Line

The operating line is found from Eq. (14.20), which gives:

$$y_A = \frac{L_M}{G_M} x_A + \left(\frac{G_{M,2}}{G_M} y_{A,2} - \frac{L_{M,2}}{G_M} x_{A,2}\right) \tag{14.82}$$

or:

$$y_A = \frac{L}{G} x_A + \left(\frac{G_2}{G} y_{A,2} - \frac{L_2}{G} x_{A,2}\right) \tag{14.83}$$

For the nondilute case considered here, L/G is not a constant but will vary from one end of the column to the other so that the operating line will no longer be straight on the

y-x plot. In order to handle this variation, it is convenient to redefine the composition in terms of mole ratios rather than mole fractions. For the gas phase, the mole ratio, Y_A, is defined as the moles of the absorbable component A in the gas divided by the moles of inert carrier gas. For the liquid phase, the mole ratio, X_A, is defined as the moles of absorbed A in the liquid divided by the moles of A-free absorbing liquid. The molar velocities of inert gas, G_s, and of A-free absorbent liquid, L_s, are related to the corresponding gas and liquid molar velocities by the following equations:

$$G_s = (1 - y_A)G \tag{14.84}$$

$$L_s = (1 - x_A)L \tag{14.85}$$

As a result, the mole ratios are related to the mole fractions by the following expressions:

$$Y_A = \frac{\text{moles of A}}{\text{moles of inert gas}} = \frac{y_A G}{G_s} = \frac{y_A}{1 - y_A} \tag{14.86}$$

$$X_A = \frac{\text{moles of A}}{\text{moles of absorbent liquid}} = \frac{x_A L}{L_s} = \frac{x_A}{1 - x_A} \tag{14.87}$$

Note that if we know y_A and x_A, we can easily calculate Y_A and X_A from Eqs. (14.86) and (14.87). We can also go backwards and calculate y_A and x_A from known values of Y_A and X_A as follows:

$$y_A = \frac{Y_A}{1 + Y_A} \tag{14.88}$$

$$x_A = \frac{X_A}{1 + X_A} \tag{14.89}$$

The advantage of using mole ratios instead of mole fractions is that the material balance can now be written more simply as:

$$G_s Y_A + L_s X_{A,2} = G_s Y_{A,2} + L_s X_A \tag{14.90}$$

or:

$$Y_A = \frac{L_s}{G_s} X_A + \left(Y_{A,2} - \frac{L_s}{G_s} X_{A,2} \right) \tag{14.91}$$

Since the inert gas and the A-free absorbent liquid velocities are constant, this is again a straight line when plotted on mole ratio scales. Now, given the equilibrium line, the gas mole fractions, and the mole fraction in the entering liquid, we can easily find the minimum value of L_s/G_s. Note that, regardless of whether it is straight or curved, on the mole fraction plot, y-x, the equilibrium line becomes curved on the mole ratio plot, Y-X. This procedure is illustrated in Example 14.4.

Example 14.4. Determination of $(L_s/G_s)_{min}$ for non-dilute conditions. Let's consider the case of an absorption column to be designed to remove 99% of the ammonia (A) present in an air stream with a mole fraction $y_{A,1} = y_{NH_3,1} = 0.25$ using pure water as the absorbent at 10 atm and 333 K. These are the same conditions as Example 14.3, except that the feed gas now contains a much higher mole fraction of ammonia.

The equilibrium line is given approximately by the polynomial $y_{NH_3} = 0.355x_{NH_3} + 1.602x^2_{NH_3} - 1.357x^3_{NH_3}$. We first convert the equilibrium line from the mole fractions basis to the mole ratios basis using Eqs. (14.86) to (14.89). Figure 14.11 (top) shows the result of this conversion. The inlet-gas mole ratio is:

$$Y_{A,1} = \frac{y_{A,1}}{1-y_{A,1}} = \frac{0.25}{1-0.25} = 0.333$$

For 99% removal, we will have $Y_{A,2} = (1-0.99)\times Y_{A,1} = 0.01\times 0.333 = 0.00333$. This corresponds to a mole fraction $y_{A,2} = Y_{A,2}/(1+Y_{A,2}) = 0.00332$. We can see that, at the dilute end of the column (i.e., at the top), the mole ratio and the mole fraction are very nearly the same. However, they are significantly different from each other at the concentrated end of the column (i.e., the bottom). We can now plot the operating line, which is straight on the *Y*-*X* plot, and vary L_s/G_s by pivoting around the point with coordinates $X_{A,2} = 0$ and $Y_{A,2} = 0.00333$ to find $(L_s/G_s)_{min} = 0.835$. The actual L_s/G_s will have to be greater than this value. If we pick, for example, $L_s/G_s = 1.4\times 0.835 = 1.17$ we obtain the solid line in Fig. 14.11 (top). Finally, we can go back to the mole fraction plot where the operating line becomes:

$$y_A = \frac{\frac{L_s}{G_s}\frac{x_A}{1-x_A} + \left(Y_{A,2} - \frac{L_s}{G_s}X_{A,2}\right)}{1+\frac{L_s}{G_s}\frac{x_A}{1-x_A} + \left(Y_{A,2} - \frac{L_s}{G_s}X_{A,2}\right)} \tag{14.92}$$

which is plotted in Fig. 14.11 (bottom). Note that this line is now curved since *L/G* varies along the column.

14.3.3 Packing Height

To determine the height of packing required for a chosen L_s/G_s, we return to the material balance for the differential slice of packing shown in Fig. 14.8. In terms of inert gas molar velocity and gas-phase mole ratio, the material balance now becomes:

$$G_s dY_A = -ak'_y(y_A - y_{A,s})dz = -\frac{ak_y}{(1-y_A)_{lm}}(y_A - y_{A,s})dz \tag{14.93}$$

This time it is convenient to work in terms of the gas-side driving force rather than the overall driving force. Before we can integrate, we need to express the left-hand side of Eq. (14.93) in terms of y_A according to the following derivation:

$$dY_A = d\left(\frac{y_A}{1-y_A}\right) = \frac{1-y_A+y_A}{(1-y_A)^2}dy_A = \frac{dy_A}{(1-y_A)^2} \tag{14.94}$$

Substituting this result in Eq. (14.93) gives:

$$G_s\frac{dy_A}{(1-y_A)^2} = -\frac{ak_y}{(1-y_A)_{lm}}(y_A - y_{A,s})dz \tag{14.95}$$

The final step is to separate the variables z and y_A in this equation, which gives:

$$dz = -\frac{G_s}{(1-y_A)ak_y}\frac{(1-y_A)_{lm}}{(1-y_A)(y_A-y_{A,s})}dy_A \tag{14.96}$$

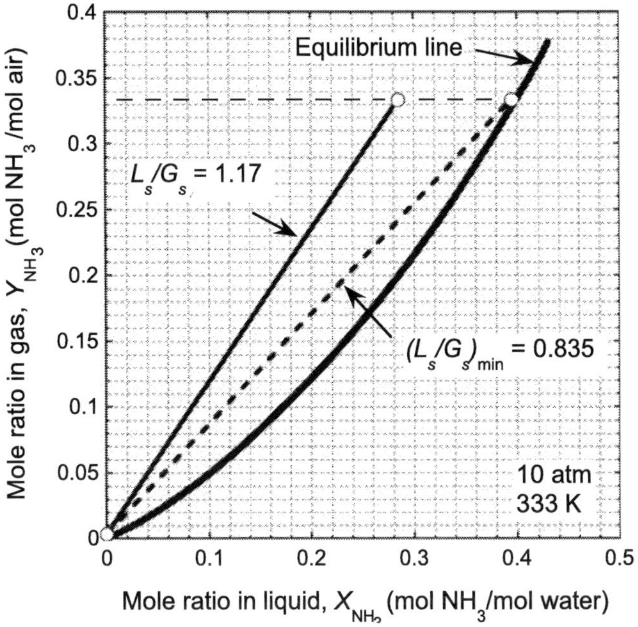

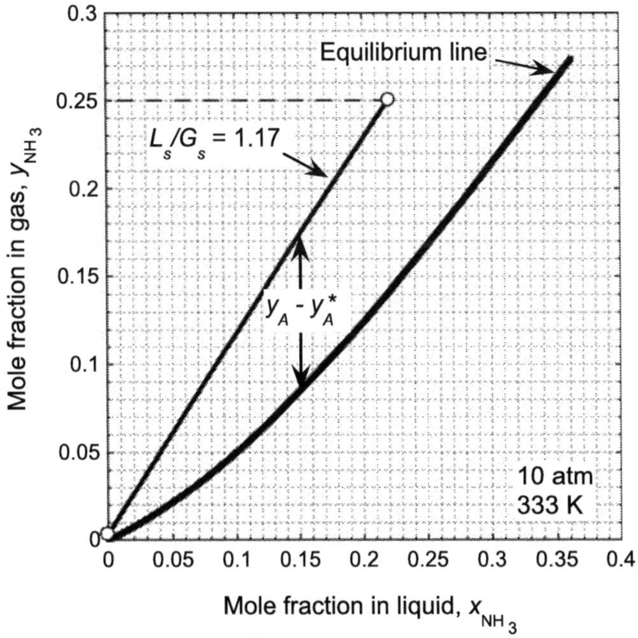

Figure 14.11 Mole ratio plot (top) and mole fraction plot (bottom) for Example 14.4.

Integrating this equation across the height of packing, Z, we obtain:

$$Z=\int_{y_{A,2}}^{y_{A,1}} \frac{G_s}{(1-y_A)ak_y}\frac{(1-y_A)_{lm}}{(1-y_A)(y_A-y_{A,s})}dy_A \tag{14.97}$$

Note that, since $G_s/(1-y_A)=G$, the first of the two ratios under the integral in Eq. (14.97) is actually $H_G=G/ak_y$. In turn, since the latter is, in practice, relatively constant with y_A, we can write the final result as follows:

$$Z=H_G N_G \tag{14.98}$$

where:

$$N_G=\int_{y_{A,2}}^{y_{A,1}} \frac{(1-y_A)_{lm}}{(1-y_A)(y_A-y_{A,s})}dy_A \tag{14.99}$$

is the number of gas-phase mass transfer units. This integral is not straightforward and numerical integration is typically needed. For this purpose, we can divide the interval $y_{A,1}-y_{A,2}$ into a suitable number of steps and evaluate the integrand at each y_A. This will require determining, at each step, the interfacial mole fraction, $y_{A,s}$, from Eq. (14.81) and can obviously get quite cumbersome if done graphically. A special case that results in a simpler calculation is when the gas-side resistance is dominant. In this case, the tie line is vertical and $y_{A,s}$ is equal to y_A^*. The latter will be the gas-phase mole fraction in equilibrium with the bulk liquid.

Note that an analogous derivation based on a liquid-side mole fraction driving force will lead to the following expression for the packing height:

$$Z=H_L N_L \tag{14.100}$$

where:

$$N_L=\int_{x_{A,2}}^{x_{A,1}} \frac{(1-x_A)_{lm}}{(1-x_A)(x_{A,s}-x_A)}dx_A \tag{14.101}$$

is the number of liquid-phase mass transfer units. In this case, if the liquid-side resistance is dominant, $x_{A,s}$ becomes equal to x_A^*, which will be the liquid mole fraction in equilibrium with the bulk gas. Example 14.5 illustrates use of these relationships to calculate the packing height under non-dilute conditions.

Example 14.5. Determination of N_G for non-dilute conditions. Let's continue Example 14.4 and calculate N_G assuming that the gas-side resistance is dominant. We already saw in Example 14.4 that this is a reasonable assumption for this system because of the high solubility of ammonia in water. Since, in this case, $y_{A,s}=y_A^*$ (with $A=NH_3$), the integrand in Eq. (14.99) becomes:

$$\frac{(1-y_A)_{lm}}{(1-y_A)(y_A-y_A^*)}$$

where:

$$(1-y_A)_{lm}=\frac{(1-y_A)-(1-y_A^*)}{\ln\left(\frac{1-y_A}{1-y_A^*}\right)}$$

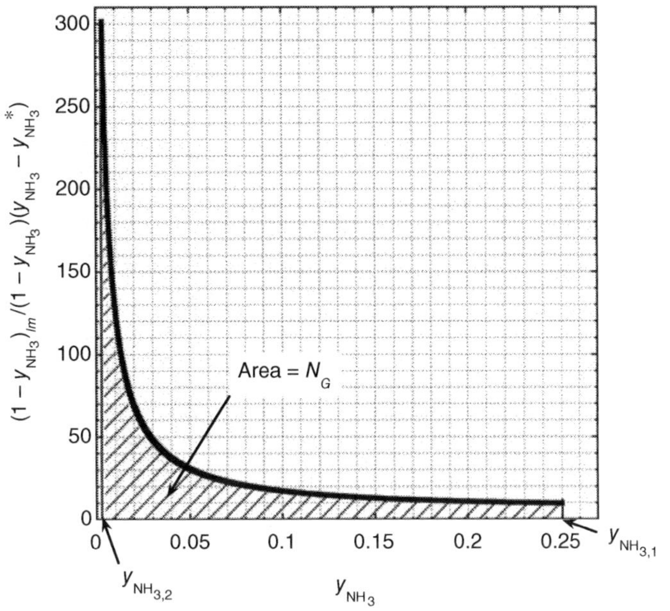

FIGURE 14.12 Calculation of N_G for Example 14.5.

y_A^* is equilibrium with x_A, which, in turn, is found on the operating line. From Eq. (14.92) we obtain:

$$x_A = \frac{\frac{G_s}{L_s}\frac{y_A}{1-y_A}+\left(X_{A,2}-\frac{G_s}{L_s}Y_{A,2}\right)}{1+\frac{G_s}{L_s}\frac{y_A}{1-y_A}+\left(X_{A_3,2}-\frac{G_s}{L_s}Y_{A,2}\right)}$$

The integrand is shown in Fig. 14.12 and the area evaluated numerically using the trapezoidal rule between $y_{A,2} = 0.00333$ and $y_{A,1} = 0.25$ gives $N_G = 6.5$. Note that the curve becomes very steep as we approach $y_{A,2}$, which makes it difficult to calculate the area precisely. We can improve the numerical calculation by dividing the column into two hypothetical parts: one from the bottom to a point where $y_A = 0.02$ and the other from this point on to the top of the column. For the bottom part, we calculate the area numerically, which gives $N_{G,\text{bottom}} = 4.31$. For the top part, since the mole fractions are less than 0.02, both the equilibrium line and the operating line are essentially linear. We can thus use the Colburn equation in this region. The initial slope of the equilibrium line is $K = 0.355$ while the slope of the operating is $L_s/G_s = 1.17$. Thus, for this portion of the column we have $\gamma = 0.355/1.17 = 0.304$. With this value, from Eq. (14.56) we obtain $N_{G,\text{top}} \sim N_{o,G} = 2.16$. Thus, for the entire column, $N_G = 4.31 + 2.16 = 6.47$. It is interesting to observe that it will take 4.31 transfer units to remove the first 92% of the ammonia (going from 25% at the bottom to 2%) and 2.16 transfer units to remove the last 8% (going from 2% to 0.33% at the top). This means that removing the 8% of the ammonia will require about 1/3 of the total transfer units or, equivalently, 1/3 of the packing height. This occurs because, at the low values of the mole fractions near the top of the column, mass transfer becomes very slow, requiring a substantial incremental height to achieve the relatively small amount of additional absorption.

14.4 Column Diameter

For given gas and liquid flow rates, the diameter of a counter-current absorption or stripping column is determined by the fluid handling capability of the particular packing used. Packing type, size, and shape will impose hydrodynamic limitations and

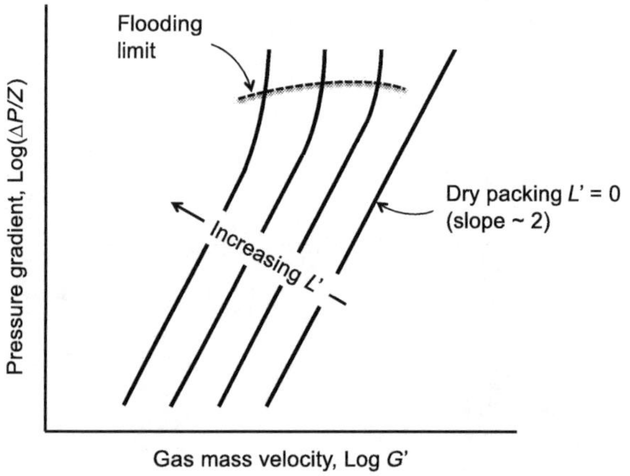

FIGURE 14.13 Pressure-flow curves for a counter-current packed absorption column on log-log scales. The curves take a sharp turn upwards at the onset of flooding.

affect the column pressure drop. In order to understand the origin of these limitations, it is instructive to consider first the pressure-flow curves sketched in Fig. 14.13, which depicts the experimental behavior expected for a typical packed column. Let's imagine that we start with a dry column (i.e., with no liquid flow) with set diameter and packing height and gradually increase the gas flow rate. The pressure gradient, $\Delta P/Z$, will gradually have to increase as we increase the gas flow rate since pressure is what actually drives the gas flow in the column. For random packings, since flow of the gas is likely turbulent, we can expect that $\Delta P/Z$ as a function of G' will follow the Burke-Plummer equation [Bird et al., *Transport Phenomena*, Wiley, 2007, p.191], which is given by:

$$\frac{\Delta P}{Z} = b\frac{1-\varepsilon}{\varepsilon^3}\frac{1}{d_p}\frac{(G')^2}{\rho_G} \tag{14.102}$$

where ε is the packing void fraction, d_p is the size of the packing pieces, and b is a numerical constant that depends on the shape of the packing. If the packing consisted of spheres, b would be equal to 1.75. The actual value will depend strongly on the particular shape and design of the packing pieces. Equation (14.102) shows that $\Delta P/Z$ will vary approximately with the square of the gas flow rate. Thus, the slope of the dry-packing pressure-flow curve on the log-log scales of Fig. 14.13 will be around 2. Because of the ε^3 and d_p terms in the denominator, Eq. (14.102) also shows that a larger void fraction and a larger packing size will result in lower values of $\Delta P/Z$ since, in both cases, the gas will experience a lower resistance to flow. Let's now imagine that we repeat the pressure-flow measurement with a finite flow rate of liquid. The liquid, driven by gravity, will flow over the surfaces of the packing pieces forming laminar falling films, as sketched in Fig. 14.1. A certain amount of liquid will be held within the column, which will reduce the space available for gas flow. As a result, we can expect that the pressure drop will be greater than for the dry packing case but that it will still follow Eq. (14.102) with a slope of about 2 on the log-log scales. As we continue to increase the gas flow, however, we will eventually reach a point where the pressure drop in the column will begin to counteract the gravity force that drives the liquid flow.

As a result, more liquid will accumulate in the packing, thereby further reducing the space available for gas flow and causing a sharper increase in the pressure drop. At some gas flow rate, the pressure forces will exactly balance the gravity force exerted on the liquid. When this happens, the liquid will no longer be able to flow down, resulting in an unstable condition that is called "flooding." As shown schematically in Fig. 14.13, repeating the pressure-flow measurements at increasing liquid flow rates will result in greater pressure drops with an onset of flooding at increasingly lower gas flow rates. In general, in order to attain a stable operation, we want to design columns whose diameter is large enough to prevent flooding. With a larger diameter, for a given gas flow rate, G' will be smaller, which, in turn, will result in a smaller pressure drop and, thus, shift the onset of flooding.

We can develop a very simple flooding model by considering the forces experienced by a laminar liquid film falling on the surface of a packing piece as sketched in Fig. 14.1. This model is simplistic for an actual packed column and we do not expect it to provide the means for a quantitative prediction. However, we will see that it can form the basis for a useful semi-empirical correlation. The forces at play in our system are viscous drag at the packing surface, F_D, gravity, F_G, buoyancy, F_B, and the force due to the pressure gradient along the packing, F_P. If the height, thickness, and width of the falling film are d_p, Δ_L, and W, respectively, we will have:

$$F_D = W d_p \mu_L \frac{v_{L,\max}}{\Delta_L}$$

$$F_G = W \Delta_L d_p \rho_L g$$

$$F_B = W \Delta_L d_p \rho_G g$$

$$F_P = W \Delta_L d_p \frac{\Delta P}{Z}$$

where $v_{L,\max}$ is the liquid velocity at the gas-liquid interface and μ_L is the liquid viscosity. Balancing these forces, we obtain:

$$F_D = F_G - F_B - F_P$$

or, substituting the corresponding expressions for F_D, F_G, F_B, and F_P,

$$\mu_L \frac{v_{L,\max}}{\Delta_L} = \Delta_L(\rho_L - \rho_G)g - \Delta_L \frac{\Delta P}{Z}$$

At the onset of flooding, the right-hand side of this equation will approach zero leading to the equality $\Delta P/Z = (\rho_L - \rho_G)g$. Combining this result with Eq. (14.102) yields:

$$\frac{(G')^2}{\rho_G(\rho_L - \rho_G)g} \frac{b}{d_p} = \frac{\varepsilon^3}{1-\varepsilon}$$

Now, for a more realistic system than that sketched in Fig. 14.1, we can expect that the term having to do with ε on the right-hand side of this equation will actually decrease as we increase the liquid flow rate. Empirically, this decrease is found to be correlated with the following variable:

$$F_{LG} = \frac{L'}{G'}\left(\frac{\rho_G}{\rho_L}\right)^{1/2} \tag{14.103}$$

Thus, for actual packings, we express flooding conditions in terms of the following equation:

$$G'\left[\frac{F_p}{\rho_G(\rho_L-\rho_G)g}\right]^{1/2} = f\left(\frac{\Delta P}{Z}, F_{LG}\right) \tag{14.104}$$

where f is an empirical function of $\Delta P/Z$ and F_{LG}. The parameter F_p appearing in this equation is called the packing factor and will depend on the size and specific design of the packing pieces. Typical units will be m^{-1}. For random packings, Section 14 in *Perry's Chemical Engineers' Handbook*, 9th edition, recommends the correlation provided by Strigle, *Packed Tower Design and Applications*, Gulf Publishing Company, 1994. Figure 14.14 shows this correlation in the form suggested by Eq. (14.104) based on the data in Fig. 14-55 of Perry's Handbook. Note that the y-axis in Fig. 14.14 has been made dimensionless and includes a viscosity correction factor $(\nu_L/\nu_{water})^{0.05}$ where ν_L is the kinematic viscosity of the liquid and ν_{water} is the kinematic viscosity of water at 20°C (= 1 cS). Flooding conditions are found in Fig. 14.14 by setting $\Delta P/Z$ equal to the value obtained from the following dimensional expression:

$$\left(\frac{\Delta P}{Z}\right)_{flooding} = 0.052(F_p)^{0.7} \tag{14.105}$$

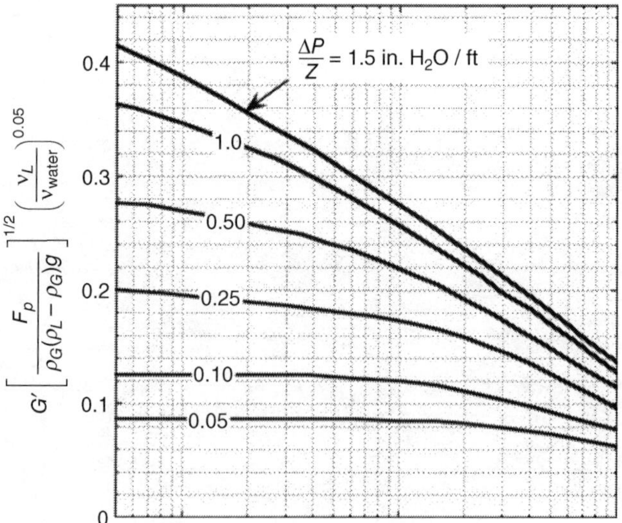

FIGURE 14.14 Generalized flooding and pressure drop correlation for random packings. The ratio ν_L/ν_{water} is the kinematic viscosity of the liquid divided by the kinematic viscosity of water at 20°C (1 cS). Divide in. H_2O/ft by 1.22×10^{-3} to get Pa/m. (Based on data in Figure 14-55 of *Perry's Chemical Engineers' Handbook*, 9th edition.)

where $\Delta P/Z$ is in inches of H_2O per ft of packing height and F_p is in m^{-1}. Values of F_p for use with this equation are found in Table 14-13 of *Perry's Chemical Engineers' Handbook*, 9th edition. Example 14.6 illustrates the use of this correlation to predict the column diameter based on the onset of flooding.

Example 14.6. Estimation of the diameter of a counter-current packed absorption column based on the onset of flooding. Let's use the method described in this section to determine the diameter needed for a column packed with 50-mm metal Intalox Saddles designed to absorb a component from an air stream into water at 1 atm and 298 K. The gas and liquid molar flow rates for this example are $G_M = 0.01$ kmol/s and $L_M = 0.06$ kmol/s. For this type and size of packing, Table 14-13 in Perry's Handbook gives $F_p = 59\ m^{-1}$. Thus, from Eq. (14.105), at the onset of flooding we will have:

$$\left(\frac{\Delta P}{Z}\right)_{\text{flooding}} = 0.052(59)^{0.7} = 0.90 \text{ in. } H_2O/\text{ft of packing}$$

Next, we calculate F_{LG}. The densities of air and water for these conditions are, respectively, $\rho_G = 1.2\ kg/m^3$ and $\rho_L = 1{,}000\ kg/m^3$. Since $G' = M_{r,G}G$ and $L' = M_{r,L}L$, where $M_{r,G}$ and $M_{r,L}$ are the molecular mass of the gas and of the liquid, respectively, we obtain:

$$F_{LG} = \frac{M_{r,L}}{M_{r,G}}\frac{L_M}{G_M}\left(\frac{\rho_G}{\rho_L}\right)^{1/2} = \frac{18}{29}\frac{0.06}{0.010}\left(\frac{1.2}{1{,}000}\right)^{1/2} = 0.13$$

From Fig. 14.14, at $F_{LG} = 0.13$ and $\Delta P/Z = 0.90$, we read:

$$G'\left[\frac{F_p}{\rho_G(\rho_L - \rho_G)g}\right]^{1/2}\left(\frac{\nu_L}{\nu_{\text{water}}}\right)^{0.05} \sim 0.24$$

Thus:

$$G'_{\text{flooding}} = 0.24 \times \left[\frac{1.2(1{,}000 - 1.2)9.8}{59}\right]^{1/2} = 3.4\ kg/m^2\ s$$

Since we want to operate well below flooding, we will design the column diameter so that the actual G' is a certain percentage of G'_{flooding}. For example, if we operate at 50% of the flooding gas velocity, we will have $G' = 0.5 \times G'_{\text{flooding}} = 1.7\ kg/m^2$ s. Finally, since $G = G_M/S$, where $S = \pi D_c^2/4$ is the column cross-sectional area, the column diameter, D_c, will be calculated as:

$$D_c = \sqrt{\frac{4}{\pi}\frac{G_M M_{r,G}}{G'}} = \sqrt{\frac{4}{\pi}\frac{0.01 \times 29}{1.7}} = 0.47\ m$$

We can also estimate the pressure drop for the chosen design value of G' at 50% of flooding. For these conditions, the value on the y-axis of Fig. 14.14 becomes:

$$G'\left[\frac{F_p}{\rho_G(\rho_L - \rho_G)g}\right]^{1/2}\left(\frac{\nu}{\nu_{\text{water}}}\right)^{0.05} \sim 0.12$$

which gives $\Delta P/Z \sim 0.1$ in. H_2O/ft or 82 Pa/m. We would need to multiply this value by the packing height, Z, to obtain the total pressure drop.

Note that the same diameter design procedure illustrated above applies whether we have dilute conditions, as we have assumed here, or not. For non-dilute cases, it is usually safer to select the diameter for an absorber using the conditions at the bottom of the column where the flow rates are greatest. For a stripper, instead, the diameter would preferably be determined using the conditions at the top of the column. The same general approach also applies to structured packings, but a different correlation would be used to find G' (e.g., see Figure 14-56 in *Perry's Chemical Engineers' Handbook*, 9th edition). The main difference between the design for structured and random packings will be that, typically, for the same F_{LG}, higher flooding velocities and lower pressures will be found for structured packings compared to random packings.

14.5 Problems

P.14.1. A gas absorption tower is used to absorb sulfur dioxide from air into water at 20°C and 2.0 atm absolute pressure. At the top of the tower, the water is 100% pure and the gas contains 5.0 mol % SO_2. At the bottom of the tower, the water contains 1.0 mol% SO_2 while the gas contains 25 mol% SO_2. The individual mass transfer coefficients are $k_x = 5.4$ mol/m^2 s and $k_y = 0.41$ mol/m^2 s for the liquid and the gas, respectively, and are constant throughout the tower. The equilibrium line is described by the data given in the following table. Find the overall mass transfer coefficients K_x and K_y and the interfacial composition at the top and at the bottom of the tower. Equilibrium data for the system SO_2 in water at 20°C are as follows (based on data in Table 2-13 of *Perry's Chemical Engineers' Handbook*, 9th edition):

p_{SO_2} (mmHg)	x_{SO_2} (Mole Fraction)
26.0	0.00141
58.4	0.00281
129	0.00563
202	0.00844
277	0.0113
353	0.0141
430	0.0169

P.14.2. Air containing 1.6% by volume sulfur dioxide is scrubbed with pure water in a packed column with 1.5 m^2 cross-sectional area and 3.5 m height, packed with No. 2 plastic Super Intalox saddles operated at a pressure of 1 atm. The gas flow rate is 0.058 kmol/s and the liquid flow rate is 2.2 kmol/s. The outlet sulfur dioxide mole fraction is 0.003. The equilibrium line is given by $y_A = 40.0x_A$ at the temperature and pressure at which the column is operated. Based on the data above, determine the values of: (a) $(L/G)/(L/G)_{min}$; (b) $N_{o,G}$; (c) $H_{o,G}$; and (d) aK_y.

P.14.3. An absorption tower was designed and built to remove 80% of a pollutant A from air fed into the bottom of the tower at a mole fraction of 0.01. The absorbing liquid is pure water and the liquid flow rate is 1.5 times the minimum possible. The equilibrium line is straight with a slope $K = 1.0$ (mol/mol). (a) Make a graph of the equilibrium and operating lines. (b) Determine the number of transfer units, $N_{o,G}$, available in the column. (c) A new law is passed requiring that 90% of the A fed be removed instead of the 80% for which the column was originally designed. Assuming that the feed gas flow rate, inlet composition, and operating temperature and pressure remain the same, by how much should the liquid flow rate be changed? (d) Another option for part (c) is to keep the liquid flow rate the same but to increase the operating pressure. Assuming that the absorption process is controlled by the gas-side resistance and that H_G is approximately independent of pressure, by how much should the operating pressure change? Note that K is a function of pressure.

P.14.4. A wastewater stream flowing at the rate of 600 GPM (gallons per minute) and containing 10 ppm (parts per million by weight) benzene is to be stripped with air in a packed column operating at 25°C and 2 atm absolute pressure. The water leaving the column is to contain 0.5 ppm benzene. The packing is 1-in. Berl saddles. The solubility of liquid benzene in water at 25°C is 0.180 g per 100 g of water. It has been suggested to use a flow rate of air of 1,000 SCFM (standard cubic feet per minute), at which the following values of the mass transfer coefficients are $ak_c = 0.80\ \text{s}^{-1}$ for the gas side and $ak_c = 0.067\ \text{s}^{-1}$ for the liquid side. Determine: (a) The minimum air flow rate. (b) The number of transfer units, $N_{o,G}$, required. (c) The overall mass transfer coefficient, aK_y, in units of mol/(m^3 s). (d) The volume of packing required in m^3.

P.14.5. The performance of a packed absorption tower is to be evaluated from the following experimental results. The test tower is 6 ft in diameter and is packed to a depth of 10 ft with 1-in. metal Pall rings. Test runs are made for the absorption of hexane vapor from air into a non-volatile hydrocarbon oil. In the tests, gas and liquid streams enter the tower at 140°F and absorption occurs isothermally. The equilibrium constant $K = y_A/x_A$ for hexane in the oil is 0.79 on a mole fraction basis at 140°F and 1 atm. The molecular weight of the oil is 200 kg/kmol. Using the data below determine: (a) The height of an overall gas-phase transfer units, $H_{o,G}$, at 1 atm and at 2 atm. (b) The individual heights of a transfer unit, H_G and H_L assuming that they are independent of pressure. Note that K is a function of pressure. (c) Determine the mass transfer coefficients ak_y and ak_x.

Parameter	Test 1	Test 2
Gas flow rate, lbmol/h	493	493
Liquid flow rate, lbmol/h	540	540
Absolute pressure, atm	1.0	2.0
Mole fraction of hexane in gas entering	0.01	0.01
Mole fraction of hexane in gas leaving	0.00208	0.00065
Mole fraction of hexane in liquid entering	0.0004	0.0004

P.14.6. A gas absorption column is to be designed to absorb SO_2 from air using pure water as the absorbent. The inlet air contains 5.0 mol% SO_2 and flow at the rate of 30.0 mol/s and it is desired to remove 95% of the SO_2. The column will be packed with 1-in. metal Pall rings and will be operated at 10°C and 10.0 atm absolute pressure with a water flow rate 1.2 times greater than the minimum and at 60% of flooding. Equilibrium data for the SO_2-water system at 10°C are given below. Taking into account the curvature of the equilibrium line, determine: (a) The minimum and actual L/G.

(b) The number of overall gas-phase mass transfer units, $N_{o,G}$. (c) The column diameter required to operate at 60% of flooding. (d) The values of H_G and H_L and of $H_{o,G}$ at the top and bottom of the column. (e) The height of packing required.

p_{SO_2} (mmHg)	Liquid Composition in g of SO_2 per 100 g of H_2O
17.0	0.5
39.5	1
88.5	2
139	3
192	4
245	5
299	6
407	8

Solubility of SO_2 in Water at 10°C Based on Table 2-13 in *Perry's Chemical Engineers' Handbook*, 9th Edition

P.14.7. It is desired to recover ammonia from an air stream at 1 atm by absorption in water in a tower packed with 2-in. plastic Flexi-saddles operating at 30°C. The inlet gas contains 20 mol% ammonia and 99% of the ammonia is to be absorbed. Pure water is fed to the top of the tower at a rate 20% greater than the minimum rate. Find the height and diameter of a tower designed to operate at one-half of the flooding velocity needed to treat 1,360 kg/h of feed gas. Note that in this problem the mole fractions are definitely high enough that both equilibrium and operating lines are curved.

p_{NH_3} (mmHg)	Liquid Composition in g of NH_3 per 100 g of H_2O
11.5	1.2
15.3	1.6
19.3	2.0
24.4	2.5
29.6	3.0
40.1	4.0
51.0	5.0
79.7	7.5
110	10
179	15
260	20
352	25

Solubility of NH_3 in Water at 30°C Based on Table 2-13 in *Perry's Chemical Engineers' Handbook*, 9th Edition

P.14.8. A gas absorption column is to be designed to remove 90.0% of the ammonia from an air stream containing 2.0 mol% ammonia. The absorbent is pure water. The column is to be packed with 50 mm metal Pall rings and will be operated at 30°C and 1 atm with a design gas-mass velocity of 800 lb/(h ft^2) and with $L/G = 1.35$. For these conditions, the equilibrium constant is $K = 1.20$. (a) Calculate the height of packing required. (b) Determine the percentage of flooding for these conditions. (c) The column is built and works according to specifications. However, after a few months it is decided to retrofit the column with new packing in order to attain 95.0% removal of the ammonia using the same height of packing. All other conditions are to remain the same. Select a new packing type and size that will be suitable for this retrofit. Make sure that flooding will not occur in the retrofitted column.

CHAPTER 15

Coupled Mass Transfer Processes

15.1 Introduction

In this chapter, we consider situations where multiple mass transfer fluxes are coupled to each other or where diffusion is coupled to other phenomena. Examples are multicomponent diffusion, where the mole fraction gradient of one species affects the flux of other species, mass transfer of charged species, where an electrical potential gradient drives electrophoretic transport, and simultaneous heat and mass transfer, where temperature and mole fraction gradients are jointly responsible for the overall energy transport process. These examples are important in a variety of applications. While the foundation for the analysis of these situations is the same diffusion mechanism that we have considered in Chaps. 8 to 14, coupling requires that we introduce additional concepts that are covered in the following sections.

15.2 Multicomponent Gas-Phase Diffusion

Let's consider diffusion in a multicomponent gas mixture for conditions where the ideal gas law is applicable. In a mixtures with n_c components there will be n_c mass transfer fluxes and $n_c(n_c-1)/2$ binary diffusivities, $\mathcal{D}_{i,j}$. For example, in a three-component mixture comprising components A, B, and C, we will have $\mathcal{D}_{A,B}$, $\mathcal{D}_{A,C}$, and $\mathcal{D}_{B,C}$. At a given temperature and pressure, each binary diffusivity can be predicted by the Chapman–Enskog theory, Eq. (10.13), or from the empirical equation of Fuller et al., Eq. (10.17). For a multicomponent system, the relationship between these diffusivities and the mass transfer fluxes is given by the following Stefan–Maxwell equation:

$$\nabla y_i = \frac{1}{C}\sum_{\substack{j=1\\ j\neq i}}^{n_c}\left[\frac{y_i N_j - y_j N_i}{\mathcal{D}_{i,j}}\right] \tag{15.1}$$

where ∇y_i is the spatial gradient in the mole fraction of component i and $C = P/\mathcal{R}T$ is the gas molar density. Equation (15.1) is also based on the kinetic theory of gases. For a binary mixture of components A and B ($n_c = 2$), Eq. (15.1) reduces to the following:

$$\nabla y_A = \frac{1}{C}\frac{y_A N_B - y_B N_A}{\mathcal{D}_{A,B}} \tag{15.2}$$

Since $y_B = 1 - y_A$, we can also write this equation as follows:

$$\nabla y_A = \frac{1}{C}\frac{y_A N_B - (1-y_A)N_A}{\mathcal{D}_{A,B}} = \frac{y_A(N_A + N_B) - N_A}{C\mathcal{D}_{A,B}} \tag{15.3}$$

or, after some rearrangement,

$$N_A = y_A(N_A + N_B) - C\mathcal{D}_{A,B}\nabla y_A \tag{15.4}$$

which is the same as Eq. (9.47). For a mixture with n_c components with $n_c > 2$, we can write the following analogous expression:

$$N_i = y_i \sum_{j=1}^{n_c} N_j - C\mathcal{D}_{i,m}\nabla y_i \tag{15.5}$$

where $\mathcal{D}_{i,m}$ is defined as the mixture diffusivity of component i. Combining Eqs. (15.1) and (15.5) gives the following expression for $\mathcal{D}_{i,m}$:

$$\mathcal{D}_{i,m} = \frac{N_i - y_i \sum_{j=1}^{n_c} N_j}{\sum_{\substack{j=1 \\ j\neq i}}^{n_c} \left[\frac{y_j N_i - y_i N_j}{\mathcal{D}_{i,j}}\right]} \tag{15.6}$$

Note that, despite their similar form, Eqs. (15.4) and (15.5) are in fact very different from each other. For a binary gas mixture, $\mathcal{D}_{A,B}$ is a physical property independent of composition. Thus, considering, for example, a steady-state diffusion problem, Eq. (15.4) can be integrated directly once the relationship between N_A and N_B is known. On the other hand, for a mixture of n_c components with $n_c > 2$, $\mathcal{D}_{i,m}$ in Eq. (15.6) will vary with both composition and with the fluxes of all components in the mixture. Thus, solving diffusion problems for multicomponent mixtures will require that we consider simultaneously all of the fluxes. A special case is when the fluxes are zero for all components except for component i. This corresponds to diffusion of component i through a stagnant mixture of gases and is encountered, for example, when a gaseous species is absorbed into a liquid from an inert gas mixture or when a membrane allows only one species to pass through. In this case, with $N_j = 0$ for all components except i, Eq. (15.6) yields:

$$\mathcal{D}_{i,m} = \frac{1 - y_i}{\sum_{\substack{j=1 \\ j\neq i}}^{n_c} \left[\frac{y_j}{\mathcal{D}_{i,j}}\right]} \tag{15.7}$$

Example 15.1 illustrates an application of these relationships.

Example 15.1. Diffusivity of a component in a stagnant gas mixture. Let's consider a gas mixture containing 10 mol% NH_3 (A), 60 mol% N_2 (B), and 30 mol% H_2 (C) at 300 K and 1 atm absolute pressure and find the mixture diffusivity of NH_3 assuming that N_2 and H_2 are stagnant. We first estimate the binary diffusivities from the equation of Fuller et al., Eq. (10.17), which is given by:

$$\mathcal{D}_{i,j}\ (\text{cm}^2/\text{s}) = 1.0 \times 10^{-3} \times \frac{T^{1.75}\left(\frac{1}{M_{r,i}} + \frac{1}{M_{r,j}}\right)^{1/2}}{P[(\Sigma v_i)^{1/3} + (\Sigma v_j)^{1/3}]^2}$$

The following values are obtained from this equation using Σv values in cm^3/mol derived from Table 10.4:

		M, (g/mol)	Σv (cm^3/mol)	Binary Diffusivity (cm^2/s)
A	NH_3	17	14.9	$\mathcal{D}_{A,B} = 0.258$
B	N_2	28	17.9	$\mathcal{D}_{B,C} = 0.770$
C	H_2	2	7.07	$\mathcal{D}_{A,C} = 0.843$

With these values, Eq. (15.7) gives:

$$\mathcal{D}_{A,m} = \frac{1-y_A}{\dfrac{y_B}{\mathcal{D}_{A,B}}+\dfrac{y_C}{\mathcal{D}_{A,C}}} = \frac{1-y_A}{\dfrac{y_B}{0.258}+\dfrac{y_C}{0.843}} = \frac{1-0.1}{\dfrac{0.6}{0.258}+\dfrac{0.3}{0.843}} = 0.336\ \text{cm}^2/\text{s}$$

which is intermediate between the binary diffusivities of the mixture components.

Let's now consider the more general case of steady-state diffusion in a film delimited by two planes separated by a distance Δ in a three-component system comprised of components A, B, and C. In this case, Eq. (15.1) yields the following two equations:

$$C\nabla y_A = \frac{y_A N_B - y_B N_A}{\mathcal{D}_{A,B}} + \frac{y_A N_C - y_C N_A}{\mathcal{D}_{A,C}} \tag{15.8}$$

$$C\nabla y_B = \frac{y_B N_A - y_A N_B}{\mathcal{D}_{A,B}} + \frac{y_B N_C - y_C N_B}{\mathcal{D}_{B,C}} \tag{15.9}$$

A third relationship of this type is not necessary and would actually not be independent of these two since, everywhere, we will have $y_A + y_B + y_C = 1$. Within the film, assuming one-dimensional steady-state diffusion along an x-axis perpendicular to the two planes that delimit the film, the mole fraction gradients are $\nabla y_A = dy_A/dx$ and $\nabla y_B = dy_B/dx$. The fluxes are constant as a function of x. A practically interesting special case is when $N_C = 0$; i.e., when components A and B diffuse in a stagnant gas C. This situation is encountered, for example, when our film is bounded by a membrane impervious to C or, otherwise, when a solubility barrier prevents its diffusion. In this case, Eqs. (15.8) and (15.9) simplify to:

$$C\frac{dy_A}{dx} = \frac{y_A N_B - y_B N_A}{\mathcal{D}_{A,B}} - \frac{y_C N_A}{\mathcal{D}_{A,C}} \tag{15.10}$$

$$C\frac{dy_B}{dx} = \frac{y_B N_A - y_A N_B}{\mathcal{D}_{A,B}} - \frac{y_C N_B}{\mathcal{D}_{B,C}} \tag{15.11}$$

Adding Eqs. (15.10) and (15.11) together yields the following result:

$$C\frac{d(y_A + y_B)}{dx} = -\left(\frac{N_A}{\mathcal{D}_{A,C}} + \frac{N_B}{\mathcal{D}_{B,C}}\right) y_C \tag{15.12}$$

or, since $d(y_A + y_B) = -dy_C$,

$$C\frac{dy_C}{dx} = \left(\frac{N_A}{\mathcal{D}_{A,C}} + \frac{N_B}{\mathcal{D}_{B,C}}\right) y_C \tag{15.13}$$

Next, since N_A and N_B are constant along x, we can separate the variables y_C and x in Eq. (15.13) and integrate the resulting equation directly from $x = 0$, where $y_A = y_{A,1}$, to $x = \Delta$, where $y_A = y_{A,2}$. This integration gives:

$$\frac{N_A}{\mathcal{D}_{A,C}} + \frac{N_B}{\mathcal{D}_{B,C}} = \frac{C}{\Delta}\ln\left(\frac{y_{C,2}}{y_{C,1}}\right) \tag{15.14}$$

where $y_{C,1}$ and $y_{C,2}$ are the mole fractions of the stagnant component at the two planes that delimit the film across which diffusion occurs. Now, if component B is also stagnant (i.e., if $N_B = 0$), Eq. (15.14) gives:

$$N_A = \frac{C\mathcal{D}_{A,C}}{\Delta}\ln\left(\frac{y_{C,2}}{y_{C,1}}\right) \tag{15.15}$$

Since, in this case, B and C are interchangeable (i.e., they are both stagnant), we will also have:

$$N_A = \frac{C\mathcal{D}_{A,B}}{\Delta}\ln\left(\frac{y_{B,2}}{y_{B,1}}\right) \tag{15.16}$$

Combining Eqs. (15.15) and (15.16) yields:

$$\frac{y_{B,2}}{y_{B,1}} = \left(\frac{y_{C,2}}{y_{C,1}}\right)^{\frac{\mathcal{D}_{A,C}}{\mathcal{D}_{A,B}}} \tag{15.17}$$

This important result shows that if $y_{B,1}$ and $y_{C,1}$ are specified on one side of the film and $y_{A,2}$ is specified on the other side, the mole fractions $y_{B,2}$ and $y_{C,2}$ will be determined by the ratio of the binary diffusivities $\mathcal{D}_{A,C}/\mathcal{D}_{A,B}$. If this ratio is not equal to 1, the two stagnant components will be redistributed attaining a ratio of mole fractions that is different on the two sides of the film. This result is not intuitive and comes about because of the coupling of diffusion fluxes predicted by the Stefan–Maxwell equation. Example 15.2 illustrates some of the implications of this result.

Example 15.2. Determination of the mass transfer flux in a ternary mixture with two stagnant components [Mathematica notebook: (M.EX.15.2)]. Let's consider again the case illustrated in Example 15.1 assuming that the $NH_3/N_2/H_2$ mixture is in contact with pure liquid water and that mass transfer of NH_3 is limited by diffusion across a gas film with a thickness Δ = 0.1 cm. If the gas-side resistance is controlling, the mole fraction of ammonia at the gas–liquid interface will be $y_{A,2} = 0$. In this case, since $y_{B,2} = 1 - y_{A,2} - y_{C,2}$, Eq. (15.17) becomes:

$$\frac{1 - y_{A,2} - y_{C,2}}{y_{B,1}} = \left(\frac{y_{C,2}}{y_{C,1}}\right)^{\frac{\mathcal{D}_{A,C}}{\mathcal{D}_{A,B}}} \tag{15.18}$$

or

$$\frac{1 - 0 - y_{C,2}}{0.6} = \left(\frac{y_{C,2}}{0.3}\right)^{\frac{0.843}{0.258}}$$

Numerical solution of this equation gives $y_{C,2} = 0.313$ and $y_{B,2} = 1 - 0.313 = 0.687$. We can see that the ratio y_B/y_C will vary from 0.6/0.3 = 2.0 in the bulk gas to 0.687/0.313 = 2.19 at the gas-liquid interface. This means that the gas at the liquid surface will be richer in N_2 compared to the bulk gas. This is actually the basis for a separation process called "sweep diffusion" where a mixture of two components is enriched by letting a third component, A, diffuse through it. Note that now that we know $y_{C,2}$, we can calculate the flux of A from Eq. (15.15). Since the gas molar density is $C = P/\mathcal{R}T = 1/(82 \times 300) = 4.07 \times 10^{-5}$ mol/cm^3, we obtain:

$$N_A = \frac{4.07 \times 10^{-5} \times 0.843}{0.1} \ln\left(\frac{0.313}{0.3}\right) = \frac{4.07 \times 10^{-5} \times 0.258}{0.1} \ln\left(\frac{0.687}{0.6}\right) = 1.42 \times 10^{-5} \text{ mol/cm}^2 \text{ s}$$

Let's now return to the more general case where only one of the three components in the mixture is stagnant; i.e., a case where A and B diffuse in stagnant C. Equation (15.14) is still valid but a second equation is needed to find N_A and N_B. With a slight rearrangement, Eqs. (15.10) and (15.11) can be re-written as follows:

$$C\frac{dy_A}{dx} = y_A\left(\frac{N_B}{\mathcal{D}_{A,B}} + \frac{N_A}{\mathcal{D}_{A,C}}\right) + y_B N_A\left(\frac{1}{\mathcal{D}_{A,C}} - \frac{1}{\mathcal{D}_{A,B}}\right) - \frac{N_A}{\mathcal{D}_{A,C}} \tag{15.19}$$

$$C\frac{dy_B}{dx} = y_A N_B\left(\frac{1}{\mathcal{D}_{B,C}} - \frac{1}{\mathcal{D}_{A,B}}\right) + y_B\left(\frac{N_A}{\mathcal{D}_{A,B}} + \frac{N_B}{\mathcal{D}_{B,C}}\right) - \frac{N_B}{\mathcal{D}_{B,C}} \tag{15.20}$$

To proceed further, we multiply both sides of Eq. (15.19) by the quantity $(N_A + N_B)/N_A$ and both sides of Eq. (15.20) by the quantity $\eta(N_A + N_B)/N_B$ where:

$$\eta = \frac{\dfrac{1}{\mathcal{D}_{A,C}} - \dfrac{1}{\mathcal{D}_{A,B}}}{\dfrac{1}{\mathcal{D}_{B,C}} - \dfrac{1}{\mathcal{D}_{A,B}}} \tag{15.21}$$

The following expressions are obtained:

$$C\frac{N_A + N_B}{N_A}\frac{dy_A}{dx} = \left[y_A\left(\frac{N_B}{\mathcal{D}_{A,B}N_A} + \frac{1}{\mathcal{D}_{A,C}}\right) + y_B\left(\frac{1}{\mathcal{D}_{A,C}} - \frac{1}{\mathcal{D}_{A,B}}\right) - \frac{1}{\mathcal{D}_{A,C}}\right](N_A + N_B) \tag{15.22}$$

and

$$C\frac{N_A + N_B}{N_B}\eta\frac{dy_B}{dx} = \left[y_A\eta\left(\frac{1}{\mathcal{D}_{B,C}} - \frac{1}{\mathcal{D}_{A,B}}\right) + y_B\eta\left(\frac{N_A}{\mathcal{D}_{A,B}N_B} + \frac{1}{\mathcal{D}_{B,C}}\right) - \frac{\eta}{\mathcal{D}_{B,C}}\right](N_A + N_B) \tag{15.23}$$

Subtracting Eq. (15.23) from Eq. (15.22), we obtain the following result after some algebraic manipulation:

$$\frac{C\mathcal{D}_{A,B}}{N_A + N_B}\frac{d}{dx}\left(y_A\frac{N_A + N_B}{N_A} - y_B\eta\frac{N_A + N_B}{N_B}\right) = y_A\frac{N_A + N_B}{N_A} - y_B\eta\frac{N_A + N_B}{N_B} + \eta - 1 \tag{15.24}$$

This equation has the form

$$\frac{C\mathcal{D}_{A,B}}{N_A + N_B}\frac{d\psi}{dx} = \psi + \eta - 1 \tag{15.25}$$

where

$$\psi = y_A\frac{N_A + N_B}{N_A} - y_B\eta\frac{N_A + N_B}{N_B} \tag{15.26}$$

is a new variable that combines both mole fractions y_A and y_B. Since η is a constant, we can separate the variables ψ and x in Eq. (15.25) and integrate the result across the thickness of the film Δ, which gives:

$$(N_A + N_B)\int_0^{\Delta} dx = C\mathcal{D}_{A,B}\int_{\psi_1}^{\psi_2} \frac{d\psi}{\psi + \eta - 1} \tag{15.27}$$

The final result, obtained by performing the integrals, is as follows:

$$(N_A + N_B) = \frac{C\mathcal{D}_{A,B}}{\Delta}\ln\left(\frac{\psi_2 + \eta - 1}{\psi_1 + \eta - 1}\right) \tag{15.28}$$

where:

$$\psi_1 = y_{A,1}\frac{N_A + N_B}{N_A} - y_{B,1}\eta\frac{N_A + N_B}{N_B} \tag{15.29}$$

and

$$\psi_2 = y_{A,2}\frac{N_A + N_B}{N_A} - y_{B,2}\eta\frac{N_A + N_B}{N_B} \tag{15.30}$$

In order to find N_A and N_B for given values of the mole fractions on the two sides of the film, Eq. (15.28) needs to be solved simultaneously with Eq. (15.14) as a system of two equations in two unknowns, which is illustrated in Example 15.3.

Example 15.3. Diffusion in a ternary mixture with one stagnant component [Mathematica notebook: (M.EX.15.3)]. This example, based on data in Sherwood et al. [*Mass Transfer*, McGraw-Hill, 1975, pp. 63–65], considers diffusional transport in a partial condenser operated with a feed gas containing ammonia (*A*), water (*B*), and hydrogen (*C*) at 3.36 atm absolute pressure. At a particular point in the condenser, the temperature is 366 K and the gas contains 30 mol% ammonia, 40 mol% water, and 30 mol% hydrogen. At the same point, the liquid contains 10 mol% ammonia and 90 mol% water with negligible hydrogen. As indicated by Sherwood et al., based on ammonia-water vapor liquid equilibrium data at this temperature and pressure, the gas at the gas–liquid interface at this point in the condenser will contain 45.5 mol% ammonia, 19.5 mol% water, and 35.0 mol% hydrogen. Let's determine the mass transfer flux of ammonia and water assuming that the rate is limited by the resistance offered by a gas film with a thickness $\Delta = 0.1$ cm. Hydrogen will be stagnant since, for these conditions, it is neither condensable nor significantly soluble in the liquid phase. The following binary diffusivities are provided by Sherwood et al. for this ternary system at 366 K and 1 atm absolute pressure:

$$\text{Ammonia/Water: } \mathcal{D}_{A,B} = 0.294 \text{ cm}^2/\text{s}$$

$$\text{Ammonia/Hydrogen: } \mathcal{D}_{A,C} = 1.14 \text{ cm}^2/\text{s}$$

$$\text{Water/Hydrogen: } \mathcal{D}_{B,C} = 1.30 \text{ cm}^2/\text{s}$$

Based on the Chapman–Enskog theory, the corresponding binary diffusivity values at 3.36 atm absolute pressure can be found by multiplying these values by the ratio 1/3.36 and are, respectively, 0.0875, 0.339, and 0.387 all in cm^2/s. For the conditions of this problem, we thus have:

$$\eta = \frac{\dfrac{1}{0.339} - \dfrac{1}{0.0875}}{\dfrac{1}{0.387} - \dfrac{1}{0.0875}} = 0.959$$

and:

$$C = \frac{P}{\mathcal{R}T} = \frac{3.36}{82 \times 366} = 1.12 \times 10^{-4} \text{ mol/cm}^3$$

Simultaneous solution of Eqs. (15.28) and (15.14) with $y_{A,1} = 0.3$, $y_{B,1} = 0.4$, $y_{C,1} = 0.3$, $y_{A,2} = 0.455$, $y_{B,2} = 0.195$, and $y_{C,2} = 0.350$ gives $N_A = 1.16 \times 10^{-5}$ mol/cm² s and $N_B = 5.35 \times 10^{-5}$ mol/cm² s. Now that we know N_A and N_B, it is easy to find the mole fraction profiles across the film. We do this by going back to Eqs. (15.13) and (15.25) and integrating from $x = 0$ to an arbitrary x, which gives:

$$y_C = y_{C,1} \exp\left[\frac{x}{C}\left(\frac{N_A}{\mathcal{D}_{A,C}} + \frac{N_B}{\mathcal{D}_{B,C}}\right)\right] \tag{15.31}$$

and

$$y_A = \frac{\psi + (1 - y_C)\eta \dfrac{N_A + N_B}{N_B}}{\dfrac{N_A + N_B}{N_A} + \eta \dfrac{N_A + N_B}{N_B}} \tag{15.32}$$

where:

$$\psi = (\psi_1 + \eta - 1)\exp\left[\frac{x}{C\mathcal{D}_{A,B}}(N_A + N_B)\right] - \eta + 1 \tag{15.33}$$

Figure 15.1 shows the mole fraction profiles calculated for this example according to Eqs. (15.31) to (15.33). As seen in this figure, none of these profiles are linear. Moreover, although the mole fraction of ammonia at the gas–liquid interface is higher than in the bulk gas (0.455 vs. 0.3), the flux of ammonia is positive; i.e., ammonia is predicted to diffuse toward the liquid against its mole fraction gradient. This is not a violation of Fick's law. Rather it is a consequence of the fact that the diffusion fluxes are coupled and that the mole fraction gradient of water affects the flux of ammonia allowing the latter to diffuse against its own mole fraction gradient. It should be noted that, in a practical application, the film thickness, Δ, would be estimated from a mass transfer coefficient according to the film model (see Sec. 13.1). In our case, if k_c is the gas-side mass transfer coefficient based, for example, on the absorption of dilute ammonia alone from a hydrogen gas, the film thickness would be estimated as $\Delta = \mathcal{D}_{A,C}/k_c$.

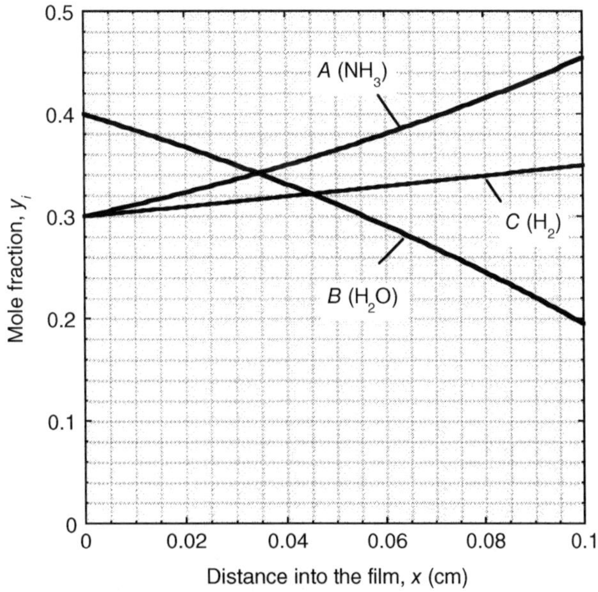

FIGURE 15.1 Mole fraction profiles across a 0.1-cm-thick gas film for ternary gas diffusion according to Example 15.3. The bulk gas is at $x = 0$ and the gas-liquid interface is at $x = 0.1$ cm.

15.3 Diffusion of Charged Species

Transport of charged species in aqueous solutions is, in general, a coupled diffusion process even under dilute conditions. In this case, coupling arises from an electrical potential gradient that drives electrophoretic transport. An electrical potential gradient can exist because it is externally applied, resulting in an electrical current. However, an electrical potential gradient can also be generated as a result of diffusional transport without an applied potential gradient.

Let's consider transport of an ion i with valence z_i in a dilute aqueous solution. Without an electrical potential gradient, the molar diffusion flux of this species will be driven by its concentration gradient and will be described by Fick's law as follows:

$$N_i = -\mathcal{D}_i' \nabla c_i \tag{15.34}$$

where $\mathcal{D}_i'$ is the limiting ion diffusivity and ∇c_i is the gradient in molar concentration of ion i. The value of $\mathcal{D}_i'$, typically expressed in units of m^2/s or cm^2/s, will depend on the properties of the diffusing ion, on the solution viscosity, and on temperature in a manner analogous to that of ordinary diffusivities. Conversely, if an electrical potential gradient $\nabla\varphi$ is applied without a concentration gradient, the molar flux of ion i will be driven by $\nabla\varphi$ and will be described by Nernst law as follows:

$$N_i = -\frac{\lambda_{m,i}}{z_i\mathcal{F}} c_i \nabla\varphi \tag{15.35}$$

where $\lambda_{m,i}$ is the molar conductance of ion i in the solution, $\mathcal{F}$ (= 96,500 Coulomb/mol) is Faraday's constant, and $\nabla\varphi$ is the electrical potential gradient in V/m. Note that in this case, the mass transfer flux of ion i will be proportional to its concentration, c_i, rather than to its concentration gradient as in Fick's law. The quantity $\lambda_{m,i}$, which, typically, will have units of Coulomb m^2/s mol V, describes the electrophoretic mobility of the ion in solution and will be a function of temperature and solution viscosity. $\lambda_{m,i}$ is related to the limiting ion diffusivity through to the following relationship known as the Nernst–Einstein equation:

$$\mathcal{D}_i' = \frac{\lambda_{m,i}\mathcal{R}T}{z_i^2\mathcal{F}^2} \tag{15.36}$$

Combining Eqs. (15.34) to (15.36) gives the following result for the overall flux of ion i when concentration and electrical potential gradients are present simultaneously:

$$N_i = -\mathcal{D}_i'\left(\nabla c_i + \frac{z_i\mathcal{F}}{\mathcal{R}T} c_i \nabla\varphi\right) \tag{15.37}$$

This relationship is known as the Nernst–Planck equation. In the following, we consider two examples of application of this equation: diffusion of a salt in a solution and counter-diffusion of two ions in an ion-exchange resin.

15.3.1 Diffusion of Electrolytes in Solution

Table 15.1 provides some representative values of $\lambda_{m,i}$ in water at infinite dilution and 298 K in units of Coulomb cm^2/s mol V along with the corresponding limiting ion diffusivities in units of cm^2/s. Note that the relationship between $\lambda_{m,i}$ and ion size is not straightforward since it is affected not only by atomic or molecular mass but also by the

Cation	$\lambda_{m,i}$ (Coulomb cm²/s mol V)	$\mathcal{D}_i'$ (10^{-5} cm²/s)	Anion	$\lambda_{m,i}$ (Coulomb cm²/s mol V)	$\mathcal{D}_i'$ (10^{-5} cm²/s)
H^+	350	9.37	OH^-	198	5.30
Li^+	38.7	1.04	Cl^-	76.3	2.04
Na^+	50.1	1.34	Br^-	78.3	2.10
K^+	73.5	1.97	I^-	76.8	2.06
Ag^+	61.9	2.67	NO_3^-	71.4	1.91
Ca^{++}	119	0.79	CH_3COO^-	40.9	1.10
Cu^{++}	108	0.72	SO_4^{--}	160	1.07
La^{+++}	209	0.62	$Fe(CN)_6^{---}$	303	0.90

$\mathcal{D}_i'$ is the Limiting Ionic Diffusivity Given by Eq. (15.36).

Table 15.1 Selected Values of Limiting Molar Conductance and Ion Diffusivity at Infinite Dilution in Water at 298 K

ion's hydration shell. Obviously, as seen from this table, protons and hydroxyl ions are transported rapidly owing to their large values of $\lambda_{m,i}$ and, correspondingly, large $\mathcal{D}_i'$. The equivalent ionic conductance, λ_i, typically expressed in units of Coulomb cm²/s equivalent V, is also frequently reported in the literature. For monovalent ions, $\lambda_{m,i}$ and λ_i are identical. However, for non-monovalent ions, the two are related by the equation $\lambda_i = \lambda_{m,i} / |z_i|$.

Let's now consider the diffusional transport of a salt $A_{z_B}B_{z_A}$ consisting of a cation A with valence z_A and an anion B with valence z_B without an applied electrical potential gradient. For this case, Eq. (15.37) yields the following flux expressions for the cation and for the anion, respectively:

$$N_A = -\mathcal{D}_A'\left(\nabla c_A + \frac{|z_A|\mathcal{F}}{\mathcal{R}T} c_A \nabla\varphi\right) \tag{15.38}$$

$$N_B = -\mathcal{D}_B'\left(\nabla c_B - \frac{|z_B|\mathcal{F}}{\mathcal{R}T} c_B \nabla\varphi\right) \tag{15.39}$$

Note that z_i will be positive for the cation and negative for the anion. $|z_i|$ denotes the absolute value. The two fluxes are coupled by the following two conditions:

(1) Electroneutrality

$$|z_A| c_A = |z_B| c_B \tag{15.40}$$

(2) No electrical current

$$|z_A| N_A = |z_B| N_B \tag{15.41}$$

Equation (15.41) is based on the fact that no work can be done without an applied electrical potential gradient. In this case, the amount of positive charge carried by the cation must be equal to the amount of negative charge carried by the anion so that the net current will be zero. Because of Eq. (15.40), we can write:

$$\nabla c_B = \frac{|z_A|}{|z_B|} \nabla c_A \tag{15.42}$$

Substituting Eqs. (15.38) and (15.39) into Eq. (15.41) yields the following result:

$$\begin{aligned} |z_A| \mathcal{D}'_A \left(\nabla c_A + \frac{\mathcal{F}}{\mathcal{R}T} |z_A| c_A \nabla \varphi \right) &= |z_B| \mathcal{D}'_B \left(\nabla c_B - \frac{|z_B| \mathcal{F}}{\mathcal{R}T} c_B \nabla \varphi \right) \\ &= |z_B| \mathcal{D}'_B \left(\frac{|z_A|}{|z_B|} \nabla c_A - \frac{|z_B| \mathcal{F}}{\mathcal{R}T} \frac{|z_A|}{|z_B|} c_A \nabla \varphi \right) \end{aligned} \tag{15.43}$$

We can now solve this equation to find the following expression for $\nabla \varphi$:

$$\nabla \varphi = \frac{\mathcal{R}T}{\mathcal{F}} \frac{\mathcal{D}'_B - \mathcal{D}'_A}{|z_A| \mathcal{D}'_A + |z_B| \mathcal{D}'_B} \frac{\nabla c_A}{c_A} \tag{15.44}$$

This important result shows that, if the limiting ion diffusivities of the cation and of the anion are different from each other, an electrical potential gradient will be established spontaneously as a result of diffusion even without an externally applied electrical potential field. This gradient will couple the mass transfer fluxes of the cation and the anion in the salt. Since we have obtained an expression for $\nabla \varphi$, we can now find the flux by substituting Eq. (15.44) into Eq. (15.38), which gives the following result:

$$\begin{aligned} N_A &= -\mathcal{D}'_A \left(\nabla c_A + |z_A| \frac{\mathcal{D}'_B - \mathcal{D}'_A}{|z_A| \mathcal{D}'_A + |z_B| \mathcal{D}'_B} \nabla c_A \right) \\ &= -\left(\mathcal{D}'_A + |z_A| \frac{\mathcal{D}'_B - \mathcal{D}'_A}{|z_A| \mathcal{D}'_A + |z_B| \mathcal{D}'_B} \right) \nabla c_A \\ &= -\frac{(|z_A| + |z_B|) \mathcal{D}'_A \mathcal{D}'_B}{|z_A| \mathcal{D}'_A + |z_B| \mathcal{D}'_B} \nabla c_A = -\mathcal{D}_{AB} \nabla c_A \end{aligned} \tag{15.45}$$

where the quantity

$$\mathcal{D}_{AB} = \frac{(|z_A| + |z_B|) \mathcal{D}'_A \mathcal{D}'_B}{|z_A| \mathcal{D}'_A + |z_B| \mathcal{D}'_B} = \frac{\dfrac{1}{|z_A|} + \dfrac{1}{|z_B|}}{\dfrac{1}{|z_A| \mathcal{D}'_A} + \dfrac{1}{|z_B| \mathcal{D}'_B}} \tag{15.46}$$

is the diffusion coefficient of the salt in the solution. The latter is often referred to as the mutual salt diffusivity. Note that $\mathcal{D}_{AB}$ (with no comma in the subscript) is a special weighted average of the individual diffusivities of the cation and anion species in the salt. In the dilute limit, $\mathcal{D}_{AB}$ will be independent of composition. Example 15.4 illustrates the application of these relationships.

Example 15.4. Diffusion coefficient of sodium chloride (NaCl) and of calcium chloride ($CaCl_2$) in water at infinite dilution and 298 K. From Table 15.1, the relevant limiting ion diffusivities are $\mathcal{D}'_{Na^+} = 1.34 \times 10^{-5}$, $\mathcal{D}'_{Ca^{++}} = 0.79 \times 10^{-5}$, and $\mathcal{D}'_{Cl^-} = 2.04 \times 10^{-5}$, all in units of cm^2/s. From Eq. (15.46) for NaCl we obtain:

$$\mathcal{D}_{NaCl} = \frac{\frac{1}{1}+\frac{1}{1}}{\frac{1}{1\times 1.34}+\frac{1}{1\times 2.04}} \times 10^{-5} = 1.62\times 10^{-5}\ cm^2/s$$

while for $CaCl_2$ we obtain:

$$\mathcal{D}_{CaCl_2} = \frac{\frac{1}{2}+\frac{1}{1}}{\frac{1}{2\times 0.79}+\frac{1}{1\times 2.04}} \times 10^{-5} = 1.34\times 10^{-5}\ cm^2/s$$

Both values compare well with experimental results; e.g., see Ribeiro et al., Electrochimica Acta 54 (2008) 192. Note that in both cases, since the fluxes are coupled by the electroneutrality and no-net current conditions, the faster diffusing chloride ion "drags along" the slower diffusing cation resulting in a mutual diffusivity of the salt that is intermediate between the cation diffusivity and the anion diffusivity. As in the case of non-electrolytes, the solution viscosity will affect the diffusivities with an inverse proportional relationship. This relationship could be used to estimate the salt diffusivity in viscous solutions. Without an applied electrical potential gradient, in conjunction with Eq. (15.45), these diffusivities provide the salt flux since, on a molar basis, $N_{Na^+} = N_{NaCl}$ and $c_{Na^+} = c_{NaCl}$ for the sodium chloride case and $N_{Ca^{++}} = N_{CaCl_2}$ and $c_{Ca^{++}} = c_{CaCl_2}$ for the calcium chloride case.

15.3.2 Ion Transport in Ion-Exchange Resins

Ion-exchange resins are used extensively in many practical applications including the removal of metal ions from contaminated water, the deionization of water, the removal of hardness (i.e., Ca^{++} and Mg^{++}) from water, the purification of fine chemicals and pharmaceuticals, and the separation of rare earths. Ion exchange resins are also used as catalysts for chemical reactions. The resins used in these applications typically consist of a water-swollen cross-linked polymer network with pendant charged functional groups linked to the polymer as sketched in Fig. 15.2. In cation-exchange resins, these functional groups are negatively charged (sulfonate groups, $-SO_3^-$, are common), while in anion-exchange resins they are positively charged (quaternary ammonium ion groups are common). In both cases, these groups cannot diffuse since they are linked to the

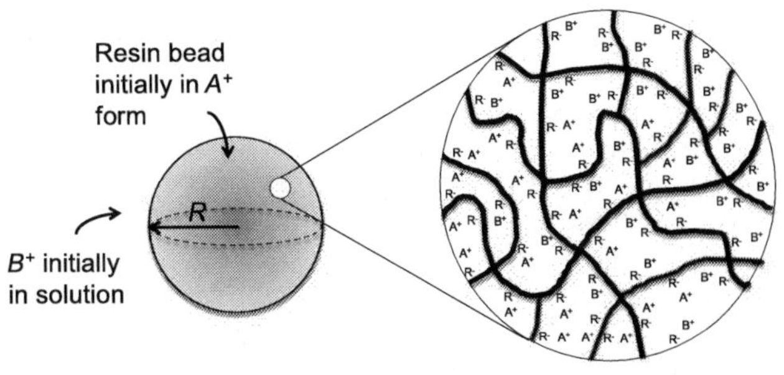

FIGURE 15.2 Schematic of a polymer-based cation exchange resin. R is the resin particle radius. A^+ and B^+ are the exchangeable counterions. R^- represents the fixed, negatively charged functional groups attached to the cross-linked polymer.

polymer network. Their charge, however, must be balanced by that of mobile cations in cation-exchange resins and by that of mobile anions in anion-exchange resins. As a result, cations present in a cation-exchange resin balancing the negatively charged groups can be exchanged for other cations in solution. Conversely, anions present in an anion-exchange resin balancing the positively charged groups can be exchanged for other anions in solution. For example, exchange of a monovalent cation A^+ initially in the resin and a monovalent cation B^+ initially in solution can occur according to the following stoichiometric relationship:

$$R^-A^+ + B^+ \rightleftharpoons R^-B^+ + A^+ \tag{15.47}$$

where R^- represents the negatively charged functional groups in the resin. An analogous relationship involving anions can be written for anion-exchange resins. In either type of resin, the exchangeable ions are referred to as the "counter-ions." Note that, because of the negatively charged functional groups, anions, referred to as "co-ions," are largely excluded from the interior of cation-exchange resins. Conversely, because of the positively charged functional groups, cations, which are now the co-ions, are largely excluded from the interior of anion-exchange resins.

In typical ion-exchange resins, the polymer mesh size is small and comparable with the size of the ions involved in the ion-exchange process. As a result, mass transfer within the resin occurs by a homogeneous diffusion mechanism where the flux is described by Eq. (15.37) which we write as follows:

$$N_i = -\bar{\mathcal{D}}_i\left(\nabla c_i + \frac{z_i\mathcal{F}}{RT}c_i\nabla\bar{\varphi}\right) \tag{15.48}$$

where $\bar{\mathcal{D}}_i$, c_i, $\nabla\bar{\varphi}$ are the ion diffusivity in the resin, the ion concentration in the resin, and the electrical potential gradient in the resin, respectively. Note that $\bar{\mathcal{D}}_i$ will depend on the properties of the diffusing ion as well as on the properties of the resin, most notably on the degree of cross-linking of the resin, which will determine its mesh size. Some representative values of $\bar{\mathcal{D}}_i$ for inorganic ions are given in Table 15.2. As seen from this table, ion diffusivities are smaller in more highly cross-linked resins and for bulkier ions. Diffusivities of organic ions in such resins are even smaller. Figure 15.3 shows data for amino acid and dipeptide cations in cross-linked poly(styrene-divinylbenzene) cation-exchange resins. The data are correlated by plotting on semi-log scales $\bar{\mathcal{D}}_A/\mathcal{D}^0_{A,W}$ vs. the product $(\hat{V}_A)^{1/3} \times x$, where $\mathcal{D}^0_{A,W}$ is the ion diffusivity in water, $\hat{V}_A$ is the LeBas volume of the diffusing cation (see Table 10.5), and x is the degree of cross-linking of the resin in wt%. Note that the correlation is exponential, which implies a strong effect of the resin cross-linking and of $(\hat{V}_A)^{1/3}$. The latter is roughly proportional to the molecular radius of the diffusing species. This strong dependence arises from the small mesh size of the polymer network in these resins, which becomes increasingly smaller as x increases, and which, in turn, results in increasing diffusional hindrance.

Let's now apply Eq. (15.48) to the prediction of the rate of exchange of two counter-ions assuming that diffusion within the resin is controlling. In practice, this will often be the case provided that the ion concentration in solution is greater than, say, 0.1 mol/L. For simplicity, we consider only the exchange of two monovalent cations A^+ and B^+ ($z_A = z_B = 1$) on a cation-exchange resin according to Eq. (15.47). For this case, when written for each counter-ion, Eq. (15.48) gives:

$$N_{A^+} = -\bar{\mathcal{D}}_{A^+}\left(\nabla c_{A^+} + \frac{\mathcal{F}}{RT}c_{A^+}\nabla\bar{\varphi}\right) \tag{15.49}$$

Resin Degree of Cross-linking, x (%)	4%	6%	8%	16%
Cation-Exchange Resins (Sulfonated)				
H^+			56.0*	
Na^+	14.1		9.4	2.40
Cs^+			13.7	3.10
Ag^+			6.42	2.75
Zn^{++}			0.63	0.14
Anion-Exchange Resins (Dimethyl Hydroxyethylamine)				
Cl^-		3.54		
Br^-	4.3	3.87	2.04	0.26
I^-		1.33		
BrO_3^-		4.55		
WO_4^{--}		1.80		

*Data from Turner et al., Chem. Eng. Sci. 21 (1966) 317.

Table 15.2 Ion Diffusivities in Cross-Linked Poly(Styrene-Divinylbenzene) Ion-Exchange Resins at 298 K. $\bar{\mathcal{D}}_A$ in 10^{-7} cm²/s. (Based on Data in Table 16-8 in *Perry's Chemical Engineers' Handbook*, 9th Edition.)

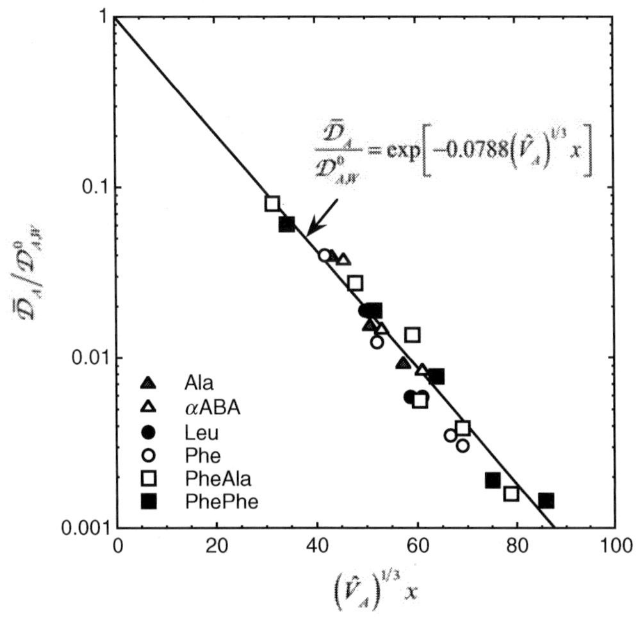

Figure 15.3 Diffusivities of monovalent amino acids and dipeptide cations in sulfonated poly(styrene-divinylbenzene) ion-exchange resins at 298 K. $\mathcal{D}^0_{A,W}$ and $\hat{V}_A$ are, respectively, the diffusivity in water and the LeBas molar volume (cm³/mol) of the diffusing species and x is the degree of cross-linking of the resin in wt%. (Based on data in Jones and Carta, Ind. Eng. Chem. Research, 32 (1993) 117.)

$$N_{B^+} = -\bar{\mathcal{D}}_{B^+}\left(\nabla c_{B^+} + \frac{\mathcal{F}}{RT} c_{B^+} \nabla \bar{\varphi}\right) \tag{15.50}$$

Note that, unlike the case of diffusion of a salt in solution, in this case, the two species, A^+ and B^+, counter-diffuse. The two fluxes are then coupled by the following constraints:

$$c_{A^+} + c_{B^+} = c_{R^-} \tag{15.51}$$

$$N_{A^+} + N_{B^+} = 0 \tag{15.52}$$

The first of these two constraints, Eq. (15.51), establishes electroneutrality. The second constraint, Eq. (15.52), imposes that there will be no net electrical current since there is no applied electrical potential. Inserting Eqs. (15.49) and (15.50) into Eq. (15.52) yields the following:

$$\begin{aligned}\bar{\mathcal{D}}_{A^+}\left(\nabla c_{A^+} + \frac{\mathcal{F}}{RT} c_{A^+} \nabla \bar{\varphi}\right) &= -\bar{\mathcal{D}}_{B^+}\left(\nabla c_{B^+} + \frac{\mathcal{F}}{RT} c_{B^+} \nabla \bar{\varphi}\right) \\ &= -\bar{\mathcal{D}}_{B^+}\left[-\nabla c_{A^+} + \frac{\mathcal{F}}{RT}(c_{R^-} - c_{A^+})\nabla \bar{\varphi}\right]\end{aligned} \tag{15.53}$$

Solving this equation for the electrical potential gradient, we obtain:

$$\nabla \bar{\varphi} = \frac{RT}{\mathcal{F}} \frac{\bar{\mathcal{D}}_{B^+} - \bar{\mathcal{D}}_{A^+}}{\bar{\mathcal{D}}_{A^+} c_{A^+} + \bar{\mathcal{D}}_{B^+}(c_{R^-} - c_{A^+})} \nabla c_{A^+} \tag{15.54}$$

This result shows that, once again, if the two ion diffusivities are different, an electrical potential gradient will exist as a result of the ion counter-diffusion process. Inserting Eq. (15.54) in Eq. (15.49) gives the following final result:

$$\begin{aligned}N_{A^+} &= -\bar{\mathcal{D}}_{A^+}\left[\nabla c_{A^+} + \frac{(\bar{\mathcal{D}}_{B^+} - \bar{\mathcal{D}}_{A^+})c_{A^+}}{\bar{\mathcal{D}}_{A^+} c_{A^+} + \bar{\mathcal{D}}_{B^+}(c_{R^-} - c_{A^+})} \nabla c_{A^+}\right] \\ &= -\bar{\mathcal{D}}_{A^+}\left[1 + \frac{(\bar{\mathcal{D}}_{B^+} - \bar{\mathcal{D}}_{A^+})c_{A^+}}{\bar{\mathcal{D}}_{A^+} c_{A^+} + \bar{\mathcal{D}}_{B^+}(c_{R^-} - c_{A^+})}\right] \nabla c_{A^+} \\ &= -\bar{\mathcal{D}}_{A^+B^+} \nabla c_{A^+}\end{aligned} \tag{15.55}$$

In this equation,

$$\bar{\mathcal{D}}_{A^+B^+} = \frac{\bar{\mathcal{D}}_{A^+}\bar{\mathcal{D}}_{B^+}}{\bar{\mathcal{D}}_{A^+}\frac{c_{A^+}}{c_{R^-}} + \bar{\mathcal{D}}_{B^+}\left(1 - \frac{c_{A^+}}{c_{R^-}}\right)} = \frac{\bar{\mathcal{D}}_{A^+}}{(\alpha - 1)\frac{c_{A^+}}{c_{R^-}} + 1} \tag{15.56}$$

is an effective diffusivity describing the ion counter-diffusion process where $\alpha = \bar{\mathcal{D}}_{A^+}/\bar{\mathcal{D}}_{B^+}$. Note that $\bar{\mathcal{D}}_{A^+B^+}$ is composition-dependent; i.e., it varies with c_{A^+}. If $c_{A^+} \to 0$ (which means that $c_{B^+} \sim c_{R^-}$), Eq. (15.56) will give $\bar{\mathcal{D}}_{A^+B^+} \sim \bar{\mathcal{D}}_{A^+}$. Conversely, if $c_{B^+} \to 0$ (which means that $c_{A^+} \sim c_{R^-}$), Eq. (15.56) will give $\bar{\mathcal{D}}_{A^+B^+} \sim \bar{\mathcal{D}}_{A^+}/\alpha = \bar{\mathcal{D}}_{B^+}$. In other words, the ion present in the smallest concentration in the resin will tend to dominate the overall transport process. This counter-intuitive phenomenon occurs because of the electrostatic coupling of the diffusion fluxes embodied through the Nernst–Planck description of ion transport. Example 15.5 illustrates some of the consequences of these relationships.

Example 15.5. Prediction of ion-exchange rates for a spherical ion-exchange particle. Ion-exchange resins used in practical applications are typically in the form of spherical beads with diameter in the range 0.5 to 1 mm. Because of the small values of the resin-phase diffusivity, diffusion in the beads is often controlling when the ion concentration in solution is not too low. Let's consider, for example, the exchange of Na^+ and H^+ on an 8% cross-linked, poly(styrene-divinylbenzene) sulfonated cation-exchange resin particle with a radius $R = 0.05$ cm. We will assume that the concentration of the fixed, negatively charged groups is $c_{R^-} = 2.6$ mmol/cm^3, which is a typical value. The temperature is 298 K. From Table 15.2, the ion diffusivity values in such a resin are $\bar{\mathcal{D}}_{H^+} = 56.0 \times 10^{-7}$ cm^2/s and $\bar{\mathcal{D}}_{Na^+} = 9.4 \times 10^{-7}$ cm^2/s. Note that both values are much smaller than the corresponding limiting ion diffusivities in solution since, within the resin, the cross-linked polymer network restricts transport. Nevertheless, just as in solution, the value for H^+ in the resin is much larger than the corresponding value for of Na^+ indicating that the former is much more mobile than the latter even in the resin. Since the particle is spherical, in order to describe the counter-diffusion process, we need to write a transient material balance for A^+ in a control volume consisting of a spherical shell within the resin particle with thickness Δr and with its spherical surfaces located at r and $r + \Delta r$. The shell volume will be $4\pi r^2 \Delta r$. The material balance for ion A^+ is as follows:

$$\text{Accumulation of } A^+ = \text{Input of } A^+ - \text{Output of } A^+$$

When applied to the spherical shell, this balance yields:

$$\frac{\partial}{\partial t}(4\pi r^2 \Delta r \times c_{A^+}) = 4\pi r^2 N_{A,r}\Big|_r - 4\pi r^2 N_{A,r}\Big|_{r+\Delta r} \tag{15.57}$$

Dividing both sides of Eq. (15.57) by $4\pi r^2 \Delta r$ and taking the limit $\Delta r \to 0$, we obtain:

$$\frac{\partial c_{A^+}}{\partial t} = -\frac{1}{r^2} \times \lim_{\Delta r \to 0}\left(\frac{r^2 N_{A,r}\Big|_{r+\Delta r} - r^2 N_{A,r}\Big|_r}{\Delta r}\right)$$

or

$$\frac{\partial c_{A^+}}{\partial t} = -\frac{1}{r^2}\frac{\partial (r^2 N_{A,r})}{\partial r} \tag{15.58}$$

The final step is to combine Eq. (15.58) with the expression for the flux given by Eq. (15.55). Since, for this case, $\nabla c_{A^+} = \partial c_{A^+}/\partial r$, we obtain:

$$\frac{\partial c_{A^+}}{\partial t} = \frac{1}{r^2}\frac{\partial}{\partial r}\left(r^2 \bar{\mathcal{D}}_{A^+B^+} \frac{\partial c_{A^+}}{\partial r}\right) \tag{15.59}$$

Initial and boundary conditions for this equation are as follows:

$$\begin{aligned} t = 0: &\quad c_{A^+} = c_{R^-} \\ r = 0: &\quad \frac{\partial c_{A^+}}{\partial r} = 0 \\ r = R: &\quad c_{A^+} = 0 \end{aligned} \tag{15.60}$$

These three conditions establish that the resin particle is initially in A^+ form (i.e., $c_{B^+} = 0$, initially), that the flux zero at the particle center (i.e., the concentration profile is radially symmetrical), and that the surface of the resin bead is immediately converted to the B^+ form at $t > 0$ (i.e., that $c_{B^+} = c_{R^-}$ at $r = R$). Note that Eq. (15.59) is different from the corresponding balance in spherical coordinates found in Table 12.1 (item C), which is valid for constant diffusivity. In the case at hand, $\bar{\mathcal{D}}_{A^+B^+}$ is a function of c_{A^+} through Eq. (15.56) and appears within the derivative $\partial/\partial r$. As a result, Eq. (15.59) is a non-linear partial differential equation which requires a numerical solution. The only exception where an analytical solution is possible will be when the two ion diffusivities are equal. In that case, there will be no electrical potential gradient and no flux coupling. As a result, we will have $\bar{\mathcal{D}}_{A^+B^+} = \bar{\mathcal{D}}_{A^+} = \bar{\mathcal{D}}_{B^+} =$ constant. In that instance, Eq. (15.59) would become:

$$\frac{\partial c_{A^+}}{\partial t} = \frac{\bar{\mathcal{D}}_{A^+}}{r^2}\frac{\partial}{\partial r}\left(r^2\frac{\partial c_{A^+}}{\partial r}\right) \tag{15.61}$$

In dimensionless form, this equation is the same as Eq. (5.66) for the analogous process of heat conduction in a sphere. The dimensionless solution is then item C in Table 5.2 with $\tau = \bar{\mathcal{D}}_{A^+}t/R^2$. For the general case, when the ion diffusivities, $\bar{\mathcal{D}}_{A^+}$ and $\bar{\mathcal{D}}_{B^+}$, differ from each other, Helfferich and Plesset [J. Chem. Phys. 28 (1958) 418] have obtained, the solution of Eq. (15.59) with initial and boundary conditions (15.60) using a finite difference numerical solution scheme. Beside the numerical solution results for several cases, Helfferich and Plesset have also provided an empirical correlation of their numerical results, which is useful for practical calculations. In dimensionless form, their correlation is given by the following expression valid for $0.1 < \alpha < 10$:

(M.E.15.62)
$$F = 1 - \frac{c_{A^+,\text{avg}}}{c_{R^-}} = \left\{1 - \exp\left[\pi^2\left(f_1(\alpha)\tau + f_2(\alpha)\tau^2 + f_3(\alpha)\tau^3\right)\right]\right\}^{1/2} \tag{15.62}$$

where $\alpha = \bar{\mathcal{D}}_{A^+}/\bar{\mathcal{D}}_{B^+}$ is the ratio of diffusivities and $\tau = \bar{\mathcal{D}}_{A^+}t/R^2$. The functions f_1, f_2, and f_3 are empirical and are defined as follows:

$$f_1(\alpha) = -(0.570 + 0.430\alpha^{0.775})^{-1} \tag{15.63}$$

$$f_2(\alpha) = (0.260 + 0.782\alpha)^{-1} \tag{15.64}$$

$$f_3(\alpha) = -(0.165 + 0.177\alpha)^{-1} \tag{15.65}$$

Note that $c_{A^+,\text{avg}}$ in Eq. (15.62) is the spherical-average concentration of ion A^+ in the resin bead, which is given by:

$$c_{A^+,\text{avg}} = \frac{3}{R^3}\int_0^R r^2 c_{A^+}(r,t)\,dr \tag{15.66}$$

We can see that F represents the fractional approach of the resin particle to saturation with B^+. F will be equal to zero at the start, when $c_{A^+} = c_{R^-}$, and will approach 1 when the resin becomes fully converted to the B^+ form. Note that this expression for the fractional approach to saturation is also valid for the exchange of two monovalent anions on an anion-exchange resin. Helfferich, *Ion Exchange*, McGraw-Hill, 1962, provides analogous expressions for the case of ion exchange of ions with unequal valence. Figure 15.4 shows the results predicted by Eqs. (15.62) to (15.65) for different values of α. Note that the curves are not symmetrical for different values of α around $\alpha = 1$. This means that the overall result will actually depend on which ion is initially in the resin.

Let's now use Eq. (15.62) to predict the time required to attain, for example, $F = 0.8$ for the exchange of Na^+ and H^+. If the resin is initially in Na^+ form (i.e., $c_{Na^+} = c_{R^-}$, initially) and is suddenly exposed to H^+, $\bar{\mathcal{D}}_{A^+} = \bar{\mathcal{D}}_{Na^+} = 9.4\times10^{-7}$ cm^2/s and $\alpha = \bar{\mathcal{D}}_{Na^+}/\bar{\mathcal{D}}_{H^+} = 0.94/5.6 = 0.168$. From Fig. 15.4 or, more precisely, from Eq. (15.62) we can find that $F = 0.8$ will be attained at $\tau = 0.079$. In turn, this corresponds to $t = \tau R^2/\bar{\mathcal{D}}_{Na^+} = 0.079\times0.05^2/9.4\times10^{-7} = 210$ s. On the other hand, if the resin is initially in H^+ form (i.e., $c_{H^+} = c_{R^-}$, initially) and is exposed to Na^+, $\bar{\mathcal{D}}_{A^+} = \bar{\mathcal{D}}_{H^+} = 56.0\times10^{-7}$ cm^2/s and $\alpha = \bar{\mathcal{D}}_{H^+}/\bar{\mathcal{D}}_{Na^+} = 5.6/0.94 =$ 5.96. In this case, from Fig. 15.4 or from Eq. (15.62) we can find that $F = 0.8$ will be attained at $\tau = 0.238$. The latter corresponds to $t = \tau R^2/\bar{\mathcal{D}}_{H^+} = 0.238\times0.05^2/56.0\times10^{-7} = 106$ s. We can see that, surprisingly, it will take less time to attain the same degree of exchange if the resin is initially equilibrated with the faster diffusing ion. This non-intuitive behavior occurs because, when the resin is initially in Na^+ form and is exposed to H^+, Na^+ will be a minority at the particle surface, where $c_{H^+} \sim c_{R^-}$. In this case, the effective diffusivity $\bar{\mathcal{D}}_{A^+B^+}$ near the particle surface will be close to the smaller $\bar{\mathcal{D}}_{Na^+}$ leading to slower rates. Conversely, when the resin is initially in H^+ form and is exposed to Na^+, H^+ will now be a minority near the particle surface where, now, $c_{Na^+} \sim c_{R^-}$. In this case, the effective diffusivity $\bar{\mathcal{D}}_{A^+B^+}$ near

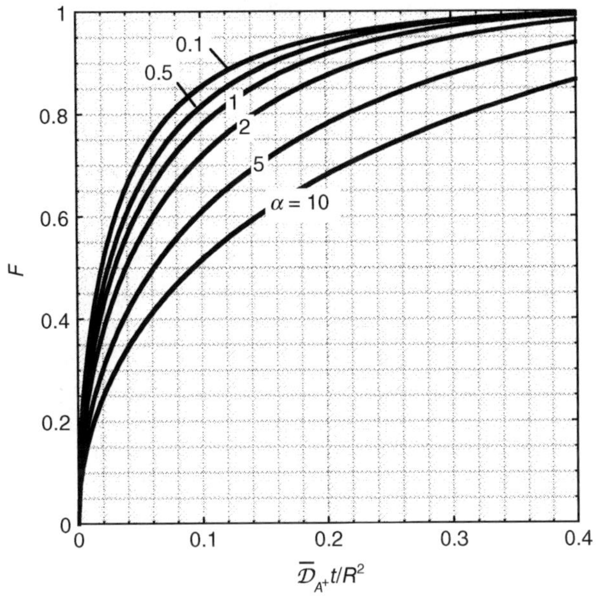

FIGURE 15.4 Solution of Eqs. (15.59) to (15.60) describing the diffusion-controlled exchange of ions A^+ and B^+ within a spherical resin particle initially saturated with A^+. F is the fractional approach to saturation with B^+. $\alpha = \bar{\mathcal{D}}_{A^+}/\bar{\mathcal{D}}_{B^+}$ is the ratio of diffusivities in the resin. Mathematica notebook: (M.F.15.4).

the particle surface will be close to the larger $\bar{\mathcal{D}}_{H^+}$ leading to faster rates. Note that if we had $\bar{\mathcal{D}}_{H^+} = \bar{\mathcal{D}}_{Na^+}$ (i.e., $\alpha = 1$), $\bar{\mathcal{D}}_{A^+B^+}$ would be a constant and the time for exchange would be exactly the same regardless of which ion is initially in the resin. The effect of α is obviously a direct consequence of the electrostatic coupling of the diffusion fluxes, which results in a composition-dependent effective diffusivity.

15.4 Simultaneous Heat and Mass Transfer

The coupling of heat and mass transfer is important in many practical applications. In this section we focus on this coupling during evaporation or condensation of a liquid or a liquid mixture in the presence of an inert gas. The main example is the air–water system, which is important in humidification, condensation, and evaporative cooling. The same principles apply, however, to many other systems. The reader of this textbook is likely to be already familiar with the concept of the "wet-bulb temperature," which is typically covered in sophomore-level material and energy balances courses. We will thus begin with an analysis of the wet-bulb thermometer but from the perspective of heat and mass transfer. We will then generalize the analysis demonstrating that the theory of the web-bulb thermometer is a special case of the general theory.

15.4.1 The Wet-Bulb Thermometer

A wet-bulb thermometer apparatus comprises a thermometer (or other temperature sensor such as a thermocouple) whose bulb is enshrouded in a cloth saturated with liquid water (A) alongside a dry-bulb thermometer. Both thermometers are placed in a

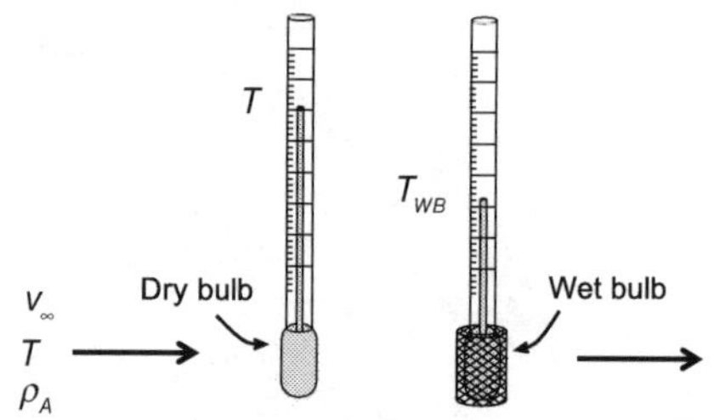

FIGURE 15.5 Schematic of dry and wet-bulb thermometers. At steady state, the wet-bulb thermometer reads a temperature T_{WB} lower than the air temperature T.

fast-moving air stream as sketched in Fig. 15.5. The air (B) flows at a velocity v_∞, has a free-stream temperature T, and contains a mass concentration of water vapor ρ_A. Note that ρ_A will have units of kg of water vapor per m^3 of moist air. The dry-bulb thermometer will quickly read the air temperature, T, upon coming to thermal equilibrium with the flowing air. On the other hand, the temperature measured by the wet-bulb thermometer will gradually drop below T, eventually settling at a steady-state value T_{WB} that we call the "wet-bulb temperature." This drop occurs as a result of the cooling caused by evaporation of water from the wet bulb. If the wet-bulb thermometer is initially at the air temperature T, water will initially evaporate quickly with a strong initial cooling effect. However, as the wet-bulb thermometer cools, the vapor pressure of water, which drives the evaporation, will decrease. As a result, the rate of evaporation will also decrease, eventually balancing the rate of convective heat transfer from the warmer air to the wet bulb. Steady state will be attained when this balance of convective heat transfer to the wet bulb and evaporative cooling is achieved. Obviously, if the air were saturated with water vapor, no evaporation and, thus, no evaporative cooling would be possible. In this case, the wet-bulb and the dry-bulb temperatures would be identical.

Empirically, the relationship between the dry-bulb temperature, the moisture content of the air, and the wet-bulb temperature is described by the psychrometric chart. Detailed charts are available in many sources. (See, for example, Fig. 12-1 of *Perry's Chemical Engineers' Handbook*, 9th edition, and Fig. 8.4-1 of Felder et al., *Elementary Principles of Chemical Processes*, Wiley, 2016, 4th edition.) Figure 15.6 shows a simplified version of the psychrometric chart where, as is customary, the moisture content is expressed in terms of the absolute humidity. This is given by the mass ratio W_A in units of kg of water vapor per kg of dry air. In this graph, the thick solid line shows the relationship between the absolute humidity and temperature in 100% saturated air; i.e., at equilibrium. This line is based on the vapor pressure of water. The dashed lines show the percentage saturation of the air, which is the relative humidity. Finally, the thin, nearly straight, diagonal lines represent conditions of constant wet-bulb temperature. Provided that flow of the air past the wet-bulb thermometer is turbulent, empirically, we will find that the relationship between T, W_A, and T_{WB} is independent of the exact value of the flow rate. From Fig. 15.6 we can see that, for example, decreasing the absolute humidity at constant T will decrease the wet-bulb temperature. In other words, the drier the air is, the greater the difference between dry-bulb and wet-bulb temperatures will be since the evaporative cooling effect will become more pronounced.

Now, suppose that we wanted to predict *a priori* the wet-bulb temperature for given values of T and W_A. In order to make this prediction, we would need to know the rate

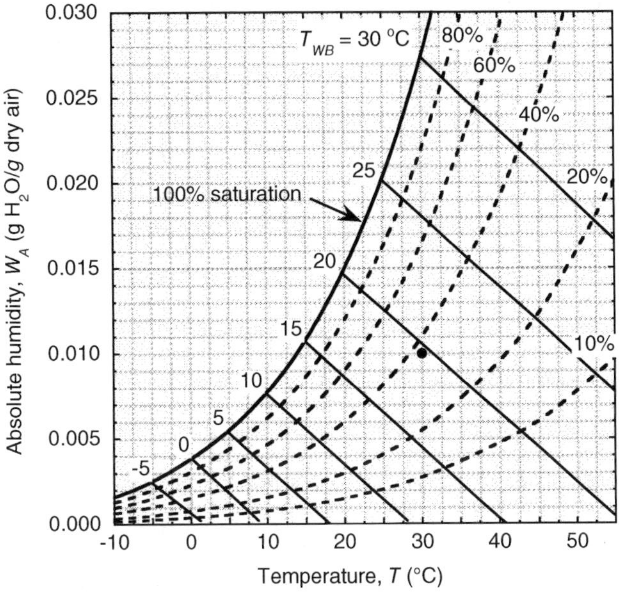

FIGURE 15.6 Psychrometric chart for the air–water system at P = 101,300 Pa. The dot represents the conditions of Example 15.6.

of convective heat transfer from the air to the wet bulb, which will depend on the unknown wet-bulb temperature. Additionally, we will need to know the rate of mass transfer, which will determine the rate of evaporative cooling. The latter will also depend on the unknown wet-bulb temperature, which, in turn, will determine the vapor pressure of water and, thus, the driving force for evaporation. Obviously, we can solve the coupled heat and mass transfer problem but not each problem independently of the other. Next, let's derive the quantitative relationship to predict the wet-bulb temperature assuming steady state. The rate of convective heat transfer from the air to the wet bulb is given by the following equation:

$$Sq = Sh(T - T_{WB}) \tag{15.67}$$

where q is the convective heat transfer flux, S is the surface area of the wet bulb, and h is the convective heat transfer coefficient. Analogously, the rate of mass transfer of water vapor from the wet bulb to the flowing air can be expressed as follows:

$$SN_A = Sk_c(C_{A,s} - C_A) \tag{15.68}$$

where N_A is the molar flux of water vapor, k_c is the convective mass transfer coefficient, and $C_{A,s}$ is the molar concentration of water vapor at the interface between the air and the wet bulb. As already noted, the latter will depend on the vapor pressure of water and, thus, will be a function of the wet-bulb temperature T_{WB}. Note that both h and k_c will, in general, be functions of the Reynolds number and, thus, of the air velocity. Assuming that heat and mass transfer occur only at the surface of the wet bulb exposed to the air, equating the rate of convective heat transfer to the thermometer to the rate of energy transfer due to evaporation of water from the wet bulb to the flowing air yields the following relationship:

$$Sh(T - T_{WB}) = Sk_c(C_{A,s} - C_A) \times M_{r,A} \times \Delta\hat{H}_v \tag{15.69}$$

where $\Delta\hat{H}_v$ is the latent heat of vaporization of water (J/kg) and $M_{r,A}$ is the molecular mass of water (kg/mol). Rearrangement of Eq. (15.69) yields the following result:

$$T_{WB} = T - \frac{k_c \Delta\hat{H}_v M_{r,A}}{h}(C_{A,s} - C_A) \tag{15.70}$$

From this equation, we can see that T_{WB} will depend only on the ratio k_c/h rather on each coefficient individually. Based on the Chilton–Colburn analogy, which is valid for the turbulent flow conditions expected for this case, according to Eq. (13.117), we have:

$$\mathrm{St_H}(\mathrm{Pr})^{2/3} = \mathrm{St_D}(\mathrm{Sc})^{2/3} \tag{15.71}$$

or

$$\frac{h}{\rho C_{p,G} v_\infty}(\mathrm{Pr})^{2/3} = \frac{k_c}{v_\infty}(\mathrm{Sc})^{2/3} \tag{15.72}$$

where Pr and Sc are, respectively, the Prandtl and Schmidt numbers of the moist air, ρ is its mass density, and $C_{p,G}$ is its heat capacity. By combining Eqs. (15.70) and (15.72), we obtain the following result:

$$T_{WB} = T - \frac{\Delta\hat{H}_v M_{r,A}}{\rho C_{p,G}}(\mathrm{Le})^{2/3}(C_{A,s} - C_A) \tag{15.73}$$

where Le = Pr/Sc is the Lewis number. Note that for the air/water vapor system, both Pr and Sc are close to 1. As a result, Le will also be close to 1. In practice, a value of Le = 1.15 is more accurate and can be used for more precise calculations. $C_{A,s}$ depends on the vapor pressure of water p_A^* at $T = T_{WB}$ and is given by:

$$C_{A,s} = y_{A,s}C = \frac{p_A^*(T_{WB})}{RT_{WB}} \tag{15.74}$$

Example 15.6 illustrates the use of these relationships to predict the wet-bulb temperature.

Example 15.6. Prediction of the wet-bulb temperature [Mathematica notebook: (M.EX.15.6)]. Let's use what we have learned so far to predict the wet-bulb temperature of air containing 0.01 g water/g of dry air at 30°C and 1 atm absolute pressure. Since the moisture content is low, the density of the moist air is close to the density of dry air at the same temperature and pressure. In this case, since $M_{r,A}C_A = \rho_A$, the absolute moisture content is $W_A \sim \rho_A/\rho$. Equation (15.73) can then be written as follows:

$$T_{WB} \sim T - \frac{\Delta\hat{H}_v}{C_{p,G}}(\mathrm{Le})^{2/3}(W_{A,s} - W_A) \tag{15.75}$$

$W_{A,s}$ will depend on the vapor pressure of water p_A^* at $T = T_{WB}$ according to the following relationship:

$$W_{A,s} \sim \frac{M_{r,H_2O}}{M_{r,\mathrm{air}}}\frac{p_A^*(T_{WB})}{P} = \frac{18}{29}\frac{p_A^*(T_{WB})}{P} \tag{15.76}$$

In this case, we can re-write Eq. (15.75) as follows:

$$W_A = \left[W_{A,s} + \frac{T_{WB} C_{p,G}}{\Delta \hat{H}_v (\text{Le})^{2/3}} \right] - \frac{T_{WB} C_{p,G}}{\Delta \hat{H}_v (\text{Le})^{2/3}} T \tag{15.77}$$

If we regard $W_{A,s}$ and T_{WB} as constants, this result predicts a straight-line relationship between W_A and T, which is consistent with the psychrometric chart. The slope of this line is expected to be about constant except for the slight dependence of $C_{p,G}$, $\Delta \hat{H}_v$, and Le on temperature and moisture content.

In order to calculate the actual value of T_{WB}, we need to know the relationship between the vapor pressure of water and temperature. Based on the Antoine equation (see, for example, Table B.4 of Felder et al., *Elementary Principles of Chemical Processes*, Wiley, 2016, 4th edition), the vapor pressure of water is given by:

$$p_A^*(T_{WB}) = \exp\left(18.3 - \frac{3{,}816}{T_{WB} + 227} \right) \tag{15.78}$$

with p_A^* is in mmHg and T_{WB} in °C. The remaining parameters are $\Delta \hat{H}_v$ = 2,400 kJ/kg and $C_{p,G}$ ~ 1.0 kJ/kg K, assuming that the properties of dry air and moist air are close to each other. The final result is obtained by combining Eqs. (15.75) to (15.78), which yields:

$$\begin{aligned} T_{WB} &\sim T - \frac{\Delta \hat{H}_v}{C_{p,G}} (\text{Le})^{2/3} \left[\frac{18}{29} \frac{1}{760} \exp\left(18.3 - \frac{3{,}816}{T_{WB} + 227} \right) - W_A \right] \\ &= 30 - 2{,}400(1.15)^{2/3} \left[\frac{18}{29} \frac{1}{760} \exp\left(18.3 - \frac{3{,}816}{T_{WB} + 227} \right) - 0.01 \right] \end{aligned} \tag{15.79}$$

Numerical solution of this equation gives T_{WB} = 19.8°C. From Fig. 15.6, we can read T_{WB} ~ 19.5°C, which is in excellent agreement with the predicted value.

15.4.2 Generalized Expressions for Simultaneous Heat and Mass Transfer

In Sec. 15.4.1, we have analyzed the problem of simultaneous heat and mass transfer at the surface of an evaporating liquid by assuming, implicitly, that the expression for the rate of convective thermal energy transport could be written according to Eq. (15.67) independently of the mass transfer rate. This is not entirely correct, however, since mass transfer will result in enthalpy transport. In this section, we re-analyze this situation by considering in greater detail the problem of simultaneous heat and mass transfer. For simplicity, we consider transport across a thin, flat film whose thickness represents a boundary layer at the liquid–gas interface, as sketched in Fig. 15.7. We assume that only A diffuses in an inert gas B. Note that neither the mole fraction profile nor the temperature profile will be straight lines in general. On the one hand, the mole fraction profile will not be straight because of the bulk-flow term in Fick's law (*cf.* Example 11.3). On the other hand, the temperature profile will not be straight because of the coupling of heat and mass transfer. In order to derive expressions for these profiles and, hence, find the rates of coupled heat and mass transfer, let's consider steady-state differential material and energy balances within the gas film. For simplicity, we assume that the relevant physical properties (density, diffusivity, and thermal conductivity) can be treated as constants. These balances are as follows:

$$\frac{dN_A}{dx} = 0 \tag{15.80}$$

$$\frac{dq_e}{dx} = 0 \tag{15.81}$$

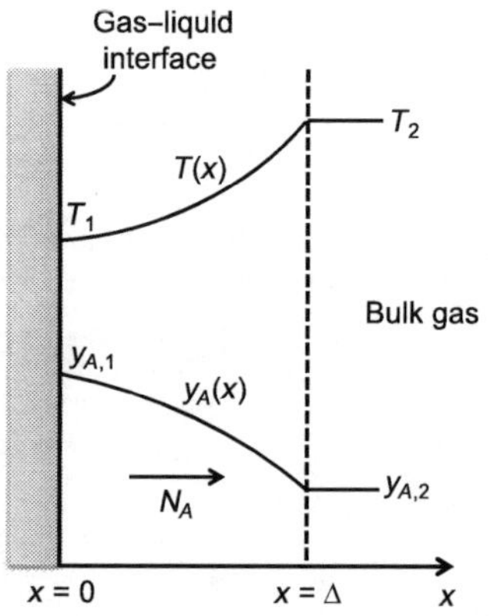

FIGURE 15.7 Schematic of temperature and mole fraction profiles for simultaneous heat and mass transfer at a gas–liquid interface.

where N_A is the mass transfer flux and q_e is the total energy flux. Based on Fick's law, for diffusion of A through stagnant B we have:

$$N_A = -\frac{\mathcal{D}_{A,B}C}{1-y_A}\frac{dy_A}{dx} \tag{15.82}$$

Substituting this expression into Eq. (15.80) and integrating between $x = 0$ and $x = \Delta$ yields the following result:

$$N_A = \frac{\mathcal{D}_{A,B}C}{\Delta}\ln\left(\frac{1-y_{A,2}}{1-y_{A,1}}\right) = k_c C\ln\left(\frac{1-y_{A,2}}{1-y_{A,1}}\right) \tag{15.83}$$

where $k_c = \mathcal{D}_{A,B}/\Delta$ is the convective mass transfer coefficient. The mole fraction profile can be found by integrating Eq. (15.82) between $x = 0$ and an arbitrary x, which gives:

$$y_A = 1 - \frac{1-y_{A,1}}{\exp\left(\dfrac{x}{\Delta}\ln\dfrac{1-y_{A,1}}{1-y_{A,2}}\right)} \tag{15.84}$$

Note that, just like in the case of the wet-bulb thermometer, we will typically know the bulk temperature T_2 and composition $y_{A,2}$, but not their interfacial values T_1 and $y_{A,1}$. The latter will have to be determined from an energy balance along with the appropriate vapor pressure relationship.

Let's now turn our attention to Eq. (15.81). In this equation, q_e is the total energy flux which is given by the sum of heat conduction, resulting from the temperature gradient in the film, and the enthalpy transport associated with mass transfer of A. Choosing T_2 as our reference temperature, we can write this sum as follows:

$$q_e = q + n_A C_{p,A}(T - T_2) = -k\frac{dT}{dx} + n_A C_{p,A}(T - T_2) \tag{15.85}$$

where k is the thermal conductivity of the gas mixture, n_A is the mass flux of A, and $C_{p,A}$ the heat capacity of gaseous A. Note that if k is in units of W/m K and n_A is in units of kg/m^2 s, $C_{p,A}$ will be in units of J/kg K. Since, based on Eq. (15.81), q_e is constant along

x, we can separate the variables x and T in Eq. (15.85) and integrate between $x = 0$ and $x = \Delta$, which gives:

$$\int_{T_1}^{T_2} \frac{dT}{n_A C_{p,A}(T - T_2) - q_e} = \int_0^{\Delta} \frac{dx}{k} \tag{15.86}$$

Carrying out the integrals we obtain:

$$\ln\left[\frac{q}{q_e - n_A C_{p,A}(T_1 - T_2)}\right] = \frac{n_A C_{p,A}}{k/\Delta} = \frac{n_A C_{p,A}}{h} = C_0 \tag{15.87}$$

where $h = k/\Delta$ is the heat transfer coefficient. Solving Eq. (15.87) for q_e gives the final result:

$$q_e = h(T_1 - T_2)\left[\frac{C_0 \exp(C_0)}{\exp(C_0) - 1}\right] \tag{15.88}$$

The temperature profile across the film can be found by changing the upper integration limits in Eq. (15.86) to arbitrary T and x values. The conductive heat transfer flux at the surface is:

$$q\big|_{x=0} = q_e - n_A C_{p,A}(T_1 - T_2) = h(T_1 - T_2)\left[\frac{C_0}{\exp(C_0) - 1}\right] \tag{15.89}$$

The term in brackets on the right-hand side of Eq. (15.98):

$$\eta = \frac{C_0}{\exp(C_0) - 1} \tag{15.90}$$

is called the "Ackermann correction." It is applied to the heat transfer coefficient to describe the effects of simultaneous mass transfer. Its magnitude will depend on the value of $n_A C_{p,A}/h$. For a system where component A is evaporated from the surface at $x = 0$, n_A will be positive and the correction factor will be less than 1. Conversely, for a system where A is condensed from the gas phase, n_A will be negative and the correction factor will be greater than 1. In either case, if $n_A C_{p,A}/h$ is small, the correction factor will approach one and the heat-transfer rate will approach the limiting value $q = h(T_1 - T_2)$.

Let's now consider the determination of the surface temperature T_1 in the case of an evaporating liquid from an insulated pool, just like we did in our analysis of the web-bulb thermometer. An energy balance for the liquid yields:

$$-q_e = \Delta\hat{H}_v n_A \tag{15.91}$$

where $\Delta\hat{H}_v$ is the latent heat of vaporization of A (in J/kg). Combining this equation with Eq. (15.88), we obtain:

$$h(T_1 - T_2)\left[\frac{C_0 \exp(C_0)}{\exp(C_0) - 1}\right] = -\Delta\hat{H}_v n_A \tag{15.92}$$

or

$$(T_1 - T_2)\left[\frac{C_0 \exp(C_0)}{\exp(C_0) - 1}\right] = -\frac{\Delta\hat{H}_v}{C_{p,A}}\frac{n_A C_{p,A}}{h} = -\frac{\Delta\hat{H}_v}{C_{p,A}} C_0 \tag{15.93}$$

With some rearrangement, this equation yields:

$$T_1 = T_2 - \frac{\Delta \hat{H}_v}{C_{p,A}} \frac{\exp(C_0) - 1}{\exp(C_0)} \tag{15.94}$$

Since $n_A = M_{r,A} \times N_A$, the term $C_0 = n_A C_{p,A}/h$ can be written as follows:

$$\frac{n_A C_{p,A}}{h} = M_{r,A} \frac{k_c C}{h} C_{p,A} \ln\left(\frac{1 - y_{A,2}}{1 - y_{A,1}}\right) \tag{15.95}$$

As seen in the Sec. 15.4.1, based on the Chilton–Colburn analogy, we have:

$$\frac{k_c}{h} = \frac{(\mathrm{Le})^{2/3}}{\rho C_{p,G}} \tag{15.96}$$

Thus, Eq. (15.95) becomes:

$$\begin{aligned} \frac{n_A C_{p,A}}{h} &= M_{r,A} \frac{C}{\rho} \frac{C_{p,A}}{C_{p,G}} (\mathrm{Le})^{2/3} \ln\left(\frac{1 - y_{A,2}}{1 - y_{A,1}}\right) \\ &= \frac{M_{r,A}}{M_{r,G}} \frac{C_{p,A}}{C_{p,G}} (\mathrm{Le})^{2/3} \ln\left(\frac{1 - y_{A,2}}{1 - y_{A,1}}\right) \end{aligned} \tag{15.97}$$

Note that $M_{r,G}$ and $C_{p,G}$ are the molecular mass and heat capacity of the gas mixture. In order to find the unknown temperature T_1, we need to solve Eq. (15.94) together with Eq. (15.97) along with an expression for the vapor pressure p_A^* as a function of temperature. The latter will give $y_{A,1} = p_A^*(T_1)/P$. Examples 15.7 and 15.8 demonstrate this calculation for two different cases.

Example 15.7. Temperature of a droplet of liquid *n*-butane evaporating in air at 325 K and 1 atm absolute temperature. This example is based on data provided by Middleman, *An Introduction to Mass and Heat Transfer*, Wiley, 1997, pp. 622–625. For *n*-butane ($M_{r,A} = 0.058$ kg/mol), we have $\Delta \hat{H}_v = 386$ kJ/kg and $C_{p,A} = 1.35$ kJ/kg K. According to Middleman, Sc = 1.5 for *n*-butane in air at 325 K. Finally, the vapor pressure of *n*-butane is given by:

$$\ln p_A^* = 15.68 - \frac{2155}{T - 34.4}$$

with p_A^* in mmHg and T in K. Thus, at 1 atm absolute pressure:

$$y_{A,1} = \frac{1}{760} \exp\left(15.68 - \frac{2155}{T_1 - 34.4}\right)$$

For air at 325 K, we have Pr = 0.7 and $C_p \sim 1.0$ kJ/kg K. Thus, Le = 0.7/1.5 = 0.47 and $(\mathrm{Le})^{2/3} = 0.60$. Inserting these values into Eqs. (15.94) and (15.97) gives:

$$T_1 = T_2 - \frac{\Delta \hat{H}_v}{C_{p,A}} \frac{\exp(C_0) - 1}{\exp(C_0)} = 325 - \frac{386}{1.35} \frac{\exp(C_0) - 1}{\exp(C_0)}$$

and

$$\frac{n_A C_{p,A}}{h} = \frac{58}{29} \frac{1.35}{1.0} (0.47)^{2/3} \ln\left(\frac{1 - 0}{1 - y_{A,1}}\right) = 1.63 \ln\left(\frac{1}{1 - y_{A,1}}\right)$$

Simultaneous solution of the three equations above gives $T_1 = 237$ K, $y_{A,1} = 0.204$, and $n_A C_{p,A}/h = 0.372$. Note that the droplet is predicted to be much cooler than the surrounding air because of the rapid evaporation of the volatile liquid n-butane. In this case, the value of $n_A C_{p,A}/h$ is high enough to make the Ackermann correction, Eq. (15.90), significantly different from 1 (actually about 0.8) indicating that there will be a substantial effect of mass transfer on the heat transfer flux. Note, however, that even at 237 K, the vapor pressure of n-butane will still be very high, which will result in a very rapid rate of evaporation and a very rapid disappearance of the droplet.

Example 15.8. Evaporation of paint solvent. Let's estimate the time needed to dry an oil-based paint applied to a wide flat surface that is 1 m long, assuming that the solvent has the properties of n-heptane. The layer of paint is 1 mm thick and air at 300 K flows over the surface at a velocity of 1.2 m/s. We will assume that the substrate onto which the paint is applied is insulated. In other words, all heat transfer will be at the air–paint interface and heat conduction to/from the substrate will be neglected. The physical properties needed for this problem are the air density, viscosity, thermal conductivity, heat capacity, the diffusivity of the n-heptane/air binary, the heat capacity of gaseous n-heptane, and the latent heat of vaporization of n-heptane. We will estimate all of these properties at 300 K, assuming that the surface temperature will not be too different from the air temperature. For air at 300 K and 1 atm absolute pressure we have (see Appendix A):

$$\rho = 1.18 \text{ kg/m}^3$$

$$\mu = 1.8 \times 10^{-5} \text{ Pa s}$$

$$k = 0.026 \text{ W/m K}$$

$$C_{p,G} = 1.0 \text{ kJ/kg K}$$

We can estimate the diffusivity of the n-heptane/air binary from the equation of Fuller et al., Eq. (10.17). For n-heptane, using additive volumes from Table 10.4 or from Table 5-13 in *Perry's Chemical Engineers' Handbook*, 9th edition, we have $\Sigma v_A = 7 \times 16.5 + 16 \times 1.98 = 147$ while for air we have $\Sigma v_B = 20.1$. Thus,

$$\mathcal{D}_{A,B} = 0.001 \frac{T^{1.75}}{P} \frac{\left(\dfrac{1}{M_{r,A}} + \dfrac{1}{M_{r,B}}\right)^{1/2}}{[(\Sigma v_A)^{1/3} + (\Sigma v_B)^{1/3}]^2} = 0.001 \frac{300^{1.75}}{1} \frac{\left(\dfrac{1}{100} + \dfrac{1}{29}\right)^{1/2}}{[(147)^{1/3} + (20.1)^{1/3}]^2} = 0.071 \text{ cm}^2/\text{s}$$

From Tables 2-69 and 2-75 of *Perry's Chemical Engineers' Handbook*, 9th edition, for n-heptane we can find:

$$\Delta \hat{H}_v = 360 \text{ kJ/kg}$$

and

$$C_{p,A} = 1.66 \text{ kJ/kg K}$$

We can now calculate the Prandtl and Schmidt numbers as follows:

$$\text{Pr} = \frac{\mu C_p}{k} = \frac{1.8 \times 10^{-5} \times 1000}{0.026} = 0.69$$

$$\text{Sc} = \frac{\mu}{\rho \mathcal{D}_{A,B}} = \frac{1.8 \times 10^{-5}}{1.18 \times (0.071 \times 10^{-4})} = 2.15$$

Thus, $(\text{Le})^{2/3} = (0.69/2.15)^{2/3} = 0.469$. Next, we need the vapor pressure on n-heptane. Between 185 and 296 K, Carruth and Kobayashi [J. Chem. Eng. Data, 18 (1973) 115] give:

$$\log_{10} p_A^* = 4.81803 - \frac{1,635.409}{T - 27.338}$$

with p_A^* in bar and T in K. Equations (15.94) and (15.97) for this case become:

$$T_1 = T_2 - \frac{\Delta\hat{H}_v}{C_{p,A}}\frac{\exp(C_0)-1}{\exp(C_0)} = 300 - \frac{360}{1.66}\frac{\exp(C_0)-1}{\exp(C_0)}$$

and

$$\frac{n_A C_{p,A}}{h} = \frac{100}{29}\frac{1.66}{1.0} \times 0.469 \ln\left(\frac{1-0}{1-y_{A,1}}\right) = 2.68 \ln\left(\frac{1}{1-y_{A,1}}\right)$$

Simultaneous solution of these two equations gives $T_1 = 284$ K, $y_{A,1} = 0.028$, $n_A C_{p,A}/h = 0.076$, and $\eta = 0.96$. Obviously, the Ackermann correction is very small in this case. We can now find the mass transfer rate. We first calculate the Reynolds number:

$$\text{Re} = \frac{\rho v_\infty L}{\mu} = \frac{1.18 \times 1.2 \times 1.0}{1.8 \times 10^{-5}} = 78{,}670$$

Based on the Nusselt number correlation in Table 6.1, item A, the Sherwood number for this case is

$$\text{Sh} = \frac{k_c L}{\mathcal{D}_{A,B}} = 0.664\,\text{Re}^{1/2}\,\text{Sc}^{1/3} = 0.664 \times (78{,}670)^{1/2}(2.15)^{1/3} = 240$$

Thus,

$$k_c = 240\frac{\mathcal{D}_{A,B}}{L} = 240\frac{0.071 \times 10^{-4}}{1.0} = 0.0017 \text{ m/s}$$

Since the molar gas density is $C = 1.18/0.029 = 40.7 \text{ mol/m}^3$, the mass flux of n-heptane will be:

$$n_A = M_{r,A} k_c C \ln\left(\frac{1}{1-y_{A,1}}\right) = 100 \times 0.0017 \times 40.7 \ln\left(\frac{1}{1-0.028}\right) = 0.196 \text{ g/m}^2\text{ s}$$

The density of liquid n-heptane is 684 kg/m³. Thus, for a 0.001 m thick layer of liquid, we can estimate that it will take $684/0.196 = 3{,}490$ s for the solvent to evaporate completely. Obviously, the time will likely be shorter if conductive heat transfer from the underlying substrate to the paint layer is considered.

Several simplifications are possible for this problem. Firstly, since the temperature and the vapor pressure are relatively low, the mole fraction of n-heptane vapor is low enough that we can approximate the flux as follows (i.e., for diffusion under dilute conditions):

$$N_A = k_c C(y_{A,1} - y_{A,2}) \tag{15.98}$$

With this approximation, the calculated value of T_1 will be practically the same as for the more rigorous expression. Also, about the same result will be obtained if we neglect the Ackermann correction. The result in this case will be the same as Eq. (15.73) and is:

$$\begin{aligned} T_1 &= T_2 - \frac{\Delta\hat{H}_v M_{r,A}}{\rho C_{p,G}}(\text{Le})^{2/3}(C_{A,1} - C_{A,2}) \\ &= 300 - \frac{360 \times 0.1}{1.18 \times 1.0} \times 0.469 \times 40.7(y_{A,1} - 0) \\ &= 300 - 582 y_{A,1} \end{aligned} \tag{15.99}$$

Solving this equation along with the expression for the vapor pressure gives $T_1 = 283$ K, which is close to the exact value. This occurs because, for this problem, the mass transfer rate is not so high as to affect heat transfer in a very significant way.

15.4.3 Evaporative Cooling Towers

In this section, we consider the analysis of evaporative water-cooling towers as an example of application of the theory of coupled heat and mass transfer processes to the analysis and design of an industrial operation. Water-cooling towers are used extensively in manufacturing facilities and in power generation plants as well as in large-scale air conditioning systems. Their purpose is to cool service water that has been warmed up in various process uses and to return it to the original service temperature. This is obtained, principally, as a result of the evaporative cooling that occurs when the warm water returning from the plant is contacted with an air stream. Because of the large value of the latent heat of vaporization of water, vaporizing a small fraction of the water fed will generally be sufficient to attain the desired cooling. Obviously, since the cooling mechanism relies on evaporation, cooling towers will work best when the ambient air is dry. Other means of cooling will often be required in geographical areas where the ambient air is warm and moist, which will hinder evaporation.

Contact between the incoming air and the warm water fed to a cooling tower can be either counter-current, with the water flowing down and the air flowing up, or cross-current, with the water flowing down and the air flowing in a cross-wise direction. In either case, the equipment is similar to packed absorption or stripping columns. However, the packing, called "fill" in this case, will often consist of slats made of wood, plastic, or metal, instead of packing pieces. The purpose of the fill is also to promote a large gas–liquid interfacial area which will facilitate evaporation, but the slat structure will typically allow larger flows of air than is possible in columns packed with small packing pieces. The air flow can be driven by fans, usually mounted at the top of the tower, or by natural convection driven by the density difference between the cooler and drier air entering at the bottom of the tower and the warmer and moister air leaving at the top. Hyperboloid column designs are often used to promote air flow in such natural draft cooling towers.

Our focus in this section is the design of a forced draft counter-current cooling tower as sketched in Fig. 15.8. More detailed analyses, including the design of cross-flow cooling towers can be found, for example, in Sherwood et al., *Mass Transfer*, McGraw-Hill, 1975, pp. 280–295.

In our analysis of the cooling tower, we want to track the temperature of the liquid, T_L, along with the temperature, T, and moisture content of the air. The latter is measured by the absolute humidity, which is given by the mass ratio W_A typically in units of kg of water vapor per kg of dry air. Let's define G_s' as the mass velocity of dry air (kg dry air/m^2 s) and L' as the liquid water mass velocity (kg water/m^2 s). Both are obtained by dividing the corresponding mass flow rates by the cross-sectional area of the cooling tower, S, which we assume is constant. We also define the quantity

$$C_{p,s} = C_{p,\text{air}} + W_A C_{p,A} \tag{15.100}$$

as the "humid air heat capacity," where $C_{p,\text{air}}$ is the heat capacity of dry air and $C_{p,A}$ is the heat capacity of water vapor. Note that the units of $C_{p,s}$ will be J per kg of dry air per K. In practice, since the moisture content W_A is usually not so high, we will have $C_{p,s} \sim C_{p,\text{air}}$. Based on the definition of $C_{p,s}$, the specific enthalpy of humid air, $\hat{h}$, at a temperature T and with a moisture content W_A can be written as follows with a reference temperature of 0°C:

$$\hat{h} = C_{p,s}T + W_A \times \Delta\hat{H}_v \tag{15.101}$$

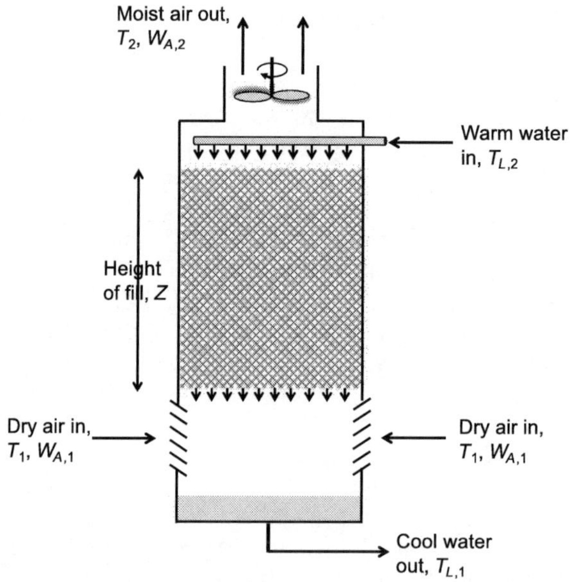

FIGURE 15.8 Schematic of a forced-draft, counter-current water-cooling tower. Air flow is induced by a fan placed at the top of the tower. The "fill" facilitates formation of a large gas–liquid interfacial area.

Note that, again, the units of $\hat{h}$ will be J per kg of dry air. Thus, the reference gas flow rate will be that of dry air on a moisture-free basis. In our analysis, we will track the conditions in the tower in a plot of $\hat{h}$ vs. the liquid temperature just like for the design of absorption or stripping columns we tracked the gas-phase composition as a function of the liquid phase composition. Two main lines will appear on this plot, as sketched in Fig. 15.9. The first is the equilibrium line, which is obtained by plotting the values of $\hat{h}$ of saturated air; i.e., the enthalpy of saturated air in equilibrium with water at T_L. We use the symbol

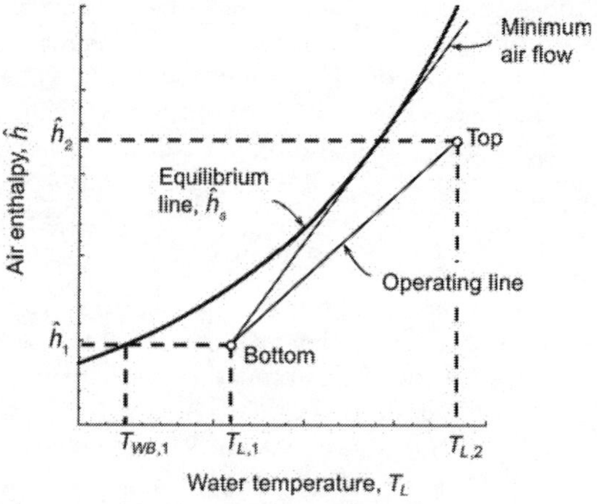

FIGURE 15.9 Equilibrium and operating lines for a counter-current water-cooling tower.

$\hat{h}_s$ to denote this enthalpy. The second line appearing on the plot is an operating line. This line is obtained from the following energy balance written for the cooling tower for a control volume extending from the bottom of the column to a point at a height z:

$$G_s'\hat{h}_1 + L'C_{p,L}T_L = G_s'\hat{h} + L'C_{p,L}T_{L,1} \tag{15.102}$$

With some rearrangement, Eq. (15.102) can be written as follows:

$$\hat{h} = \frac{L'C_{p,L}}{G_s'}T_L + \left(\hat{h}_1 - \frac{L'C_{p,L}}{G_s'}T_{L,1}\right) \tag{15.103}$$

Note that the slope of this operating line is $L'C_{p,L}/G_s'$, which will have units of J/kg of dry air K. As seen in Fig. 15.9, because of the upwards curvature of the equilibrium line, there will often be a maximum value possible for $L'C_{p,L}/G_s'$ and, thus, a minimum possible air flow rate, that occurs when the operating line becomes tangent to the equilibrium line. The actual value of $L'C_{p,L}/G_s'$ will need to be smaller than this maximum and, thus, the air flow rate will have to be larger than the corresponding minimum in order to obtain a tower of finite height.

Next, we turn our attention to the rate equations, which are obtained by writing differential material and energy balances along the height of the tower, z. We write these balances neglecting the Ackermann correction, which is usually small for the air–water system. If a is the air–water interfacial area per unit volume, the gas-phase balances for a differential section of the column with height dz are as follows:

$$G_s'C_{p,s}dT = h(T_s - T)adz \tag{15.104}$$

$$G_s'dW_A = k_W(W_{A,s} - W_A)adz \tag{15.105}$$

The energy balance for the liquid phase in the same section yields:

$$L'C_{p,L}dT_L = h(T - T_s) + k_W(W_A - W_{A,s}) \times \Delta\hat{H}_v \tag{15.106}$$

In these equations, T_s is the temperature at the air–liquid interface and $W_{A,s}$ is the absolute humidity (or mass ratio in kg water/kg dry air) in equilibrium at T_s. Note that in Eq. (15.105) we have expressed the rate of mass transfer in terms of a mass transfer coefficient k_W multiplied by a driving force written as the difference between the mass ratio (or absolute humidity) $W_{A,s}$ at saturation and the actual mass ratio or absolute humidity of the bulk air, W_A. In practice, since conditions are typically relatively dilute, we will have $k_W \sim \rho k_c$. Multiplying Eq. (15.105) by $\Delta\hat{H}_v$ and adding the result to Eq. (15.104) yields the following result:

$$G_s'(C_{p,s}dT + \Delta\hat{H}_v dW_A) = k_W\left[\frac{h}{k_W C_{p,s}}C_{p,s}(T_s - T) + (W_{A,s} - W_A) \times \Delta\hat{H}_v\right]adz \tag{15.107}$$

It can be verified that for the air–water system, since Le ~ 1, the ratio $h/k_W C_{p,s}$ will be close to 1. In this case, considering Eq. (15.101), Eq. (15.107) will simplify to:

$$G_s'd\hat{h} = k_W(\hat{h}_s - \hat{h})adz \tag{15.108}$$

Integrating Eq. (15.108) over the height of the tower fill gives the final result:

$$Z = \frac{G_s'}{k_W a}\int_{\hat{h}_1}^{\hat{h}_2} \frac{d\hat{h}}{\hat{h}_s - \hat{h}} \tag{15.109}$$

This result has the familiar form:

$$Z = H_G N_{G,\hat{h}} \tag{15.110}$$

where

$$H_G = \frac{G_s'}{k_W a} \tag{15.111}$$

is the height of a gas-phase mass transfer unit and

$$N_{G,\hat{h}} = \int_{\hat{h}_1}^{\hat{h}_2} \frac{d\hat{h}}{\hat{h}_s - \hat{h}} \tag{15.112}$$

is the number of gas-enthalpy transfer units. It is indeed remarkable that we end up with an expression that combines mass transfer and enthalpy transfer units. This result is obtained as a consequence of the coupling of heat and mass transfer processes but is rigorously applicable only to the air–water system for which Le ~ 1.

In order to calculate $N_{G,\hat{h}}$ from Eq. (15.112), we need $\hat{h}_s$, which is the enthalpy of humid air at the surface temperature T_s. In general, this would be obtained by considering the liquid-side heat transfer resistance in combination with Eq. (15.106). In practice, however, the liquid-side heat transfer resistance is typically negligible, which means that the surface temperature will be essentially the same as the temperature of the bulk liquid, or $T_s = T_L$. The calculation of $N_{G,\hat{h}}$ for this case is demonstrated in Example 15.9. Values of H_G are obtained from correlations that are specific to the cooling tower equipment as found, for example, in Sherwood et al., *Mass Transfer*, McGraw-Hill, 1975, Section 7.11. For slat-packed cooling towers, Sherwood et al. report experimental values of H_G between 2 and 15 ft depending on the ratio of air and water flow rates.

Example 15.9. Calculation of the height of a water-cooling tower [Mathematica notebook: (M.EX.15.9)]. Let's calculate the height of the fill in a cooling tower needed to cool water from 45°C to 25°C with air entering at 20°C with a relative humidity of 40%. Figure 15.10 provides the enthalpy-temperature diagram for the air-water system at 1 atm absolute pressure based on data in Table 12-4 of *Perry's Chemical Engineer's Handbook*, 9th edition. Enthalpy reference states are air at –17.8°C (0°F) and liquid water at 0°C. Over the temperature range 10°C to –45°C, the equilibrium line is represented, approximately, by the following empirical equation:

$$\hat{h}_s = 10.652 + 4.3843(T_L) - 0.10951(T_L)^2 + 0.0027003(T_L)^3 \tag{15.113}$$

with $\hat{h}_s$ in kJ/kg dry air and T_L in °C. Based on Fig. 15.6, the wet-bulb temperature of the entering air is 12.5°C. The corresponding humid enthalpy $\hat{h}$ is equal to the value at saturation since the constant wet-bulb temperature lines and the constant enthalpy lines are essentially coincident for the air–water system. Thus, from Eq. (15.113) we have $\hat{h}_1 = \hat{h}_s(T_{L,1}) = 53.6$ kJ/kg dry air. As seen in Fig. 15.10, the operating line becomes tangent to the equilibrium line at $T_L \sim 40$°C, which gives $(L'C_{p,L}/G_s')_{\max} =$ 8.62 kJ/kg°C. This corresponds to the minimum value of G_s'. Let's assume that we actually operate with an air flow rate that is 20% higher the minimum. In this case, $L'C_{p,L}/G_s'$ will be 8.62/(1 + 0.2) = 7.18 kJ/kg°C. The corresponding operating line is plotted in Fig. 15.10. With an inlet water temperature of 45°C, the outlet air enthalpy will be $\hat{h}_2 = 197$ kJ/kg dry air. The integrand in the expression for $N_{G,\hat{h}}$, Eq. (15.112), is calculated as follows. For a given $\hat{h}$, we calculate the temperature T_L from the operating line as:

$$T_L = T_{L,1} + \frac{\hat{h} - \hat{h}_1}{L'C_{p,L}/G_s'} = 25 + \frac{\hat{h} - 53.6}{7.18} \tag{15.114}$$

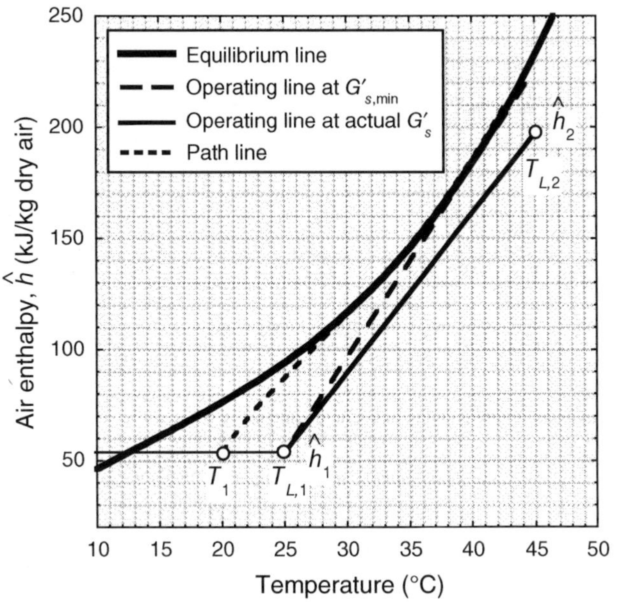

FIGURE 15.10 Equilibrium, operating lines, and path line for Example 15.9.

This temperature is then used to calculate $\hat{h}_s$ from Eq. (15.113). Finally, the integral of $(\hat{h}_s - \hat{h})^{-1}$ is evaluated numerically. The result is $N_{G,\hat{h}} = 5.75$. The tower height would be obtained multiplying this value times H_G.

Next, we consider the calculation of the air temperature along the tower. Note that although we know the air enthalpy $\hat{h}$ at each point along the operating line and, thus, along the height of the tower, the actual air temperature and humidity will depend on a different calculation, which requires that we return to the differential balances. Dividing Eq. (15.108) by Eq. (15.104) and considering that $k_W C_{p,s}/h \sim 1$ for the air–water system, we obtain:

$$\frac{d\hat{h}}{dT} = \frac{\hat{h}_s - \hat{h}}{T_s - T} = \frac{\hat{h}_s(T_L) - \hat{h}}{T_L - T} \tag{15.115}$$

This equation can be integrated starting at the point $T = T_1$ where $\hat{h} = \hat{h}_1$ and $T_L = T_{L,1}$ (i.e., at the bottom of the tower) to find the gas temperature along the height of the tower. The $\hat{h}$ vs. T line obtained from this equation is called the "path line" and is also plotted in Fig. 15.10. To calculate it numerically, we use the following recurrence formula:

$$\hat{h}_{i+1} = \hat{h}_i + \frac{\hat{h}_s(T_{L,i}) - \hat{h}_i}{T_{L,i} - T_i}\Delta T \tag{15.116}$$

where ΔT is a small increment in gas temperature. Starting with $i = 1$, $T_{L,i}$ is obtained from the operating line, Eq. (15.114), at each value of $\hat{h}$. As seen in Fig. 15.10, the path line in this case quickly approaches the equilibrium line indicating that the air will stay essentially at the wet-bulb temperature and will leave the tower at near saturated conditions. Note that, for these conditions, fog (i.e., condensation in the bulk air) might form at some point along the tower height.

A final calculation of interest for this problem is how much water will be lost to evaporation to obtain the desired cooling. As see in Fig. 15.10, the air leaving the tower will be essentially saturated at 42°C. From the psychrometric chart or Table 12.4 in Perry's Handbook, the moisture content of this air will be 0.052 kg water/kg dry air. The moisture content of the entering air is 0.006 kg water/kg dry air. By material balance, the rate of water evaporation will thus be proportional to $G'_s(0.052 - 0.006)$. For the conditions of this problem, $L'C_{p,L}/G'_s = 7.18$ kJ/kg°C. Since, for liquid water, $C_{p,L} = 4.2$ kJ/kg°C, we will have $L'/G'_s = 7.18/4.184 = 1.72$ kg water/kg dry air. As a result, the amount of water evaporated per kg of water fed will be $0.052 - 0.006)/1.72 = 0.027$ kg/kg. Thus, evaporation of a mere 2.7% of the water fed will result in cooling from 45°C to 25°C.

15.5 Problems

P.15.1. The following problem is based on data and analyses of ternary gas diffusion found in Toor [AIChE J. 3 (1957) 198] for the system CO_2 (A)/H_2O (B)/H_2 (C) at 40°C and 155 mmHg absolute pressure. (a) Estimate the binary diffusion coefficients $\mathcal{D}_{A,B}$, $\mathcal{D}_{B,C}$, and $\mathcal{D}_{A,C}$ for these conditions. (b) Consider a situation where at a certain point in a separation unit steady-state diffusion of this ternary mixture occurs through a stagnant gas film with thickness $\Delta = 0.1$ cm. A solubility barrier prevents C from moving while A and B can freely diffuse. On one side of the film we have $y_{A,1} = 0$, $y_{B,1} = 0.7$ and $y_{C,1} = 0.3$. On the other side of the film, we have $y_{C,2} = 0.2$. Obtain the steady-state diffusion fluxes of A and B as a function of $y_{B,2}$ for $0 < y_{B,2} < 0.7$. Plot the values of $N_i RT\Delta/P$ for each component vs. $y_{B,2}$ in this range. Note that $y_{A,2} = 1 - y_{B,2} - 0.2$. (c) Find the value of $y_{B,2}$ for which the flux of B is zero. Explain what happens between this value and $y_{B,2} = 0.7$. Note that in this range, the gradient in $y_{B,2}$ is negative; i.e., $y_{B,1} > y_{B,2}$.

P.15.2. A diaphragm diffusion cell is used to determine diffusivities in aqueous solutions by measuring the mass transfer flux across an inert macroporous slab that separates two well-mixed compartments as shown in Fig. P.15.2. In this particular case, the slab is 1.0 mm thick and has a porosity $\varepsilon_p = 0.5$ and a tortuosity factor $\tau_p = 2.0$ (*cf.* Sec. 10.3.3). The temperature is 298 K. Assuming that diffusion through the macroporous slab is controlling: (a) Predict the mass transfer flux of K^+ and Cl^- if compartment 1 is filled with a 0.1 M KCl solution and compartment 2 is filled with pure water. (b) Predict the mass transfer flux of K^+ and H^+ if compartment 1 is filled with a 0.1 M KCl solution and compartment 2 is filled with a 0.1 M HCl solution.

P.15.3. Lysine is an essential amino acid used in many applications, including animal and human foods and pharmaceuticals. A sulfonated poly(styrene-divinylbenzene) cation-exchange resin with an 8% degree of cross-linking is used to bind lysine at pH 6 from an aqueous solution at 25°C. At this pH, lysine diffuses in the resin in monovalent cationic form. The resin particle diameter is 600 μm. (a) Estimate the diffusivity of lysine in water, $\mathcal{D}^0_{Lys,W}$, and the ion diffusivity of monovalent lysine cations, $\bar{\mathcal{D}}_{Lys^+}$, in this resin. (b) Estimate the time required to convert the resin particle to 80% monovalent lysine form if the resin starts 100% saturated in NH_4^+ form and is immersed in a large volume of a lysine solution. (c) Estimate the time required to convert the resin particle to 80% NH_4^+ form if the resin starts 100% saturated in monovalent lysine in and is immersed in a large volume of a NH_4Cl solution. (d) Repeat the calculation in (b) and (c) neglecting electrostatic

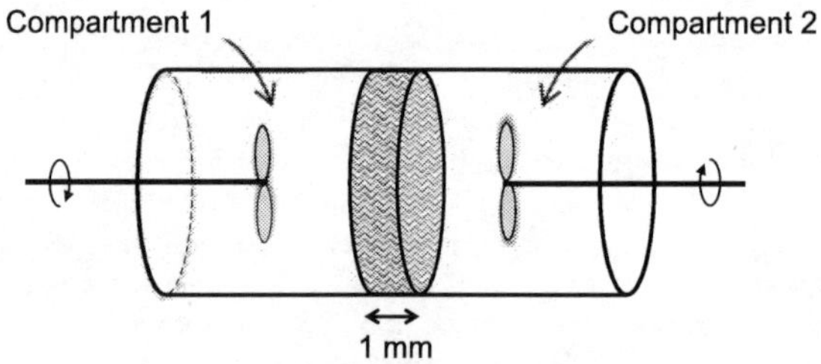

FIGURE P.15.2

coupling of diffusion fluxes and assuming that the effective diffusivity is a constant equal to $\bar{\mathcal{D}}_{Lys^+}$. The ion diffusivity of NH_4^+ in this resin has been reported to be $\bar{\mathcal{D}}_{NH_4^+} = 0.18 \times 10^{-7}$ cm²/s [Nagai and Carta, Sep. Sci. Technol. 39 (2004) 3711].

P.15.4. Spherical particles are used in many practical applications including adsorption and ion exchange. Mass transfer in such particles can be described starting with a differential material balance in spherical coordinates. If the intraparticle diffusivity is constant, starting with an initially uniform concentration $c_{A,0}$ and with a uniform surface concentration $c_{A,s}$, as shown in Fig. P.15.3, the average concentration in the particle $c_{A,\text{avg}}$ is given in dimensionless form by item C in Table 5.2. With reference to Fig. P.15.4, an approximate solution of this mass transfer problem can be found using a pseudo-steady-state approximation by assuming that the mass transfer resistance in the sphere is represented by diffusion through a solid film with thickness $\Delta = R/5$ where R is the actual particle radius. With a constant diffusivity, $\bar{\mathcal{D}}_A$, the diffusion flux of A across this film is:

$$N_A = -\bar{\mathcal{D}}_A \frac{dc_A}{dx}$$

Assuming pseudo-steady-state, N_A is constant across the film. Thus, integrating this equation across the film we obtain:

$$N_A = \frac{\bar{\mathcal{D}}_A}{R/5}(c_{A,s} - c_{A,\text{avg}})$$

A material balance yields the following equation:

$$\frac{4\pi R^3}{3}\frac{dc_{A,\text{avg}}}{dt} = 4\pi R^2 \frac{\bar{\mathcal{D}}_A}{R/5}(c_{A,s} - c_{A,\text{avg}})$$

or

$$\frac{dc_{A,\text{avg}}}{dt} = \frac{15\bar{\mathcal{D}}_A}{R^2}(c_{A,s} - c_{A,\text{avg}})$$

Separating the variables and integrating gives:

$$\int_{c_{A,0}}^{c_{A,\text{avg}}} \frac{dc_{A,\text{avg}}}{c_{A,s} - c_{\text{avg}}} = \frac{15\bar{\mathcal{D}}_A}{R^2}\int_0^t dt$$

The final result is:

$$Y_{\text{avg}} = \frac{c_{A,\text{avg}} - c_{A,s}}{c_{A,0} - c_{A,s}} = \exp\left(-\frac{15\bar{\mathcal{D}}_A}{R^2}t\right) = \exp(-15\tau)$$

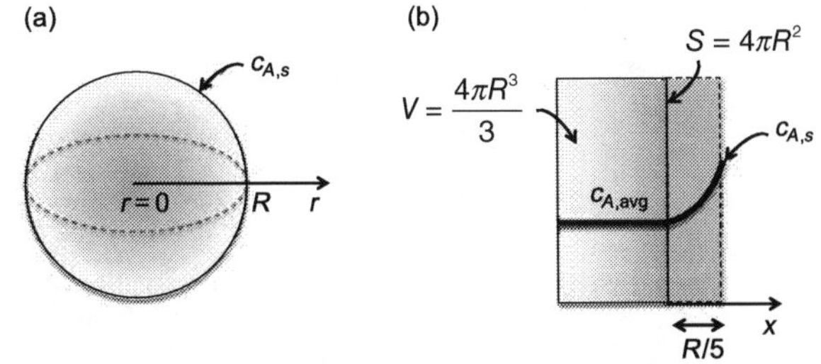

FIGURE P.15.4

where $\tau = \bar{D}_A t/R^2$. Although this expression is only approximate, this simplified approach has the advantage that it can be easily extended to cases where the diffusivity is composition dependence. In the adsorption field, this is called the linear driving force approximation (LDF) and is used extensively in lieu of the more complex description of transport in spherical coordinates. (a) Compare the result of the film model approximation above for the case of constant $\bar{D}_A$ with the exact solution by item C in Table 5.2 by plotting Y_{avg} vs. τ in both cases. (b) Extend the film model approximation analysis to the ion exchange of two ions A and B with diffusivities $\bar{D}_{A^+}$ and $\bar{D}_{B^+}$ assuming $c_{A,0} = c_{R^-}$ and $c_{A,s} = 0$. In this case, the flux is given by:

$$N_A = -\frac{\bar{D}_{A^+}}{(\alpha - 1)\dfrac{c_{A^+}}{c_{R^-}} + 1} \frac{dc_{A^+}}{dx}$$

(c) Compare the results of your extension with the exact solution for diffusion in a spherical ion-exchange particle (Eq. (15.61)) by plotting $F = 1 - c_{A^+,avg}/c_{R^-}$ vs. $D_{A^+}t/R^2$ in the two cases for $\alpha = 0.1$ and $\alpha = 10$.

P.15.5. A thin porous metal plate, 0.3 m long and very wide, is placed in a stream of dry air at 500°C and 1 atm absolute pressure, flowing at a velocity of 15 m/s parallel to the 0.3 m dimension. Liquid water is supplied continuously to the plate and evaporates at the plate surface to keep the plate cool. (a) Determine the steady-state temperature of the plate assuming that the liquid water is supplied at the same temperature at which the plate is kept. (b) Determine the rate at which water has to be supplied to the plate per square meter of plate area.

P.15.6. A forced-draft cooling tower is to be designed to cool water from 35°C to 15°C in a dry, cool climate where the air is at 20°C with a relative humidity of 30%. The design air mass velocity is to be $G'_s = 1.6$ kg dry air/m² s with an L'/G'_s value 1.25 times smaller than the maximum. For these conditions, H_G is estimated to be 1.5 m. (a) Determine the maximum and actual values of L'/G'_s and the height of fill required. (b) Suppose that the tower is built with the height calculated in part (a) but that, all of a sudden, the relative humidity of the inlet air increases to 50%, while its temperature remains the same. Assuming that G'_s and H_G remain the same, determine by what percentage the water flow rate will have to decrease if it is still desired to have an outlet water temperature of 15°C.

CHAPTER 16

Mass Transfer with Reaction

16.1 Introduction

The simultaneous occurrence of diffusional mass transfer and chemical reactions is important in a broad range of applications including catalysis and chemical reactors, fuel cells, emission control devices for automotive applications, and bioreactors. The treatment of mass transfer in these systems can be found in many chemical reaction and biochemical engineering textbooks. Simultaneous mass transfer and chemical reaction is also important in many different types of gas-absorption operations where a poorly or moderately soluble gas is absorbed into a reactive liquid. This situation, which is the main topic of this chapter, is encountered, for instance, in the removal of certain gaseous pollutants from air and other gaseous streams, in the absorption of carbon dioxide in alkaline solutions as a means of carbon capture, in the absorption of hydrogen sulfide into amine solutions for natural gas sweetening, in the absorption of oxygen into aqueous solutions in water treatment systems, and in the absorption of nitrogen oxides in the manufacture of nitric acid. Many of the underlying principles are the same regardless of the specific application. Thus, the more focused analyses in this chapter can be used for a conceptual understanding of many different systems. The foundation for these analyses can be found in the classical textbooks by Astarita, *Mass Transfer with Chemical Reaction*, Elsevier, 1967, and by Sherwood et al., *Mass Transfer*, McGraw-Hill, 1975, Chapter 8.

16.2 Absorption with a Bimolecular Reaction

Let's consider the absorption of a gaseous component A into a liquid that contains a species B that reacts with dissolved A according to the following stoichiometry:

$$A + \upsilon B \rightarrow P \tag{16.1}$$

where υ is a stoichiometric coefficient and P represents one or more reaction products. We will assume that the reaction is irreversible and that B is not volatile. In other words, we assume that B will be confined to the liquid phase. We also assume that the reaction occurs in the liquid only between the A that is physically dissolved and B. In this case, the rate of the chemical reaction will generally depend on the local concentration of dissolved A and on the concentration of B. For example, the rate of reaction could be represented by the following second-order kinetics expression:

$$R_A = -kc_Ac_B \tag{16.2}$$

where k is a second-order rate constant and c_A and c_B are the molar concentrations of A and B in the liquid, respectively. Typical units of R_A will be mol/m^3 s or mol/L s. In this case, k will have units of m^3/mol s or L/mol s.

In the simplest formulation, we can use the film model to represent mass transfer in the liquid by defining a liquid-film thickness $\Delta = \mathcal{D}_A/k_c$ where $\mathcal{D}_A$ is the diffusivity of A in the liquid and k_c is the mass transfer coefficient. In practice, k_c can be determined experimentally by measuring the rate of absorption of A in the absence of chemical reaction, or it can be estimated from an appropriate correlation for mass transfer coefficients. For example, if the absorption process occurs in a packed absorption column, we could use the Onda method, Eq. (14.64), to estimate k_x. In turn, we can calculate $k_c = k_x/c_L$, where c_L is the molar density of the liquid.

Let's consider first the steady-state absorption of a component A from a gas into a liquid in the stirred gas-liquid contactor sketched in Fig. 16.1. The liquid is fed at volumetric flow rate $\dot{V}$ and contains a concentration c_B of a component B that reacts with the A dissolved in the liquid. The gas-liquid interfacial area is S, and the volume of liquid in the reactor is V. We will assume that the bulk liquid is mixed and that mass transfer of A is limited by diffusion within a thin film at the gas-liquid interface. We will also assume that conditions are dilute. As illustrated in Fig. 16.1, if we look at the concentration profiles of

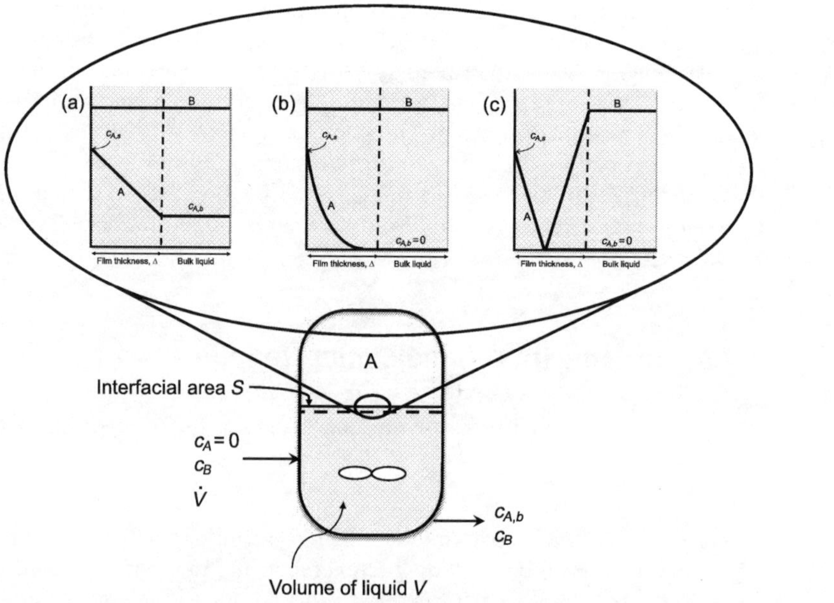

FIGURE 16.1 Concentration profiles of A and B near the gas–liquid interface for the absorption of A with the reaction $A + \upsilon B \rightarrow P$ based on the film model. (a) Profiles for a slow reaction, (b) profiles for a fast reaction, (c) profiles for an instantaneous reaction. $\Delta = \mathcal{D}_A/k_c$ is the thickness of the mass-transfer film at the gas-liquid interface.

dissolved A and of reactant B within this film, we can envision several different scenarios depending on the relative magnitude of reaction and diffusion rates. Three different scenarios illustrated in Fig. 16.1 are as follows:

(a) Slow reaction. If the reaction kinetics is slow, the reaction will not occur appreciably within the film. In this case, A will diffuse across the film and reach the bulk liquid where it will finally react with B. In this case, as we have learned in previous chapters, the rate of mass transfer of A will be given by the following equation:

$$SN_A = Sk_c(c_{A,s} - c_{A,b}) \tag{16.3}$$

where $c_{A,s}$ is the concentration of A dissolved in the liquid at the gas-liquid interface and $c_{A,b}$ is the concentration of A present in the bulk liquid. $c_{A,s}$ will depend on the solubility of gaseous A in the liquid and will be a function of the partial pressure of A in the gas. If the liquid is well mixed, $c_{A,b}$ will also be the concentration of dissolved A in the liquid leaving the vessel. In this case, the rate at which A will be consumed in the bulk liquid is given by the following equation:

$$VR_A = Vkc_{A,b}c_{B,b} \tag{16.4}$$

where here $c_{B,b}$ is the concentration of B in the bulk liquid. A material balance on component A in the liquid gives the following relationship:

$$\text{Rate of absorption of } A = \text{Rate of consumption of } A + \text{Output rate of } A$$

For simplicity, in this analysis we assume that the concentration of B in the bulk liquid, $c_{B,b}$, is the same as in the feed. This will be the case, for instance, if B is in excess. In this case, the material balance for component A can be written as follows:

$$Sk_c(c_{A,s} - c_{A,b}) = Vkc_{A,s}c_{B,b} + \dot{V}c_{A,b} \tag{16.5}$$

Solving this equation for $c_{A,b}$ gives the following expression:

$$c_{A,b} = \frac{Sk_c/\dot{V}}{1 + Sk_c/\dot{V} + Vkc_{B,b}/\dot{V}} c_{A,s} \tag{16.6}$$

The final result is obtained by substituting Eq. (16.6) into Eq. (16.3), which yields the following expression for the rate of absorption:

$$SN_A = \frac{1 + Vkc_{B,b}/\dot{V}}{1 + Sk_c/\dot{V} + Vkc_{B,b}/\dot{V}} Sk_c c_{A,s} \tag{16.7}$$

Note that the effect of reaction in this case is simply to reduce the concentration of component A in the bulk liquid. Mass transfer and reaction will then be in series, rather than occurring simultaneously. If the reaction rate increases, the term $Vkc_{B,b}/\dot{V}$ in Eq. (16.6) will also increase. As a consequence, $c_{A,b}$ will decrease.

At some point, $c_{A,b}$ will become zero. In this case, the rate of absorption will be simply:

$$SN_A = Sk_c c_{A,s} \tag{16.8}$$

For these conditions, the rate of absorption, and thus, the overall rate of reaction, will be completely controlled by the rate of mass transfer. The reaction kinetics will no longer have an effect other than to keep the concentration of A at zero in the bulk liquid and, thus, to maximize the driving force for mass transfer.

(b) **Fast reaction**. In this case, the reaction is not only fast enough to keep the concentration of component A in the bulk liquid at zero, as in case (a), but it is also fast enough that A is consumed appreciably within the diffusion film. In other words, A will be consumed before having a chance to make it to the bulk liquid. Since, in this case, the reaction will occur while A diffuses, mass transfer and reaction will be coupled. The concentration profile of A in the film will then no longer be linear and the concentration of A will drop more steeply within the film compared to the slow reaction case. Since the rate of mass transfer is proportional to the concentration gradient of A, the steeper sloping profile will also result in faster mass transfer. Thus, in this case, the reaction will not only keep the bulk concentration of A at zero, but it will also enhance the rate at which A will be absorbed relative to the value predicted by Eq. (16.8). In order to describe this effect, we define an enhancement factor, ϕ, as follows:

$$\phi = \frac{\text{Mass transfer rate with reaction in the film}}{\text{Mass transfer rate without reaction in the film}} = \frac{SN_A}{Sk_c c_{A,s}} = \frac{N_A}{k_c c_{A,s}} \geq 1 \tag{16.9}$$

where SN_A is the rate of mass transfer with reaction and N_A is the corresponding flux. According to this definition of ϕ, the rate of mass transfer will be given by:

$$SN_A = S\phi k_c c_{A,s} \tag{16.10}$$

Predicting ϕ and, thus, predicting the absorption rate will require that we analyze the problem of simultaneous reaction and diffusion in the film.

(c) **Instantaneous reaction**. In this case, the reaction is not only fast enough to consume component A within the film, but it is so fast that the concentration of B will also be reduced significantly within the film. If, as illustrated by case (c) in Fig. 16.1, the reaction between A and B is assumed to be instantaneous (i.e., if it occurs at an infinite rate), the concentrations of both A and B will reach zero at a plane located at some distance δ from the gas-liquid interface, smaller than the actual film thickness Δ. In this case, component A will diffuse from the gas-liquid interface toward this plane while component B will diffuse toward the same plane from the bulk liquid. The reaction will occur only where the two components A and B meet. The absorption rate will also be enhanced relative to the slow reaction case and will be faster than that predicted by Eq. (16.8). Predicting ϕ and, thus, the absorption rate will also require that we analyze the problem of diffusion of both A and B within the film.

16.3 Expressions for the Enhancement Factor Based on the Film Model

The differential material balances describing steady-state diffusion and reaction in a flat film of thickness Δ are readily derived from Table 12.1 assuming constant diffusivities and dilute conditions. Two material balances will be needed in general: one for the gaseous component A dissolved in the liquid and the other for the liquid-phase reactant B. On a molar basis, from item A in Table 12.1, we obtain:

$$0 = \mathcal{D}_A \frac{d^2 c_A}{dx^2} + R_A \tag{16.11}$$

and

$$0 = \mathcal{D}_B \frac{d^2 c_B}{dx^2} + R_B \tag{16.12}$$

where $\mathcal{D}_A$ and $\mathcal{D}_B$ are, respectively, the diffusivities of A and B in the liquid. Because of the reaction stoichiometry, we will have $R_B = \upsilon R_A$. Thus, assuming that the rate is represented by Eq. (16.2), Eqs. (16.11) and (16.12) become:

$$\mathcal{D}_A \frac{d^2 c_A}{dx^2} = k c_A c_B \tag{16.13}$$

$$\mathcal{D}_B \frac{d^2 c_B}{dx^2} = \upsilon k c_A c_B \tag{16.14}$$

The boundary conditions for these equations can be derived as follows. At the gas–liquid interface, where $x = 0$, the concentration of dissolved A will be in equilibrium with the gas and, thus, c_A will be equal to $c_{A,s}$. Conversely, if B is not volatile, its flux and, thus, its concentration gradient will be zero at $x = 0$. At the edge of the film adjacent to the bulk liquid, where $x = \Delta$, c_A will be zero if the reaction is fast while c_B will be equal to its value in the bulk liquid. As a result, the boundary conditions for Eqs. (16.13) and (16.14) are as follows:

$$\begin{aligned} x = 0:&\quad c_A = c_{A,s},\quad \frac{dc_B}{dx} = 0 \\ x = \Delta:&\quad c_A = 0,\quad c_B = c_{B,b} \end{aligned} \tag{16.15}$$

It is instructive to consider first the limiting solutions of this set of equations and boundary conditions, which are applicable over particular ranges of the parameter values. We refer to the conditions where these limiting solutions are applicable as "reaction regimes." After establishing these reaction regimes, we will then consider the full solution of these equations.

16.3.1 Fast Pseudo-First-Order Reaction Regime

In this regime, the reaction is fast enough that component A is consumed within the film as in case (b) of Fig. 16.1. At the same time, the concentration of B is high enough that its variation within the film will be negligible. In other words, in this regime,

$c_B \sim c_{B,b}$ throughout the film. In this case, Eq. (16.13) alone is sufficient to describe mass transfer of A and can be written as follows:

$$\mathcal{D}_A \frac{d^2 c_A}{dx^2} = k_I c_A \tag{16.16}$$

where $k_I = k c_{B,b}$ is a pseudo-first-order rate constant with units of s^{-1}. The value of k_I will depend, of course, on the concentration of B in the bulk liquid. The boundary conditions for Eq. (16.16) are just:

$$\begin{aligned} x = 0: &\quad c_A = c_{A,s} \\ x = \Delta: &\quad c_A = 0 \end{aligned} \tag{16.17}$$

In dimensionless form, Eqs. (16.16) and (16.17) can be written as follows:

$$\frac{d^2\theta_A}{d\xi^2} = N^2\theta_A \tag{16.18}$$

$$\begin{aligned} \xi = 0: &\quad \theta_A = 1 \\ \xi = 1: &\quad \theta_A = 0 \end{aligned} \tag{16.19}$$

where $\theta_A = c_A/c_{A,s}$ is the dimensionless concentration of dissolved A, $\xi = x/\Delta$ is the dimensionless distance from the gas-liquid interface, and N is a reaction/diffusion modulus given by:

$$N = \sqrt{\frac{k_I \Delta^2}{\mathcal{D}_A}} = \sqrt{\frac{k_I \mathcal{D}_A}{k_c^2}} \tag{16.20}$$

Remarkably, even though the physical meaning is completely different, in this dimensionless form, Eq. (16.18) is exactly the same as the equation that describes heat transfer in a straight fin, Eq. (4.15). The first of the two boundary conditions is also the same as Eq. (4.16). Thus, mathematically, the only difference between the equations describing these two processes is the boundary condition at $\xi = 1$ where, for the fin, we assumed $d\theta/d\xi = 0$ according to Eq. (4.17), while, for mass transfer with reaction, we have assumed $\theta_A = 0$. The solution of Eqs. (16.18) and (16.19) is as follows:

$$\theta_A = \frac{\sinh[N(1-\xi)]}{\sinh(N)} \tag{16.21}$$

Figure 16.2 shows the dimensionless concentration profiles calculated from Eq. (16.21) for different values of N. We can see that when N is zero or very small (say, $N = 0.1$), this equation predicts a linear or nearly linear concentration profile of A. In this case, little, if any, consumption of A occurs within the film. As N increases, however, we see that the A concentration profile becomes nonlinear with an increasingly steeper slope at the gas–liquid interface ($\xi = 0$).

Beside the concentration profiles, we are interested, of course, in the mass transfer rate and, hence, the flux of A at $x = 0$. The latter can be obtained from Fick's law and is given by:

$$N_A = -\mathcal{D}_A \left.\frac{dc_A}{dx}\right|_{x=0} \tag{16.22}$$

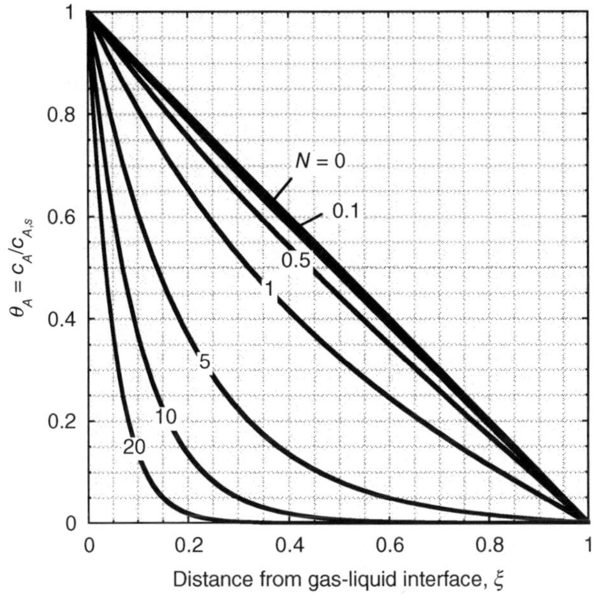

FIGURE 16.2 Concentration profiles of component *A* in the film predicted for a fast pseudo-first-order reaction for different values of $N = \sqrt{k_l\Delta^2/\mathcal{D}_A} = \sqrt{k_l\mathcal{D}_A/k_c^2}$. Mathematica notebook: (M.F.16.2).

From this equation, we can see that as the concentration profile at $x = 0$ becomes steeper, the flux increases because of the larger magnitude of $(dc_A/dx)_{x=0}$. In terms of the dimensionless variables θ_A and ξ, Eq. (16.22) can be written as follows:

$$N_A = -\frac{\mathcal{D}_A c_{A,s}}{\Delta}\frac{d\theta_A}{d\xi}\bigg|_{\xi=0} \tag{16.23}$$

Finally, taking the derivative $d\theta_A/d\xi$ from Eq. (16.21) we obtain:

$$\begin{aligned} N_A &= -\frac{\mathcal{D}_A c_{A,s}}{\Delta}\frac{d}{d\xi}\left(\frac{\sinh[N(1-\xi)]}{\sinh(N)}\right)\bigg|_{\xi=0} = \frac{\mathcal{D}_A c_{A,s} N}{\Delta}\frac{\cosh(N)}{\sinh(N)} \\ &= \frac{\mathcal{D}_A c_{A,s}}{\Delta}\frac{N}{\tanh(N)} \end{aligned} \tag{16.24}$$

We can now find an expression for the enhancement factor. Since, without reaction in the film, the mass-transfer flux is just $\mathcal{D}_A c_{A,s}/\Delta$, from Eq. (16.24) we have:

$$\phi = \frac{N_A}{\mathcal{D}_A c_{A,s}/\Delta} = \frac{N}{\tanh(N)} \tag{16.25}$$

Interestingly, this expression for ϕ is exactly the reciprocal of the function that describes the heat transfer efficiency of a straight fin in terms of the fin modulus, Eq. (4.27). This is, thus, another example where, remarkably, two processes that are physically very different end up being described by analogous functions.

Figure 16.3 shows a plot of Eq. (16.25) on log-log scales. The physical meaning of N can now be understood. As seen in Fig. 16.3, small values of N correspond to values of ϕ close to 1, while large values of N correspond to large values of ϕ. Based on Eq. (16.20), N^2 can be written as the ratio of the quantities $\Delta^2/\mathcal{D}_A$ and $1/k_I$. Both of these quantities have dimensions of time. The former is the time scale for diffusion of component A across the film (*cf.* Sec. 12.3.2) while the latter is the time scale for the reaction of A. As a result, N^2 can be viewed as the ratio of diffusion and reaction times. If the reaction is slow, the reaction time is long. In this case, N^2 and, thus, N, will be small and little if any enhancement of the rate of absorption of A will occur. Conversely, if the time scale for diffusion across the film is long compared to the reaction time, N^2 and, thus, N will be large. In this case, A will be consumed within the film resulting in a large enhancement of the rate of absorption.

We can see from Fig. 16.3 that for $N > 2.5$, Eq. (16.25) is closely approximated by:

$$\phi \sim N \tag{16.26}$$

In this case, the mass-transfer flux is given by

$$N_A = \frac{\mathcal{D}_A}{\Delta} c_{A,s}\phi \sim \frac{\mathcal{D}_A}{\Delta} c_{A,s} N = c_{A,s}\sqrt{k_I \mathcal{D}_A} \tag{16.27}$$

Note that this result is independent of the film thickness and, thus, independent of the mass transfer coefficient k_c. Mathematically, this is a consequence of the fact that the hyperbolic tangent of a large number is very close to one. Physically, this occurs because, when N is large, the reaction is fast compared to diffusion. In this case, the concentration

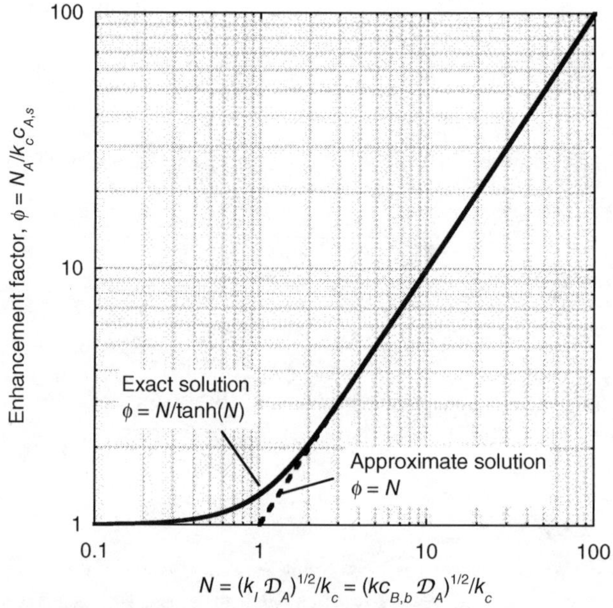

FIGURE 16.3 Enhancement factor for mass transfer with reaction in the fast pseudo-first-order reaction regime based on the film model for mass transfer.

of dissolved A will be reduced to zero very close to the gas-liquid interface, well before A will reach Δ. As a result, for these conditions, it will not really matter what the actual values of Δ or, equivalently, k_c are.

16.3.2 Instantaneous Reaction Regime

As sketched for case (c) of Fig. 16.1, when the reaction can be considered to be instantaneous (i.e., infinitely fast), both c_A and c_B will approach zero at a distance $x = \delta$ from the gas–liquid interface at a plane where the two species will meet and react. We refer to this as the reaction plane or reaction front. If the reaction is instantaneous, A and B cannot coexist. As a result, between the gas-liquid interface and the reaction plane, no B will be present and there will be only diffusion of A. Conversely, between the reaction plane and the bulk liquid, no A will be present and there will be only diffusion of B. In this case, the following flux expressions can be written for A and B, respectively:

$$N_A = \frac{\mathcal{D}_A}{\delta} c_{A,s} \tag{16.28}$$

$$N_B = \frac{\mathcal{D}_B}{\delta - \Delta} c_{B,b} \tag{16.29}$$

Note that N_A will be positive, since diffusion of A will be from the gas-liquid interface at $x = 0$ to the reaction plane at $x = \delta$. Conversely, N_B will be negative, since diffusion of B will be from the bulk liquid at $x = \Delta$ to the reaction plane at $x = \delta$. In order to satisfy the stoichiometry of Eq. (16.1), these two fluxes must be balanced as follows:

$$N_A = -\frac{1}{\upsilon} N_B \tag{16.30}$$

Combining this stoichiometric constraint with Eqs. (16.28) and (16.29) we obtain:

$$\upsilon \frac{\mathcal{D}_A}{\delta} c_{A,s} = \frac{\mathcal{D}_B}{\Delta - \delta} c_{A,b} \tag{16.31}$$

Solving this equation for δ we can find the location of the reaction plane, which is:

$$\frac{\delta}{\Delta} = \frac{1}{1 + \mathcal{D}_B c_{B,b} / \upsilon \mathcal{D}_A c_{A,s}} \tag{16.32}$$

The final result for the flux of A is obtained by substituting Eq. (16.32) into Eq. (16.28), which gives:

$$N_A = \frac{\mathcal{D}_A}{\Delta} c_{A,s} \left(1 + \frac{\mathcal{D}_B c_{B,b}}{\upsilon \mathcal{D}_A c_{A,s}} \right) \tag{16.33}$$

The enhancement factor in this regime is thus:

$$\phi = \frac{N_A}{\mathcal{D}_A c_{A,s} / \Delta} = 1 + \frac{\mathcal{D}_B c_{B,b}}{\upsilon \mathcal{D}_A c_{A,s}} \tag{16.34}$$

Note that Eq. (16.34) gives the maximum possible value of ϕ obtained when the reaction is infinitely fast.

16.3.3 Exact Solution

A solution of the simultaneous diffusion-reaction problem can be obtained for the general case by numerically solving Eqs. (16.13) to (16.15). In dimensionless form, these equations and boundary conditions can be written as follows:

$$\frac{d^2\theta_A}{d\xi^2} = N^2\theta_A\theta_B \tag{16.35}$$

$$\frac{d^2\theta_B}{d\xi^2} = \frac{N^2}{\beta}\theta_A\theta_B \tag{16.36}$$

$$\begin{aligned} &\xi = 0: \quad \theta_A = 1, \quad \frac{d\theta_B}{d\xi} = 0 \\ &\xi = \Delta: \quad \theta_A = 0, \quad \theta_B = 1 \end{aligned} \tag{16.37}$$

where $\theta_A = c_A/c_{A,s}$ and $\theta_B = c_B/c_{B,b}$ are the dimensionless concentrations of components A and B, respectively, $\xi = x/\Delta$ is the dimensionless distance in the film, $\beta = \mathcal{D}_B c_{B,b}/\upsilon\mathcal{D}_A c_{A,s}$, and $N = \sqrt{kc_{B,b}\Delta^2/\mathcal{D}_A}$. The solution of Eq. (16.35) to (16.37) will depend on the two parameters N and β that appear in the equations. Once the solution is obtained (which is not so trivial because this is a boundary value problem), the mass-transfer flux of A and the enhancement factor will be obtained from the following two equations:

$$N_A = -\frac{\mathcal{D}_A c_{A,s}}{\Delta}\frac{d\theta_A}{d\xi}\bigg|_{\xi=0} \tag{16.38}$$

and

$$\phi = \frac{N_A}{\mathcal{D}_A c_{A,s}/\Delta} = -\frac{d\theta_A}{d\xi}\bigg|_{\xi=0} \tag{16.39}$$

Sherwood et al. [*Mass Transfer*, McGraw-Hill, 1975, p. 330] provide the following approximate analytical expression for ϕ which is in good agreement with the numerical solution of these equations:

(M.16.40)

$$\phi = \frac{N\sqrt{1-(\phi-1)/\beta}}{\tanh[N\sqrt{1-(\phi-1)/\beta}]} \tag{16.40}$$

Although this equation is implicit in ϕ, it is still simpler to use than the numerical solution, which becomes challenging at high values of N. Figure 16.4 shows the enhancement factor calculated from Eq. (16.40) for a range of values of N and β. As seen from this figure, for a given value of β, the enhancement factor increases with N, as expected, because larger values of N correspond to a faster reaction and, thus, to a greater consumption of A near the gas-liquid interface. However, we can see that, for a given β, ϕ does not continue to increase with N to infinity. Rather, ϕ is limited to a plateau value, which, as seen in Fig. 16.4, is given by $\phi \sim 1+\beta$. This value coincides with Eq. (16.34), which is predicted for an instantaneous reaction. Example 16.1 compares the exact solution for ϕ with the limiting cases of fast pseudo-first-order and instantaneous reaction.

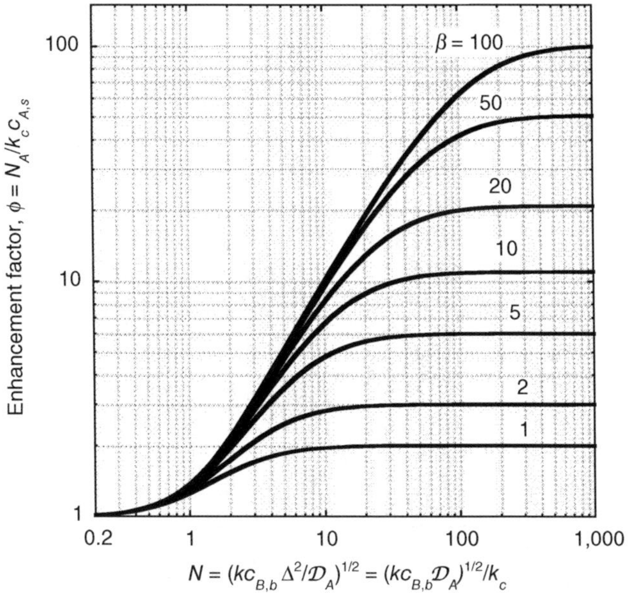

Figure 16.4 Enhancement factor for a bimolecular reaction near the gas-liquid interface. $\beta = \mathcal{D}_B c_{B,b}/\upsilon\mathcal{D}_A c_{A,s}$ based on the film model for mass transfer. Mathematica notebook: (M.E.16.40).

Example 16.1. Comparison of exact solution and limiting expressions for the enhancement factor. Let's consider, as an example, a case where $\beta = 20$. For the fast pseudo-first-order reaction regime, we have $\phi = N/\tanh(N)$ where $N = \sqrt{kc_{B,b}\mathcal{D}_A/k_c^2}$. For $N > 2.5$, this result is well approximated by $\phi \sim N$. Conversely, for the instantaneous reaction regime, we have $\phi = 1 + \beta$, which, in this case, is equal to 21. Figure 16.5 compares the exact solution for $\beta = 20$ with these two limiting cases of fast pseudo-first-order and instantaneous reaction over a range of values of N. As seen in this figure, there is an excellent agreement between the exact solution and those describing the limiting cases. For this case, the exact solution is nearly the same as that for the fast pseudo-first-order limit when $N < 10$. On the other hand, the exact solution is nearly the same as the instantaneous reaction limit when $N > 70$.

As seen from Example 16.1, in practice, the limiting solutions established for the fast pseudo-first-order and instantaneous reaction regimes provide reasonable estimates for ϕ over their respective ranges of validity. If we look at the intersections between the lines predicted by the limiting cases in Fig. 16.5, we can find the range of conditions that define the different reaction regimes. These ranges are defined in Table 16.1 along with the corresponding limiting expressions for ϕ. To use this table in practice, we first estimate N and β. We then determine in which reaction regime we fall according to the approximate range given. Within these ranges, we can use the corresponding approximate solution. Outside these ranges, we need to use the exact numerical solution, such as Eq. (16.40). Example 16.2 illustrates the use of these relationships.

Example 16.2. Mass transfer during the absorption of carbon dioxide in water and in sodium hydroxide solutions. Let's estimate the rate of absorption of CO_2 (component A) from air in pure water and in a 0.05 M ($= 5 \times 10^{-5}$ mol/cm^3) aqueous sodium hydroxide solution at 293 K and 1 atm total pressure for a case where the liquid-side mass transfer coefficient is $k_c = 0.02$ cm/s. Because of the low solubility of CO_2 in water, we will assume that the liquid-side resistance is controlling. In this case, the mole fraction of CO_2 in the gas will determine directly the concentration of dissolved CO_2 at the gas-liquid interface.

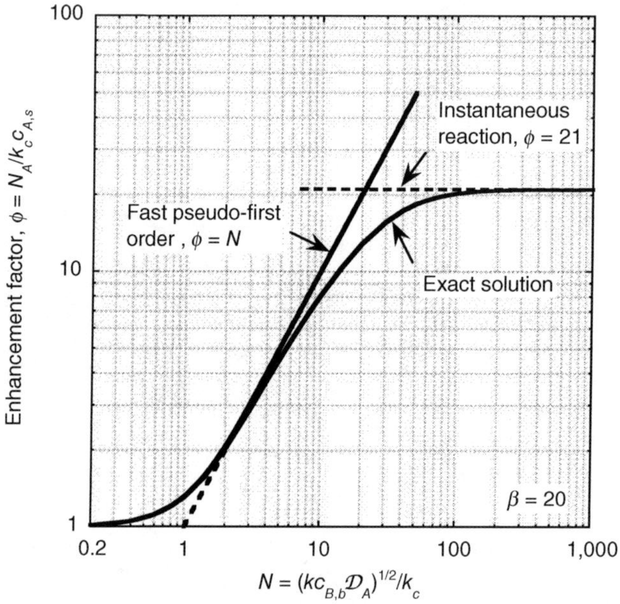

FIGURE 16.5 Comparison of exact and limiting solutions for a bimolecular reaction with $\beta = 20$ for Example 16.1.

Reaction Regime	Approximate Range	Enhancement Factor
Slow reaction	$N << 1$	$\phi \sim 1$
Fast pseudo-first order	$1 << N << \beta$	$\phi \sim N$
Instantaneous	$N >> \beta$	$\phi \sim 1 + \beta$

TABLE 16.1 Reaction Regimes for Mass Transfer with a Bimolecular Reaction

We will consider three different cases: a first case with a low gas-phase mole factions of CO_2, $y_A = 0.01$, a second case with a high gas-phase mole fraction, $y_A = 1.0$, and a third case with an intermediate gas-phase mole fraction, $y_A = 0.3$. Based on Example 13.3, at 293 K, the diffusivity of CO_2 in water is $\mathcal{D}_A = 1.7 \times 10^{-5}$ cm^2/s while the solubility of pure CO_2 in water at 1 atm pressure is 4.0×10^{-5} mol/cm^3. Thus, at 1 atm total pressure and a mole fraction y_A, the concentration of dissolved CO_2 right at the gas-liquid interface will be $c_{A,s} = 4.0 \times 10^{-5} \times y_A$ in units of mol/cm^3. Dissolved CO_2 reacts in alkaline solutions according to the following mechanism:

$$CO_2 + OH^- \rightarrow HCO_3^- \tag{16.41}$$

$$HCO_3^- + OH^- \rightarrow CO_3^{--} + H_2O \tag{16.42}$$

The overall reaction is thus:

$$CO_2 + 2OH^- \rightarrow CO_3^{--} + H_2O \tag{16.43}$$

Because reaction (16.42) is very fast, the first step, reaction (16.41), in the overall reaction will actually be controlling. As a result, the overall reaction rate will be described by the following equation:

$$R_A = -kc_A c_B \tag{16.44}$$

where $c_B = c_{OH^-}$ is the concentration of OH^-. Based on data in Sherwood et al. [*Mass Transfer*, McGraw-Hill, 1975, p. 363, Fig. 8.23], the second-order rate constant for this reaction at 293 K is $k = 5.4 \times 10^6$ cm³/mol s.

Let's consider first the case of absorption in pure water (i.e., without reaction). In this case, the mass-transfer flux will be

$$N_A = k_c c_{A,s} = 0.02 \times 4.0 \times 10^{-5} \times y_A$$

which gives values of $N_A = 8.0 \times 10^{-9}$ mol/cm² s when $y_A = 0.01$ and $N_A = 8.0 \times 10^{-7}$ mol/cm² s when $y_A = 1.0$. Next, let's consider the case of absorption in 0.05 M NaOH. In this case, we need to calculate the enhancement factor, which is determined by the values of the two parameters N and β. For N we have:

$$N = \sqrt{\frac{k c_{B,b} \mathcal{D}_A}{k_c^2}} = \sqrt{\frac{5.4 \times 10^6 \times 5.0 \times 10^{-5} \times 1.7 \times 10^{-5}}{0.02^2}} = 3.39$$

In order to calculate β, we need to estimate $\mathcal{D}_B = \mathcal{D}_{OH^-}$. From Table 15.1, after correcting for the effect of temperature and slightly higher viscosity at 293 K, we find that the limiting ionic diffusivity of OH^- in water is 4.6×10^{-5} cm²/s. In order to find $\mathcal{D}_{OH^-}$, we need to consider the effects of coupling of diffusion fluxes on OH^-, CO_3^{--}, and Na^+ according to the methods in Sec. 15.3. On this basis, at 293 K, Nijsing et al. [Chem. Eng. Sci. 10 (1959) 88] found a value of $\mathcal{D}_{OH^-} = 2.8 \times 10^{-5}$ cm²/s, which, not surprisingly, is intermediate between the limiting ionic diffusivities of the different ions in solution. Using this value of $\mathcal{D}_{OH^-}$, we obtain:

$$\beta = \frac{\mathcal{D}_B c_{B,b}}{\upsilon \mathcal{D}_A c_{A,s}} = \frac{2.8 \times 10^{-5} \times 5.0 \times 10^{-5}}{2 \times 1.7 \times 10^{-5} \times (4.0 \times 10^{-5} y_A)} = \frac{1.03}{y_A}$$

Thus, β will be equal to 103 when $y_A = 0.01$ and to 1.03 when $y_A = 1.0$. The calculation of ϕ will depend on the value of y_A according to the following two cases:

1. For $y_A = 0.01$ (i.e., for a very dilute gas), we will have $\beta = 103$, which is much larger than N. Based on Table 16.1, we will be in the fast, pseudo-first-order reaction regime. CO_2 will be consumed within the film but the OH^- concentration will remain essentially uniform and about equal to its value in the bulk liquid. Accordingly, in this case, we will have $\phi \sim N = 3.39$. The corresponding flux will be $N_A = \phi k_c c_{A,s} = 3.39 \times 8.0 \times 10^{-9} = 2.7 \times 10^{-8}$ mol/cm² s.

2. For $y_A = 1.0$, we will have $\beta = 1.03$, which is much smaller than N. Based on Table 16.1, we will be in the instantaneous reaction regime. Both CO_2 and OH^- will be consumed within the film, and the reaction will occur only at the reaction plane. Accordingly, in this case, we will have $\phi \sim 1 + \beta = 1 + 1.03 = 2.03$. The corresponding flux will be $N_A = \phi k_c c_{A,s} = 2.03 \times 8.0 \times 10^{-7} = 1.62 \times 10^{-6}$ mol/cm² s. Note that, in this case, the mass-transfer flux of CO_2 will no longer be proportional to y_A (as it was in case 1), since, now, y_A will appear both in the value of $c_{A,s}$ and in the expression for ϕ.

Let's now consider the third case where the CO_2 gas-phase mole fraction has an intermediate value $y_A = 0.3$. In this case, we will have $\beta = 3.43$, which is comparable in magnitude to N. Based on Table 16.1, we will now be in the transition between the fast pseudo-first-order and the instantaneous reaction regimes. In this case, we will obtain ϕ from Fig. 16.4 or, more precisely, from Eq. (16.40). The latter gives $\phi = 2.55$. The corresponding flux will be $N_A = \phi k_c c_{A,s} = 2.55 \times 0.02 \times 1.2 \times 10^{-5} = 6.1 \times 10^{-7}$ mol/cm² s. Figure 16.6 shows the dimensionless concentration profiles in the film predicted by the numerical solution of Eqs. (16.35) to (16.37) for the two cases with $y_A = 0.01$ and for $y_A = 0.3$. In both cases, as indicated by the curvature of the A (CO_2) concentration profile, θ_A, CO_2 will be consumed within the film, resulting in the enhanced mass transfer rate that we have predicted. However, different results are obtained for B. For $y_A = 0.01$, the B (OH^-) concentration profile, θ_B, is predicted to remain almost flat across the width of the film at $\theta_B \sim 1$. Conversely, for $y_A = 0.3$, B is predicted to also be consumed

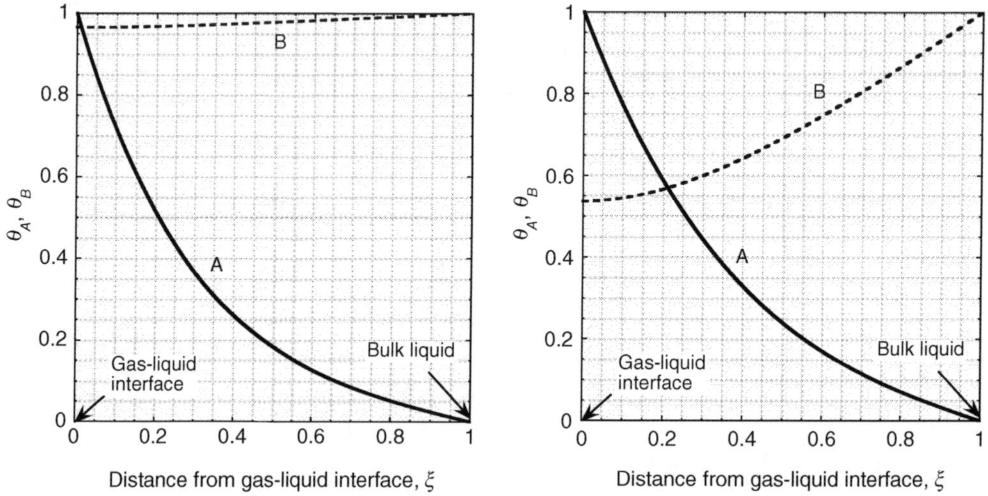

FIGURE 16.6 Concentration profiles in the film at the gas-liquid interface for Example 16.2 with $y_A = 0.01$, corresponding to $N = 3.39$ and $\beta = 103$ (left), and with $y_A = 0.3$, corresponding to $N = 3.39$ and $\beta = 3.43$ (right). Mathematica notebook: (M.F.16.6).

significantly near the gas-liquid interface at $\xi = 0$. In turn, this consumption reduces the reaction rate near the surface and, thus, limits the enhancement. The reason why a lower β leads to greater consumption of B at the surface is that lower values of β correspond to lower values of $c_{B,b}$ so that the reaction reduces B to a greater extent than when β is larger.

16.4 Expressions for the Enhancement Factor Based on the Penetration Model

The penetration model or penetration theory was introduced in Sec. 13.2.2 to describe the rate of mass transfer at the interface between a falling liquid film and an adjacent gas when the depth of penetration of the absorbed component in the liquid is small. In some cases, the penetration model may be a more realistic description of mass transfer compared to the simpler film model, such as when dealing with mass transfer in packed gas-absorption columns and other gas-liquid contactors. According to the penetration model, Eq. (13.84), in the absence of chemical reaction, the average flux of an absorbed component A at the gas-liquid interface is described by the following equation:

$$N_{A,\text{avg}} = \sqrt{\frac{4\mathcal{D}_A}{\pi t_E}}(c_{A,s} - c_{A,b}) \tag{16.45}$$

where t_E is the time of exposure of the liquid surface to the gas. Accordingly, the mass transfer coefficient is given by:

$$k_c = \sqrt{\frac{4\mathcal{D}_A}{\pi t_E}} \tag{16.46}$$

Although derived for transient diffusion in a laminar falling film, the penetration model can also be applied to more complex situations if t_E is treated as an empirical parameter as illustrated in Example 13.3. In this case, the idea is to use the model as a framework to predict mass transfer when, for example, a reaction takes place in the liquid. For instance, we can apply this model to the situation illustrated in Fig. 16.1 to describe the absorption of component A reacting with a liquid-phase reactant B. For this situation, t_E could be determined from Eq. (16.46) based on an experimentally determined value of k_c to develop a transient model for simultaneous diffusion and reaction. We can recognize again the possibility of three extreme scenarios depending on the relative magnitude of the rate of reaction and the rate of diffusion. The three cases are analogous to those conceptualized based on the film model and correspond to slow reaction, fast reaction, and instantaneous reaction scenarios. Figure 16.7 illustrates the expected concentration profiles of A and B corresponding to these three cases. The only substantial difference between these profiles and those expected based on the film model is that, regardless of reaction, the penetration model profiles will be transient and not described by straight lines.

The transient differential balances describing diffusion and reaction in the liquid are derived again from item A in Table 12.1 and are as follows, on a molar basis:

$$\frac{\partial c_A}{\partial t} = \mathcal{D}_A \frac{\partial^2 c_A}{\partial x^2} + R_A \tag{16.47}$$

$$\frac{\partial c_B}{\partial t} = \mathcal{D}_B \frac{\partial^2 c_B}{\partial x^2} + R_B \tag{16.48}$$

Considering again a bimolecular reaction occurring according to Eqs. (16.1) and (16.2), these equations become:

$$\frac{\partial c_A}{\partial t} = \mathcal{D}_A \frac{\partial^2 c_A}{\partial x^2} - kc_A c_B \tag{16.49}$$

$$\frac{\partial c_B}{\partial t} = \mathcal{D}_B \frac{\partial^2 c_B}{\partial x^2} - \upsilon kc_A c_B \tag{16.50}$$

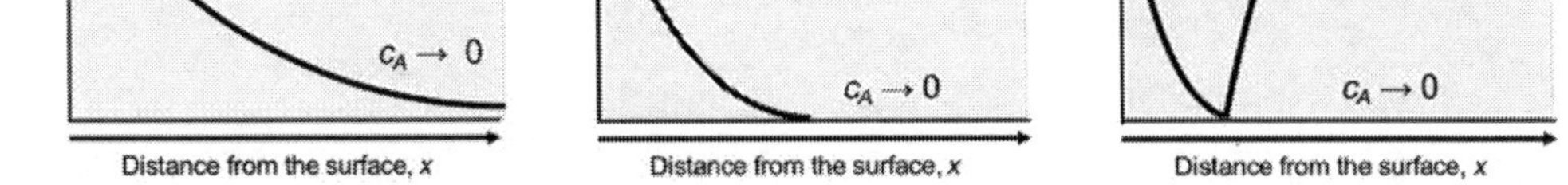

FIGURE 16.7 Concentration profiles of components A and B near the gas-liquid interface for the absorption of A with the bimolecular reaction $A + \upsilon B \to P$ based on the penetration model for mass transfer with a time of exposure t_E. (a) Profiles for a slow reaction, (b) profiles for a fast reaction, (c) profiles for an instantaneous reaction.

Note that these are now partial differential equations. If we assume that we start with no A in the liquid and that B is not volatile, initial and boundary conditions for these equations will be as follows:

$$\begin{aligned} t=0: &\quad c_A=0,\ c_B=c_{B,b} \\ x=0: &\quad c_A=c_{A,s},\ \frac{\partial c_B}{\partial x}=0 \\ x\to\infty: &\quad c_A\to 0,\ c_B\to c_{B,b} \end{aligned} \tag{16.51}$$

Let's now consider the three scenarios shown in Fig. 16.7. In the first case (a), the reaction is too slow to occur significantly near the gas-liquid interface during the time of exposure and, thus, does not affect the rate of mass transfer. In this case, Eq. (16.49) with $k=0$ is sufficient to describe the rate of absorption of A. From the solution of this equation we obtain:

$$N_{A,\text{avg}}=\sqrt{\frac{4\mathcal{D}_A}{\pi t_E}}\times c_{A,s} \tag{16.52}$$

This will be the rate of purely physical absorption of A. In the second case (b), the reaction is fast enough to consume A appreciably near the gas-liquid interface during the time of exposure. However, conditions are such that the concentration of B is still not significantly different from its value in the bulk liquid. Thus, everywhere we will have $c_B=c_{B,b}$. In this case, Eq. (16.49) and its relevant boundary conditions can be written as follows:

$$\frac{\partial c_A}{\partial t}=\mathcal{D}_A\frac{\partial^2 c_A}{\partial x^2}-k_I c_A \tag{16.53}$$

$$\begin{aligned} t=0: &\quad c_A=0 \\ x=0: &\quad c_A=c_{A,s} \\ x\to\infty: &\quad c_A\to 0 \end{aligned} \tag{16.54}$$

where $k_I=kc_{B,b}$ is again a pseudo-first-order rate constant. We do not need an equation for B in this case, since $c_B=c_{B,b}$. Analytical solution of Eq. (16.53) with the initial and boundary conditions given by Eq. (16.54) gives the following result for the flux of A at the gas-liquid interface at a particular time t:

$$N_A=\left[\text{erf}(k_I t)^{1/2}+\frac{\exp(-k_I t)}{(\pi k_I t)^{1/2}}\right](k_I\mathcal{D}_A)^{1/2}c_{A,s} \tag{16.55}$$

We are interested in finding the average flux during the time of exposure t_E so that we can compare it with Eq. (16.52) and find the enhancement factor. This average flux can be found by integrating Eq. (16.55) as follows:

$$N_{A,\text{avg}}=\frac{1}{t_E}\int_0^{t_E}N_A(t)dt \tag{16.56}$$

Substituting Eq. (16.55) into this expression and integrating gives the following result:

$$N_{A,\text{avg}} = \frac{1}{t_E}\int_0^{t_E}\left[\text{erf}(k_I t)^{1/2} + \frac{\exp(-k_I t)}{(\pi k_I t)^{1/2}}\right](k_I \mathcal{D}_A)^{1/2} c_{A,s}\,dt$$
$$= \left[\left(1+\frac{1}{2k_I t_E}\right)\text{erf}(k_I t_E)^{1/2} + \frac{\exp(-k_I t_E)}{(\pi k_I t_E)^{1/2}}\right](k_I \mathcal{D}_A)^{1/2} c_{A,s} \tag{16.57}$$

Therefore, the enhancement factor is:

$$\phi = \frac{N_{A,\text{avg}}}{k_c c_{A,s}} = \left[\left(1+\frac{1}{2k_I t_E}\right)\text{erf}(k_I t_E)^{1/2} + \frac{\exp(-k_I t_E)}{(\pi k_I t_E)^{1/2}}\right]\left(\frac{k_I \mathcal{D}_A}{k_c^2}\right)^{1/2} \tag{16.58}$$

Note that because of Eq. (16.46), the product $k_I t_E$ that appears in this equation is actually equal to $4k_I\mathcal{D}_A/\pi k_c^2 = 4N^2/\pi$ with $N = \sqrt{k_I\mathcal{D}_A/k_c^2}$. As a result, the enhancement factor can be expressed simply in terms of N as follows:

$$\phi = \left[\left(1+\frac{1}{8N^2/\pi}\right)\text{erf}\left(\frac{4N^2}{\pi}\right)^{1/2} + \frac{\exp(-4N^2/\pi)}{(4N^2)^{1/2}}\right]N \tag{16.59}$$

Let's examine this result. First, we find that although the penetration model is conceptually and mathematically different from the film model, the same parameter N will determine the magnitude of the enhancement factor. Second, although the mathematical result appears very different from that obtained based on the film model, Eq. (16.25), there are, in fact, striking similarities that can be seen by plotting these equations. Figure 16.8 compares the effects of N on ϕ predicted by Eq. (16.59), obtained from the penetration

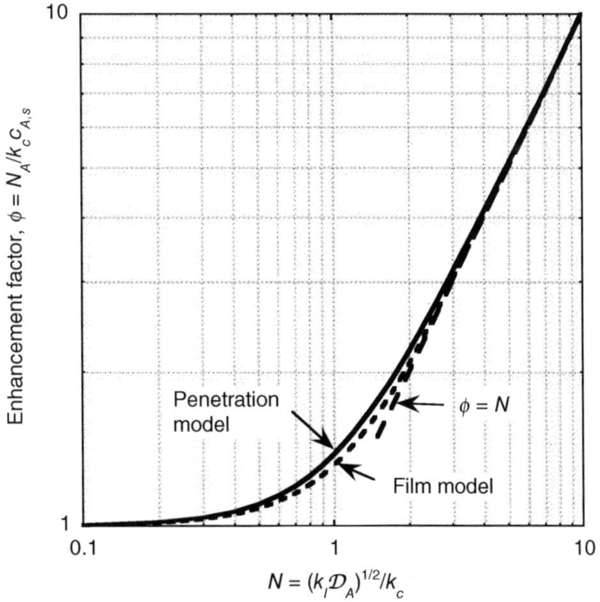

FIGURE 16.8 Comparison of film and penetration model predictions of the enhancement factor for mass transfer with a pseudo-first-order reaction.

model, with that predicted by Eq. (16.25), which was obtained from the film model. As seen in this figure, the two predictions are in remarkably close agreement. A major conclusion is that the simpler film model will generally be sufficient for practical calculations. In both cases, we can see that at values of $N > 2.5$, the solution is approximated very well by $\phi = N$. This can be seen also directly from Eq. (16.59). For large values of N, the term erf(N) will approach 1, while the terms with N in the denominator will approach 0. As a result, the whole term in brackets in Eq. (16.59) will approach 1. This agreement between the film model and the penetration model is not fortuitous. If we take a look at the original equations from which these results are derived, Eq. (16.16) for the film model and Eq. (16.53) for the penetration model, we see that the only difference between the two is the existence of the accumulation term $\partial c_A / \partial t$ in Eq. (16.53). This term quickly approaches zero as the time of exposure increases and a balance between diffusion of A into the liquid and its reaction with B is established leading to a steady-state concentration profile near the gas-liquid interface.

Let's now consider the third case (c) in Fig. 16.7 where the reaction can be considered to be instantaneous. In this case, while their concentration profiles are no longer linear, A and B still meet at a reaction plane located at a distance δ from the gas-liquid interface. The difference between the steady-state film model and the penetration model is that, in the penetration model, δ will, in general, vary with time. The solution of the penetration model for this case gives the following expression for the mass-transfer flux:

$$N_{A,\text{avg}} = \left(\sqrt{\frac{\mathcal{D}_A}{\mathcal{D}_B}} + \sqrt{\frac{\mathcal{D}_B}{\mathcal{D}_A}} \frac{c_{B,b}}{\upsilon c_{A,s}} \right) k_c c_{A,s} \tag{16.60}$$

The enhancement factor predicted for this case will thus be:

$$\phi = \frac{N_{A,\text{avg}}}{k_c c_{A,s}} = \sqrt{\frac{\mathcal{D}_A}{\mathcal{D}_B}} + \sqrt{\frac{\mathcal{D}_B}{\mathcal{D}_A}} \frac{c_{B,b}}{\upsilon c_{A,s}} \tag{16.61}$$

Obviously, if the diffusivities $\mathcal{D}_A$ and $\mathcal{D}_B$ are equal to each other, this result is exactly the same as that found for the film model, Eq. (16.34). In other cases, unless the diffusivities are very different from each other, film and penetration model results will not differ dramatically from each other since the square root of the diffusivity ratio will tend to dampen the effects of these differences. Example 16.3 illustrates a practical use of these relationships.

Another model used to describe mass transfer at gas–liquid interfaces is the surface renewal model introduced by Danckwerts in 1951 [Ind. Eng. Chem. 43 (1951) 1460]. This model assumes that there is a distribution of times of exposures of liquid elements at the gas-liquid interface that is described by the normal probability distribution function. The model predicts a mass transfer coefficient k_c that depends on the fractional rate of renewal of liquid elements at the surface. Even this model, when used to predict the enhancement factor for a bimolecular reaction, gives results that are nearly indistinguishable from those obtained from the simpler film model.

Example 16.3. Absorption of nitrogen dioxide in a laminar water jet. Kramers et al. [Chem. Eng. Sci. 14 (1961) 115] have studied the absorption of NO_2/N_2O_4 gas mixtures in water using a laminar jet apparatus at 20°C. In this apparatus, shown schematically in Fig. 16.9, liquid water flows out of an orifice into a gas phase forming a cylindrical jet with a diameter D. The flow rate of the liquid and the length of the jet will determine the time of exposure of the liquid to the surrounding gas phase. If the liquid jet is

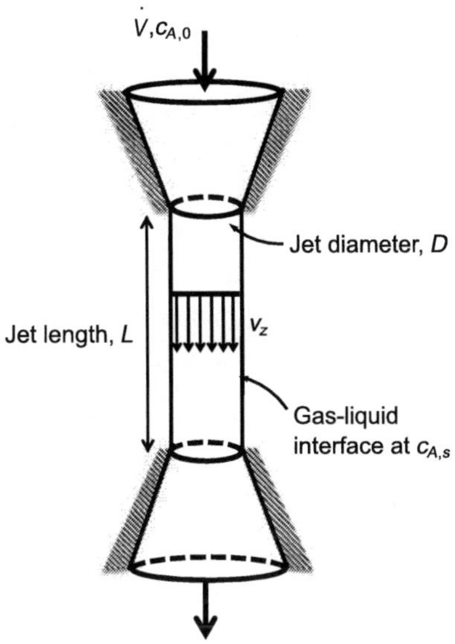

Figure 16.9 Laminar jet gas-absorption apparatus for Example 16.3.

laminar, mass transfer will occur by diffusion alone. In the experiments of Kramers et al. the gas phase contained a mixture of NO_2 and its dimer N_2O_4 in equilibrium with each other. While NO_2 does not dissolve or react appreciably with water, N_2O_4 formed in the gas phase dissolves in water and reacts according to the following reaction:

$$N_2O_4 + H_2O \rightarrow HNO_3 + HNO_2$$

If HNO_3 and HNO_2 do not accumulate significantly in the liquid, the reaction can be treated as being practically irreversible. In this case, since water is in excess, the reaction rate is described by the following equation:

$$R_{N_2O_4} = -k_l c_{N_2O_4}$$

where k_l is a first-order rate constant and $c_{N_2O_4}$ is the concentration of dissolved N_2O_4. At the gas-liquid interface, $c_{N_2O_4,s}$ will be related to the partial pressure of N_2O_4 in the gas, $p_{N_2O_4}$, by Henry's law, which gives:

$$c_{N_2O_4,s} = \frac{c}{H} p_{N_2O_4}$$

where c is the molar density of water and H is the Henry's law constant in units of pressure. Table 16.2 gives some of the results of the measurements of Kramers et al. obtained at $p_{N_2O_4} \sim 0.14$ atm using jets of different length with a diameter $D = 1$ mm and a liquid flow rate $\dot{V} = 2.56\ \text{cm}^3/\text{s}$.

Note that the rate of absorption of N_2O_4 observed in these experiments increases, as expected, with the length of the jet since a longer jet provides a greater gas-liquid interfacial area. However, as seen in this table, at the same time, the average flux decreases as the jet length is increased. Let's use the penetration model to explain this experimentally observed behavior and to determine both the solubility of N_2O_4 and the rate constant k_l from these data. Since the jet is cylindrical, the average mass-transfer flux of N_2O_4, given in Table 16.2, is calculated from the rate absorption based on the area of the gas-liquid interface, πDL, as follows:

$$N_{N_2O_4,\text{avg}} = \frac{\text{Rate of absorption}}{\pi DL}$$

Length of the Jet, L (cm)	Observed Rate of Absorption of N_2O_4 (mol/s)	Calculated Average Flux, $N_{N_2O_4,avg}$ (mol/cm² s)	Calculated Time of Exposure, t_E (s)
2.03	0.93×10^{-5}	1.45×10^{-5}	0.0062
4.00	1.57×10^{-5}	1.25×10^{-5}	0.012
8.01	2.92×10^{-5}	1.16×10^{-5}	0.025

TABLE 16.2 Experimental Data from Kramers et al. for the Absorption of N_2O_4 in Water in 1-mm-Diameter Laminar Water Jets with a Flow Rate of 2.56 cm³/s and $p_{N_2O_4} = 0.14$ atm

where L is the length of the liquid jet exposed to the gas. Neglecting friction at the gas-liquid interface, the velocity within the jet will be uniform and equal to the flow rate divided by the cross-sectional area of the jet according to:

$$v_z = \frac{\dot{V}}{\pi D^2/4} = \frac{2.56}{\pi \times 0.1^2/4} = 326 \text{ cm/s}$$

Thus, for this case, the time of exposure of the liquid to the gas will be:

$$t_E = \frac{L}{v_z} = \frac{L}{326}$$

where L is in cm and t_E is in s. The calculated values of t_E are shown in Table 16.2.

Based on the penetration model and, hence, assuming that the depth of penetration of N_2O_4 is small compared to the diameter of the liquid jet, Eq. (16.57) gives the following model for the case at hand:

$$N_{N_2O_4,avg} = \left[\left(1 + \frac{1}{2k_l t_E}\right)\text{erf}(k_l t_E)^{1/2} + \frac{\exp(-k_l t_E)}{(\pi k_l t_E)^{1/2}}\right](k_l \mathcal{D}_{N_2O_4})^{1/2} \frac{c}{H} p_{N_2O_4}$$

The diffusivity of N_2O_4 can be estimated from the Wilke-Chang equation, Eq. (10.24). Using a LeBas volume $\hat{V}_{N_2O_4} = 63.8$ cm³/mol, the diffusivity in water at 293 K is estimated to be $\mathcal{D}_{N_2O_4} = 1.23 \times 10^{-5}$ cm²/s. Since we know t_E and $p_{N_2O_4}$, our model for $N_{N_2O_4,avg}$ now contains only two unknown parameters, k_l and H. We could determine these parameters from a non-linear regression of the model equation to the experimental values of $N_{N_2O_4,avg}$. Alternatively, however, it is useful to note that if $k_l t_E > 1$, the error function term in brackets in this equation will become very close to 1 while the second term in brackets will become very close to 0. As a result, if $k_l t_E > 1$ the expression for the flux will be well approximated by:

$$N_{N_2O_4,avg} \sim \left(1 + \frac{1}{2k_l t_E}\right)(k_l \mathcal{D}_{N_2O_4})^{1/2} \frac{c}{H} p_{N_2O_4}$$

In this case, plotting the experimental values of $N_{N_2O_4,avg}$ vs. $1/t_E$ we will expect a straight line with slope equal to the quantity $(k_l \mathcal{D}_{N_2O_4})^{1/2} c p_{N_2O_4}/2k_l H$ and intercept equal to the quantity $(k_l \mathcal{D}_{N_2O_4})^{1/2} c p_{N_2O_4}/H$. The ratio of intercept and slope is thus expceted to be equal to $2k_l$. The Henry's law constant H can then be calculated from either the slope or the intercept. Figure 16.10 shows this plot. The corresponding parameter values, obtained by linear regression, are $k_l = 220$ s^{-1} and $c/H = 1.45 \times 10^{-3}$ mol/cm³ atm or $H = 0.0556/1.45 \times 10^{-3} = 38.3$ atm. We can verify that, based on the value of k_l obtained, for the conditions of these experiments, the product $k_l t_E$ was indeed larger than 1, which justifies using the approximation above.

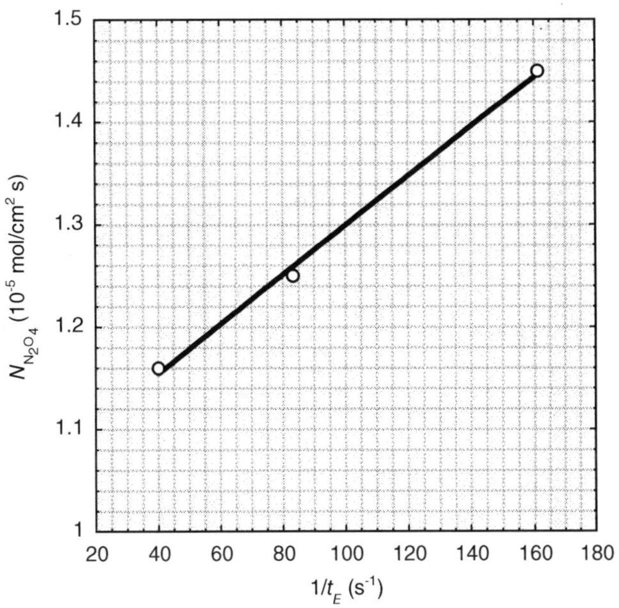

FIGURE 16.10 Experimental data for Example 16.3 (symbols) compared with the approximate expression for the flux.

16.5 Mass Transfer with Reaction with Generalized Kinetics

The developments in the previous sections are strictly applicable only to the irreversible kinetics described by Eq. (16.2). Nevertheless, the key concepts of slow, fast, and instantaneous reactions are still applicable to different reaction kinetics. In some cases, simple modifications of the theory developed based on Eq. (16.2) can be used to describe different situations. For instance, consider the case where the reaction kinetics is represented by the following nth order expression:

$$R_A = -k_n c_A^n \tag{16.62}$$

where k_n is an nth order rate constant. We can refer to this as a pseudo nth order kinetics since, in practice, k_n could still be a function of the concentration of a liquid-phase reactant B. With this rate equation, we could go through the solution of the corresponding equation describing diffusion and reaction and find a new expression for the enhancement factor. On the other hand, it has been shown that if $n > 1$ the solution obtained with first-order kinetics (i.e., with $n = 1$) can be used to describe diffusion coupled with the nth order reaction kinetics, Eq. (16.25), using the following modified expression for N:

$$N = \sqrt{\frac{2}{n+1} \frac{k_n (c_{A,s})^{n-1} \mathcal{D}_A}{(k_c)^2}} \tag{16.63}$$

If N is large, we will have:

$$\phi \sim N \tag{16.64}$$

Thus, the flux will be given by:

$$N_A \sim c_{A,s}\sqrt{\frac{2}{n+1}k_n(c_{A,s})^{n-1}\mathcal{D}_A} \tag{16.65}$$

This result implies that if $n > 1$, the mass-transfer flux will no longer be proportional to the concentration of A at the gas-liquid interface, as would be expected for a first-order reaction, but will vary, instead, with $(c_{A,s})^{(n+1)/2}$. Note that these relationships do not apply to zero-order reactions (i.e., with $n = 0$). Such reactions can be handled as described by Astarita, Chem. Eng. Sci. 17 (1962) 708.

For more complicated reaction kinetics, including those involving the practically important case of simultaneous absorption and reaction of two gases, and that of reversible reactions, the reader of this book is referred to specialized textbooks such as Sherwood et al., *Mass Transfer*, McGraw-Hill, 1975, Chap. 8, and Astarita, *Mass Transfer with Chemical Reaction*, Elsevier, 1967.

16.6 Mass Transfer with Reaction in a Packed Gas-Absorption Column

In Chap. 14, we addressed the calculation of the height of packing required for the absorption of a gaseous component A in a counter-current gas-absorption column in the absence of reaction; i.e., for systems where the amount and rate of absorption are limited by the physical dissolution of A into the liquid. For dilute conditions and with a linear equilibrium line describing the solubility of A in the liquid, we discovered that this calculation requires two distinct considerations. The first is the determination of the minimum liquid flow rate required. The second is the calculation of the packing height based on the number of transfer units and the height of a transfer unit. If instead of relying just on physical absorption, we use, as the absorbent, a reactive liquid or a liquid containing a reactive component B, both of these considerations are affected. First, if the reaction is irreversible and the concentration of B in the liquid remains above zero, the concentration of dissolved A in the bulk liquid will be kept at zero. As a result, the equilibrium line will no longer be limiting. Second, if the reaction is sufficiently fast, the rate of mass transfer of A will be enhanced relative to that observed for purely physical absorption. As a result, the height of a transfer unit will also be smaller than in the absence of reaction.

Let's consider the absorption column sketched in Fig. 16.11 assuming that conditions are dilute (i.e., y_A is, say, equal to 0.05 or less) and that the equilibrium line describing the physical dissolution of A in the liquid is straight and given by:

$$y_A = Kx_A \tag{16.66}$$

where y_A and x_A are the mole fractions of A in the gas and liquid phase, respectively. In this equation, $K = H(T)/P$, where $H(T)$ is the Henry's law constant in units of pressure. If the solubility of A in the liquid is low (i.e., if K is relatively large), the liquid-phase mass transfer resistance will often be controlling. In this case, the concentration of A dissolved in the

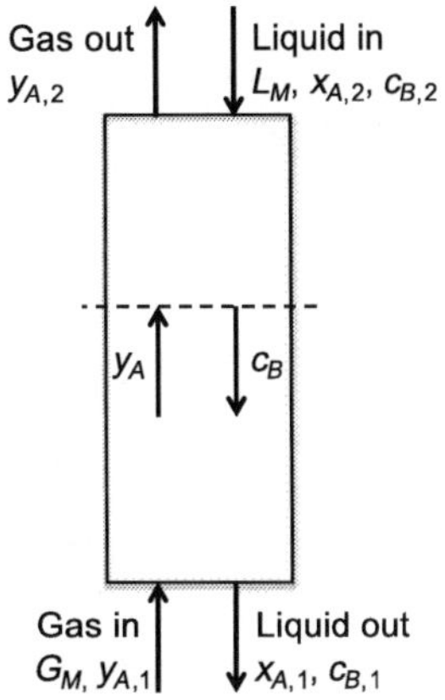

FIGURE 16.11 Schematic of a gas-absorption column for non-reactive and reactive absorption. For the reactive case, $x_{A,1} = x_{A,2} = 0$ if the reaction is irreversible.

liquid at the gas-liquid interface will be in equilibrium with the gas. Since $x_A = c_A/c_L$, where c_L is the molar density of the liquid, from Eq. (16.66), we obtain:

$$c_{A,s} = \frac{y_A}{K} c_L \tag{16.67}$$

c_L can be obtained from the mass density of the liquid remembering that $c_L = \rho_L / M_{r,L}$ where $M_{r,L}$ is the molecular mass of the liquid. As shown in Fig. 16.11, $y_{A,1}$ is the mole fraction of the absorbable component A in the feed gas while $c_{B,2}$ is the reactant concentration in the absorbent liquid fed to the column. Without reaction, e.g., if $c_{B,2} = 0$, we can calculate the minimum L/G from Eq. (14.26), which gives:

$$\left(\frac{L}{G}\right)_{\min} = \frac{y_{A,1} - y_{A,2}}{\dfrac{y_{A,1}}{K} - x_{A,2}} \tag{16.68}$$

The actual L/G will have to be larger than this minimum value. Next, we can calculate $N_{o,G}$ either from Eq. (14.55) or, more conveniently, from the Colburn equation, Eq. (14.56). The latter is:

$$N_{o,G} = \frac{1}{1-\gamma} \ln\left[(1-\gamma)\frac{y_{A,1} - Kx_{A,2}}{y_{A,2} - Kx_{A,2}} + \gamma\right] \tag{16.69}$$

where:

$$\gamma = \frac{KG}{L} = \frac{K}{L/G} \tag{16.70}$$

is the stripping factor. The next step is to calculate $H_{o,G} = \gamma H_L$, where $H_L = L/ak_x$ is the height of a liquid-phase mass transfer unit. Note that we neglect H_G here since we have assumed that the liquid-side resistance is controlling. H_L can be calculated, for example, from the Onda method, using Eq. (14.62) to estimate a and Eq. (14.64) to estimate k_x. Finally, we obtain the packing height as:

$$Z = N_{o,G} H_{o,G} \tag{16.71}$$

Let's now consider what happens if the dissolved A reacts irreversibly with a liquid-phase reactant B according to the stoichiometry and kinetics given by Eqs. (16.1) and (16.2), respectively. In this case, since B is consumed as A is absorbed, we need a material balance linking the consumption of B to that of A. A material balance from the top of the column to a point in the column at a height z from the bottom where the gas-phase mole fraction of A is y_A and the liquid-phase concentration of B is c_B, yields the following equation:

$$G_M(y_A - y_{A,2}) = \frac{1}{\upsilon}\frac{L_M}{c_L}(c_{B,2} - c_B) \tag{16.72}$$

Note that, for simplicity, we are using the symbol c_B in lieu of $c_{B,b}$ to denote the bulk concentration of the liquid-phase reactant. With some rearrangement, Eq. (16.72) can be written as follows:

$$c_B = c_{B,2} - \frac{\upsilon c_L G_M}{L_M}(y_A - y_{A,2}) = c_{B,2} - \frac{\upsilon c_L G}{L}(y_A - y_{A,2}) \tag{16.73}$$

This equation links directly the local values of the reactant concentration in the liquid, c_B, and the gas-phase mole fraction of component A, y_A.

Next, we consider the rate of absorption. If the reaction is fast enough to occur near the gas-liquid interface, mass transfer of A will be enhanced relative to the case of no reaction. Thus, the flux will be given by:

$$N_A = k_c c_{A,s}\phi(N,\beta) \tag{16.74}$$

where $\phi(N,\beta)$ is the enhancement factor, which, in turn, will be a function of the two parameters $N = \sqrt{kc_B\mathcal{D}_A/k_c^2}$ and $\beta = \mathcal{D}_B c_B/\upsilon\mathcal{D}_A c_{A,s}$. As already noted, the mass transfer coefficient k_c is related to k_x by $k_c = k_x/c_L$. Let's now write a differential balance for component A in the gas phase across a differential section of the column exactly like we did for the non-reactive case. With reference to Fig. 14.8, we can write:

$$G_M dy_A = -SaN_A dz = -Sak_c c_{A,s}\phi dz \tag{16.75}$$

where S is the column cross section and a is the gas-liquid interfacial area per unit column volume. Since $G = G_M/S$, we obtain:

$$dz = -\frac{Gc_L}{ak_x}\frac{dy_A}{c_{A,s}\phi} \tag{16.76}$$

Finally, if the liquid-side resistance is dominant, the dissolved concentration of A at the gas-liquid interface, $c_{A,s}$, will be locally in equilibrium with the bulk gas. In this case, $c_{A,s}$ and y_A will be related to each other by Eq. (16.67). Substituting this relationship into Eq. (16.76) we obtain:

$$dz = -\frac{GK}{ak_x}\frac{dy_A}{y_A\phi} \tag{16.77}$$

The packing height is then found by integrating Eq. (16.77) from one end of the column to the other, which gives:

$$Z = \frac{GK}{ak_x}\int_{y_{A,2}}^{y_{A,1}}\frac{dy_A}{y_A\phi} = H_L\frac{GK}{L}\int_{y_{A,2}}^{y_{A,1}}\frac{dy_A}{y_A\phi} \tag{16.78}$$

In order to integrate Eq. (16.78), we need to obtain ϕ at each point from Eq. (16.40) or from Fig. 16.4 as a function of N and β. Note that since, in general, c_B will vary along the packing height according to Eq. (16.73), both of these parameters will also vary along the length of the column. We can express these parameters as follows:

$$N=\sqrt{\frac{kc_B\mathcal{D}_A}{k_c^2}}=\sqrt{\frac{k\mathcal{D}_A\left[c_{B,2}-\frac{\upsilon c_L G}{L}(y_A-y_{A,2})\right]}{k_c^2}} \tag{16.79}$$

$$\beta=\frac{\mathcal{D}_B c_B}{\upsilon\mathcal{D}_A c_{A,s}}=\frac{\mathcal{D}_B\left[c_{B,2}-\frac{\upsilon c_L G}{L}(y_A-y_{A,2})\right]}{\upsilon\mathcal{D}_A y_A c_L/K} \tag{16.80}$$

In the special case where absorption occurs in the fast pseudo-first-order reaction regime throughout the height of the packing, we will have $\phi \sim N$. In this case, Eq. (16.78) will become:

$$\begin{aligned}Z&=\frac{GK}{ak_x}\int_{y_{A,2}}^{y_{A,1}}\frac{dy_A}{y_A\left\{k\mathcal{D}_A\left[c_{B,2}-\frac{\upsilon c_L G}{L}(y_A-y_{A,2})\right]\Big/k_c^2\right\}^{1/2}}\\&=\frac{GK}{ac_L\sqrt{k\mathcal{D}_A}}\int_{y_{A,2}}^{y_{A,1}}\frac{dy_A}{y_A\left[c_{B,2}-\frac{\upsilon c_L G}{L}(y_A-y_{A,2})\right]^{1/2}}\end{aligned} \tag{16.81}$$

Note that, in this reaction regime, the packing height will become independent of the liquid-side mass transfer coefficient, k_x or k_c. This will occur because, in this case, the flux will just be $N_A=k_c c_{A,s}\sqrt{kc_B\mathcal{D}_A/k_c^2}=c_{A,s}\sqrt{kc_B\mathcal{D}_A}$, which is independent of k_x or k_c. The gas-liquid interfacial area a will, however, continue to play a critical role in determining the mass transfer rate and, thus, the packing height. Example 16.4 illustrates an application of these relationships.

Example 16.4. Packing height for the absorption of CO_2 in water and in NaOH solutions [Mathematica notebook: (M.EX.16.4)]. Building on Example 16.2, let's consider the calculation of the height of packing needed for an absorption column packed with 2-in. metal Pall rings designed to remove CO_2 (A) from air at 293 K and 10.0 atm absolute pressure operated with a gas molar velocity $G=90$ mol/m² s (0.009 mol/cm² s). The feed air contains 5 mol% CO_2 ($y_{A,1}=0.05$) and we want to remove 95% of the CO_2. Neglecting the change in the gas total molar flow rate, the mole fraction of CO_2 in the air leaving the column will be $y_{A,2}=0.05\times(1-0.95)=0.0025$. We will assume that there is no dissolved CO_2 in the absorbent liquid entering the column. The physical solubility of CO_2 at 293 K and 1 atm absolute pressure is 4.0×10^{-5} mol/cm³. Thus, assuming that Henry's law is valid, at 10 atm absolute pressure we will have $c_{A,s}=4.0\times10^{-4}\times y_A$ in units of mol/cm³. The distribution coefficient, $K=y_A/x_A$, at this pressure and temperature, will thus be $K=c_L/4.0\times10^{-4}=0.0556/4.0\times10^{-4}=138$ where $c_L=0.0556$ mol/cm³ is the molar density of water. Since CO_2 is only sparingly soluble in water for these conditions, we will assume that the liquid-side resistance is dominant. The solutions for the two different cases of absorption in water and in a NaOH solution are as follows.

(a) Absorption in water without reaction. If there is no reaction, we calculate the $(L/G)_{\min}$ from Eq. (16.68), which gives:

$$\left(\frac{L}{G}\right)_{\min}=\frac{0.05-0.0025}{\frac{0.05}{138}-0}=131$$

Let's take the actual L/G to be 20% larger than this minimum, so that we will have L/G = 157 and L = 14,100 mol/m² s. Next, we can calculate $N_{o,G}$ from the Colburn equation using $\gamma = KG/L = 138 \times 90/14{,}100 = 0.881$, which gives:

$$N_{o,G} = \frac{1}{1-0.881}\ln\left[(1-0.881)\frac{0.05-0}{0.0025-0}+0.881\right] = 9.93$$

The next step is to estimate the gas-liquid interfacial area and the liquid-side mass transfer coefficient from the Onda method, Eqs. (14.62) and (14.64). The mass velocities are $G' = G \times M_{r,G} = 90 \times 29/1{,}000 = 2.61$ kg/m² s and $L' = L \times M_{r,L} = 14{,}100 \times 18/1{,}000 = 254$ kg/m² s. For 2-in. metal Pall rings, Table 14.13 in *Perry's Chemical Engineers' Handbook*, 9th edition, gives $a_p = 105$ m²/m³. With these values, from Eqs. (14.62) and (14.64) we obtain $a = 103$ m²/m³ and $k_x = 88.2$ mol/m² s. In turn, these values give $H_L = 1.54$ m and $H_{o,G} \sim \gamma H_L = 0.881 \times 1.54 =$ 1.36 m. The final result is:

$$Z = H_{o,G}N_{o,G} = 1.36 \times 9.93 = 13.5 \text{ m}$$

(b) Absorption with reaction. If we use a NaOH solution as the absorbent liquid, the CO_2 concentration in the bulk liquid will be kept at essentially zero since the reaction is practically irreversible for these conditions. This means that we will be able to use a much smaller liquid flow rate compared to the case without reaction. For example, if we choose $c_{B,2}$ = 0.5 mol/L (= 5.0×10^{-4} mol/cm³) and a liquid molar velocity of 1,410 mol/m² s (10 times smaller than in the case of no reaction), from Eq. (16.73), the outlet concentration of OH^- in the liquid leaving the column will be:

$$c_{B,1} = c_{B,2} - \frac{\upsilon c_L G}{L}(y_{A,1} - y_{A,2}) = 5.0 \times 10^{-4} - \frac{2 \times 0.0556 \times 90}{1{,}410}(0.05 - 0.0025) = 1.63 \times 10^{-4} \text{ mol/cm}^3$$

This value is still much higher than the dissolved CO_2 concentration at the gas-liquid interface at the bottom of the column, which is $c_{A,s} = 4.0 \times 10^{-4} \times 0.05 = 2.0 \times 10^{-5}$ mol/cm³. This will ensure that there will still be a large driving force for CO_2 absorption even at the bottom of the column since depletion of OH^- will be moderate. Next, let's establish the reaction regime. At the liquid mass velocity of 1,410 mol/m² s that we have chosen, the Onda method predicts $a = 86.5$ m²/m³ (0.865 cm²/cm³) and $k_x =$ 21.4 mol/m² s or 2.14×10^{-3} mol/cm² s. In turn, we will have $k_c = k_x/c_L = 2.14 \times 10^{-3}/0.0556 = 0.0385$ cm/s. We can now calculate N and β. At the bottom of the column, the values of these parameters will be:

$$N = \sqrt{\frac{kc_{B,1}\mathcal{D}_A}{k_c^2}} = \sqrt{\frac{5.4 \times 10^6 \times 1.63 \times 10^{-4} \times 1.7 \times 10^{-5}}{(0.0385)^2}} = 3.21$$

and:

$$\beta = \frac{\mathcal{D}_B c_{B,1}}{\upsilon \mathcal{D}_A c_{A,s,1}} = \frac{2.8 \times 10^{-5} \times 1.63 \times 10^{-4}}{2 \times 1.7 \times 10^{-5} \times 2.0 \times 10^{-5}} = 6.86$$

while at the top of the column, we will have:

$$N = \sqrt{\frac{kc_{B,2}\mathcal{D}_A}{k_c^2}} = \sqrt{\frac{5.4 \times 10^6 \times 5.0 \times 10^{-4} \times 1.7 \times 10^{-5}}{(0.0385)^2}} = 5.56$$

and:

$$\beta = \frac{\mathcal{D}_B c_{B,2}}{\upsilon \mathcal{D}_A c_{A,s,2}} = \frac{2.8 \times 10^{-5} \times 5.0 \times 10^{-4}}{2 \times 1.7 \times 10^{-5} \times 1.0 \times 10^{-6}} = 412$$

As seen from these values, at both ends of the column we will have $1 < N < \beta$. Thus, based on Table 16.1, we will be in the fast pseudo-first-reaction regime throughout the height of the packing. In this case, we will be able to approximate ϕ as being equal to N. The final result for the packing height is obtained from Eq. (16.81). The integral in this equation, evaluated numerically, has a value of 152 in units of $(cm^3/mol)^{1/2}$. Thus, the packing height will be:

$$Z = \frac{90 \times 10^{-4} \times 138}{0.865 \times 0.0556\sqrt{5.4 \times 10^6 \times 1.7 \times 10^{-5}}} \times 152 = 410 \text{ cm} = 4.1 \text{ m}$$

As seen from this example, the reaction with NaOH allows using just 1/10th of the flow of absorbent liquid and less than 1/3rd of the packing height compared to the case without reaction. This will happen because, on the one hand, the reaction will remove the physical equilibrium limitation due to the low solubility of CO_2 while, on the other hand, it will enhance the rate of mass transfer by virtue of its occurrence near the gas-liquid interface.

16.7 Problems

P.16.1. The following data were obtained for the absorption of a sparingly soluble gas A at 20°C into an aqueous solution of a non-volatile reactant B in a laminar jet apparatus as described in Example 16.3. The time of exposure was 0.1 s and the irreversible reaction $A + B \rightarrow P$ occurred. The reaction kinetics is first order in A and first order in B, the solubility of A is 1.0×10^{-3} mol/cm³ atm, and the diffusivity of B in water at infinite dilution is 2.0×10^{-5} cm²/s. The gas-side resistance is negligible. Based on these data determine: (a) The diffusivity of A in pure water. (b) The rate constant for the chemical reaction. Make sure you verify the reaction regime. (c) The enhancement factor and the specific rate of absorption of A (N_A) at 20°C if the partial pressure of A is increased to 1.0 atm, the concentration of B is 4.0×10^{-3} mol/cm³, and the time of exposure is 0.01 s.

Experiment #	1	2	3
Time of exposure, t_E (s)	0.1	0.1	0.2
Partial pressure of A, p_A (atm)	0.01	0.01	0.01
Concentration of B, $c_{B,b}$ (mol/cm³)	0	2.0×10^{-3}	2.0×10^{-3}
Specific rate of absorption, N_A (mol/cm² s)	0.18×10^{-6}	2.1×10^{-6}	2.1×10^{-6}

P.16.2. A counter-current gas-absorption column is to be designed to remove CO_2 from an inert gas stream by reaction with an amine solution at 20°C. At this temperature, CO_2 reacts with the amine according to the following irreversible reaction:

$$CO_2 + 2RNH_2 \rightarrow RNHCOO^- + RNH_3^+$$

The reaction kinetics is the first order in CO_2 and the first order in the amine, and the second-order rate constant is 10,200 L/mol s. The gas entering the column contains 10 mol% CO_2 while the gas leaving the column contains 0.1 mol% CO_2. The concentration of RNH_2 in the absorbent liquid is 1.75 mol/L at the top of the column and 1.00 mol/L at the bottom. The absolute pressure in the column is 10.0 atm. The following data can be used for this problem:

Solubility of CO_2 in the amine solution = 2.5×10^{-5} mol/cm³ atm

Liquid-side mass transfer coefficient = 0.02 cm/s

Gas-liquid interfacial area = 1.5 cm²/cm³

Diffusivity of CO_2 in the amine solution = 1.4×10^{-5} cm²/s

Diffusivity of RNH_2 in solution = 0.70×10^{-5} cm²/s

Using these data and neglecting the gas-side mass transfer resistance, determine the rate of absorption of CO_2 per unit column volume at the top and the bottom of the column.

P.16.3. Measurements of the rate of absorption of CO_2 in water and in aqueous NaOH solutions are to be used to determine the mass transfer coefficient, k_c, and the gas-liquid interfacial area, a, in a counter-current laboratory-scale column packed with a new proprietary packing. The column has a diameter of 0.15 m, contains 0.75 m of packing, and is operated at 30°C and 1 atm absolute pressure. Air containing 5.0 mol% CO_2 is supplied at the rate of 0.31 mol/s. The liquid is supplied at the rate of 4.2 mol/s. Two experiments are conducted. In the first, the liquid fed is pure water while in the second it is an aqueous solution containing 3 mol/L NaOH. At steady-state, the following data are obtained:

Experiment	NaOH Concentration (mol/L)	Mole Fraction of CO_2 in the Outlet Gas	Mole Fraction of CO_2 in the Outlet Liquid
1	0	~0.05	2.1×10^{-5}
2	3.0	0.01	ND

(a) Determine the product of the gas-liquid interfacial area and the liquid-side mass transfer coefficient ak_c in units of s^{-1} from experiment 1. (b) Determine the gas-liquid interfacial area a in units of m² per m³ of column from experiment 2 assuming that absorption occurs in the fast pseudo-first-order reaction regime. Verify that this assumption is appropriate. (c) Combine the results in (a) and (b) to obtain the liquid-side mass transfer coefficient in units of m/s. The following additional data can be used for this problem: physical solubility of CO_2 in water and NaOH solutions = 5.0×10^{-4} mol CO_2/mol atm; diffusivity of dissolved $CO_2 = 1.3 \times 10^{-9}$ m²/s; rate constant for reaction of CO_2 with $OH^- = 2 \times 10^4$ L/mol s. The gas-side resistance and the change in OH^- concentration due to absorption of CO_2 in experiment 2 can be neglected.

P.16.4. Hydrogen sulfide (H_2S) is to be removed from natural gas by absorption into amine solutions in a process known as natural gas sweetening. Experiments were done to measure the rate of absorption of pure H_2S at 30°C and 1 atm absolute pressure in a laminar liquid jet apparatus (see Example 16.3) using either water or an aqueous solution containing 1.0×10^{-3} mol/cm³ of an amine. The jet is 5 cm long and has a dimeter of 0.1 cm. The liquid flow rate is 4.0 cm³/s. H_2S reacts with the amine according to the reaction:

$$H_2S + R_2NH \rightarrow R_2NH_2^+ + HS^-$$

The reaction is essentially infinitely fast and the physical solubility of H_2S in water and in the amine solution is 8.0×10^{-5} mol/cm³ atm. The rate of absorption of H_2S was determined to be 0.14 cm³/s for the water jet and 1.5 cm³/s for the amine jet. Based on these data: (a) Determine the mass transfer coefficient and diffusivity of H_2S in water. (b) Determine the enhancement factor for absorption in the amine solution and the diffusivity of the amine. (c) Suppose that the results are used to predict the rate of absorption in a gas-absorption column operated at 10 atm absolute pressure and 30°C at a point where the mole fraction of H_2S in the gas phase is 0.01, the amine concentration in the liquid is 1.0×10^{-3} mol/cm³, and the liquid-side and gas-side mass transfer coefficients are $ak_x = 6.0 \times 10^{-4}$ mol/cm³ s and $ak_y = 1.5 \times 10^{-3}$ mol/cm³ s, respectively. What is the rate of absorption in mol/cm³ s for these conditions if the gas-side resistance is neglected? What is the rate of absorption in mol/cm³ s for these conditions if the gas-side resistance is considered?

P.16.5. Oxygen reacts irreversibly in aqueous sodium sulfate solutions according to the following reaction kinetics:

$$R_{O_2} = k_2(C_{O_2})^2$$

where k_2 is a second-order rate constant. The reaction is catalyzed by cobalt ions in solution. Thus, k_2 depends on the sodium sulfite concentration as well as on the concentration of cobalt. For 0.8 mol/L sodium sulfite solutions at 30°C, based on data in Sherwood et al. [*Mass Transfer*, McGraw-Hill, 1975, p. 373–375] we obtain $k_2 = 2.5 \times 10^{16} C_{Co}$ where k_2 is in cm^3/mol s and C_{Co} is the cobalt concentration in mol/cm^3. The physical solubility of O_2 in 0.8 mol/L sodium sulfite is 0.58×10^{-6} mol/cm^3 atm and the diffusivity is 1.2×10^{-5} cm^2/s. (a) Estimate the rate of absorption of pure O_2 at 1 atm in 0.8 mol/L sodium sulfite in a single-sphere absorber as illustrated in Example 13.3 if the solution contains no cobalt. The sphere radius is $R = 2$ cm and the flow rate of the liquid is 1.0 cm^3/s. (b) Estimate the rate of absorption for the same conditions in (a) if the solution contains a cobalt concentration of 0.6×10^{-7} mol/cm^3. Verify that the reaction regime is fast pseudo-first-order.

Appendix A

Physical Properties of Liquid Water and Air

The analysis and solution of problems involving heat and mass transfer processes require extensive physical property data. Representative data and estimation methods for thermal conductivities and diffusivities were provided in Chaps. 2 and 10, respectively. However, additional thermophysical properties are needed to solve a variety of problems. The physical properties of pure substances are widely available nowadays in a variety of printed and electronic resources. Of particular relevance are the data collections in the Design Institute for Physical Properties (DIPPR) of the American Institute of Chemical Engineers (AIChE), DIPPR 801 Critically Evaluated Gold Standard™ Database, Design Institute for Physical Properties, AIChE, New York, NY, USA (2016), which contains data for more than 2,000 compounds. Selected data are also readily available in Section 2 of *Perry's Chemical Engineers' Handbook*, 9th edition, both in print and electronically through the Access Engineering Library. For the solution of practical problems, the reader of this book is thus generally referred to these extensive and widely available compilations of physical property data. However, selected data for liquid water and for air at 1 atm absolute pressure are also given in the following tables since these systems are encountered so frequently. Wolfram Mathematica notebooks are provided in this book's companion website to calculate the properties listed in these tables as a function of temperature.

Inspection of the data included in Tables A.1 and A.2 shows that several approximations are possible, especially when they are used for the estimation of transport coefficients, which is, itself, affected by considerable uncertainties. We can see that for liquid water, density, heat capacity, thermal conductivity, and latent heat of vaporization vary by less than 20% over the temperature range 273–373 K. Thus, using average values of $\rho = 980$ kg/m^3, $C_p = 4{,}200$ J/kg K, $k = 0.63$ W/m K, and $\Delta\hat{H}_v = 2.4 \times 10^6$ J/kg will be satisfactory in many cases within this range of temperatures. Conversely, both the viscosity, μ, and, especially, the vapor pressure, p^*, will vary substantially with temperature requiring the use of precise values. Note that the Prandtl number for water also varies substantially over this temperature range primarily as a result of the effect of temperature on viscosity. For air at 1 atm absolute pressure, between 273 and 973 K the density, ρ, decreases in inverse proportion with absolute temperature, while C_p and Pr are nearly constant with temperature. Over a more modest temperature range between 273 and 373 K at 1 atm absolute pressure, density, heat capacity, viscosity, thermal conductivity, and Prandtl number of air vary by less than 32% around average values of $\rho = 1.1$ kg/m^3, $C_p = 1{,}000$ J/kg K, $\mu = 2.0 \times 10^{-5}$ Pa s, $k = 0.028$ W/m K, and Pr = 0.71. These average values can be used, with caution, for quick, order-of-magnitude estimates even outside this narrow temperature range.

T (K)	ρ (kg/m^3)	C_p (J/kg K)	μ (Pa s)	k (W/m K)	Pr	$\Delta\hat{H}_v$ (J/kg)	p^* (Pa)
273.2	993.3	4,226	0.001693	0.5672	12.61	2.496×10^6	612
278.2	993.5	4,214	0.001476	0.5757	10.80	2.485×10^6	875
283.2	993.3	4,204	0.001295	0.5839	9.32	2.474×10^6	1,232
288.2	992.8	4,195	0.001144	0.5917	8.11	2.463×10^6	1,711
293.2	991.9	4,189	0.001016	0.5992	7.10	2.452×10^6	2,347
298.2	990.8	4,183	0.000908	0.6064	6.26	2.441×10^6	3,180
303.2	989.4	4,179	0.000815	0.6132	5.56	2.429×10^6	4,260
308.2	987.9	4,177	0.000736	0.6198	4.96	2.418×10^6	5,646
313.2	986.1	4,175	0.000668	0.6260	4.45	2.406×10^6	7,406
318.2	984.1	4,175	0.000608	0.6319	4.02	2.395×10^6	9,621
323.2	981.9	4,176	0.000557	0.6375	3.65	2.383×10^6	12,382
328.2	979.6	4,177	0.000512	0.6428	3.32	2.371×10^6	15,798
333.2	977.2	4,179	0.000472	0.6477	3.05	2.359×10^6	19,986
338.2	974.6	4,182	0.000437	0.6524	2.80	2.347×10^6	25,086
343.2	971.8	4,186	0.000406	0.6567	2.59	2.335×10^6	31,248
348.2	968.9	4,190	0.000378	0.6608	2.40	2.322×10^6	38,645
353.2	965.9	4,195	0.000354	0.6645	2.23	2.310×10^6	47,464
358.2	962.8	4,200	0.000332	0.6679	2.09	2.297×10^6	57,915
363.2	959.5	4,206	0.000312	0.6711	1.96	2.284×10^6	70,225
368.2	956.1	4,212	0.000294	0.6739	1.84	2.271×10^6	84,644
373.2	952.6	4,218	0.000278	0.6764	1.74	2.258×10^6	101,441

TABLE A.1 Selected Physical Properties of Liquid Water between 273.2 K and 373.2 K. [Based on Data in Tables 2-8, 2-32, 2-69, 2-72, 2-135, and 2-147 of *Perry's Chemical Engineers' Handbook*, 9th edition. Mathematica notebook: (M.T.A.1)]

T (K)	ρ (kg/m^3)	C_p (J/kg K)	μ (Pa s)	k (W/m K)	Pr	$g\hat{\beta}\rho^2/\mu^2$ (1/m^3 K)
273.2	1.2919	1,001	1.724×10^{-5}	0.02416	0.714	2.02×10^{8}
298.2	1.1836	1,001	1.846×10^{-5}	0.02598	0.711	1.35×10^{8}
323.2	1.0920	1,002	1.963×10^{-5}	0.02775	0.709	9.39×10^{7}
348.2	1.0136	1,004	2.075×10^{-5}	0.02948	0.706	6.72×10^{7}
373.2	0.9457	1,006	2.184×10^{-5}	0.03118	0.704	4.93×10^{7}
398.2	0.8863	1,008	2.288×10^{-5}	0.03285	0.702	3.69×10^{7}
423.2	0.8340	1,012	2.390×10^{-5}	0.03450	0.701	2.82×10^{7}
448.2	0.7875	1,015	2.488×10^{-5}	0.03612	0.699	2.19×10^{7}
473.2	0.7459	1,019	2.584×10^{-5}	0.03771	0.698	1.73×10^{7}
498.2	0.7084	1,024	2.677×10^{-5}	0.03929	0.698	1.38×10^{7}
523.2	0.6746	1,029	2.767×10^{-5}	0.04085	0.697	1.11×10^{7}
548.2	0.6438	1,034	2.855×10^{-5}	0.04238	0.697	9.09×10^{6}
573.2	0.6157	1,040	2.941×10^{-5}	0.04391	0.697	7.50×10^{6}
598.2	0.5900	1,046	3.026×10^{-5}	0.04541	0.697	6.23×10^{6}
623.2	0.5663	1,052	3.108×10^{-5}	0.04690	0.697	5.23×10^{6}
648.2	0.5445	1,058	3.188×10^{-5}	0.04838	0.697	4.41×10^{6}
673.2	0.5243	1,064	3.267×10^{-5}	0.04984	0.697	3.75×10^{6}
698.2	0.5055	1,070	3.344×10^{-5}	0.05129	0.697	3.21×10^{6}
723.2	0.4880	1,076	3.420×10^{-5}	0.05273	0.698	2.76×10^{6}
748.2	0.4717	1,082	3.494×10^{-5}	0.05416	0.698	2.39×10^{6}
773.2	0.4565	1,088	3.567×10^{-5}	0.05557	0.698	2.08×10^{6}
798.2	0.4422	1,093	3.639×10^{-5}	0.05698	0.698	1.81×10^{6}
823.2	0.4287	1,099	3.709×10^{-5}	0.05837	0.698	1.59×10^{6}
848.2	0.4161	1,105	3.778×10^{-5}	0.05976	0.698	1.40×10^{6}
873.2	0.4042	1,110	3.846×10^{-5}	0.06113	0.699	1.24×10^{6}
898.2	0.3929	1,115	3.914×10^{-5}	0.06250	0.698	1.10×10^{6}
923.2	0.3823	1,121	3.980×10^{-5}	0.06386	0.698	9.80×10^{5}
948.2	0.3722	1,126	4.045×10^{-5}	0.06520	0.698	8.76×10^{5}
973.2	0.3627	1,131	4.109×10^{-5}	0.06655	0.698	7.85×10^{5}

Table A.2 Selected Physical Properties of Air at 1 atm Absolute Pressure Between 273.2 K and 973.2 K. [Based on Data in Tables 2-75, 2-138, and 2-145 of *Perry's Chemical Engineers' Handbook*, 9th edition. Mathematica notebook: (M.T.A.2)]

Appendix B

Unit Conversion Factors and Physical Constants

Comprehensive tables of unit conversion factors can be found in many printed and on-line sources, including Section 1 in *Perry's Chemical Engineers' Handbook*, 9th edition. The following tables provide some commonly used factors and physical constants that are encountered in the examples and chapter problems included in this book.

Area	ft^2	× 0.09290 =	m^2
Density	lb/ft^3	× 16.02 =	kg/m^3
Energy	BTU	× 1,055 =	J
Force	lb ft/s^2	× 0.1383 =	N (kg m/s^2)
Heat capacity	BTU/lb°F kcal/kg K	× 4,186 = × 4,186 =	J/kg K
Heat transfer coefficient	BTU/h ft^2°F	× 5.678 =	W/m^2 K
Kinematic viscosity, thermal diffusivity, diffusivity	ft^2/h	× 0.00002581 =	m^2/s
Length	ft	× 0.3048 =	m
Mass	lb	× 0.4536 =	kg
Moles	lbmol	× 0.4536 = × 453.6 =	kmol mol
Molecular mass	lb/lbmol, g/mol	× 1 = × 0.001 =	kg/kmol kg/mol
Power	BTU/h	× 0.2931 =	W (J/s)
Pressure, shear stress	lbf/in.^{2} (psi) atm mmHg at 0°C in. H_2O at 60°F	× 6,895 = × 101,300 = × 133.3 = × 248.8 =	Pa (kg/m s^2)
Thermal conductivity	BTU/h ft°F	× 1.731 =	W/m K
Time	h	× 3,600 =	s
Velocity	ft/h	× 0.00008467 =	m/s
Viscosity	lb/ft h cp (centipoise)	× 0.0004134 = × 0.001 =	Pa s (kg/m s)
Volume	ft^3 L (liter)	× 0.02832 = × 0.001 =	m^3

TABLE B.1 Commonly Used Unit Conversion Factors

Degrees Fahrenheit (°F) to Celsius (°C)	°C = (°F − 32)/1.8
Degrees Celsius (°C) to Kelvin (K)	K = °C + 273.2
Degrees Fahrenheit (°F) to Kelvin (K)	K = (°F − 32)/1.8 + 273.2

*Note: An increment of 1°C is equal to an increment of 1 K and to an increment of 1.8°F.

TABLE B.2 Commonly Used Temperature Unit Conversions*

Avogadro's number	$\mathcal{N}$	6.022×10^{23} molecules/mol
Boltzmann constant	k_B	1.381×10^{-23} J/molecule K
Faraday's constant	$\mathcal{F}$	96,500 Coulomb/mol
Gravity constant	g	9.807 m/s^2 32.17 ft/s^2
Ideal gas constant	$\mathcal{R}$	8.314 J/mol K 0.08206 atm L/mol K 82.06 atm cm^3/mol K
Stefan-Boltzmann constant	σ	5.670×10^{-8} W/m^2 K^4

Table B.3 Commonly Used Physical Constants

Index

Note: Page numbers followed by *f* denote figures; page numbers followed by *t* denote tables.

D

G

H

R

V

W

Z

Made in the USA
Monee, IL
13 September 2026

62976826R00238